Ford Granada
Service and Repair Manual

A K Legg LAE MIMI

(481–1AB10)

Models covered

Ford Granada L, LX, GL, GLS, S, Ghia, Ghia X and Injection
Saloon and Estate including special and limited edition models
1993 cc ohc, 2293 cc V6 and 2792 cc V6 petrol engines
Covers fuel injection and models with "S" pack modifications

Does not cover Diesel engines, or "new" Granada introduced March 1985

ABCDE
FGHIJ
KLMNO
PQRST

© Haynes Publishing 1998

A book in the **Haynes Service and Repair Manual Series**

All rights reserved. No part of this book may be reproduced or transmitted in any form or by any means, electronic or mechanical, including photocopying, recording or by any information storage or retrieval system, without permission in writing from the copyright holder.

ISBN 1 85960 084 0

British Library Cataloguing in Publication Data
A catalogue record for this book is available from the British Library.

Printed by **J H Haynes & Co Ltd, Sparkford, Nr Yeovil, Somerset BA22 7JJ, England**

Haynes Publishing
Sparkford, Nr Yeovil, Somerset BA22 7JJ, England

Haynes North America, Inc
861 Lawrence Drive, Newbury Park, California 91320, USA

Editions Haynes S.A.
Tour Aurore - La Défense 2, 18 Place des Reflets,
92975 PARIS LA DEFENSE Cedex, France

Haynes Publishing Nordiska AB
Box 1504, 751 45 UPPSALA, Sweden

Contents

LIVING WITH YOUR FORD GRANADA

Introduction	Page	0•4
Acknowledgements	Page	0•4
Safety First!	Page	0•5
Conversion factors	Page	0•6

MOT Test Checks

Checks carried out from the driver's seat	Page	0•7
Checks carried out with the vehicle on the ground	Page	0•8
Checks carried out with the vehicle raised	Page	0•9
Checks carried out on your vehicle's exhaust emission system	Page	0•10

MAINTENANCE

Routine maintenance

General dimensions, weights and capacities	Page	0•11
Jacking and towing	Page	0•12
Buying spare parts and vehicle identification numbers	Page	0•12
Routine maintenance (also see Chapter 13)	Page	0•13
Recommended lubricants and fluids	Page	0•14

Contents

REPAIRS & OVERHAUL

Engine and Associated Systems

Engine (also see Chapter 13)	Chapter 1
Cooling system (also see Chapter 13)	Chapter 2
Fuel and exhaust systems (also see Chapter 13)	Chapter 3
Ignition system (also see Chapter 13)	Chapter 4

Transmission

Clutch (also see Chapter 13)	Chapter 5
Manual and automatic transmission (also see Chapter 13)	Chapter 6
Propeller shaft (also see Chapter 13)	Chapter 7
Rear axle (also see Chapter 13)	Chapter 8

Brakes

Braking system (also see Chapter 13)	Chapter 9

Electrical

Electrical system (also see Chapter 13)	Chapter 10

Suspension and steering

Suspension and steering (also see Chapter 13)	Chapter 11

Body Equipment

Bodywork and fittings (also see Chapter 13)	Chapter 12

Additional information

Supplement: Revisions and information on later models	Chapter 13

Wiring diagrams

Refer to	Page WD•1

Reference

Fault finding	Page REF•1
Tools and working facilities	Page REF•3
General repair procedures	Page REF•5

Index

Refer to	Page IND•1

Introduction

Introduction to the Ford Granada

Introduced in September 1977, the new Granada replaces the previous Consul, Granada range which was introduced in March 1972. Major changes include a bodyshell sharing aerodynamic and styling similarities with the Mk IV Cortina range, and a lightweight German based V6 engine of 2.3 and 2.8 litres. Further design refinements have been added including improved suspension and braking, restyled interior and an optional high performance petrol injection 2.8 litre engine.

The base Granada L version is powered by the well proven 2.0 litre ohc engine. All other models utilize the V6 power unit in 2.3 or 2.8 litre capacity.

A wide range of options are available which include automatic transmission, central door locking system, electric windows, sliding roof, air-conditioning, headlight washers, special interior trim and high performance suspension packs.

In 1979 the Granada 2.8i S and 2.8i GL were discontinued to be replaced by the better equipped 2.8i GLS. Other changes include the central door locking system now standard equipment on the GL and Ghia model and halogen fog lamps, headlight washers and electric rear windows standard on the Ghia.

Engine modifications for the 1980 model year gave improved power and economy whilst retaining the well-tried basic design. Certain trim changes were also made to improve further the standard of this highly successful range of cars.

For the 1982 model year, improvements included detail changes to the bumpers, rear lights and grille. A long-life exhaust system was also fitted, and modifications made to the power assisted steering, suspension, seats, controls and locks. Spoilers were fitted to both front and rear of the 2.8 Injection models.

Progress since 1982 has concentrated mainly on improving the already opulent trim and outstanding comfort of the top line models, at the same time raising the standard of equipment fitted to basic models. Details of equipment fitted to later models will be found in Chapter 13 at the end of the manual.

Acknowledgements

Thanks are due to Champion Spark Plug who supplied spark plug information. Certain illustrations are the copyright of the Ford Motor Company Limited, and are used with their permission. Thanks are also due to Sykes-Pickavant, who provided many of the workshop tools, and all those at Sparkford who assisted in the production of this manual.

We take great pride in the accuracy of information given in this manual, but vehicle manufacturers make alterations and design changes during the production run of a particular vehicle of which they do not inform us. No liability can be accepted by the authors or publishers for loss, damage or injury caused by any errors in, or omissions from, the information given.

Ford Granada Saloon

Safety First! 0•5

Working on your car can be dangerous. This page shows just some of the potential risks and hazards, with the aim of creating a safety-conscious attitude.

General hazards

Scalding
• Don't remove the radiator or expansion tank cap while the engine is hot.
• Engine oil, automatic transmission fluid or power steering fluid may also be dangerously hot if the engine has recently been running.

Burning
• Beware of burns from the exhaust system and from any part of the engine. Brake discs and drums can also be extremely hot immediately after use.

Crushing
• When working under or near a raised vehicle, always supplement the jack with axle stands, or use drive-on ramps. **Never venture under a car which is only supported by a jack.** 
• Take care if loosening or tightening high-torque nuts when the vehicle is on stands. Initial loosening and final tightening should be done with the wheels on the ground.

Fire
• Fuel is highly flammable; fuel vapour is explosive.
• Don't let fuel spill onto a hot engine.
• Do not smoke or allow naked lights (including pilot lights) anywhere near a vehicle being worked on. Also beware of creating sparks (electrically or by use of tools).
• Fuel vapour is heavier than air, so don't work on the fuel system with the vehicle over an inspection pit.
• Another cause of fire is an electrical overload or short-circuit. Take care when repairing or modifying the vehicle wiring.
• Keep a fire extinguisher handy, of a type suitable for use on fuel and electrical fires.

Electric shock
• Ignition HT voltage can be dangerous, especially to people with heart problems or a pacemaker. Don't work on or near the ignition system with the engine running or the ignition switched on. 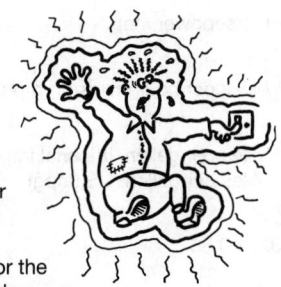
• Mains voltage is also dangerous. Make sure that any mains-operated equipment is correctly earthed. Mains power points should be protected by a residual current device (RCD) circuit breaker.

Fume or gas intoxication
• Exhaust fumes are poisonous; they often contain carbon monoxide, which is rapidly fatal if inhaled. Never run the engine in a confined space such as a garage with the doors shut. 
• Fuel vapour is also poisonous, as are the vapours from some cleaning solvents and paint thinners.

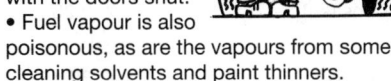

Poisonous or irritant substances
• Avoid skin contact with battery acid and with any fuel, fluid or lubricant, especially antifreeze, brake hydraulic fluid and Diesel fuel. Don't syphon them by mouth. If such a substance is swallowed or gets into the eyes, seek medical advice.
• Prolonged contact with used engine oil can cause skin cancer. Wear gloves or use a barrier cream if necessary. Change out of oil-soaked clothes and do not keep oily rags in your pocket.
• Air conditioning refrigerant forms a poisonous gas if exposed to a naked flame (including a cigarette). It can also cause skin burns on contact.

Asbestos
• Asbestos dust can cause cancer if inhaled or swallowed. Asbestos may be found in gaskets and in brake and clutch linings. When dealing with such components it is safest to assume that they contain asbestos.

Special hazards

Hydrofluoric acid
• This extremely corrosive acid is formed when certain types of synthetic rubber, found in some O-rings, oil seals, fuel hoses etc, are exposed to temperatures above 400°C. The rubber changes into a charred or sticky substance containing the acid. *Once formed, the acid remains dangerous for years. If it gets onto the skin, it may be necessary to amputate the limb concerned.*
• When dealing with a vehicle which has suffered a fire, or with components salvaged from such a vehicle, wear protective gloves and discard them after use.

The battery
• Batteries contain sulphuric acid, which attacks clothing, eyes and skin. Take care when topping-up or carrying the battery.
• The hydrogen gas given off by the battery is highly explosive. Never cause a spark or allow a naked light nearby. Be careful when connecting and disconnecting battery chargers or jump leads.

Air bags
• Air bags can cause injury if they go off accidentally. Take care when removing the steering wheel and/or facia. Special storage instructions may apply.

Diesel injection equipment
• Diesel injection pumps supply fuel at very high pressure. Take care when working on the fuel injectors and fuel pipes.

⚠ *Warning: Never expose the hands, face or any other part of the body to injector spray; the fuel can penetrate the skin with potentially fatal results.*

Remember...

DO
• Do use eye protection when using power tools, and when working under the vehicle.
• Do wear gloves or use barrier cream to protect your hands when necessary.
• Do get someone to check periodically that all is well when working alone on the vehicle.
• Do keep loose clothing and long hair well out of the way of moving mechanical parts.
• Do remove rings, wristwatch etc, before working on the vehicle – especially the electrical system.
• Do ensure that any lifting or jacking equipment has a safe working load rating adequate for the job.

DON'T
• Don't attempt to lift a heavy component which may be beyond your capability – get assistance.
• Don't rush to finish a job, or take unverified short cuts.
• Don't use ill-fitting tools which may slip and cause injury.
• Don't leave tools or parts lying around where someone can trip over them. Mop up oil and fuel spills at once.
• Don't allow children or pets to play in or near a vehicle being worked on.

Conversion Factors

Length (distance)
Inches (in)	X 25.4	= Millimetres (mm)	X 0.0394	= Inches (in)
Feet (ft)	X 0.305	= Metres (m)	X 3.281	= Feet (ft)
Miles	X 1.609	= Kilometres (km)	X 0.621	= Miles

Volume (capacity)
Cubic inches (cu in; in3)	X 16.387	= Cubic centimetres (cc; cm3)	X 0.061	= Cubic inches (cu in; in3)
Imperial pints (Imp pt)	X 0.568	= Litres (l)	X 1.76	= Imperial pints (Imp pt)
Imperial quarts (Imp qt)	X 1.137	= Litres (l)	X 0.88	= Imperial quarts (Imp qt)
Imperial quarts (Imp qt)	X 1.201	= US quarts (US qt)	X 0.833	= Imperial quarts (Imp qt)
US quarts (US qt)	X 0.946	= Litres (l)	X 1.057	= US quarts (US qt)
Imperial gallons (Imp gal)	X 4.546	= Litres (l)	X 0.22	= Imperial gallons (Imp gal)
Imperial gallons (Imp gal)	X 1.201	= US gallons (US gal)	X 0.833	= Imperial gallons (Imp gal)
US gallons (US gal)	X 3.785	= Litres (l)	X 0.264	= US gallons (US gal)

Mass (weight)
Ounces (oz)	X 28.35	= Grams (g)	X 0.035	= Ounces (oz)
Pounds (lb)	X 0.454	= Kilograms (kg)	X 2.205	= Pounds (lb)

Force
Ounces-force (ozf; oz)	X 0.278	= Newtons (N)	X 3.6	= Ounces-force (ozf; oz)
Pounds-force (lbf; lb)	X 4.448	= Newtons (N)	X 0.225	= Pounds-force (lbf; lb)
Newtons (N)	X 0.1	= Kilograms-force (kgf; kg)	X 9.81	= Newtons (N)

Pressure
Pounds-force per square inch (psi; lbf/in^2; lb/in^2)	X 0.070	= Kilograms-force per square centimetre (kgf/cm^2; kg/cm^2)	X 14.223	= Pounds-force per square inch (psi; lbf/in^2; lb/in^2)
Pounds-force per square inch (psi; lbf/in^2; lb/in^2)	X 0.068	= Atmospheres (atm)	X 14.696	= Pounds-force per square inch (psi; lbf/in^2; lb/in^2)
Pounds-force per square inch (psi; lbf/in^2; lb/in^2)	X 0.069	= Bars	X 14.5	= Pounds-force per square inch (psi; lbf/in^2; lb/in^2)
Pounds-force per square inch (psi; lbf/in^2; lb/in^2)	X 6.895	= Kilopascals (kPa)	X 0.145	= Pounds-force per square inch (psi; lbf/in^2; lb/in^2)
Kilopascals (kPa)	X 0.01	= Kilograms-force per square centimetre (kgf/cm^2; kg/cm^2)	X 98.1	= Kilopascals (kPa)
Millibar (mbar)	X 100	= Pascals (Pa)	X 0.01	= Millibar (mbar)
Millibar (mbar)	X 0.0145	= Pounds-force per square inch (psi; lbf/in^2; lb/in^2)	X 68.947	= Millibar (mbar)
Millibar (mbar)	X 0.75	= Millimetres of mercury (mmHg)	X 1.333	= Millibar (mbar)
Millibar (mbar)	X 0.401	= Inches of water (inH$_2$O)	X 2.491	= Millibar (mbar)
Millimetres of mercury (mmHg)	X 0.535	= Inches of water (inH$_2$O)	X 1.868	= Millimetres of mercury (mmHg)
Inches of water (inH$_2$O)	X 0.036	= Pounds-force per square inch (psi; lbf/in^2; lb/in^2)	X 27.68	= Inches of water (inH$_2$O)

Torque (moment of force)
Pounds-force inches (lbf in; lb in)	X 1.152	= Kilograms-force centimetre (kgf cm; kg cm)	X 0.868	= Pounds-force inches (lbf in; lb in)
Pounds-force inches (lbf in; lb in)	X 0.113	= Newton metres (Nm)	X 8.85	= Pounds-force inches (lbf in; lb in)
Pounds-force inches (lbf in; lb in)	X 0.083	= Pounds-force feet (lbf ft; lb ft)	X 12	= Pounds-force inches (lbf in; lb in)
Pounds-force feet (lbf ft; lb ft)	X 0.138	= Kilograms-force metres (kgf m; kg m)	X 7.233	= Pounds-force feet (lbf ft; lb ft)
Pounds-force feet (lbf ft; lb ft)	X 1.356	= Newton metres (Nm)	X 0.738	= Pounds-force feet (lbf ft; lb ft)
Newton metres (Nm)	X 0.102	= Kilograms-force metres (kgf m; kg m)	X 9.804	= Newton metres (Nm)

Power
Horsepower (hp)	X 745.7	= Watts (W)	X 0.0013	= Horsepower (hp)

Velocity (speed)
Miles per hour (miles/hr; mph)	X 1.609	= Kilometres per hour (km/hr; kph)	X 0.621	= Miles per hour (miles/hr; mph)

Fuel consumption*
Miles per gallon, Imperial (mpg)	X 0.354	= Kilometres per litre (km/l)	X 2.825	= Miles per gallon, Imperial (mpg)
Miles per gallon, US (mpg)	X 0.425	= Kilometres per litre (km/l)	X 2.352	= Miles per gallon, US (mpg)

Temperature
Degrees Fahrenheit = (°C x 1.8) + 32   Degrees Celsius (Degrees Centigrade; °C) = (°F - 32) x 0.56

*It is common practice to convert from miles per gallon (mpg) to litres/100 kilometres (l/100km), where mpg (Imperial) x l/100 km = 282 and mpg (US) x l/100 km = 235

MOT Test Checks

This is a guide to getting your vehicle through the MOT test. Obviously it will not be possible to examine the vehicle to the same standard as the professional MOT tester. However, working through the following checks will enable you to identify any problem areas before submitting the vehicle for the test.

Where a testable component is in borderline condition, the tester has discretion in deciding whether to pass or fail it. The basis of such discretion is whether the tester would be happy for a close relative or friend to use the vehicle with the component in that condition. If the vehicle presented is clean and evidently well cared for, the tester may be more inclined to pass a borderline component than if the vehicle is scruffy and apparently neglected.

It has only been possible to summarise the test requirements here, based on the regulations in force at the time of printing. Test standards are becoming increasingly stringent, although there are some exemptions for older vehicles. For full details obtain a copy of the Haynes publication Pass the MOT! (available from stockists of Haynes manuals).

An assistant will be needed to help carry out some of these checks.

The checks have been sub-divided into four categories, as follows:

1 Checks carried out **FROM THE DRIVER'S SEAT**

2 Checks carried out **WITH THE VEHICLE ON THE GROUND**

3 Checks carried out **WITH THE VEHICLE RAISED AND THE WHEELS FREE TO TURN**

4 Checks carried out on **YOUR VEHICLE'S EXHAUST EMISSION SYSTEM**

---

1 Checks carried out **FROM THE DRIVER'S SEAT**

Handbrake

☐ Test the operation of the handbrake. Excessive travel (too many clicks) indicates incorrect brake or cable adjustment.

☐ Check that the handbrake cannot be released by tapping the lever sideways. Check the security of the lever mountings.

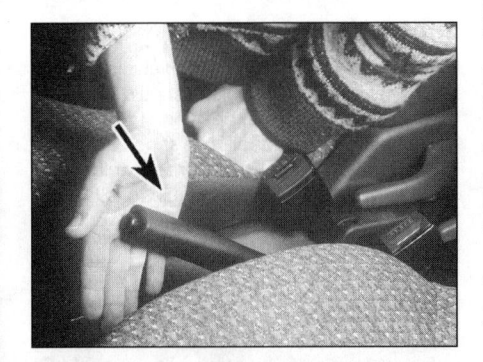

Footbrake

☐ Depress the brake pedal and check that it does not creep down to the floor, indicating a master cylinder fault. Release the pedal, wait a few seconds, then depress it again. If the pedal travels nearly to the floor before firm resistance is felt, brake adjustment or repair is necessary. If the pedal feels spongy, there is air in the hydraulic system which must be removed by bleeding.

☐ Check that the brake pedal is secure and in good condition. Check also for signs of fluid leaks on the pedal, floor or carpets, which would indicate failed seals in the brake master cylinder.

☐ Check the servo unit (when applicable) by operating the brake pedal several times, then keeping the pedal depressed and starting the engine. As the engine starts, the pedal will move down slightly. If not, the vacuum hose or the servo itself may be faulty.

Steering wheel and column

☐ Examine the steering wheel for fractures or looseness of the hub, spokes or rim.

☐ Move the steering wheel from side to side and then up and down. Check that the steering wheel is not loose on the column, indicating wear or a loose retaining nut. Continue moving the steering wheel as before, but also turn it slightly from left to right.

☐ Check that the steering wheel is not loose on the column, and that there is no abnormal movement of the steering wheel, indicating wear in the column support bearings or couplings.

Windscreen and mirrors

☐ The windscreen must be free of cracks or other significant damage within the driver's field of view. (Small stone chips are acceptable.) Rear view mirrors must be secure, intact, and capable of being adjusted.

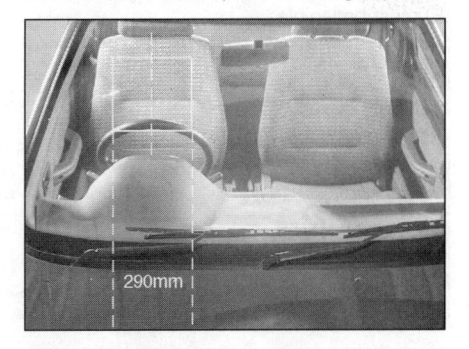

MOT Test Checks

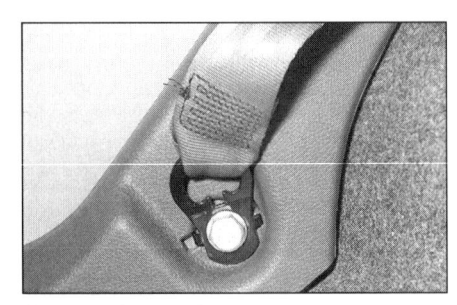

Seat belts and seats

Note: *The following checks are applicable to all seat belts, front and rear.*

☐ Examine the webbing of all the belts (including rear belts if fitted) for cuts, serious fraying or deterioration. Fasten and unfasten each belt to check the buckles. If applicable, check the retracting mechanism. Check the security of all seat belt mountings accessible from inside the vehicle.

☐ The front seats themselves must be securely attached and the backrests must lock in the upright position.

Doors

☐ Both front doors must be able to be opened and closed from outside and inside, and must latch securely when closed.

2 Checks carried out WITH THE VEHICLE ON THE GROUND

Vehicle identification

☐ Number plates must be in good condition, secure and legible, with letters and numbers correctly spaced – spacing at (A) should be twice that at (B).

☐ The VIN plate (A) and homologation plate (B) must be legible.

Electrical equipment

☐ Switch on the ignition and check the operation of the horn.

☐ Check the windscreen washers and wipers, examining the wiper blades; renew damaged or perished blades. Also check the operation of the stop-lights.

☐ Check the operation of the sidelights and number plate lights. The lenses and reflectors must be secure, clean and undamaged.

☐ Check the operation and alignment of the headlights. The headlight reflectors must not be tarnished and the lenses must be undamaged.

☐ Switch on the ignition and check the operation of the direction indicators (including the instrument panel tell-tale) and the hazard warning lights. Operation of the sidelights and stop-lights must not affect the indicators - if it does, the cause is usually a bad earth at the rear light cluster.

☐ Check the operation of the rear foglight(s), including the warning light on the instrument panel or in the switch.

Footbrake

☐ Examine the master cylinder, brake pipes and servo unit for leaks, loose mountings, corrosion or other damage.

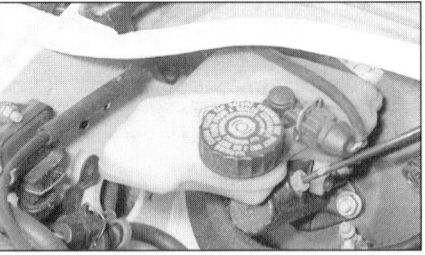

☐ The fluid reservoir must be secure and the fluid level must be between the upper (**A**) and lower (**B**) markings.

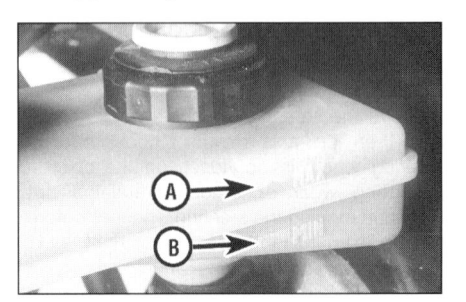

☐ Inspect both front brake flexible hoses for cracks or deterioration of the rubber. Turn the steering from lock to lock, and ensure that the hoses do not contact the wheel, tyre, or any part of the steering or suspension mechanism. With the brake pedal firmly depressed, check the hoses for bulges or leaks under pressure.

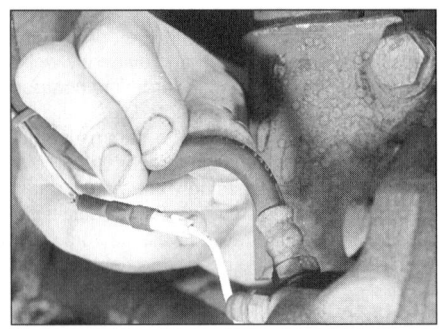

Steering and suspension

☐ Have your assistant turn the steering wheel from side to side slightly, up to the point where the steering gear just begins to transmit this movement to the roadwheels. Check for excessive free play between the steering wheel and the steering gear, indicating wear or insecurity of the steering column joints, the column-to-steering gear coupling, or the steering gear itself.

☐ Have your assistant turn the steering wheel more vigorously in each direction, so that the roadwheels just begin to turn. As this is done, examine all the steering joints, linkages, fittings and attachments. Renew any component that shows signs of wear or damage. On vehicles with power steering, check the security and condition of the steering pump, drivebelt and hoses.

☐ Check that the vehicle is standing level, and at approximately the correct ride height.

Shock absorbers

☐ Depress each corner of the vehicle in turn, then release it. The vehicle should rise and then settle in its normal position. If the vehicle continues to rise and fall, the shock absorber is defective. A shock absorber which has seized will also cause the vehicle to fail.

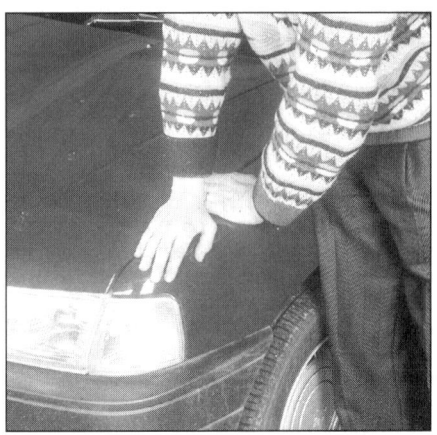

MOT Test Checks

Exhaust system

☐ Start the engine. With your assistant holding a rag over the tailpipe, check the entire system for leaks. Repair or renew leaking sections.

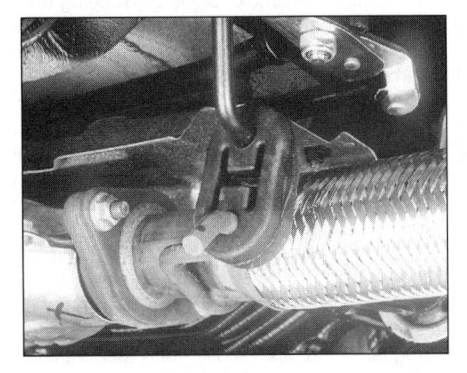

3 Checks carried out WITH THE VEHICLE RAISED AND THE WHEELS FREE TO TURN

Jack up the front and rear of the vehicle, and securely support it on axle stands. Position the stands clear of the suspension assemblies. Ensure that the wheels are clear of the ground and that the steering can be turned from lock to lock.

Steering mechanism

☐ Have your assistant turn the steering from lock to lock. Check that the steering turns smoothly, and that no part of the steering mechanism, including a wheel or tyre, fouls any brake hose or pipe or any part of the body structure.

☐ Examine the steering rack rubber gaiters for damage or insecurity of the retaining clips. If power steering is fitted, check for signs of damage or leakage of the fluid hoses, pipes or connections. Also check for excessive stiffness or binding of the steering, a missing split pin or locking device, or severe corrosion of the body structure within 30 cm of any steering component attachment point.

Front and rear suspension and wheel bearings

☐ Starting at the front right-hand side, grasp the roadwheel at the 3 o'clock and 9 o'clock positions and shake it vigorously. Check for free play or insecurity at the wheel bearings, suspension balljoints, or suspension mountings, pivots and attachments.

☐ Now grasp the wheel at the 12 o'clock and 6 o'clock positions and repeat the previous inspection. Spin the wheel, and check for roughness or tightness of the front wheel bearing.

☐ If excess free play is suspected at a component pivot point, this can be confirmed by using a large screwdriver or similar tool and levering between the mounting and the component attachment. This will confirm whether the wear is in the pivot bush, its retaining bolt, or in the mounting itself (the bolt holes can often become elongated).

☐ Carry out all the above checks at the other front wheel, and then at both rear wheels.

Springs and shock absorbers

☐ Examine the suspension struts (when applicable) for serious fluid leakage, corrosion, or damage to the casing. Also check the security of the mounting points.

☐ If coil springs are fitted, check that the spring ends locate in their seats, and that the spring is not corroded, cracked or broken.

☐ If leaf springs are fitted, check that all leaves are intact, that the axle is securely attached to each spring, and that there is no deterioration of the spring eye mountings, bushes, and shackles.

☐ The same general checks apply to vehicles fitted with other suspension types, such as torsion bars, hydraulic displacer units, etc. Ensure that all mountings and attachments are secure, that there are no signs of excessive wear, corrosion or damage, and (on hydraulic types) that there are no fluid leaks or damaged pipes.

☐ Inspect the shock absorbers for signs of serious fluid leakage. Check for wear of the mounting bushes or attachments, or damage to the body of the unit.

Driveshafts (fwd vehicles only)

☐ Rotate each front wheel in turn and inspect the constant velocity joint gaiters for splits or damage. Also check that each driveshaft is straight and undamaged.

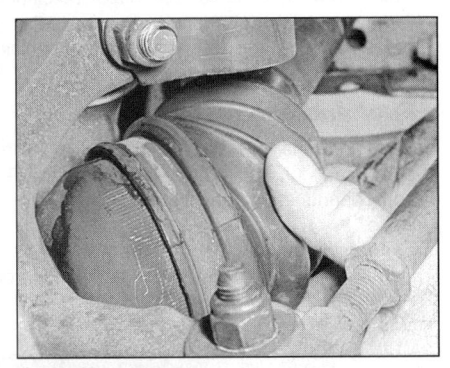

Braking system

☐ If possible without dismantling, check brake pad wear and disc condition. Ensure that the friction lining material has not worn excessively, (A) and that the discs are not fractured, pitted, scored or badly worn (B).

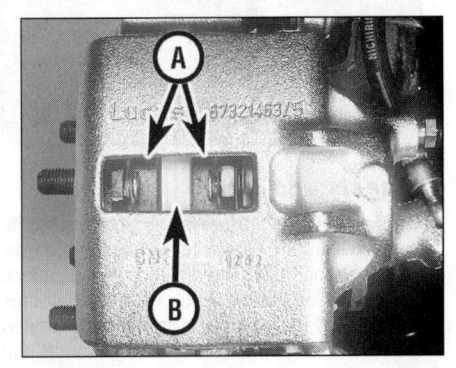

☐ Examine all the rigid brake pipes underneath the vehicle, and the flexible hose(s) at the rear. Look for corrosion, chafing or insecurity of the pipes, and for signs of bulging under pressure, chafing, splits or deterioration of the flexible hoses.

☐ Look for signs of fluid leaks at the brake calipers or on the brake backplates. Repair or renew leaking components.

☐ Slowly spin each wheel, while your assistant depresses and releases the footbrake. Ensure that each brake is operating and does not bind when the pedal is released.

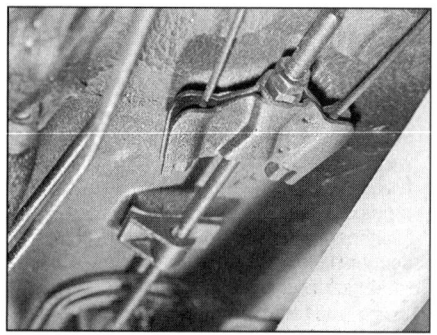

MOT Test Checks

☐ Examine the handbrake mechanism, checking for frayed or broken cables, excessive corrosion, or wear or insecurity of the linkage. Check that the mechanism works on each relevant wheel, and releases fully, without binding.

☐ It is not possible to test brake efficiency without special equipment, but a road test can be carried out later to check that the vehicle pulls up in a straight line.

Fuel and exhaust systems

☐ Inspect the fuel tank (including the filler cap), fuel pipes, hoses and unions. All components must be secure and free from leaks.

☐ Examine the exhaust system over its entire length, checking for any damaged, broken or missing mountings, security of the retaining clamps and rust or corrosion.

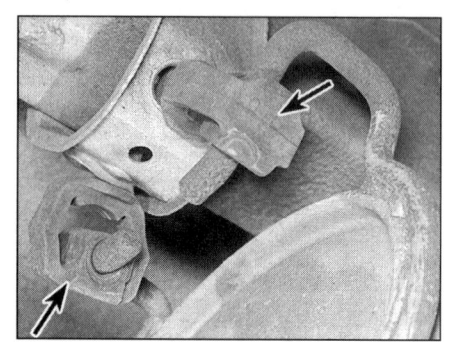

Wheels and tyres

☐ Examine the sidewalls and tread area of each tyre in turn. Check for cuts, tears, lumps, bulges, separation of the tread, and exposure of the ply or cord due to wear or damage. Check that the tyre bead is correctly seated on the wheel rim, that the valve is sound and properly seated, and that the wheel is not distorted or damaged.

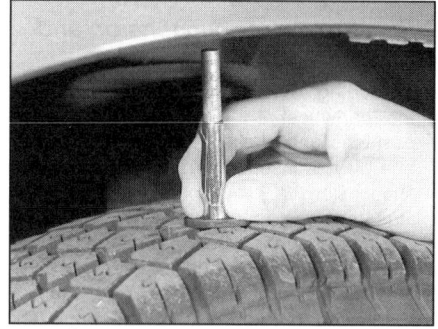

☐ Check that the tyres are of the correct size for the vehicle, that they are of the same size and type on each axle, and that the pressures are correct.

☐ Check the tyre tread depth. The legal minimum at the time of writing is 1.6 mm over at least three-quarters of the tread width. Abnormal tread wear may indicate incorrect front wheel alignment.

Body corrosion

☐ Check the condition of the entire vehicle structure for signs of corrosion in load-bearing areas. (These include chassis box sections, side sills, cross-members, pillars, and all suspension, steering, braking system and seat belt mountings and anchorages.) Any corrosion which has seriously reduced the thickness of a load-bearing area is likely to cause the vehicle to fail. In this case professional repairs are likely to be needed.

☐ Damage or corrosion which causes sharp or otherwise dangerous edges to be exposed will also cause the vehicle to fail.

4 Checks carried out on YOUR VEHICLE'S EXHAUST EMISSION SYSTEM

Petrol models

☐ Have the engine at normal operating temperature, and make sure that it is in good tune (ignition system in good order, air filter element clean, etc).

☐ Before any measurements are carried out, raise the engine speed to around 2500 rpm, and hold it at this speed for 20 seconds. Allow the engine speed to return to idle, and watch for smoke emissions from the exhaust tailpipe. If the idle speed is obviously much too high, or if dense blue or clearly-visible black smoke comes from the tailpipe for more than 5 seconds, the vehicle will fail. As a rule of thumb, blue smoke signifies oil being burnt (engine wear) while black smoke signifies unburnt fuel (dirty air cleaner element, or other carburettor or fuel system fault).

☐ An exhaust gas analyser capable of measuring carbon monoxide (CO) and hydrocarbons (HC) is now needed. If such an instrument cannot be hired or borrowed, a local garage may agree to perform the check for a small fee.

CO emissions (mixture)

☐ At the time of writing, the maximum CO level at idle is 3.5% for vehicles first used after August 1986 and 4.5% for older vehicles. From January 1996 a much tighter limit (around 0.5%) applies to catalyst-equipped vehicles first used from August 1992. If the CO level cannot be reduced far enough to pass the test (and the fuel and ignition systems are otherwise in good order) then the carburettor is badly worn, or there is some problem in the fuel injection system or catalytic converter (as applicable).

HC emissions

☐ With the CO emissions within limits, HC emissions must be no more than 1200 ppm (parts per million). If the vehicle fails this test at idle, it can be re-tested at around 2000 rpm; if the HC level is then 1200 ppm or less, this counts as a pass.

☐ Excessive HC emissions can be caused by oil being burnt, but they are more likely to be due to unburnt fuel.

Diesel models

☐ The only emission test applicable to Diesel engines is the measuring of exhaust smoke density. The test involves accelerating the engine several times to its maximum unloaded speed.

Note: *It is of the utmost importance that the engine timing belt is in good condition before the test is carried out.*

☐ Excessive smoke can be caused by a dirty air cleaner element. Otherwise, professional advice may be needed to find the cause.

Dimensions 0•11

Dimensions

	Up to 1981	1982 onwards
Overall length:		
Saloon	182.4 in (4633 mm)	184.2 in (4679 mm)
Estate	187.1 in (4752 mm)	188.9 in (4767 mm)
Overall width	70.5 in (1790 mm)	70.9 in (1800 mm)
Overall height:		
Saloon	54.3 in (1378 mm)	55.8 in (1416 mm)
Estate	54.4 in (1381 mm)	56.1 in (1424 mm)
Wheel base	109.0 in (2768 mm)	109.0 in (2768 mm)

Kerb weight (nominal)

	Saloon	Estate
Base, L and LS:		
2.0 litre engine	2772 lb (1260 kg)	2937 lb (1335 kg)
2.3 V6 engine	2915 lb (1325 kg)	3069 lb (1395 kg)
2.8 V6 engine	3025 lb (1375 kg)	-
GL and GLS:		
2.3 V6 engine	3014 lb (1370 kg)	3142 lb (1425 kg)
2.8 V6 engine	3080 lb (1400 kg)	3201 lb (1455 kg)
Ghia and Ghia with "S" pack:		
2.3 V6 engine	3058 lb (1390 kg)	-
2.8 V6 engine (carburettor)	3124 lb (1420 kg)	3263 lb (1480 kg)
2.8 V6 engine (fuel injection)	3164 lb (1435 kg)	3307 lb (1500 kg)

Capacities

Engine oil with filter change (approx):
- 4-cyl ..... 6.6 pints (3.75 litres)
- V6 ..... 7.5 pints (4.25 litres)

Engine oil without filter change (approx):
- 4-cyl ..... 5.7 pints (3.25 litres)
- V6 ..... 7.0 pints (4.00 litres)

Cooling system (approx):
- 4-cyl ..... 12.9 pints (7.3 litres)
- V6 up to 1981 ..... 18.0 pints (10.2 litres)
- V6 from 1982 ..... 15.9 pints (9.0 litres)

Fuel tank:
- Saloon ..... 14.3 gallons (65 litres)
- Estate ..... 13.7 gallons (62 litres)

Manual transmission:
- "B" type ..... 3.0 pints (1.7 litres)
- "E" type ..... 3.5 pints (2.0 litres)
- "N" type ..... 3.0 pints (1.7 litres)

Automatic transmission (from dry):
- 4-cyl and 2.3 V6 ..... 11.1 pints (6.3 litres)
- 2.8 V6 ..... 13.0 pints (7.4 litres)

Rear axle:
- Models up to 1981 ..... 3.4 pints (1.9 litres)
- 1982 models onwards ..... 2.5 pints (1.4 litres)

Jacking, Towing, Spare Parts

Jacking points

To change a wheel in an emergency, use the jack supplied with the car. Ensure that the roadwheel nuts are released before jacking up the car and make sure that the arm of the jack is fully engaged with the body bracket and that the base of the jack is standing on a firm level surface (photo).

The jack supplied with the vehicle is not suitable for use when raising the car for maintenance or repair operations. For this work use a trolley, hydraulic or screw type jack located under the front crossmember, bodyframe sidemembers or rear axle casing. Always supplement the jack with axle-stands or blocks before crawling beneath the car.

Towing points

If your car is being towed, make sure that the tow rope is attached to a towing eye or the front crossmember. If the vehicle is equipped with automatic transmission, the distance towed must not exceed 15 miles (24 km), nor the speed 30 mph (48 km/h), otherwise serious damage to the transmission may result. If these limits are likely to be exceeded, disconnect and remove the propeller shaft.

Make sure that the steering column is unlocked (position II) before the tow commences.

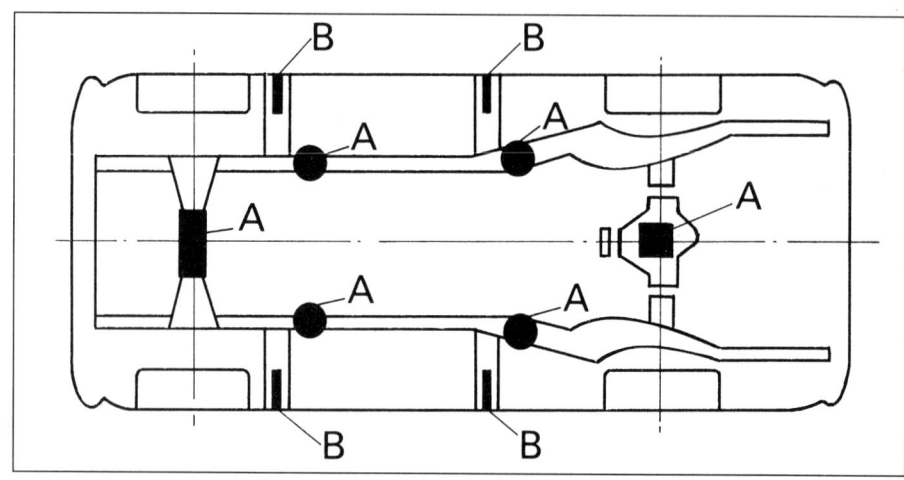

Workshop jack support points (A) wheel changing jack support points (B)

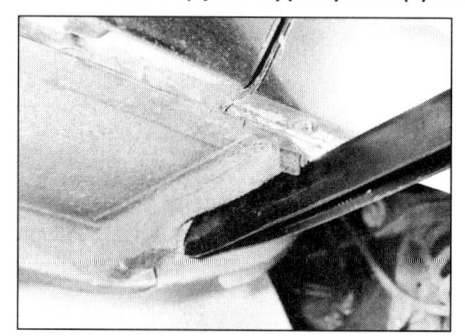

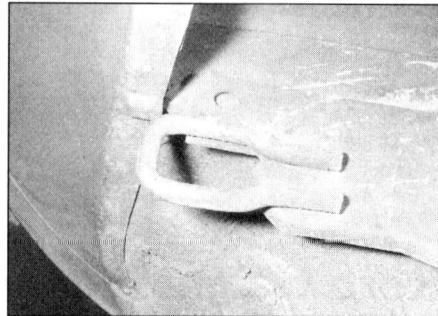

Jack in position

Rear towing hook

Buying spare parts and vehicle identification numbers

Buying spare parts

Spare parts are available from many sources, for example: Ford garages, other garages and accessory shops, and motor factors. Our advice regarding spare part sources is as follows:

Officially appointed Ford garages - This is the best source of parts which are peculiar to your car and are otherwise not generally available (eg. complete cylinder heads, internal gearbox components, badges, interior trim etc). It is also the only place at which you should buy parts if your car is still under warranty - non-Ford components may invalidate the warranty. To be sure of obtaining the correct parts it will always be necessary to give the storeman your car's vehicle identification number, and if possible, to take the "old" part along for positive identification. Remember that many parts are available on a factory exchange scheme - and parts returned should always be clean! It obviously makes good sense to go straight to the specialists on your car for this type of part for they are best equipped to supply them.

Other garages and accessory shops - These are often very good places to buy materials and components needed for the maintenance of your car (eg, oil filters, spark plugs, bulbs, fan belts, oils and greases, touch-up paint, filler paste etc). They also sell general accessories, usually have convenient opening hours, charge lower prices and can often be found not far from home.

Motor factors - Good factors will stock all of the more important components which wear out relatively quickly (eg. clutch components, pistons, valves, exhaust systems, brake cylinders/pipes/hoses/seals/shoes and pads etc). Motor factors will often provide new or reconditioned components on a part exchange basis - this can save a considerable amount of money.

Vehicle identification numbers

Although many individual parts, and in some cases sub-assemblies, fit a number of different models, it is dangerous to assume that just because they look the same, they are the same. Differences are not always easy to detect except by serial numbers. Make sure, therefore, that the appropriate identity number for the model or subassembly is known and quoted when a spare part is ordered.

The vehicle identification plate is mounted on the left-hand side of the front body panel and may be seen once the bonnet is open (photo).

The chassis number is stamped into the floor panel inside the front door sill.

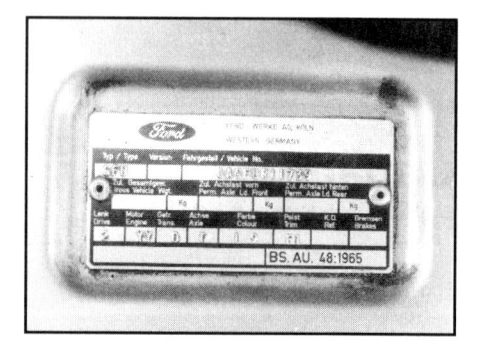

Vehicle identification plate

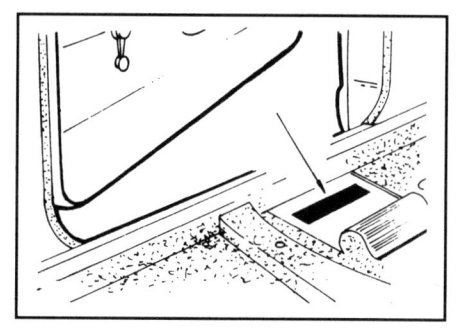

Location of chassis number

Routine Maintenance 0•13

For modifications, and information applicable to later models, see Supplement at end of manual

Maintenance is essential for ensuring safety, and desirable for the purpose of getting the best in terms of performance and economy from your car. Over the years the need for periodic lubrication - oiling, greasing and so on - has been drastically reduced, if not totally eliminated. This has unfortunately tended to lead some owners to think that because no such action is required, components either no longer exist, or will last forever. This is a serious delusion. It follows therefore that the largest initial element of maintenance is visual examination and a general sense of awareness. This may lead to repairs or renewals, but should help to avoid roadside breakdowns. Other neglect results in unreliability, increased running costs, more rapid wear and depreciation of the vehicle in general.

Every 260 miles (400 km) or weekly

Check tyre pressures and inflate if necessary - don't forget the spare
Check and top-up the engine oil (photo)
Check and top-up battery electrolyte level (photo)
Check and if necessary, top up water reservoir adding a screen wash
Check and top-up coolant level (photo)
Check operation of all lights
Check the brake fluid reservoir

Every 6000 miles (10 000 km)

Change engine oil and renew filter (photos)
Clean fuel pump filter (carburettor engines)
Check carburettor adjustment (if fitted)
Check drivebelt tension; renew if frayed (photo)
Clean and adjust spark plugs
Check, adjust or renew distributor contact breaker points (OHC engines)
Check for wear in all steering joints and condition of flexible dust excluders
Inspect brake fluid lines and hoses for leaks, damage or deterioration
Inspect front disc pads and rear brake shoe linings for wear
Check and top-up brake fluid reservoir
Lubricate door hinges and controls
Check and top-up gearbox oil level
Check the fluid level in the power steering reservoir and top-up if necessary (if fitted)
Check the power steering system for leaks (if fitted)
Check the power steering pump drivebelt (if fitted)
Check and if necessary adjust the air conditioning compressor drivebelt (where fitted)
Check and adjust valve clearances
Check and top-up rear axle oil level Move position of roadwheels (balanced off car) to even out tread wear
Clean crankcase ventilation valve
Check exhaust system for leaks or broken mountings

Every 12 000 miles (19 000 km)

Complete the 6000 mile (10 000 km) service and in addition:
Check ignition timing
Check front wheel alignment
Check headlight alignment
Renew air cleaner element
Renew fuel filter (fuel injection engines)

Every 24 000 miles (38 000 km) or yearly intervals

Drain cooling system and refill with "long-life" type antifreeze mixture. If other than "long-life" mixture is used, cooling system should be drained, flushed and refilled every autumn.
Clean, re-lubricate and adjust front hub bearings.

Every 36 000 miles (68 000 km)

Bleed hydraulic system, renew all system seals and refill with clean, fresh fluid
Renew servo unit filter.

Once a year or more often if the time can be spared

Cleaning

Examination of components requires that they are cleaned. The same applies to the body of the car, inside and out, in order that deterioration due to rust or unknown damage may be detected. Certain parts of the body frame, if rusted badly, can result in the vehicle being declared unsafe and it will not pass the annual test for roadworthiness.

Exhaust system

An exhaust system must be leakproof, and the noise level below a certain maximum. Excessive leaks may cause carbon monoxide fumes to enter the passenger compartment. Excessive noise constitutes a public nuisance. Repair or renew defective sections when symptoms are apparent.

Location of rear screen washer

Topping up engine oil level

Topping up the battery

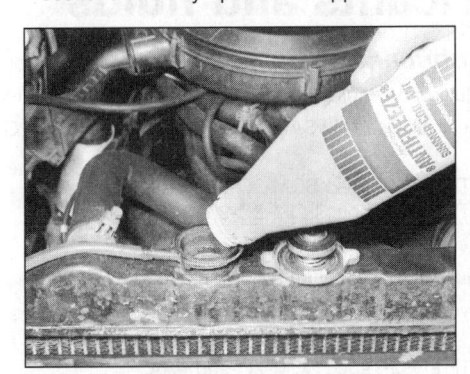

Topping up coolant level

Greasing bonnet lock

Routine Maintenance

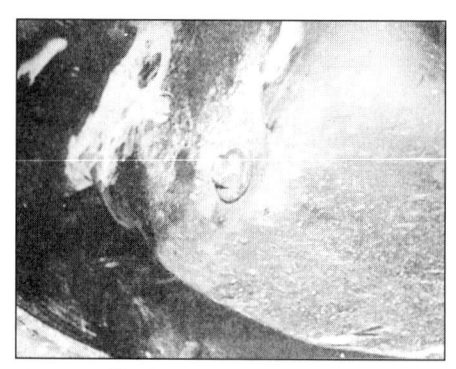

Engine sump drain plug

Change the oil filter

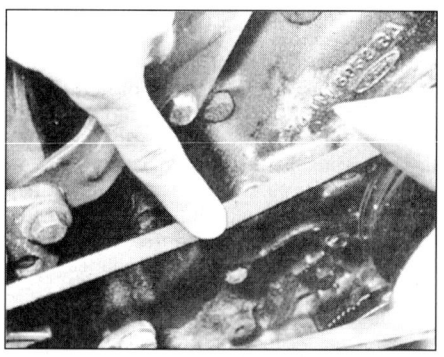

Check fan belt tension

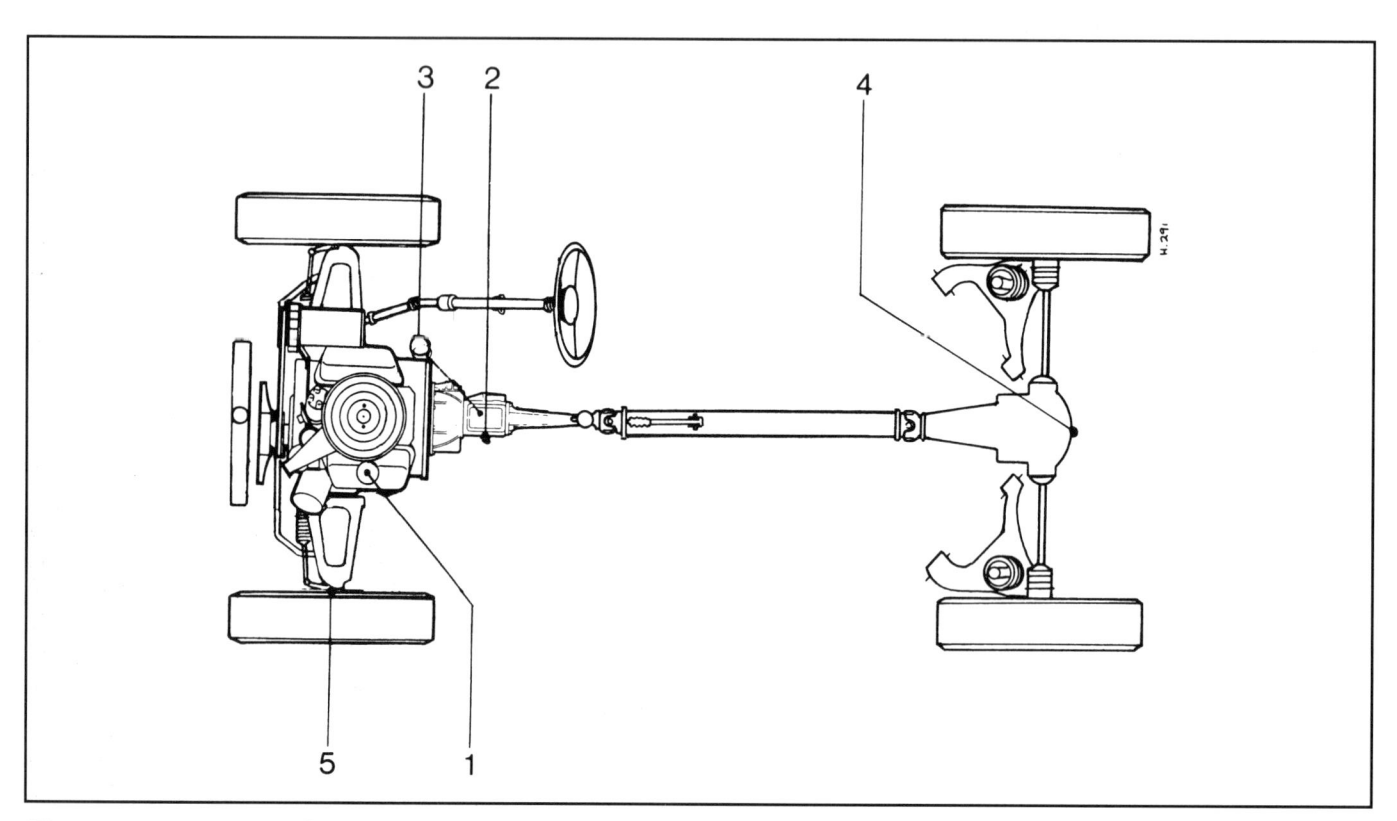

Recommended lubricants and fluids

Component or system	Lubricant type/specification
1 Engine	Multigrade engine oil, viscosity range SAE 10W/30 to 10W/50, to API SF/CC or SF/CD
2 Manual gearbox*	Gear oil, viscosity SAE 80EP, to Ford spec SQM-2C-9008-A
3 Automatic transmission	
Early models (black dipstick)	ATF to Ford spec SQM-2C-9007-AA
Later models (red dipstick)	ATF to Ford spec SQM-2C-9010-A
4 Rear axle	Gear oil, viscosity SAE 90EP, to API GL5
5 Front wheel hub bearings	Multi-purpose lithium based grease to NLGI-2
Brake fluid reservoir	Hydraulic fluid to Ford spec SAM-6C9103-A
Power steering	
Up to 1981	ATF to Ford spec SQM-2C-9007-AA
1982 on	ATF to Ford spec SQM-2C-9010-A
Cooling system	Antifreeze to Ford spec SSM-97B9103-A

*Note: See Supplement for alternative lubricant for N type transmission

Chapter 1 Engine

For modifications, and information applicable to later models, see Supplement at end of manual

Contents

Part A In-line ohc engine

Auxiliary shaft and timing cover - refitting	46
Auxiliary shaft - removal	10
Cam followers - refitting	53
Camshaft, camshaft bearings and cam followers - examination and renovation	31
Camshaft drivebelt - refitting and timing	56
Camshaft drivebelt - removal (engine in car)	16
Camshaft drivebelt tensioner and thermostat housing - refitting	55
Camshaft - refitting	52
Camshaft - removal	18
Connecting rods and gudgeon pins - examination and renovation	30
Connecting rods to crankshaft - reassembly	43
Crankshaft and main bearings - removal	15
Crankshaft - examination and renovation	26
Crankshaft pulley, sprocket and timing belt cover - removal	13
Crankshaft rear oil seal - refitting	45
Crankshaft - refitting	39
Crankshaft sprocket and pulley, and auxiliary shaft sprocket - refitting	48
Cylinder bores - examination and renovation	28
Cylinder head and piston crowns - examination and renovation	35
Cylinder head - refitting	54
Cylinder head - removal (engine in car)	8
Cylinder head - removal (engine on bench)	9
Engine components - examination for wear	25
Engine dismantling - general	7
Engine - initial start-up after major overhaul or repair	60
Engine reassembly - general	38
Engine - refitting with transmission	59
Engine - refitting without transmission	58
Engine - removal with transmission	6
Engine - removal without transmission	5
Fault diagnosis - engine	See end of Chapter
Flywheel and clutch - refitting	50
Flywheel and sump - removal	11
Flywheel ring gear - examination and renovation	34
General description	1
Gudgeon pin - removal	20
Lubrication and crankcase ventilation systems - description	22
Main and big-end bearings - examination and renovation	27
Major operations possible with engine in car	2
Major operations requiring engine removal	3
Methods of engine removal	4
Oil filter - removal and refitting	24
Oil pump and strainer - removal	12
Oil pump - dismantling, inspection and reassembly	23
Oil pump - refitting	44
Piston rings - refitting	41
Piston rings - removal	21
Pistons and connecting rods - reassembly	40
Pistons and piston rings - examination and renovation	29
Pistons, connecting rods and big-end bearings - removal	14
Pistons - refitting	42
Sump - inspection	37
Sump - refitting	47
Thermostat housing and belt tensioner - removal	19
Timing gears and belt - examination and renovation	33
Valve clearances - checking and adjustment	57
Valve guides - inspection	36
Valves and valve seats - examination and renovation	32
Valves - refitting	51
Valves - removal	17
Water pump - refitting	49

Part B V6 engines

Ancillary components - refitting	95
Camshaft and camshaft bearings - inspection and renovation	72
Camshaft and front intermediate plate - refitting	87
Connecting rods and gudgeon pins- examination and renovation	78
Crankcase ventilation system - description and maintenance	84
Crankshaft - examination and renovation	79
Crankshaft main and big-end bearings - examination and renovation	80
Crankshaft - refitting	86
Cylinder bores - inspection and renovation	74
Cylinder heads - dismantling, renovation and reassembly	73
Cylinder heads, rocker shafts and inlet manifold - refitting	93
Engine ancillaries - removal	68
Engine - dismantling	69
Engine dismantling - general	67
Engine - initial start-up after major overhaul or repair	97
Engine modifications - later models	98
Engine oil sump, water pump and crankshaft pulley - refitting	92
Engine reassembly - general	85
Engine - refitting in car	96
Engine with carburettor - removal	65
Engine with fuel injection - removal	66
Fault diagnosis - engine	See end of Chapter
Flywheel and clutch - refitting	90
Flywheel ring gear - inspection and renewal	82
General description	61
Lubrication system - description	83
Major operations requiring engine removal	63
Major operations with engine in place	62
Methods of engine removal	64
Oil pump - dismantling, inspection and reassembly	81
Oil pump - refitting	89
Pistons and connecting rods - refitting	88
Pistons and piston rings - examination and renovation	76
Piston rings - refitting	77
Piston rings - removal	75
Timing gears and timing cover - refitting	91
Valve clearances - checking and adjustment	94
Valve rockers - dismantling and reassembly	70
Valve tappets and pushrods - inspection	71

Degrees of difficulty

Easy, suitable for novice with little experience	**Fairly easy,** suitable for beginner with some experience	**Fairly difficult,** suitable for competent DIY mechanic	**Difficult,** suitable for experienced DIY mechanic	**Very difficult,** suitable for expert DIY or professional

1•2 Engine

Specifications

In-line ohc engine

General
Engine type	Four cylinder in-line, single overhead camshaft
Firing order	1-3-4-2 (No 1 at timing belt end)
Bore	3.575 in (90.8 mm)
Stroke	3.03 in (76.95 mm)
Cubic capacity	1993 cc
Maximum continuous engine speed	5850 rpm
Engine bhp (DIN)	96 at 5200 rpm
Max torque (DIN)	111 lbf ft at 3500 rpm
Compression ratio	9.2:1

Cylinder block
Cast identification marks	20
Number of main bearings	5
Cylinder bore dia. grades:	
Standard grade:	
1 - in (mm)	3.5748 to 3.5752 (90.800 to 90.810)
2 - in (mm)	3.5752 to 3.5756 (90.810 to 90.820)
3 - in (mm)	3.5756 to 3.5760 (90.820 to 90.830)
4 - in (mm)	3.5760 to 3.5764 (90.830 to 90.840)
Oversize A - in (mm)	3.5949 to 3.5953 (91.310 to 91.320)
Oversize B - in (mm)	3.5953 to 3.5957 (91.320 to 91.330)
Oversize C - in (mm)	3.5957 to 3.5961 (91.330 to 91.340)
Standard supp. in service - in (mm)	3.5760 to 3.5764 (90.830 to 90.840)
Oversize 0.5 - in (mm)	3.5957 to 3.5961 (91.330 to 91.340)
Oversize 1.0 - in (mm)	3.6154 to 3.6157 (91.830 to 91.840)
Centre main bearing width - in (mm)	1.072 to 1.070 (27.22 to 27.17)
Main bearing liners fitted (inner diameter):	
Standard - in (mm)	2.2440 to 2.2455 (57.000 to 57.038)
Undersize - in (mm):	
0.25 mm	2.2342 to 2.2357 (56.750 to 56.788)
0.50 mm	2.2244 to 2.2259 (56.500 to 56.538)
0.75 mm	2.2145 to 2.2160 (56.250 to 56.288)
1.00 mm	2.2047 to 2.2062 (56.000 to 56.038)
Main bearing bore - in (mm):	
Standard	2.3866 to 2.3873 (60.620 to 60.640)
Oversize	2.4023 to 2.4031 (61.020 to 61.040)

Crankshaft
Endfloat - in (mm)	0.0032 to 0.0110 (0.08 to 0.28)
Main bearing journal diameters - in (mm):	
Standard	2.2429 to 2.2437 (56.970 to 56.990)
Undersize:	
0.25 mm	2.2330 to 2.2338 (56.720 to 56.740)
0.50 mm	2.2232 to 2.2240 (56.470 to 56.490)
0.75 mm	2.2133 to 2.2141 (56.220 to 56.240)
1.00 mm	2.2035 to 2.2043 (55.970 to 55.990)
Thrust washer thickness - in (mm):	
Standard	0.0905 to 0.0925 (2.300 to 2.350)
Undersize	0.0984 to 0.1003 (2.500 to 2.550)
Bearing shell to main journal clearance - in (mm)	0.0004 to 0.0025 (0.010 to 0.064)
Big-end journal diameter - in (mm):	
Standard	2.0464 to 2.0472 (51.980 to 52.000)
Undersize:	
0.25 mm	2.0366 to 2.0373 (51.730 to 51.750)
0.50 mm	2.0267 to 2.0275 (51.480 to 51.500)
0.75 mm	2.0169 to 2.0177 (51.230 to 51.250)
1.00 mm	2.0070 to 2.0078 (50.980 to 51.00)

Camshaft
Drive	Toothed belt
Number of bearings	3
Thrust plate thickness - in (mm)	0.157 to 0.158 (3.98 to 4.01)
Camshaft keyway width - in (mm)	0.162 to 0.165 (4.114 to 4.184)
Cam lift - in (mm)	0.249 (6.3323)
Cam length (heel-to-toe) - in (mm)	1.427 to 1.440 (36.26 to 36.60)

Engine 1•3

Camshaft (cont.)
Journal diameter - in (mm):
   Front .... 1.6531 to 1.6539 (41.99 to 42.01)
   Centre .... 1.7562 to 1.7570 (44.61 to 44.63)
   Rear .... 1.7712 to 1.7720 (44.99 to 45.01)
Bearing inner diameter - in (mm):
   Front .... 1.6549 to 1.6557 (42.035 to 42.055)
   Centre .... 1.7580 to 1.7588 (44.655 to 44.675)
   Rear .... 1.7730 to 1.7738 (45.035 to 45.055)
Endfloat - in (mm) .... 0.004 to 0.008 (0.104 to 0.204)
Colour code .... Yellow

Auxiliary shaft
Endfloat - in (mm) .... 0.0021 to 0.0080 (0.054 to 0.204)

Pistons
Piston diameter:
 Standard grade:
   1 - in (mm) .... 3.5734 to 3.5738 (90.765 to 90.775)
   2 - in (mm) .... 3.5738 to 3.5742 (90.775 to 90.785)
   3 - in (mm) .... 3.5742 to 3.5746 (90.785 to 90.795)
   4 - in (mm) .... 3.5746 to 3.5750 (90.795 to 90.805)
 Oversize supplied in service:
   0.5 - in (mm) .... 3.5937 to 3.5947 (91.280 to 91.305)
   1.0 - in (mm) .... 3.6134 to 3.6144 (91.780 to 91.805)
Piston clearance in cylinder bore - in (mm) .... 0.001 to 0.0024 (0.025 to 0.060)
Ring gap (in situ):
   Top - in (mm) .... 0.015 to 0.023 (0.38 to 0.58)
   Centre - in (mm) .... 0.015 to 0.023 (0.38 to 0.58)
   Bottom - in (mm) .... 0.0157 to 0.055 (0.4 to 1.4)

Gudgeon pins
Length - in (mm) .... 2.83 to 2.87 (72.0 to 72.8)
Diameter:
   Red - in (mm) .... 0.94465 to 0.94476 (23.994 to 23.997)
   Blue - in (mm) .... 0.94476 to 0.94488 (23.997 to 24.000)
   Yellow - in (mm) .... 0.94488 to 0.94500 (24.000 to 24.003)
Clearance in piston - in (mm) .... 0.0003 to 0.0006 (0.008 to 0.014)
Interference in connecting rod - in (mm) .... 0.0007 to 0.0015 (0.018 to 0.039)

Connecting rods
Bore diameters - in (mm):
   Big-end .... 2.165 to 2.166 (55.00 to 55.02)
   Small-end .... 0.9434 to 0.9439 (23.964 to 23.976)
Bearing inside diameter (fitted) - in (mm):
   Standard .... 2.0474 to 2.0489 (52.006 to 52.044)
 Undersize:
   0.25 mm .... 2.0376 to 2.0391 (51.756 to 51.794)
   0.50 mm .... 2.0277 to 2.0292 (51.506 to 51.544)
   0.75 mm .... 2.0179 to 2.0194 (51.256 to 51.294)
   1.00 mm .... 2.0081 to 2.0096 (51.006 to 51.044)
Big-end bearing to journal clearance - in (mm) .... 0.0002 to 0.0023 (0.006 to 0.060)

Cylinder head
Cast identification number .... 0 (early models), 20 (later models)
Valve seat angle .... 44° 30' to 45°
Valve guide inside diameter, inlet and exhaust:
   Standard - in (mm) .... 0.3174 to 0.3184 (8.063 to 8.088)
 Oversize:
   0.2 - in (mm) .... 0.3253 to 0.3263 (8.263 to 8.288)
   0.4 - in (mm) .... 0.3332 to 0.3342 (8.463 to 8.488)
Parent bore for camshaft bearing liners:
   Front - in (mm) .... 1.6557 to 1.6549 (42.055 to 42.035)
   Centre - in (mm) .... 1.7589 to 1.7580 (44.675 to 44.655)
   Rear - in (mm) .... 1.7738 to 1.7730 (45.055 to 45.035)

Valves
Valve clearances (cold):
   Inlet - in (mm) 0.008 (0.20)
   Exhaust - in (mm) 0.010 (0.25)
Valve timing:
   Inlet valve opens .... 24° BTDC
   Inlet valve closes .... 64° ABDC
   Exhaust valve opens .... 70° BBDC
   Exhaust valve closes .... 18° ATDC

Inlet valves

Length - in (mm)	4.3563 to 4.3760 (110.65 to 111.15)
Valve head diameter - in (mm)	1.654 ± 0.008 (42 ± 0.2)
Valve stem diameter:	
Standard - in (mm)	0.3167 to 0.3159 (8.043 to 8.025)
Oversize:	
0.2 - in (mm)	0.3245 to 0.3238 (8.243 to 8.225)
0.4 in (mm)	0.3324 to 0.3317 (8.443 to 8.425)
Valve stem to guide clearance in (mm)	0.0008 to 0.0025 (0.020 to 0.063)
Valve lift - in (mm)	0.3985 (10.121)
Valve spring free length - in (mm)	1.73 (44)
Valve spring compressed length - in (mm)	0.945 (24)

Exhaust valves

Length - in (mm)	4.336 to 4.372 (110.15 to 111.05)
Head diameter - in (mm)	1.409 to 1.425 (35.80 to 36.20)
Stem diameter - in (mm):	
Standard	0.3149 to 0.3156 (7.999 to 8.017)
Oversize:	
0.2	0.3227 to 0.3235 (8.199 to 8.217)
0.4	0.3306 to 0.3313 (8.399 to 8.417)
0.6	0.3385 to 0.3392 (8.599 to 8.617)
0.8	0.3464 to 0.3471 (8.799 to 8.817)
Stem-to-guide clearance - in (mm)	0.0018 to 0.0035 (0.046 to 0.089)
Lift - in (mm)	0.3984 (10.121)
Spring free length - in (mm):	
Early models	1.7323 (44.0)
Later models	1.8504 (47.0)
Spring compressed length - in (mm):	
Early models	0.9646 (24.5)
Later models	0.9449 (24.0)

Engine lubrication data

Oil type specification	Multigrade engine oil, viscosity range SAE 10W/30 to 10W/50, to API SF/CC or SF/CD
Oil filter	Champion C102
Minimum oil pressure readings (at 80°C):	
At 750 rpm:	
Early models	14 lbf/in^2 (1.0 kgf/cm^2)
Later models	29 lbf/in^2 (2.1 kgf/cm^2)
At 2000 rpm (all models)	35 lbf/in^2 (2.5 kgf/cm^2)
Relief valve opening pressure	57 to 67 lbf/in^2 (4.0 to 4.7 kgf/cm^2)
Oil pump outer rotor to housing clearance - in (mm)	0.006 to 0.012 (0.15 to 0.30)
Oil pump inner rotor to outer rotor clearance - in (mm)	0.002 to 0.008 (0.05 to 0.20)
Oil pump rotor endfloat - in (mm)	0.0012 to 0.004 (0.03 to 0.10)

Torque wrench settings

	lbf ft	Nm
Main bearing cap bolts	70	95
Big-end bearing cap bolts	35	48
Crankshaft pulley bolt	40	55
Camshaft and auxiliary shaft sprocket bolts	33	45
Flywheel bolts	50	68
Oil pump mounting bolts	14	19
Oil pump cover bolts	9	12
Sump bolts:		
Stage 1	1.5	2
Stage 2	5	7
Stage 3 (after 20 minutes running)	7	10
Sump drain plug	18	24
Oil pressure switch	10	14
Valve adjuster ball-pin	33	45
Cylinder head bolts:		
Stage 1	15 to 30	20 to 40
Stage 2	36 to 51	49 to 69
Stage 3 (after 10 to 20 minute wait)	54 to 61	73 to 83
Stage 4 (after 15 minutes running)	70 to 85	95 to 115
Rocker cover bolts (in this sequence - for bolt identification see Fig. 1.15):		
Bolts 1 to 6	5	7
Bolts 7 and 8	1.8	2.5
Bolts 9 and 10	5	7
Bolts 7 and 8	5	7

Torque wrench settings (cont.)

	lbf ft	Nm
Timing cover	10	14
Inlet manifold	14	19
Exhaust manifold	16	22
Spark plugs	18	25

V6 engines

General

Type	6 cylinder ohv in 60° "V"
Firing order	1-4-2-5-3-6
Bore:	
2.3 litre	3.54 in (90.02 mm)
2.8 litre	3.66 in (93.02 mm)
Stroke:	
2.3 litre	2.37 in (60.14 mm)
2.8 litre	2.62 in (68.50 mm)
Cubic capacity:	
2.3 litre	2294 cc
2.8 litre	2792 cc
Compression ratio:	
2.3 litre - early models	8.75:1
2.3 litre - later models	9.00:1
2.8 litre	9.2:1
Maximum continuous engine speed:	
Later 2.8 models with fuel injection	6000 rpm
All other models	5700 rpm
Engine bhp (DIN): - figures in brackets are for later models:	
2.3 litre	106 @ 5000 rpm (114 @ 5300 rpm)
2.8 litre	133 @ 5200 rpm (135 @ 5200 rpm)
2.8 litre fuel injection	158 @ 5700 rpm (160 @ 5700 rpm)
Max torque (DIN):	
2.3 litre	130 lbf ft @ 3000 rpm
2.8 litre	162 lbf ft @ 3000 rpm
2.8 litre fuel injection	166 lbf ft @ 4300 rpm

Cylinder block

Number of main bearings	4
Bore diameter (2.3 litre):	
Standard:	
grade 1	3.5433 to 3.5437 in (90.000 to 90.010 mm)
grade 2	3.5437 to 3.5440 in (90.010 to 90.020 mm)
grade 3	3.5440 to 3.5444 in (90.020 to 90.030 mm)
grade 4	3.5444 to 3.5448 in (90.030 to 90.040 mm)
Oversize:	
grade A	3.5633 to 3.5637 in (90.510 to 90.520 mm)
grade B	3.5637 to 3.5641 in (90.520 to 90.530 mm)
grade C	3.5641 to 3.5645 in (90.530 to 90.540 mm)
Standard service	3.5444 to 3.5448 in (90.030 to 90.040 mm)
Oversize 0.5 mm	3.5641 to 3.5645 in (90.530 to 90.540 mm)
Oversize 1.0 mm	3.5838 to 3.5842 in (91.030 to 91.040 mm)
Bore diameter (2.8 litre):	
Standard:	
grade 1	3.6618 to 3.6622 in (93.010 to 93.020 mm)
grade 2	3.6622 to 3.6626 in (93.020 to 93.030 mm)
grade 3	3.6626 to 3.6630 in (93.030 to 93.040 mm)
grade 4	3.6630 to 3.6634 in (93.040 to 93.050 mm)
Oversize:	
grade A	3.6819 to 3.6823 in (93.520 to 93.530 mm)
grade B	3.6823 to 3.6826 in (93.530 to 93.540 mm)
grade C	3.6826 to 3.6830 in (93.540 to 93.550 mm)
Standard service	3.6630 to 3.6634 in (93.040 to 93.050 mm)
Oversize 0.5 mm	3.6826 to 3.6830 in (93.540 to 93.550 mm)
Oversize 1.0 mm	3.7023 to 3.7028 in (94.040 to 94.050 mm)
Centre main bearing width (2.3 litre):	
Standard	0.8488 to 0.8508 in (21.560 to 21.610 mm)
Undersize	1.054 to 1.056 in (26.771 to 26.821 mm)
Centre main bearing width (2.8 litre):	
Standard - early models	0.8488 to 0.8508 in (21.560 to 21.610 mm)
Standard - later models	0.8902 to 0.8921 in (22.610 to 22.660 mm)
Undersize - all models	1.0563 to 1.0579 in (26.829 to 26.871 mm)

1•6 Engine

Cylinder block (cont.)
Main bearing shell internal diameter (fitted 2.3 litre):
  Standard .... 2.2444 to 2.457 in (57.008 to 57.042 mm)
  Undersize:
    0.254 mm .... 2.2344 to 2.357 in (56.754 to 56.788 mm)
    0.508 mm .... 2.2244 to 2.2257 in (56.500 to 56.534 mm)
    0.762 mm .... 2.2144 to 2.2157 in (56.246 to 56.280 mm)
    1.020 mm .... 2.2044 to 2.2057 in (55.992 to 56.026 mm)
Main bearing shell internal diameter (fitted 2.8 litre):
  Standard .... 2.2448 to 2.2461 in (57.018 to 57.052 mm)
  Undersize:
    0.254 mm .... 2.2348 to 2.2361 in (56.764 to 56.798 mm)
    0.508 mm .... 2.2248 to 2.2261 in (56.510 to 56.544 mm)
    0.762 mm .... 2.2148 to 2.2161 in (56.256 to 56.290 mm)
    1.020 mm .... 2.2048 to 2.2061 in (56.002 to 56.036 mm)
Main bearing parent bore diameter:
  Standard .... 2.3866 to 2.3874 in (60.620 to 60.640 mm)
  Oversize 0.38 mm .... 2.4015 to 2.4023 in (61.000 to 61.020 mm)
Camshaft parent bore diameter:
  Front .... 1.7726 to 1.7740 in (45.025 to 45.060 mm)
  Centre 1 .... 1.7576 to 1.7590 in (44.645 to 44.680 mm)
  Centre 2 .... 1.7427 to 1.7440 in (44.265 to 44.300 mm)
  Rear .... 1.7277 to 1.7291 in (43.885 to 43.920 mm)

Crankshaft
Endfloat .... 0.003 to 0.011 in (0.080 to 0.280 mm)
Backlash .... 0.002 to 0.005 in (0.050 to 0.140 mm)
Main bearing journal diameter:
  Standard .... 2.2433 to 2.2440 in (56.980 to 57.000 mm)
  Undersize:
    0.254 mm .... 2.2333 to 2.2340 in (56.726 to 56.746 mm)
    0.508 mm .... 2.2233 to 2.2240 in (56.472 to 56.492 mm)
    0.762 mm .... 2.2133 to 2.2140 in (56.218 to 56.238 mm)
    1.020 mm .... 2.2033 to 2.2040 in (55.964 to 55.984 mm)
Centre main bearing width .... 1.0389 to 1.0409 in (26.390 to 26.440 mm)
Main bearing clearance .... 0.0002 to 0.0025 in (0.006 to 0.064 mm)
Big-end journal diameter:
  Standard .... 2.1251 to 2.1259 in (53.980 to 54.000 mm)
  Undersize:
    0.254 mm .... 2.1151 to 2.1159 in (53.726 to 53.746 mm)
    0.508 mm .... 2.1051 to 2.1059 in (53.472 to 53.492 mm)
    0.762 mm .... 2.0951 to 2.0959 in (53.218 to 53.238 mm)
    1.020 mm .... 2.0851 to 2.0859 in (52.964 to 52.984 mm)

Camshaft
Number of bearings 4
Drive .... Gear driven from crankshaft
Thrust plate thickness:
  Red .... 0.1559 to 0.1568 in (3.960 to 3.985 mm)
  Blue .... 0.1569 to 0.1579 in (3.986 to 4.011 mm)
Backlash:
  2.3 litre .... 0.002 to 0.005 in (0.05 to 0.14 mm)
  2.8 litre .... 0.007 to 0.011 in (0.17 to 0.27 mm)
Spacer thickness:
  Red .... 0.1604 to 0.1614 in (4.075 to 4.100 mm)
  Blue .... 0.1615 to 0.1624 in (4.101 to 4.125 mm)
Cam lift (except 2.8 litre fuel injection) .... 0.2545 to 0.2565 (6.465 to 6.516 mm)
Cam lift (2.8 litre fuel injection):
  Inlet .... 0.2638 in (6.700 mm)
  Exhaust:
    Early models .... 0.2622 in (6.660 mm)
    Later models .... 0.2598 in (6.600 mm)
Cam length, heel to toe:
  Early models, except fuel injection .... 1.3385 to 1.3465 in (33.998 to 34.201 mm)
  Later models, except fuel injection .... 1.3526 to 1.3581 in (34.355 to 34.495 mm)
  Fuel injection models, inlet .... 1.3266 to 1.3333 in (33.695 to 33.865 mm)
  Fuel injection models, exhaust .... 1.3226 to 1.3293 in (33.595 to 33.765 mm)
Journal diameter:
  Front .... 1.6497 to 1.6505 in (41.903 to 41.923 mm)
  Centre 1 .... 1.6347 to 1.6355 in (41.522 to 41.542 mm)
  Centre 2 .... 1.6197 to 1.6205 in (41.141 to 41.161 mm)
  Rear .... 1.6047 to 1.6055 in (40.760 to 40.780 mm)

Engine 1•7

Camshaft (cont.)
Bearing inside diameter:
- Front .... 1.6514 to 1.6522 in (41.948 to 41.968 mm)
- Centre 1 .... 1.6364 to 1.6372 in (41.567 to 41.587 mm)
- Centre 2 .... 1.6214 to 1.6222 in (41.186 to 41.206 mm)
- Rear .... 1.6064 to 1.6072 in (40.805 to 40.825 mm)

Endfloat .... 0.0008 to 0.0040 in (0.02 to 0.10 mm)

Pistons and piston rings
Piston diameter (2.3 litre):
- Standard:
  - 1 .... 3.5414 to 3.5418 in (89.952 to 89.962 mm)
  - 2 .... 3.5418 to 3.5422 in (89.962 to 89.972 mm)
  - 3 .... 3.5422 to 3.5425 in (89.972 to 89.982 mm)
  - 4 .... 3.5425 to 3.5429 in (89.982 to 89.992 mm)
- Service standard .... 3.5424 to 3.5433 in (89.978 to 90.002 mm)
  - Service oversize 0.5 mm .... 3.5621 to 3.5630 in (90.478 to 90.502 mm)
  - Service oversize 1.0 mm .... 3.5818 to 3.5827 in (90.978 to 91.002 mm)

Piston diameter (2.8 litre):
- Standard:
  - 1 .... 3.6603 to 3.6607 in (92.972 to 92.982 mm)
  - 2 .... 3.6607 to 3.6611 in (92.982 to 92.992 mm)
  - 3 .... 3.6611 to 3.6615 in (92.992 to 93.002 mm)
  - 4 .... 3.6615 to 3.6619 in (93.002 to 93.012 mm)

Clearance in bore:
- 2.3 litre .... 0.0011 to 0.0024 in (0.028 to 0.062 mm)
- 2.8 litre .... 0.0011 to 0.0019 in (0.028 to 0.048 mm)

Ring gap (fitted):
- Top .... 0.015 to 0.023 in (0.38 to 0.58 mm)
- Centre .... 0.015 to 0.023 in (0.38 to 0.58 mm)
- Bottom .... 0.015 to 0.055 in (0.38 to 1.40 mm)

Gudgeon pins
Diameter:
- Red .... 0.9446 to 0.9447 in (23.994 to 23.997 mm)
- Blue .... 0.9447 to 0.9448 in (23.997 to 24.000 mm)

Clearance in piston (floating):
- 2.3 litre .... 0.0002 to 0.0004 in (0.005 to 0.011 mm)
- 2.8 litre .... 0.0003 to 0.0005 in (0.008 to 0.014 mm)

Interference in connecting rod .... 0.0007 to 0.0017 in (0.018 to 0.042 mm)

Connecting rods
Big-end bearing internal diameter (2.3 litre):
- Standard .... 2.1262 to 2.1276 in (54.008 to 54.042 mm)
- Undersize:
  - 0.254 mm .... 2.1162 to 2.1176 in (53.754 to 53.788 mm)
  - 0.508 mm .... 2.1062 to 2.1076 in (53.500 to 53.534 mm)
  - 0.762 mm .... 2.0962 to 2.0976 in (53.246 to 53.280 mm)
  - 1.020 mm .... 2.0862 to 2.0876 in (52.992 to 53.026 mm)

Big-end bearing internal diameter (2.8 litre):
- Standard .... 2.1226 to 2.1282 in (54.016 to 54.056 mm)
- Undersize 0.254 mm .... 2.1166 to 2.1182 in (53.762 to 53.802 mm)

Bore diameter:
- Big-end .... 2.2370 to 2.2378 in (56.820 to 56.840 mm)
- Small-end .... 0.9432 to 0.9439 in (23.958 to 23.976 mm)

Crankpin/bearing shell clearance .... 0.0002 to 0.0025 in (0.006 to 0.064 mm)

Valves
Valve seat angle .... 44°30' to 45°

Valve guide bore:
- Standard .... 0.3174 to 0.3184 in (8.063 to 8.088 mm)
- Oversize 0.2 mm .... 0.3253 to 0.3262 in (8.263 to 8.288 mm)
- Oversize 0.4 mm .... 0.3331 to 0.3341 in (8.463 to 8.488 mm)

Valve clearance (cold):
- Inlet .... 0.014 in (0.35 mm)
- Exhaust .... 0.016 in (0.40 mm)

Valve timing (2.3 litre):   **Early models**   **Later models**

- Inlet valve:
  - Opens .... 20° BTDC   25° BTDC
  - Closes .... 56° ABDC   51° ABDC
- Exhaust valve:
  - Opens .... 62° BBDC   67° BBDC
  - Closes .... 14° ATDC   9° ATDC

Valves (cont.)

	Early models and all fuel injection	Later models except fuel injection
Valve timing (2.8 litre):		
Inlet valve:		
Opens	24° BTDC	20° BTDC
Closes	72° ABDC	56° ABDC
Exhaust valve:		
Opens	73° BBDC	62° BBDC
Closes	25° ATDC	14° ATDC
Tappet diameter	0.8736 to 0.8740 in (22.190 to 22.202 mm)	
Tappet clearance in housing	0.0009 to 0.0023 in (0.023 to 0.060 mm)	

Inlet valves

Length .... 4.133 to 4.181 in (105.000 to 106.200 mm)
Head diameter:
 2.3 litre - early models .... 1.261 to 1.275 in (32.030 to 32.410 mm)
 2.3 litre - later models .... 1.561 to 1.577 in (39.670 to 40.060 mm)
 2.8 litre .... 1.644 to 1.660 in (41.850 to 42.240 mm)
Stem diameter:
 Standard .... 0.315 to 0.316 in (8.025 to 8.043 mm)
 Oversize 0.2 mm .... 0.323 to 0.324 in (8.225 to 8.243 mm)
 Oversize 0.4 mm .... 0.331 to 0.332 in (8.425 to 8.443 mm)
 Oversize 0.6 mm .... 0.339 to 0.340 in (8.625 to 8.643 mm)
 Oversize 0.8 mm .... 0.347 to 0.348 in (8.825 to 8.843 mm)
Stem/guide clearance .... 0.0007 to 0.0024 in (0.020 to 0.063 mm)
Valve lift (except 2.8 litre fuel injection) .... 0.3555 to 0.3586 in (9.03 to 9.11mm)
Valve lift (2.8 litre fuel injection) .... 0.3705 in (9.43 mm)
Valve spring free length:
 All 2.3 litre and early 2.8 litre (except fuel injection) ...... 1.909 in (48.5 mm)
 Later 2.8 litre and all fuel injection models .... 1.988 in (50.5 mm)

Exhaust valves

Length .... 4.141 to 4.181 in (105.200 to 106.200 mm)
Head diameter:
 2.3 litre - early models .... 1.261 to 1.275 in (32.030 to 32.410 mm)
 2.3 litre - later models .... 1.331 to 1.346 in (33.830 to 34.210 mm)
 2.8 litre - all models .... 1.408 to 1.423 in (35.830 to 36.210 mm)
Stem diameter:
 Standard .... 0.314 to 0.315 in (7.999 to 8.017 mm)
 Oversize 0.2 mm .... 0.322 to 0.323 (8.199 to 8.217 mm)
 Oversize 0.4 mm .... 0.330 to 0.331 in (8.399 to 8.417 mm)
 Oversize 0.6 mm .... 0.338 to 0.339 in (8.599 to 8.617 mm)
 Oversize 0.8 mm .... 0.346 to 0.347 in (8.799 to 8.817 mm)
Stem/guide clearance .... 0.0018 to 0.0035 in (0.046 to 0.089 mm)
Valve lift (except 2.8 litre fuel injection) .... 0.3548 to 0.3580 in (9.01 to 9.09 mm)
Valve lift (2.8 litre fuel injection) .... 0.3627 in (9.23 mm)
Valve spring free length:
 All 2.3 litre and early 2.8 litre models (except fuel injection) .... 1.909 in (48.5 mm)
 Later 2.8 litre and all fuel injection models .... 1.988 in (50.5 mm)

Engine lubrication data

Oil type/specification .... Multigrade engine oil, viscosity range SAE 10W/30 to 10W/50, to API SF/CC or SF/CD
Oil filter .... Champion C102
Oil change capacity approx:
 With filter renewal .... 7.5 pints (4.25 litres)
 Without filter renewal .... 7.0 pints (4.0 litres)
Minimum oil pressure:
 At 750 rpm .... 14.5 lbf/in^2 (1.0 kgf/cm^2)
 At 2000 rpm .... 40.6 lbf/in^2 (2.8 kgf/cm^2)
Pressure relief valve opens at .... 58 to 68 lbf/in^2 (4.0 to 4.7 kgf/cm^2)
Warning light operates at .... 4.5 to 8.5 lbf/in^2 (0.3 to 0.6 kgf/cm^2)
Oil pump:
 Rotor/casing clearance .... 0.0059 to 0.0118 in (0.15 to 0.30 mm)
 Inner/outer rotor clearance .... 0.0019 to 0.0078 in (0.05 to 0.20 mm)
Endfloat rotor/casing .... 0.0011 to 0.0039 in (0.03 to 0.10 mm)

Torque wrench settings

	lbf ft	Nm
Main bearing caps	70	95
Big-end caps	23	31
Crankshaft pulley vibration damper central bolt (if fitted):		
2.3 litre manual without air conditioning	33	45
All other models (19 mm bolt)	85	115
Camshaft gear	33	45

Engine 1•9

Torque wrench settings

	lbf ft	Nm
Flywheel	49	67
Timing cover	13	18
Oil pump	11	15
Oil pump cover	7	10
Rocker shaft	46	63
Oil sump		
First stage	4	6
After 15 minutes at 1000 rpm	6	8
Oil sump drain plug	17	23
Oil pressure switch	10	14
Cylinder head bolts (Hexagon head bolts only):		
First stage	30 to 40	39 to 54
Second stage	40 to 50	54 to 69
Third stage (after 10 to 20 minutes wait)	70 to 85	95 to 115
After 15 minutes running at 1000 rpm	70 to 85	95 to 115
Rocker cover	4	6
Inlet manifold (carburettor engines):		
First stage	5	7
Second stage	8	11
Third stage	13	18
Fourth stage	17	23
After 15 minutes running at 1000 rpm	17	23
Inlet manifold - (fuel injection engines)	13	18
Pressure plate to flywheel	13	18
Fuel pump	13	18
Water pump	6	8
Spark plug	26	35
Exhaust manifolds:		
Pre-February 1978	17	23
Post February 1978	20	27

PART A
IN-LINE OHC ENGINE

1 General description

The 2.0 litre in-line ohc engine cylinder head is of the crossflow design. As flat top pistons are used the combustion chambers are contained in the cylinder head. The combined crankcase and cylinder block is made of cast iron and houses the pistons and crankshaft. Attached to the underside of the crankcase is a pressed steel sump which acts as a reservoir for the engine oil. Full information on the lubricating system will be found in Section 22.

The cast iron cylinder head is mounted on top of the cylinder block and acts as a support for the overhead camshaft. The slightly angled valves operate directly in the cylinder head and are controlled by the camshaft via cam followers. The camshaft is operated by a toothed reinforced composite rubber belt from the crankshaft. To eliminate backlash and prevent slackness of the belt a spring loaded tensioner in the form of a jockey wheel is in contact with the back of the belt. It serves two further functions, to keep the belt away from the water pump and also to increase the contact area of the camshaft and crankshaft sprockets.

The drivebelt also drives the auxiliary shaft sprocket and it is from this shaft that the oil pump, distributor and fuel pump operate.

The inlet manifold is mounted on the left-hand side of the cylinder head and to this the carburettor is fitted. A water jacket is

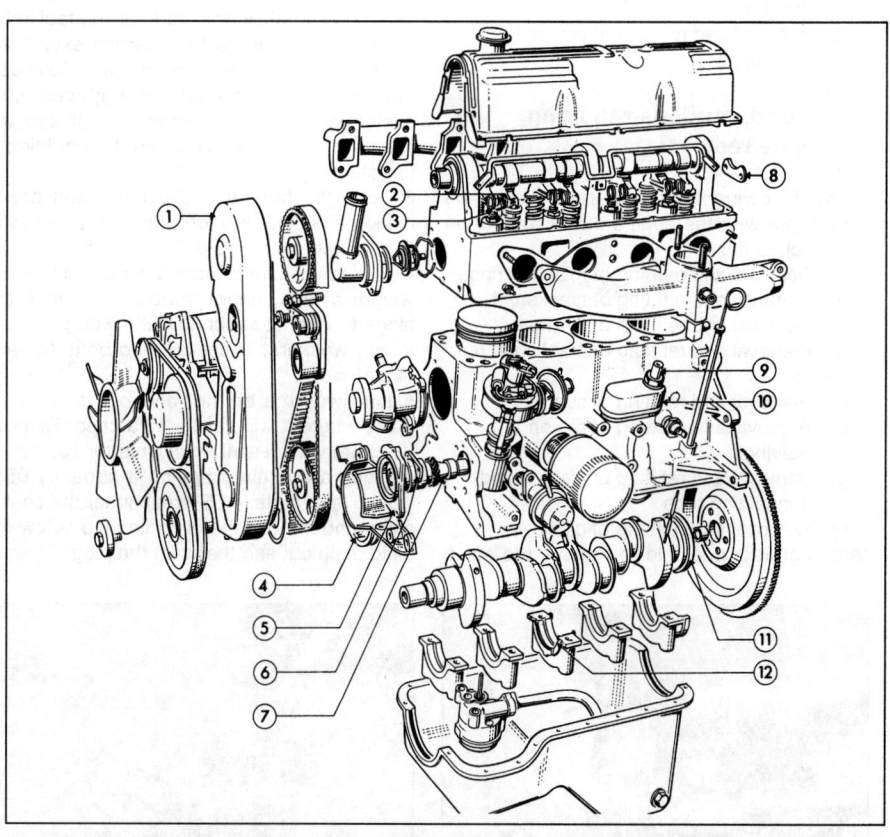

Fig. 1.1 Exploded view of the main engine components (Sec 1)

1 Belt cover
2 Cam follower
3 Cam follower spring
4 Crankshaft timing cover
5 Auxiliary shaft cover
6 Auxiliary shaft thrust plate
7 Auxiliary shaft
8 Thrust plate
9 Ventilation valve
10 Oil separator
11 Oil seal
12 Thrust washer

1•10 Engine

incorporated in the inlet manifold so that the petrol/air charge may be at the correct temperature before entering the combustion chambers.

The exhaust manifold is mounted on the right-hand side of the cylinder head and connects to a single downpipe and silencer system.

Aluminium alloy pistons are connected to the crankshaft by H-section forged steel connecting rods and gudgeon pins. The gudgeon pin is a press fit in the small end of the connecting rod but a floating fit in the piston boss. Two compression rings and one scraper ring, all located above the gudgeon pin, are fitted.

The forged crankshaft runs in five main bearings and endfloat is accommodated by fitting thrust washers either side of the centre main bearings.

2 Major operations possible with engine in car

The following major operations can be carried out to the engine with it in place:
a) Removal and refitting of camshaft
b) Removal and refitting of cylinder head
c) Removal and refitting of camshaft drivebelt
d) Removal and refitting of engine front mountings

3 Major operations requiring engine removal

The following major operations can be carried out with the engine out of the car on the bench or floor:
a) Removal and refitting of main bearings
b) Removal and refitting of crankshaft
c) Removal and refitting of flywheel
d) Removal and refitting of crankshaft rear oil seal
e) Removal and refitting of sump*
f) Removal and refitting of big-end bearings
g) Removal and refitting of pistons and connecting rods
h) Removal and refitting of oil pump

*Although it is possible to remove the sump pan if the engine is raised, the steering shaft disconnected and the crossmember lowered, the amount of work involved is not considered worthwhile.

4 Methods of engine removal

The engine may be lifted out either on its own or in unit with the transmission. On models fitted with automatic transmission it is recommended that the engine be lifted out on its own, unless a substantial crane or overhead hoist is available, because of the weight factor. If the engine and transmission are removed as a unit they have to be lifted out at a very steep angle, so make sure that there is sufficient lifting height available.

5 Engine - removal without transmission

1 The do-it-yourself owner should be able to remove the power unit fairly easily in about 4 hours. It is essential to have a good hoist and two axle-stands if an inspection pit is not available.
2 The sequence of operations listed in this Section is not critical as the position of the person undertaking the work, or the tool in his hand, will determine to a certain extent the order in which the work is tackled. Obviously the power unit cannot be removed until everything is disconnected from it and the following sequence will ensure that nothing is forgotten.
3 Open the bonnet and using a soft pencil mark the outline position of both the hinges at the bonnet to act as a datum for refitting.
4 With the help of a second person to take the weight of the bonnet undo and remove the hinge to bonnet securing bolts with plain and spring washers. There are two bolts to each hinge.
5 Lift away the bonnet and put it in a safe place where it will not be scratched. Remove the battery as described in Chapter 10.
6 Place a container having a capacity of at least 8 Imp. pints (4.55 litres) under the engine sump and remove the oil drain plug. Allow the oil to drain out and then refit the plug.

7 Refer to Chapter 3, and remove the air cleaner assembly from the top of the carburettor.
8 Mark the HT leads so that they may be refitted in their original positions and detach from the spark plugs.
9 Release the HT lead from the centre of the ignition coil.
10 Spring back the clips securing the distributor cap to the distributor body and lift off the distributor cap and leads.
11 Refer to Chapter 2 and drain the cooling system and remove the radiator.
12 Slacken the clip that secures the heater hose to the heater unit. Pull off the hose.
13 Slacken the clip that secures the hoses to the automatic choke and pull off the two hoses.
14 Slacken the clip securing the water hose to the adaptor elbow on the side of the inlet manifold and pull off the hose.
15 Slacken the clip that secures the fuel feed pipe to the carburettor float chamber and pull off the hose. Plug the end to stop dirt ingress or fuel loss due to syphoning.
16 Detach the throttle control inner cable from the operating rod (photo) and place out of the way.
17 Unscrew the throttle control outer cable securing nut and detach the cable from the mounting bracket (photo).
18 Remove the vacuum pipes from the manifold to brake servo unit and also the transmission vacuum hose. (Automatic transmission only).
19 Undo and remove the two nuts securing the exhaust down pipe clamp plate to the manifold. Slide the clamp plate down the pipe.
20 Detach the temperature gauge cable from the sender unit in the inlet manifold side of the cylinder head.
21 Pull the crankcase ventilation valve and hose from the oil separator on the left-hand side of the cylinder block.
22 Detach the oil pressure warning light cable from the switch located below the oil separator.
23 Detach the multi-plug connector from the rear of the alternator.
24 Make a note of the electrical connections to the starter motor solenoid and detach the cables (photo).

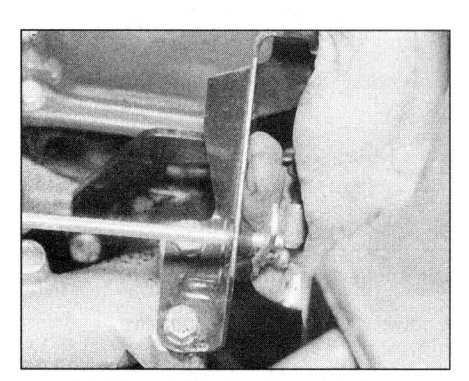

5.16 Detaching throttle cable from operating rod

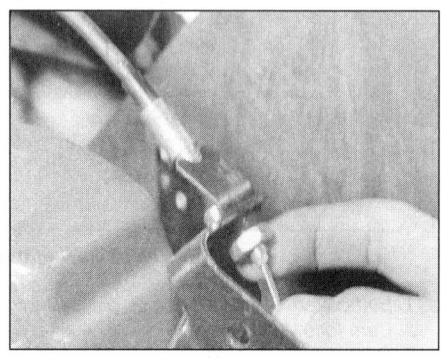

5.17 Detaching throttle cable assembly from mounting bracket

5.24 Location of cables at rear of starter solenoid

25 Raise the car at the front and support with axle-stands or blocks.
26 From underneath the car remove the starter motor.
27 Unscrew the clutch/converter housing lower bolts.
28 On automatic transmission models prise the rubber grommet from the cut away section of the starter motor aperture (certain models only) and then, using a suitable socket and extension, unscrew and remove the intermediate drive plate to torque converter retaining bolts.
29 Place a jack or support under the transmission and remove the remaining clutch/converter housing to engine bolts.
30 Check that apart from the engine mountings, the respective components and fittings to the engine are disconnected and positioned out of the way.
31 Place the lifting sling in position round the engine and adjust the lifting tackle so that it is just supporting the weight of the engine.
32 Unscrew the engine mounting lower retaining nuts and remove with flat washers.
33 On removal of the engine from automatic transmission models it is essential that the torque converter is retained on the transmission. While the engine is being removed, press the torque converter to the rear and after the engine has been removed secure the converter to the transmission with a piece of wood and metal bar. (See photo 66.35 Part B of this Chapter). The transmission should also be supported by ropes attached to a bar positioned across the engine compartment on the inner wing panels.
34 Carefully raise the engine from the car ensuring that it doesn't snag any of the surrounding components. When the sump is clear of the front grille panel the car can be wheeled out of the way, but make sure that the transmission is supported as just described.

6 Engine - removal with transmission

1 If automatic transmission is fitted it is advisable, due to the combined weight to remove the automatic transmission and engine separately as described in this Chapter and Chapter 6.
2 If manual transmission is fitted proceed as follows.
3 Disconnect the engine attachments as described in the previous Section.
4 Refer to Chapter 7 and remove the propeller shaft.
5 Undo and remove the nuts and bolts securing the exhaust downpipe to the manifold. Push the downpipe outward and tie out of the way of the gearbox.
6 If a centre console is fitted, this must be removed from inside the car. Directions will be found in Chapter 6.
7 Lift up the gear change lever gaiter and carefully remove the spring retainer.
8 Using a screwdriver prise open the locking tab and draw the gear change lever upward.

9 Refer to Chapter 5 and detach the clutch cable from the release arm.
10 Using a pair of circlip or pointed pliers, compress the circlip retaining the speedometer cable to the side of the gearbox casing. Withdraw the speedometer cable assembly.
11 If a reversing light is fitted disconnect the lead from the gearbox mounted switch.
12 Disconnect the crossmember from the gearbox and then detach the crossmember from the underside of the body by undoing and removing the retaining bolts and washers.
13 The engine and gearbox assembly should now be ready for removal. Check that no hoses, cables or other items have been left connected and all is clear so that lifting out can begin.
14 Lower the gearbox and raise the engine, drawing the unit forwards as much as possible.
15 It will now be necessary to tilt the unit at a steep angle and lift upward through the engine compartment. If there is not sufficient clearance it will be necessary to jack-up the front of the car and support on stands.
16 Continue raising the unit until the gearbox can be lifted over the front body member.
17 Move the car rearward or the unit forward until clear of the engine compartment and lower the unit to the floor. Suitably support so that it does not roll over.

7 Engine dismantling - general

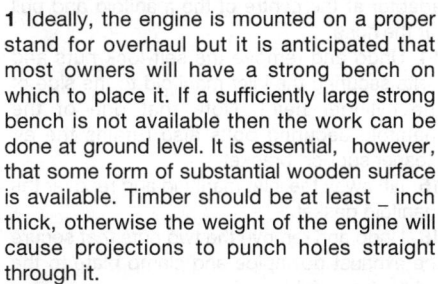

1 Ideally, the engine is mounted on a proper stand for overhaul but it is anticipated that most owners will have a strong bench on which to place it. If a sufficiently large strong bench is not available then the work can be done at ground level. It is essential, however, that some form of substantial wooden surface is available. Timber should be at least _ inch thick, otherwise the weight of the engine will cause projections to punch holes straight through it.
2 It will save a great deal of time later if the engine is thoroughly cleaned down on the exterior before any dismantling begins. This can be done by using paraffin and a stiff brush or more easily, probably, by the use of a water soluble solvent which can be brushed on and then the dirt swilled off with a water jet. This will dispose of all the heavy grease and grit once and for all so that later cleaning of individual components will be a relatively clean process and the paraffin bath will not become contaminated with abrasive metal.
3 As the engine is stripped down, clean each part as it comes off. Try to avoid immersing parts with oilways in paraffin as pockets of liquid could remain and cause oil dilution in the critical first few revolutions after reassembly. Clean oilways with wire, or preferably, an air jet.
4 It is helpful to obtain a few blocks of wood to support the engine while it is in the process of dismantling.

HAYNES HINT *Where possible, avoid damaging gaskets on removal, especially if new ones have not been obtained. They can be used as patterns if new ones have to be specially cut.*

5 Start dismantling at the top of the engine and then turn the block over and deal with the sump and crankshaft etc, afterwards.
6 Nuts and bolts should be refitted in their locations where possible to avoid confusion later. As an alternative keep each group of nuts and bolts (all the timing gear cover bolts for example) together in a jar or tin. This is particularly important since the various components are retained and secured with a mixture of AF and metric nuts and bolts.
7 Many items which are removed must be refitted in the same position, if they are not being renewed. These include valves, rocker arms, tappets, pistons, pushrods, bearings and connecting rods. Some of these are marked on assembly to avoid any possibility of mixing them up during overhaul. Others are not, and it is a great help if adequate preparation is made in advance to classify these parts. Suitably labelled tins or jars and, for small items, egg trays, tobacco tins and so on, can be used. The time spent in this operation will be amply repaid later.
8 Before beginning a complete overhaul or if the engine is being exchanged for a works reconditioned unit the following items should be removed:
All nuts and bolts associated with the following
Fuel system components
Carburettor
Inlet manifold
Exhaust manifolds
Fuel pump
Fuel lines
Ignition system components
Spark plugs
Distributor
Electrical system components
Alternator and mounting brackets
Starter motor
Cooling system components
Fan and hub
Water pump
Thermostat housing and thermostat
Water temperature sender unit
Engine
Crankcase ventilation tube
Oil filter element
Oil pressure sender unit
Oil level dipstick
Oil filler cap
Engine mounting brackets
Clutch
Clutch pressure plate assembly
Clutch friction plate assembly

Some of these items have to be removed for individual servicing or renewal periodically and details can be found under the appropriate Chapter.

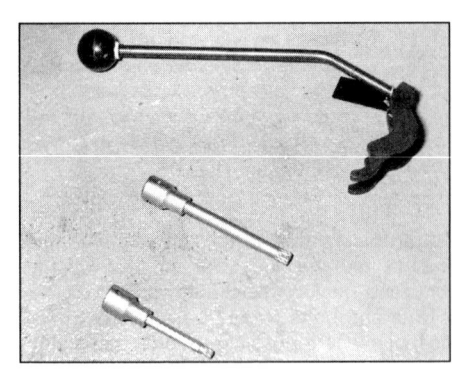

1•12 Engine

7.9 The three special tools necessary for dismantling

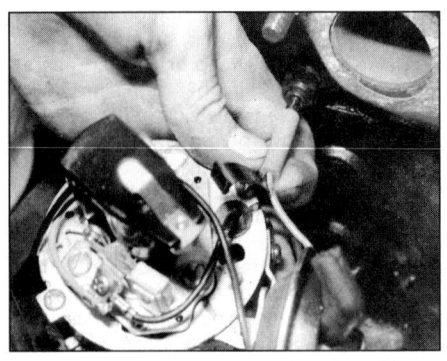

8.18 Thermal transmitter electric cable detachment

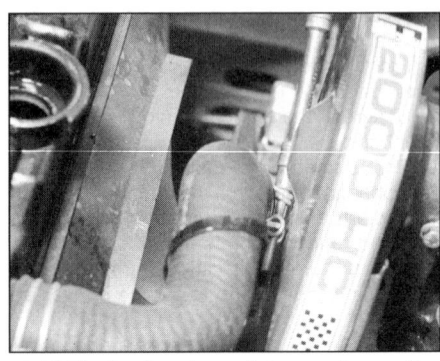

8.19 Slackening the radiator top hose clip

9 In addition it should be noted that three special tools are required to complete the service operation of the cylinder head, the belt tensioner mounting plate and the oil pump and the valves (photo). These are two wrenches to fit the internally splined bolts of the cylinder head, oil pump and camshaft belt tensioner mounting plate and a special valve spring compressor. The first two items are readily available from Ford dealers or motor supply stores while a universal type compressor will serve as a substitute for the official tool if slightly modified and used with caution.

8 Cylinder head - removal (engine in car)

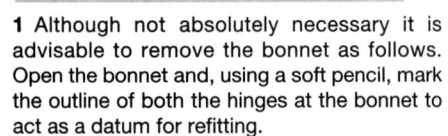

1 Although not absolutely necessary it is advisable to remove the bonnet as follows. Open the bonnet and, using a soft pencil, mark the outline of both the hinges at the bonnet to act as a datum for refitting.
2 With the help of a second person to take the weight of the bonnet undo and remove the hinge to bonnet securing bolts. There are two bolts to each hinge.
3 Lift away the bonnet and put in a safe place where it will not be scratched.
4 Disconnect the battery cables.
5 Refer to Chapter 3 and remove the air cleaner assembly from the top of the carburettor.
6 Mark the HT leads so that they may be refitted in their original positions and detach from the spark plugs.

7 Pull off the HT lead from the top of the ignition coil. Spring back the clips securing the distributor cap to the distributor body and lift off the cap and leads.
8 Refer to Chapter 2 and drain the cooling system.
9 Slacken the clip that secures the hoses to the automatic choke and pull off the hoses.
10 Slacken the clip that secures the fuel feed and return pipes to the carburettor float chamber and pull off the hoses. Plug the ends to stop dirt ingress or fuel loss.
11 Remove the distributor vacuum advance pipe and brake servo vacuum pipe from the manifold.
12 Slacken the clip securing the hose to the inlet manifold branch pipe adaptor and pull off the hose.
13 Slacken the clip securing the hose to the adaptor at the centre of the manifold and pull off the hose.
14 Undo and remove the self-lock nuts and bolts securing the inlet manifold to the side of the cylinder head. Note that one of the manifold securing bolts also retains the air cleaner support bracket.
15 Lift away the inlet manifold and recover the manifold gasket.
16 Undo and remove the two nuts that secure the exhaust downpipe and clamp plate to the exhaust manifold.
17 Slide the clamp plate down the exhaust pipe.
18 Detach the thermal transmitter electric cable from the inlet manifold side of the cylinder head (photo).

19 Slacken the radiator top hose clips and completely remove the hose (photo).
20 Undo and remove the bolts, spring and plain washers that secure the top cover to the cylinder head (photo).
21 Lift away the top cover (photo).
22 Undo and remove the two self-locking nuts that secure the heat deflector plate to the top of the exhaust manifold. Lift away the deflector plate (photo).
23 Undo and remove the bolts, spring and plain washers that secure the toothed drivebelt guard (photo).
24 Lift away the guard (photo).
25 Release the tension from the drivebelt by slackening the spring loaded roller mounting plate securing bolt (photo).
26 Lift the toothed drivebelt from the camshaft sprocket (photo).
27 Using the special tool (21-002) together with a socket wrench (photo), slacken the cylinder head securing bolts in a diagonal and progressive manner until all are free from tension. Remove in sequence (Fig. 1.2) the ten bolts, noting that because of the special shape of the bolt head, no washers are used. Unfortunately there is no other tool suitable to slot into the bolt head so do not attempt to improvise, which will only cause damage to the bolt.
28 The cylinder head may now be removed by lifting upwards (photo). If the head is stuck, try to rock it to break the seal. Under no circumstances try to prise it apart from the cylinder block with a screwdriver or cold chisel, as damage may be done to the faces of

8.20 Removing the top cover securing bolts . . .

8.21 . . . and lifting off the top cover

8.22 Removing the heat deflector plate

Engine 1•13

8.23 Removing the belt guard securing bolts

8.24 Belt guard removal

8.25 Releasing belt tensioner mounting plate securing bolt

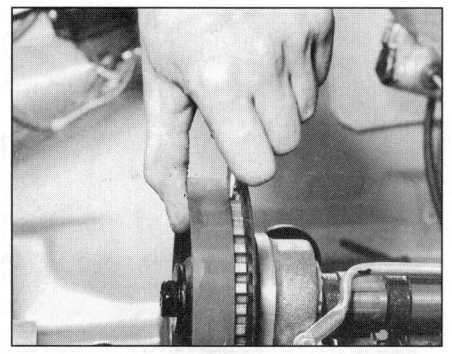

8.26 Removing the belt from the camshaft sprocket

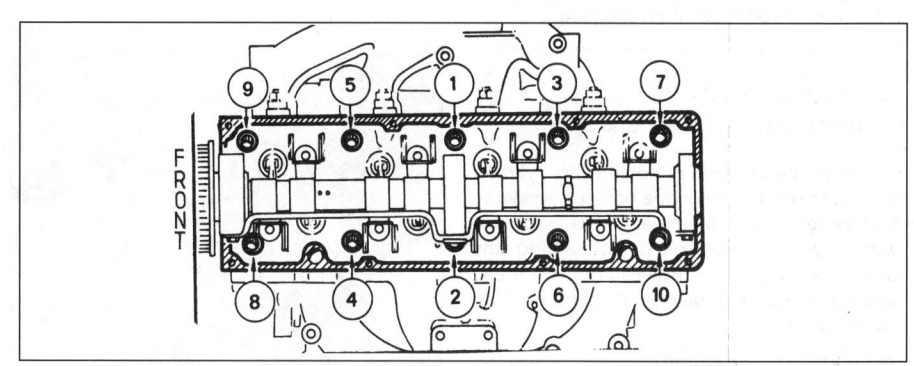

Fig. 1.2 Correct order for slackening or tightening cylinder head bolts (Sec 8)

8.27 Slackening the cylinder head securing bolts

8.28a Cylinder head removal

8.28b Engine with cylinder head removed

10.1 Auxiliary shaft sprocket securing bolt removal

10.2 Remove the auxiliary shaft timing cover securing bolts . . .

10.3 . . . and lift off the timing cover

1•14 Engine

the cylinder head and block. Under no circumstances hit the head directly with a metal hammer as this may cause the casting to fracture. Several sharp taps with the hammer, at the same time pulling upwards, should free the head. Lift the head off and place to one side (photo).

> **HAYNES HiNT** *If the head will not readily free, temporarily refit the battery and turn the engine over using the starter motor, as the compression in the cylinders will often break the cylinder head joint. If this fails to work, strike the head sharply with a plastic headed or wooden hammer, or with a metal hammer with an interposed piece of wood to cushion the blow.*

9 Cylinder head - removal (engine on bench)

The procedure for removing the cylinder head with the engine on the bench is similar to that for removal when the engine is in the car, with the exception of disconnecting the controls and services. Refer to Section 8 and follow the sequence given in paragraphs 20 to 28, inclusive.

10 Auxiliary shaft - removal

1 Using a metal bar lock the shaft sprocket, and with an open ended spanner undo and remove the bolt and washer that secures the sprocket to the shaft (photo).
2 Undo and remove the three bolts and spring washers that secure the shaft timing cover to the cylinder block (photo).
3 Lift away the timing cover (photo).
4 Undo and remove the two crosshead screws that secure the shaft thrust plate to the cylinder block (photo).
5 Lift away the thrust plate (photo).
6 The shaft may be drawn forwards and then lifted away (photo).

11 Flywheel and sump - removal

1 With the clutch removed, as described in Chapter 5, lock the flywheel using a screwdriver in mesh with the starter ring gear and undo the six bolts that secure the flywheel to the crankshaft in a diagonal and progressive manner (photo). Lift away the bolts.
2 Mark the relative position of the flywheel and crankshaft and then lift away the flywheel (photo).
3 Undo the remaining engine backplate securing bolts and ease the backplate from the two dowels. Lift away the backplate (photo).
4 Undo and remove the bolts that secure the sump to the underside of the crankcase (photo).
5 Lift away the sump and its gasket (photo).

10.4 Unscrew the auxiliary shaft thrust plate securing screws . . .

10.5 . . . and remove the thrust plate

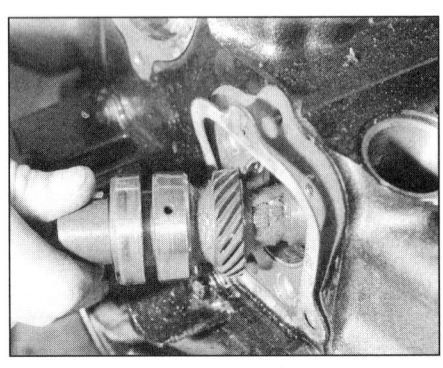

10.6 Withdrawal of auxiliary shaft

11.1 Remove the flywheel securing bolts . .

11.2 . . . and lift away the flywheel

11.3 Backplate removal

11.4 Removing the sump securing bolts

11.5 Lifting away sump

Engine 1•15

12.2 Removal of oil pump securing bolts

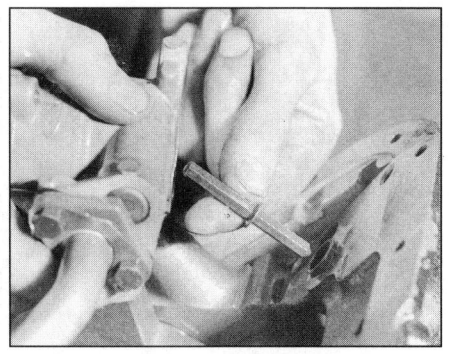

12.3 Lifting out oil pump, pick-up pipe and driveshaft

13.3 Removal of sprocket from crankshaft

12 Oil pump and strainer - removal

1 Undo and remove the screw and spring washer that secures the oil pump pick-up pipe support bracket to the crankcase.
2 Using special tool (21-012) undo the two special bolts that secure the oil pump to the underside of the crankcase. Unfortunately, there is no other tool suitable to slot into the screw head so do not attempt to improvise, which will only cause damage to the screw (photo).
3 Lift away the oil pump and strainer assembly (photo).
4 Carefully lift away the oil pump drive making a special note of which way round it is fitted.

13 Crankshaft pulley, sprocket and timing belt cover - removal

1 Lock the crankshaft using a block of soft wood placed between a crankshaft web and the crankcase. Then, using a socket and suitable extension, undo the bolt that secures the crankshaft pulley. Recover the large diameter plain washer.
2 Using two large screwdrivers, ease the pulley from the crankshaft. Recover the large diameter thrust washer.
3 Again using the screwdrivers ease the sprocket from the crankshaft (photo).
4 Undo and remove the bolts and spring washers that secure the timing belt cover to the front of the crankcase.
5 Lift away the timing belt cover and gasket (photo).

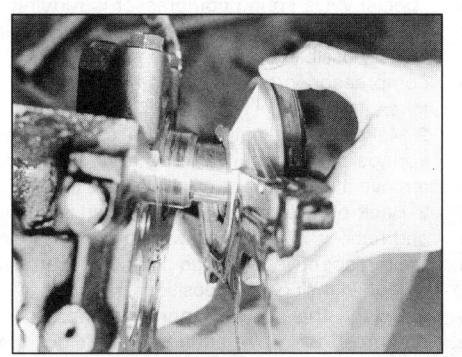

14 Pistons, connecting rods and big-end bearings - removal

1 Note that the pistons have an arrow marked on the crown showing the forward facing side (photo). Inspect the big-end bearing caps and connecting rods to make sure identification marks are visible. This is to ensure that the correct end caps are fitted to the correct connecting rods and the connecting rods placed in their respective bores.
2 Undo the big-end nuts and place to one side in the order in which they were removed.
3 Remove the big-end caps, taking care to keep them in the right order and the correct way round (photo).

> **HAYNES HiNT** *Ensure that the shell bearings are kept with their correct connecting rods unless the rods or shells are to be renewed.*

4 If the big-end caps are difficult to remove, they may be gently tapped with a soft hammer.
5 To remove the shell bearings, press the bearing opposite the groove in both the connecting rod and its cap, and the bearing will slide out easily.
6 Withdraw the pistons and connecting rods upwards and ensure they are kept in the correct order for refitting in the same bores as they were originally fitted.

15 Crankshaft and main bearings - removal

With the engine removed from the car and separated from the transmission, and the drivebelt, crankshaft pulley and sprocket, flywheel and backplate, oil pump, big-end bearings and pistons all dismantled, proceed to remove the crankshaft and main bearings.

1 Make sure that identification marks are visible on the main bearing caps, so that they

13.5 Removal of timing cover and gasket

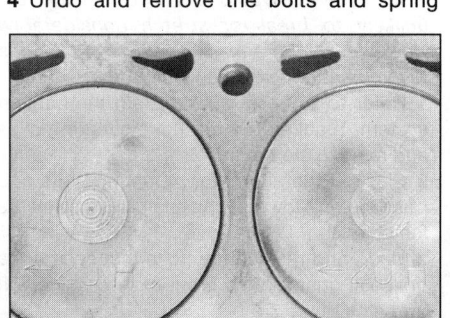

14.1 Piston identification marks stamped on crown

14.3 Lifting away big-end cap

15.1 Main bearing cap identification marks

1•16 Engine

15.3 Lifting away No 2 main bearing cap

15.4 Rear main bearing cap removal

15.8 Lifting away crankshaft rear oil seal

may be fitted in their original positions and also the correct way round (photo).
2 Undo by one turn at a time the bolts which hold the five bearing caps.
3 Lift away each main bearing cap and the bottom half of each bearing shell, taking care to keep the bearing shell in the right caps (photo).
4 When removing the rear main bearing end cap, note that this also retains the crankshaft rear oil seal (photo).
5 When removing the centre main bearing, note the bottom semicircular halves of the thrust washers, one half lying on either side of the main bearing. Lay them with the centre main bearing along the correct side.
6 As the centre and rear bearings end caps are accurately located by dowels, it may be

15.9 Cylinder block with crankshaft removed

necessary to gently tap the caps to release them.
7 Slightly rotate the crankshaft to free the upper halves of the bearing shells and thrust washers which can be extracted and placed carefully over the correct bearing cap.
8 Carefully lift away the crankshaft rear oil seal (photo).
9 Remove the crankshaft by lifting it away from the crankcase (photo).

16 Camshaft drivebelt - removal (engine in car)

It is possible to remove the camshaft drivebelt with the engine in situ but experience is such that this type of belt is very reliable and unlikely to break or stretch considerably. However, during a major engine overhaul it is recommended that a new belt is fitted.

1 Refer to Chapter 2, and drain the cooling system. Slacken the top hose securing clips and remove the top hose.
2 Slacken the alternator mounting bolts and push the unit towards the engine. Lift away the fan belt.
3 Undo and remove the bolts that secure the drivebelt guard to the Front of the engine. Lift away the guard.
4 Slacken the belt tensioner mounting plate securing bolt and release the tension on the belt.

5 Place the car in gear (manual gearbox only), and apply the brakes firmly. Undo and remove the bolt and plain washer that secure the crankshaft pulley to the nose of the crankshaft. On vehicles fitted with automatic transmission, the starter must be removed and the ring gear jammed to prevent the crankshaft rotating.
6 Using two screwdrivers carefully ease off the pulley (photo).
7 Recover the plain large diameter thrust washer.
8 The drivebelt may now be lifted away (photo).

17 Valves - removal

1 To enable the valves to be removed a special valve spring compressor is required. This has a part number of 21-005. However it is just possible to use a universal valve spring compressor provided extreme caution is taken.
2 Make a special note of how the cam follower springs are fitted and, using a screwdriver, remove these from the cam followers (photos).
3 Back off fully the cam follower adjustment and remove the cam followers. Keep these in their respective order so that they can be refitted in their original positions.
4 Using the valve spring compressor, compress the valve springs and lift out the collets (photo).

16.6 Crankshaft pulley removal

16.8 Timing belt removal

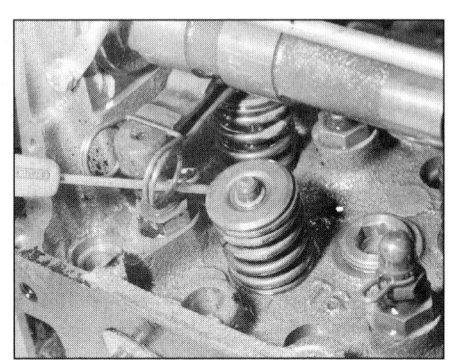

17.2 Cam follower spring removal

Engine 1•17

17.4 Compressing valve spring

18.1 Removing camshaft lubrication pipe (later models have a modified oil pipe)

18.2 Camshaft lubrication pipe oil holes (early type)

5 Remove the spring cap and spring and, using a screwdriver, prise the oil retainer caps out of their seats. Remove each valve and keep it in its respective order unless it is so badly worn that it is to be renewed. Also keep the valve springs, cups etc, in the correct order.
6 If necessary unscrew the ball head bolts.

> **HAYNES HINT** *If the valves are going to be used again, place them in a sheet of card having eight numbered holes corresponding with the relative positions of the valves when fitted.*

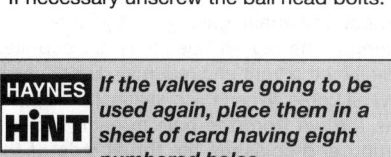

18 Camshaft - removal

It is not necessary to remove the engine from the car in order to remove the camshaft. However, it will be necessary to remove the cylinder head first (Section 8) as the camshaft has to be withdrawn from the rear.
Before removing the camshaft, check it for wear as given in Section 31, paragraph 5.
1 Undo and remove the bolts, and spring washers and bracket that secure the camshaft lubrication pipe. Lift away the pipe (photo).
2 Carefully inspect the fine oil drillings in the pipe to make sure that none are blocked (photo).

3 Using a metal bar, lock the camshaft drive sprocket then undo and remove the sprocket securing bolt and washer (photo).
4 Using a soft-faced hammer or screwdriver, ease the sprocket from the camshaft (photo).
5 Undo and remove the two bolts and spring washers that secure the camshaft thrust plate to the rear bearing support (photo).
6 Lift away the thrust plate noting which way round it is fitted (photo).
7 Remove the cam follower springs and then the cam followers as detailed in Section 17, paragraphs 2 and 3.
8 The camshaft may now be removed by using a soft-faced hammer and tapping rearwards. Take care not to cut the fingers when the camshaft is being handled as the sides of the lobes can be sharp (photo).

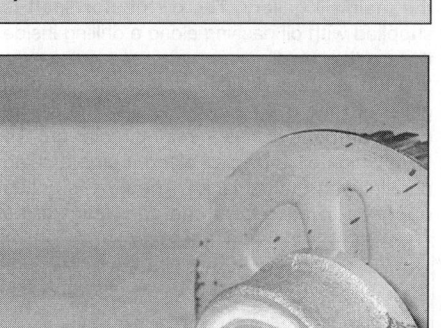

18.3 Using a metal bar to lock the camshaft sprocket

18.4 Removing the camshaft sprocket

18.5 Undo the camshaft thrust plate securing bolts . . .

18.6 . . . and lift off the thrust plate

18.8 Tapping the camshaft through bearings

18.9 Camshaft removal

1•18 Engine

9 Lift the camshaft carefully through the bearing inserts as the lobes can easily damage the soft metal bearing surfaces (photo).
10 If the oil seal has hardened or become damaged, it may be removed by prising it out with a screwdriver (photo).

19 Thermostat housing and belt tensioner - removal

1 Removal of these parts will usually only be necessary if the cylinder head is to be completely dismantled.
2 Undo and remove the two bolts and spring washers that secure the thermostat housing to the front face of the cylinder head.
3 Lift away the thermostat housing and recover its gasket (photo).

18.10 Camshaft oil seal removal

19.3 Thermostat housing removal

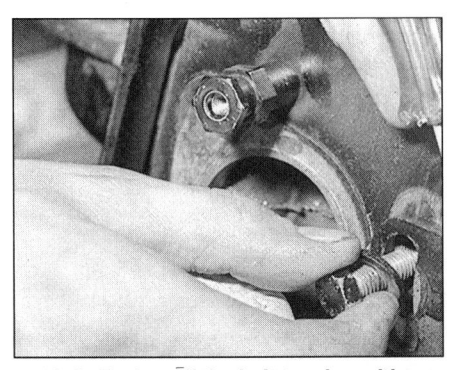

19.4b Easing off the belt tension with a screwdriver

4 Undo and remove the bolt and spring washer that secures the belt tensioner to the cylinder head. It will be necessary to override the tension using a screwdriver as a lever (photos).
5 Using tool number 21-012 (the tool for removal of the oil pump securing bolts) unscrew the tensioner mounting plate and spring shaped bolt and lift away the tensioner assembly (photo).

20 Gudgeon pin - removal

Interference type fit gudgeon pins are used and it is important that no damage is caused during removal and refitting. Because of this, should it be necessary to fit new pistons, take the parts along to the local Ford garage who will have the special equipment to do this job.

21 Piston rings - removal

1 To remove the piston rings, slide them carefully over the top of the piston, taking care not to scratch the aluminium alloy; never slide them off the bottom of the piston skirt. It is very easy to break the cast iron piston rings if they are pulled off roughly, so this operation should be done with extreme care. It is helpful to make use of old feeler gauges.

19.4a Removal of belt tensioner mounting plate securing bolt

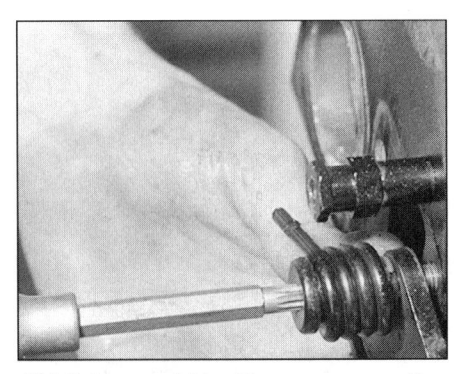

19.5 Using special tool to remove mounting plate and spring securing bolt from belt tensioner

2 Lift one end of the piston ring to be removed out of its groove and insert under it the end of the feeler gauge.
3 Turn the feeler gauge slowly round the piston and, as the ring comes out of its groove, apply slight upward pressure so that it rests on the land above. It can then be eased off the piston, with the feeler gauge stopping it from slipping into an empty groove if it is any but the top piston ring that is being removed.
4 The piston rings must always be removed from the top of the piston.

22 Lubrication and crankcase ventilation systems - description

1 The pressed steel oil sump is attached to the underside of the crankcase and acts as a reservoir for the engine oil. The oil pump draws oil through a strainer located under the oil surface, passes it along a short passage and into the full flow oil filter. The freshly filtered oil flows from the centre of the filter element and enters the main gallery. Five small drillings connect the main gallery to the five main bearings. The big-end bearings are supplied with oil by the front and rear main bearings via skew oil bores. When the crankshaft is rotating oil is thrown from the hole in each big-end bearing and splashes the thrust side of the piston and bore.
2 The auxiliary shaft is lubricated directly from the main oil gallery. The distributor shaft is supplied with oil passing along a drilling inside the auxiliary shaft.
3 A further three drillings connect the main oil gallery to the overhead camshaft. The centre camshaft bearing has a semi-circular groove from which oil is passed along a pipe running parallel with the camshaft. The pipe is drilled opposite to each cam and cam follower so

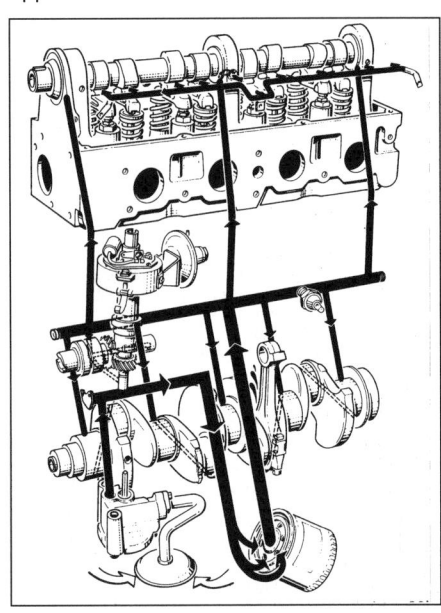

Fig. 1.3 Circulation of lubricant through the engine (Sec 22)

Engine 1•19

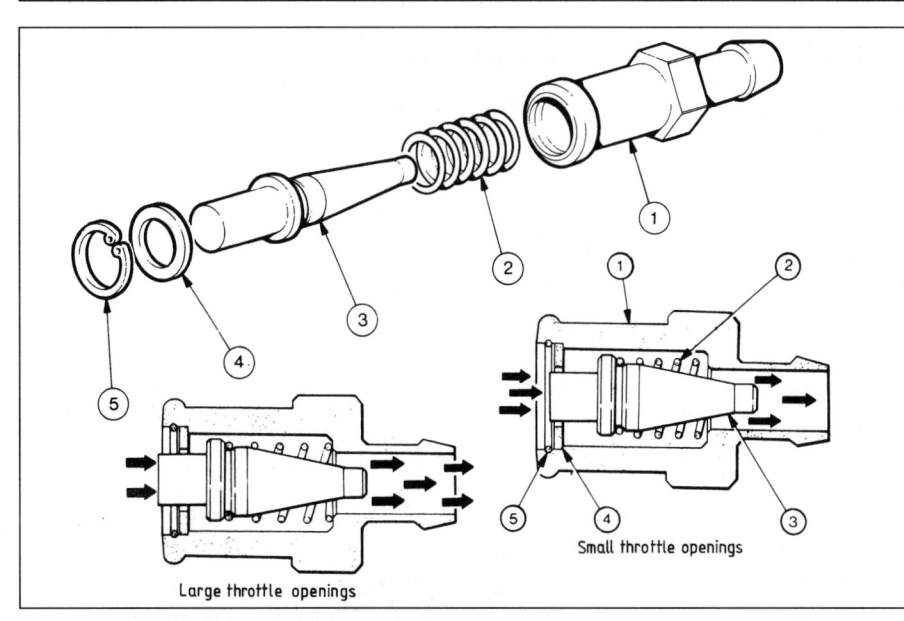

Fig. 1.4 The PCV valve showing large and small throttle opening positions (Sec 22)
1 Valve body 2 Spring 3 Piston 4 Washer 5 Circlip

providing the lubrication to the sump, via large drillings in the cylinder head and cylinder block.

4 The engine oil dipstick is on the left-hand side of the engine, and indicates the maximum and minimum oil levels.

5 A semi-enclosed engine ventilation system is used to control crankcase vapour. It is controlled by the amount of air drawn in by the engine when running and the throughput of the regulator valve (Fig. 1.4). The system is known as the PCV system (Positive Crankcase Ventilation) and the advantage of the system is that should the "blowby" exceed the capacity of the PCV valve, excess fumes are fed into the engine through the air cleaner. This is caused by the rise in crankcase pressure which creates a reverse flow in the air intake pipe.

6 Periodically, pull the valve and hose from the rubber grommet of the oil separator and inspect the valve for free movement. If it is sticky in action or is choked with sludge, dismantle it and clean the components.

7 Occasionally check the security and condition of the system connecting hoses.

23 Oil pump - dismantling, inspection and reassembly

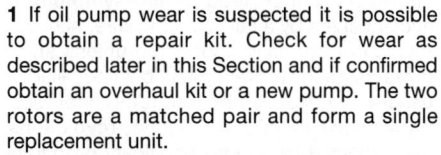

1 If oil pump wear is suspected it is possible to obtain a repair kit. Check for wear as described later in this Section and if confirmed obtain an overhaul kit or a new pump. The two rotors are a matched pair and form a single replacement unit.

2 Undo and remove the two bolts and spring washers that secure the intake cowl to the oil pump body. Lift away the cowl and its gasket (Fig.1.5).

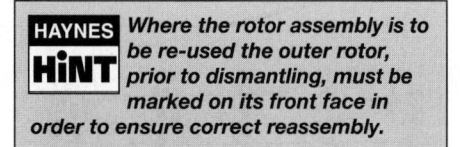

Haynes Hint: *Where the rotor assembly is to be re-used the outer rotor, prior to dismantling, must be marked on its front face in order to ensure correct reassembly.*

3 Note the relative position of the oil pump cover and body and then undo and remove the three bolts and spring washers. Lift away the cover.

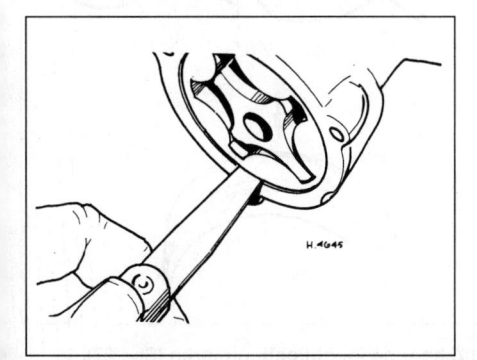

Fig. 1.6. Checking oil pump outer rotor to body clearance (Sec 23)

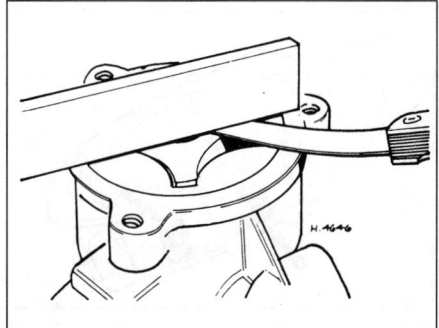

Fig. 1.7 Checking oil pump endfloat (Sec 23)

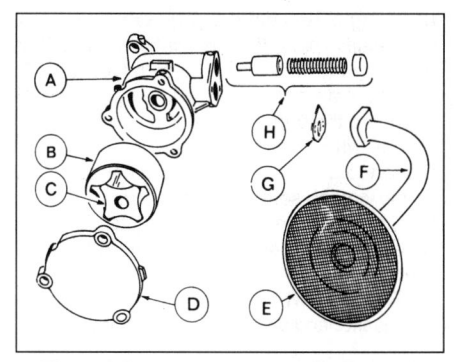

Fig. 1.5 Oil pump components (Sec 23)

A Pump body   E Strainer
B Outer rotor   F Suction pipe
C Inner rotor   G Gasket
D Cover   H Pressure relief valve

4 Carefully remove the rotors from the housing.

5 Using a centre punch, tap a hole in the centre of the pressure relief valve sealing plug, and make a note to obtain a new one.

6 Screw in a self-tapping screw and, using an open-ended spanner, withdraw the sealing plug.

7 Thoroughly clean all parts in petrol or paraffin and wipe dry using a non-fluffy rag. The necessary clearances may now be checked using a machined straight-edge (a good steel rule) and a set of feeler gauges. The critical clearances are between the lobes of the centre rotor and convex faces of the outer rotor; between the outer rotor and the pump body: and between both rotors and the end cover plate.

8 The rotor lobe clearance may be checked using feeler gauges and should be within the limits specified.

9 The clearance between the outer rotor and pump body should be within the limits specified (Fig. 1.6).

10 The endfloat clearance may be measured by placing a steel straight edge across the end of the pump and measuring the gap between the rotors and the straight edge. The gap in either rotor should be within the limits specified as shown in Fig. 1.7.

11 If the only excessive clearances are endfloat it is possible to reduce them by removing the rotors and lapping the face of the body on a flat bed until the necessary clearances are obtained. It must be emphasised, however, that the face of the body must remain perfectly flat and square to the axis of the rotor spindle otherwise the clearances will not be equal and the end cover will not be a pressure tight fit to the body. It is worth trying, of course, if the pump is in need of renewal anyway but unless done properly, it could seriously jeopardise the rest of the overhaul. Any variations in the other two clearances should be overcome with a new unit.

12 With all parts scrupulously clean, first refit the relief valve and spring and lightly lubricate with engine oil.

1•20 Engine

13 Using a suitable diameter drift, drive in a new sealing plug, flat side outwards until it is flush with the intake cowl bearing face.
14 Lubricate both rotors with engine oil and insert into the body. Fit the oil pump cover and secure with the three bolts in a diagonal sequence and progressive manner to a final torque wrench setting as specified.
15 Fit the intermediate shaft into the rotor driveshaft and make sure that the rotor turns freely.
16 Fit the cowl to the pump body, using a new gasket and secure with two bolts.

24 Oil filter - removal and refitting

1 The oil filter element is of the disposable element type and is located on the left-hand side of the cylinder block.
2 To remove, simply unscrew the old unit but be prepared for oil spillage and place a container or tray underneath when removing.
3 Wipe clean the mating surface at the cylinder block and the surrounding area. Before fitting the new element, smear the rubber O-ring on the new filter with a small amount of clean engine oil. Screw the new filter into position firmly by hand but do not overtighten. When the engine is restarted, check around the filter to ensure that it has sealed correctly.

25 Engine components - examination for wear

1 When the engine has been stripped down and all parts properly cleaned, decisions have to made as to what needs renewal and the following sections tell the examiner what to look for. In any border line case it is always best to decide in favour of a new part. Even if a part may still be serviceable, its life will have been reduced by wear, and the degree of trouble needed to renew it in the future must be taken into consideration. However, these things are relative and it depends on whether a quick "survival" job is being done or whether the car as a whole is being regarded as having many thousands of miles of useful and economical life remaining.

26 Crankshaft - examination and renovation

1 Examine the main bearing journals and the crankpins, and if there are any scratches or score marks, the shaft will need regrinding. Such conditions will nearly always be accompanied by similar deterioration in the machine bearing shells.
2 Each bearing journal should also be round and can be checked with a micrometer or caliper gauge around the periphery at several points. If there is more than 0.001 inch (0.0254 mm) of ovality, regrinding is necessary.
3 A main Ford agent or motor engineering specialist will be able to decide to what extent regrinding is necessary and supply the special undersize shell bearings to match whatever may need grinding off.
4 Before taking the crankshaft for regrinding, check the cylinder bores and pistons as it may be advantageous to have all relevant components attended to at the same time.
5 The crankshaft oilways must be cleared. This can be done by probing with wire or by blowing through with an air line. Then insert the nozzle of an oil gun into the respective oilways, each time blanking off the previous hole, and squirt oil through the shaft. It should emerge through the next hole. Any oilway blockage must obviously be cleaned out prior to refitting the crankshaft.
6 Refer to Fig. 1.8 and Section 27 for the crankshaft and main bearing identification code information.

27 Main and big-end bearings - examination and renovation

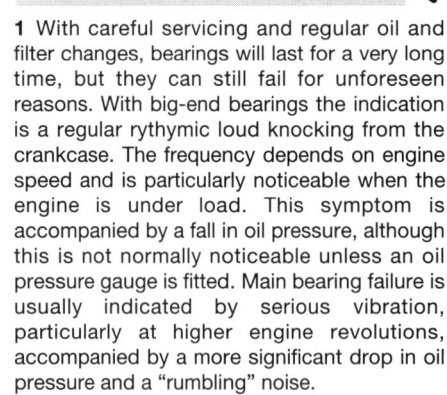

1 With careful servicing and regular oil and filter changes, bearings will last for a very long time, but they can still fail for unforeseen reasons. With big-end bearings the indication is a regular rythymic loud knocking from the crankcase. The frequency depends on engine speed and is particularly noticeable when the engine is under load. This symptom is accompanied by a fall in oil pressure, although this is not normally noticeable unless an oil pressure gauge is fitted. Main bearing failure is usually indicated by serious vibration, particularly at higher engine revolutions, accompanied by a more significant drop in oil pressure and a "rumbling" noise.
2 Bearing shells in good condition have bearing surfaces with a smooth, even matt silver/grey colour all over. Worn bearings will show patches of a different colour when the bearing metal has worn away and exposed the underlay. Damaged bearings will be pitted or scored. If the crankshaft is in good condition it is merely a question of obtaining another set of bearings the same size. A reground crankshaft will need new bearing shells as a matter of course.
3 The original bore in the cylinder block may have been standard or 0.015 inch (0.38 mm) oversize and in the latter instance this will be indicated by the bearing caps which will be marked with white paint.
4 The original size of the crankshaft main bearing journals may have been standard or 0.010 inch (0.25 mm) undersize. If the journals were undersize originally, this will be indicated by a green stripe on the first balance weight.
5 The original size of the big-end journal was either standard or 0.010 inch (0.25 mm) undersize. If undersize, the corresponding journal web will be marked with a green spot.

> **HAYNES HINT** *If the crankshaft is not being reground, but the bearings are to be renewed, take the old ones along to your supplier and this will act as a check that you are getting the correct size bearings. Undersize bearings are marked as such on the reverse face.*

6 Further information regarding main and big-end bearing identification is shown in Fig. 1.8.

28 Cylinder bores - examination and renovation

1 A new cylinder is perfectly round and the walls parallel throughout its length. The action of the piston tends to wear the walls at right angles to the gudgeon pin due to side thrust.

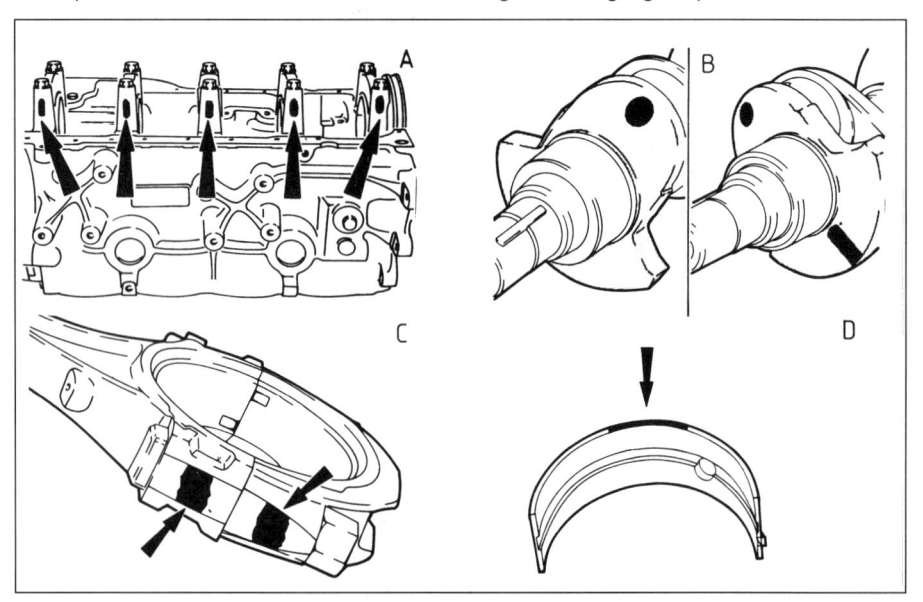

Fig. 1.8 Crankshaft, connecting rod and bearing identification marks (Sec 27)

A *Main bearing identification* C *Connecting rod identification* D *Bearing identification*
B *Crankshaft identification*

This wear takes place principally on that section of the cylinder swept by the piston rings.

2 It is possible to get an indication of bore wear by removing the cylinder heads with the engine still in the car. With the piston down in the bore first signs of wear can be seen and felt just below the top of the bore where the top piston ring reaches and there will be a noticeable lip (other than normal carbon build-up). If there is no lip it is fairly reasonable to expect that bore wear is not severe and any lack of compression or excessive oil consumption is due to worn or broken piston rings or pistons (See Section 35).

3 If it is possible to obtain a bore measuring micrometer, measure the bore in the thrust plane below the lip and again at the bottom of the cylinder in the same plane. If the difference is more than 0.003 inch (0.0762 mm) then a rebore is necessary. Similarly a difference of 0.003 inch (0.0762 mm) or more across the bore diameter is a sign of ovality calling for a rebore.

4 Any bore which is significantly scratched or scored will need reboring. This symptom usually indicates that the piston or rings are damaged. Even if only one cylinder is in need of reboring it will still be necessary for all cylinders to be bored and fitted with new oversize pistons and rings. Your Ford agent or local motor engineering specialist will be able to rebore and obtain the necessary matched pistons. If the crankshaft is undergoing regrinding also, it is a good idea to let the same firm renovate and reassemble the crankshaft and pistons to the block. A reputable firm normally gives a guarantee for such work. In cases where engines have been rebored already to their maximum, cylinder liners are available which may be fitted. In such cases the same reboring processes have to be followed and the services of a specialist engineering firm are required.

29 Pistons and piston rings - examination and renovation

1 Worn pistons and rings can usually be diagnosed when the symptoms of excessive oil consumption and low compression occur and are sometimes, though not always, associated with worn cylinder bores. Compression testers that fit onto the spark plug holes are available and these can indicate where low pressure is occurring. Wear usually accelerates the more it is left so when the symptoms occur early action can possibly save the expense of a rebore.

2 Another symptom of piston wear is piston-slap - a knocking noise from the crankcase, not to be confused with big-end bearing failure. It can be heard clearly at low engine speed when there is no load (idling for example) and is much less audible when the engine speed increases. Piston wear usually occurs in the skirt or lower end of the piston and is indicated by vertical streaks in the worn area which is always on the thrust side. It can also be seen where the skirt thickness is different.

3 Piston ring wear can be checked by first removing the rings from the pistons as described in Section 21. Then place the rings in the cylinder bores from the top, pushing them down about 1½ inches (38 mm) with the head of the piston (from which the rings have been removed) so that they rest square in the cylinder. Then measure the gap at the ends of the ring with a feeler gauge. If it exceeds 0.020 inch (0.508 mm) for the two top compression rings, or 0.015 inch (0.381 mm) for the lower oil control ring then they need renewal.

4 The grooves in which the rings locate in the piston can also become enlarged in use. The clearance between ring and piston, in the groove, should not exceed 0.004 inch (0.102 mm) for the top two compression rings and 0.003 inch (0.0762 mm) for the lower oil control ring.

5 However, it is rare that a piston is only worn in the ring grooves and the need to renew them for this fault alone is hardly ever encountered. Wherever pistons are renewed the weight of the four piston/connecting rod assemblies should be kept within the limit variation of 8 gms. to maintain engine balance.

30 Connecting rods and gudgeon pins - examination and renovation

1 Gudgeon pins are a shrink fit into the connecting rods. Neither of these would normally need renewal unless the pistons were being changed, in which case the new pistons would automatically be supplied with new gudgeon pins.

2 Connecting rods are not subject to wear but in extreme circumstances such as engine seizure they could be distorted. Such conditions may be visually apparent but where doubt exists they should be checked for alignment and if necessary renewed or straightened. The bearing caps should also be examined for indications of filing down which may have been attempted in the mistaken idea that bearing slackness could be remedied in this way. If there are such signs then the connecting rods should be renewed.

31 Camshaft, camshaft bearings and cam followers - examination and renovation

1 The camshaft bearing bushes should be examined for signs of scoring and pitting. If they need renewal they will have to be dealt with professionally as, although it may be relatively easy to remove the old bushes, the correct fitting of new ones requires special tools. If they are not fitted evenly and square from the very start they can be distorted thus causing localised wear in a very short time. See your Ford dealer or local engineering specialist for this work.

2 The camshaft itself may show signs of wear on the bearing journals, or cam lobes. The main decision to take is what degree of wear justifies renewal, which is costly. Any signs of scoring or damage to the bearing journals cannot be removed by regrinding. Renewal of the whole camshaft is the only solution.

> **HAYNES HINT** *When overhauling the valve gear, check that oil is being ejected from the nozzles onto the cam followers. Turn the engine on the starter to observe this.*

3 The cam lobes themselves may show signs of ridging or pitting on the high points. If ridging is light then it may be possible to smooth it out with fine emery. The cam lobes however, are surface hardened and once this is penetrated wear will be very rapid thereafter.

4 The faces of the cam followers which bear on the camshaft should show no signs of pitting, scoring or other forms of wear. They should not be a loose sloppy fit on the ballheaded bolt. Inspect the face which bears onto the valve stem and if pitted the cam follower must be renewed.

5 If the valve gear is known to be noisy but no readily detectable wear can be found on the camshaft, cam followers or ball-pins, a check should be made of the camshaft lobe bases for variation in the clearance between the base faces and the cam followers. The cylinder head must be assembled and the valve clearances correctly set for this check - see Section 57.

6 Measure the cam-to-follower clearance as shown in Fig.1.14, and also with the cam peak located at about 30° to 50° each side of vertical. Carry out this check with each of the cams in turn. Should the clearance in any of these positions vary by more than 0.015 in (0.04 mm) the camshaft must be renewed.

32 Valves and valve seats - examination and renovation

1 With the valves removed from the cylinder heads examine the heads for signs of cracking, burning away and pitting of the edge where it seats in the head. The seats of the valves in the cylinder head should also be examined for the same signs. Usually it is the valve that deteriorates first but if a bad valve is

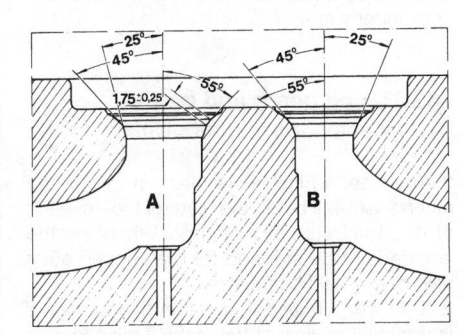

Fig. 1.9 Valve seat angles (Sec 32)

not rectified the seat will suffer and this is more difficult to repair.

2 Provided there are no obvious signs of serious pitting, the valve should be ground with its seat. This may be done by placing a smear of carborundum paste on the edge of the valve and using a suction type valve holder, grinding the valve in situ. This is done with a semi-rotary action, rotating the handle of the valve holder between the hands and lifting it occasionally to redistribute the traces of paste. Use a coarse paste to start with graduating to a fine paste. As soon as a matt grey unbroken line appears on both the valve and seat the valve is "ground in".

> **HAYNES HiNT** *All traces of carbon should also be cleaned from the head and neck of the valve stem. A wire brush mounted in a power drill is a quick and effective way of doing this.*

3 If the valve requires renewal it should be ground into the seat in the same way as an old valve.
4 Another form of valve wear can occur on the stem where it runs in the guide in the cylinder head. This can be detected by trying to rock the valve from side to side. If there is any movement at all it is an indication that the valve stem or guide is worn. Check the stem first with a micrometer at points along and around its length and if they are not within the specified size new valves will probably solve the problem. If the guides are worn, however, they will need reboring for oversize valves or for fitting guide inserts. The valve seats will need recutting to ensure they are concentric with the stems. This work should be given to your Ford dealer or local engineering works.
5 When valve seats are badly burnt or pitted, requiring renewal, inserts may be fitted - or renewed if already fitted once before - and once again this is a specialist task to be carried out by a suitable engineering firm.
6 When all valve grinding is completed it is essential that every trace of grinding paste is removed from the valves and ports in the cylinder head. This should be done by thorough washing in petrol or paraffin and blowing out with a jet of air. If particles of carborundum should work their way into the engine they would cause havoc with bearings or cylinder walls.

33 Timing gears and belt - examination and renovation

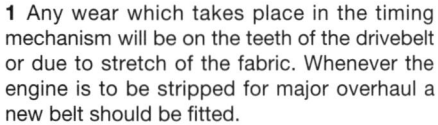

1 Any wear which takes place in the timing mechanism will be on the teeth of the drivebelt or due to stretch of the fabric. Whenever the engine is to be stripped for major overhaul a new belt should be fitted.
2 It is very unusual for the timing gears (sprockets) to wear at the teeth. If the securing bolts/nuts have been loose it is possible for the keyway or hub bore to wear. Check these two points and if damage or wear is evident a new gear must be obtained.

34 Flywheel ring gear - examination and renovation

1 If the ring gear is badly worn or has missing teeth it should be renewed. The old ring can be removed from the flywheel by cutting a notch between two teeth with a hacksaw and then splitting it with a cold chisel.
2 To fit a new ring gear requires heating the ring to 400°F (204°C). This can be done by polishing four equal spaced sections of the gear laying it on a suitable heat resistant surface (such as fire bricks) and heating it evenly with a blow lamp or torch until the polished areas turn a light yellow tint. Do not overheat or the hard wearing properties will be lost. The gear has a chamfered inner edge which should go against the shoulder when put on the flywheel. When hot enough place the gear in position quickly, tapping it home, if necessary, and let it cool naturally without quenching in any way.

35 Cylinder head and piston crowns - examination and renovation

1 When a cylinder head is removed, either in the course of overhaul or for inspection of bores or valve condition when the engine is in the car, it is normal to remove all carbon deposits from the piston crown and head.
2 This is best done with a cup shaped wire brush and an electric drill and is fairly straightforward when the engine is dismantled and the pistons removed. Sometimes hard spots of carbon are not easily removed except by a scraper. When cleaning the pistons with a scraper, take care not to damage the surface of the piston in any way.
3 When the engine is in the car certain precautions must be taken when decarbonising the piston crowns in order to prevent dislodged pieces of carbon falling into the interior of the engine which could cause damage to cylinder bores, pistons and rings - or if allowed into the water passage - damage to the water pump. Turn the engine so that the piston being worked on is at the top of its stroke and then mask off the adjacent cylinder bores and all surrounding water jacket orifices with paper and adhesive tape. Press grease into the gap all round the piston to keep carbon particles out and then scrape all carbon away by hand carefully. Do not use a power drill and wire brush when the engine is in the car as it will be virtually impossible to keep all the carbon dust clear of the engine. When completed carefully clear out the grease round the rim of the piston with a matchstick or something similar - bringing any carbon particles with it. Repeat the process on the other piston crown. It is not recommended that a ring of carbon is left round the edge of the piston on the theory that it will aid oil consumption. This was valid in the earlier days of long stroke low revving engines but modern engines, fuels and lubricants cause less carbon deposits anyway, and any left behind tends merely to cause hot spots.

36 Valve guides - inspection

1 Examine each valve guide internal bore for signs of scoring and wear. Insert the respective valves and if they are a loose fit in the guides and/or they can be rocked laterally, either the valve stem or internal bore of the guide is badly worn.
2 Try a new valve in the guides and if it is still loose, then the guides must be reamed oversize and new valves fitted. This is a task that is best left to your Ford dealer.

37 Sump - inspection

Wash out the sump in petrol and wipe dry. Carefully inspect for signs of damage or distortion. Renew if required. Scrape away all of the old gasket and sealant solution from the cylinder block mating face.

38 Engine reassembly - general

All engine components must be cleaned of old oil, sludge and old gaskets. The work area should be cleaned also prior to reassembly. In addition to the normal range of essential tools the following must be available before reassembling the engine:
  a) Complete set of new gaskets
  b) Supply of clean rags
  c) Clean can full of engine oil of the correct graded
  d) Torque wrench
  e) All new spare parts as required
  f) A tube of gasket sealant solution

39 Crankshaft - refitting

Ensure that the crankcase is thoroughly clean and all oilways are clear. A thin twist drill on a piece of wire is useful for cleaning them out. If possible blow them out with compressed air. Treat the crankshaft in the same fashion, and then inject engine oil into the crankshaft oilways. Commence work on rebuilding the engine by refitting the crankshaft and main bearings.

1 Wipe the bearing shell location in the crankcase with a soft, non-fluffy rag.
2 Wipe the crankshaft journals with a soft, non-fluffy rag.
3 If the old main bearing shells are to be renewed (not to do so is a false economy unless they are virtually new) fit the five upper

Engine 1•23

39.3 Inserting bearing shells into crankcase

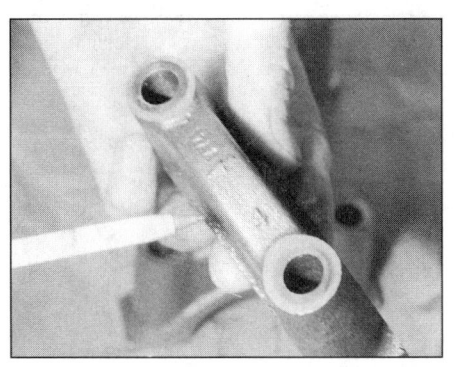

39.4 Main bearing cap identification marks

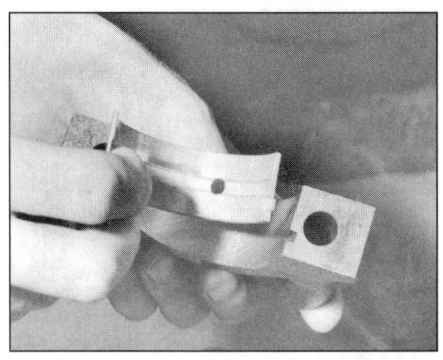

39.6 Fitting bearing shell to main bearing cap

halves of the main bearing shells to their location in the crankcase (photo).
4 Identify each main bearing cap and place in order. The number is cast onto the cap and with intermediate caps an arrow is also marked so that the cap is fitted the correct way round (photo).
5 Wipe the end cap bearing shell location with a soft non-fluffy rag.
6 Fit the bearing half shell into each main bearing cap (photo).
7 Fit the bearing half shell into each location in the crankcase.
8 Apply a little grease to each side of the centre main bearing so as to retain the thrust washers (photo).
9 Fit the upper halves of the thrust washers into their grooves either side of the main bearing. The slots must face outwards (photo).
10 Lubricate the crankshaft journals and the upper and lower main bearing shells with engine oil (photo).
11 Carefully lower the crankshaft into the crankcase (photo).
12 Lubricate the crankshaft main bearing journals again and then fit No 1 bearing cap (photo). Fit the two securing bolts but do not tighten yet.
13 Apply a little gasket cement to the crankshaft rear main bearing end cap location (photo).
14 Next fit No 5 cap (photo). Fit the two securing bolts but as before do not tighten yet.

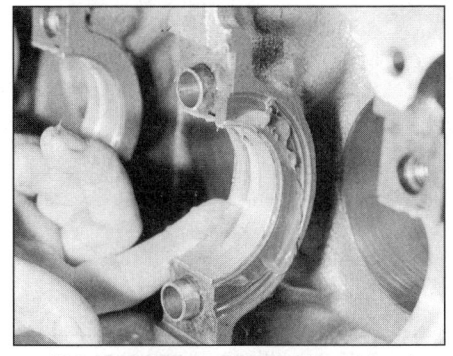

39.8 Applying grease to either side of centre main bearing

39.9 Fitting thrust washers to centre main bearing

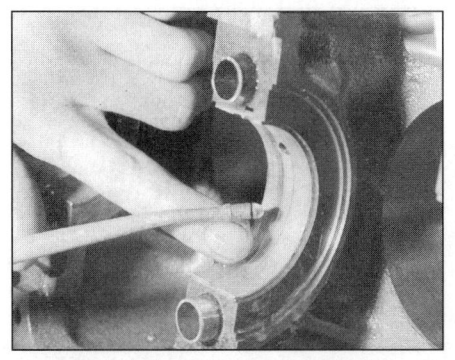

39.10 Lubricating bearing shells

39.11 Fitting crankshaft to crankcase

39.12 Refitting No 1 main bearing cap. Note identification mark

39.13 Applying gasket cement to rear main bearing cap location

39.14 Refitting rear main bearing cap

1•24 Engine

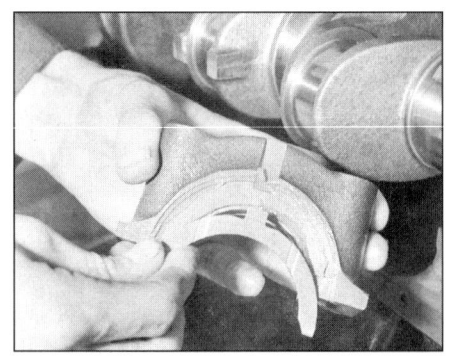

39.15 Fitting thrust washers to centre main bearing cap

39.16 All main bearing caps in position

39.17 Tightening main bearing cap securing bolts

39.18 Using feeler gauges to check crankshaft endfloat

42.3 Positioning ring gaps

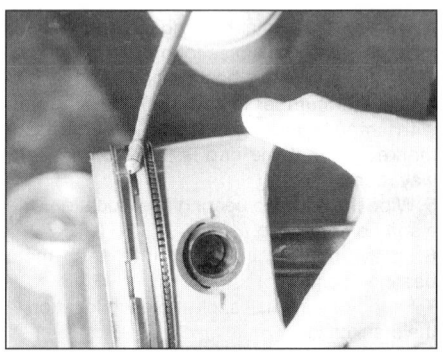

42.4 Lubricating pistons prior to fitting

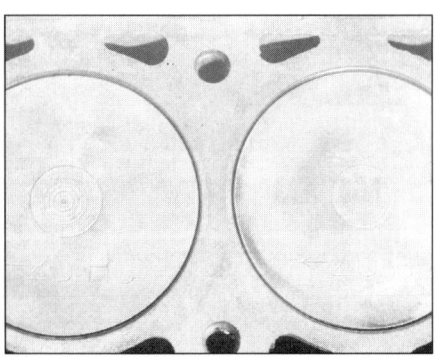

42.5 Piston identification marks

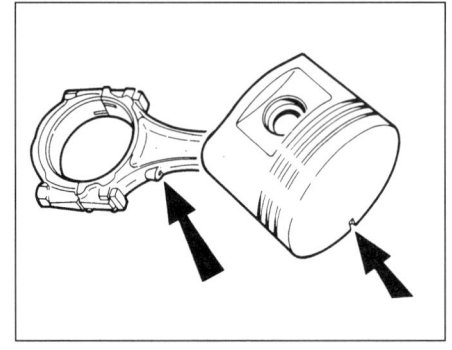

Fig. 1.10 Piston identification mark relative to connecting rod oil jet hole (Sec 42)

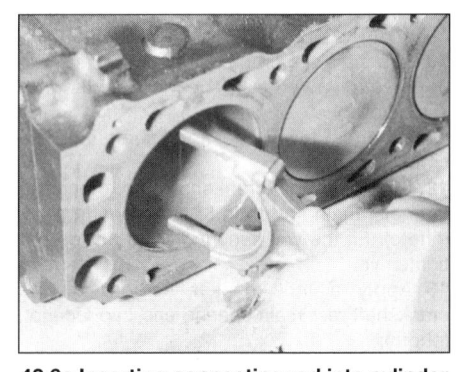

42.6a Inserting connecting rod into cylinder bore

42.6b Piston ring compressor correctly positioned

42.7 Pushing piston down bore

43.6 Refitting big-end cap securing nuts

Engine 1•25

15 Apply a little grease to either side of the centre main bearing and caps so as to retain the thrust washers. Fit the thrust washers with the tag located in the groove and the slots facing outwards (photo).
16 Fit the centre main bearing end cap and the two securing bolts. Then refit the intermediate main bearing end caps. Make sure that the arrows point towards the front of the engine (photo).
17 Lightly tighten all main cap securing bolts and then fully tighten in a progressive manner to a final torque wrench setting as specified (photo).
18 Using a screwdriver ease the crankshaft fully forwards and with feeler gauges check the clearance between the crankshaft journal side and the thrust washers. The clearance must not exceed that specified. Undersize thrust washers are available (photo).
19 Test the crankshaft for freedom of rotation. Should it be stiff to turn or possess high spots, a most careful inspection must be made with a micrometer, preferably by a qualified mechanic, to get to the root of the trouble. It is very seldom that any trouble of this nature will be experienced when fitting the crankshaft.

40 Pistons and connecting rods - reassembly

As a press type gudgeon pin is used (see Sections 20 and 30) this operation must be carried out by the local Ford garage.

41 Piston rings - refitting

1 Check that the piston ring grooves and oilways are thoroughly clean and unblocked. Pistons rings must always be fitted over the head of the piston and never from the bottom.
2 The easiest method to use when fitting rings is to wrap a feeler gauge round the top of the piston and place the rings one at a time, starting with the bottom oil control ring, over the feeler gauge.
3 The feeler gauge, complete with ring can then be slid down the piston over the other piston ring grooves until the correct groove is reached. The piston ring is then slid off the feeler gauge into the groove.
4 An alternative method is to fit the rings by holding them slightly open with the thumbs and both of the index fingers. This method requires a steady hand and great care as it is easy to open the ring too much and break it.

42 Pistons - refitting

1 The piston, complete with connecting rods, can be fitted to the cylinder bores in the following sequence: With a wad of clean non-fluffy rag wipe the cylinder bores clean.
2 The pistons, complete with connecting rods, are fitted to their bores from the top of the block.
3 Locate the piston ring gaps in the following manner (photo):

Top (compression)	150° to one side of the oil control ring helical expander gap
Centre (compression)	150° to the opposite side of the oil control ring helical expander gap
Bottom (oil control)	Helical expander: opposite the marked piston front side Intermediate rings: 1 inch (25 mm) each side of the helical expander gap

4 Lubricate the piston and rings with engine oil (photo).
5 Fit a universal piston ring compressor and prepare to insert the first piston into the bore. Make sure it is the correct piston connecting rod assembly for that particular bore, that the connecting rod is the correct way round and that the front of the piston is towards the front of the bore, ie towards the front of the engine (photo).
6 Again lubricate the piston skirt and insert into the bore up to the bottom of the piston ring compressor (photos).
7 Gently but firmly tap the piston through the piston ring compressor and into the cylinder bore with a wooden or plastic faced hammer (photo). Make certain that the rings do not jam against the block face and break.

43 Connecting rods to crankshaft - reassembly

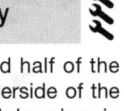

1 Wipe clean the connecting rod half of the big-end bearing cap and the underside of the shell bearing and fit the shell bearing in position with its locating tongue engaged with the corresponding cut-out in the rod.
2 If the old bearings are nearly new and are being refitted then ensure they are refitted in their correct locations on the correct rods. Note that it is false economy not to fit new bearings unless the old ones are nearly new.
3 Generously lubricate the crankpin journals with engine oil and turn the crankshaft so that the crankpin is in the most advantageous position for the connecting rod to be drawn onto it.
4 Wipe clean the connecting rod bearing cap and back of the shell bearing, and fit the shell bearing in position ensuring that the locating tongue at the back of the bearing engages with the locating groove in the connecting rod cap.
5 Generously lubricate the shell bearing and offer up the connecting rod bearing cap to the connecting rod.
6 Refit the connecting rod nuts (photo).
7 Tighten the bolts with a torque wrench set to the specified torque (photo).
8 When all the connecting rods have been fitted, rotate the crankshaft to check that everything is free, and that there are no high spots causing binding.

44 Oil pump - refitting

1 Wipe the mating faces of the oil pump and underside of the cylinder block.
2 Insert the hexagonal driveshaft into the end of the oil pump (photo).
3 Offer up the oil pump and refit the two special bolts. Using the special tool (21-020) and a torque wrench tighten the two bolts to the specified torque (photo).
4 Refit the one bolt and spring washer that secures the oil pump pick-up pipe support bracket to the crankcase.

43.7 Tightening the big-end cap securing nuts

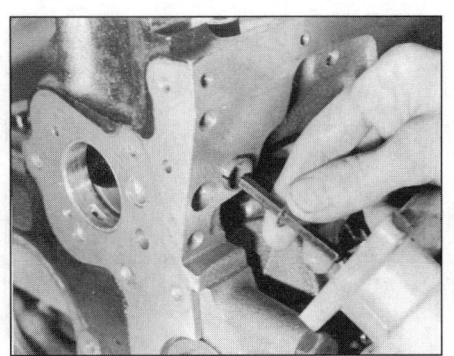

44.2 Inserting oil pump driveshaft

44.3 Tightening oil pump securing bolts

1•26 Engine

45.1 Refitting rectangular shaped seals to rear of crankshaft

45.2 Fitting seal into rear main bearing cap

45.3 Refitting crankshaft rear oil seal

Note: *When a new or overhauled oil pump is being fitted it must be filled with engine oil and turned by hand through one complete revolution prior to being fitted. This will prime the pump.*

45 Crankshaft rear oil seal - refitting

1 Apply some gasket sealant to the slot on either side of the rear main bearing end cap and insert a rectangular shaped seal (photo).
2 Apply some gasket sealant to the slot in the rear main bearing end cap and carefully insert the shaped seal (photo).
3 Lightly smear some grease on the crankshaft rear oil seal and carefully ease it over the end of the crankshaft. The spring must be inwards (photo).
4 Using a soft metal drift carefully tap the seal into position (photo).

46 Auxiliary shaft and timing cover - refitting

1 Carefully insert the auxiliary shaft into the front face of the cylinder block (photo).
2 Position the thrust plate into its groove in the auxiliary shaft - countersunk faces of the holes facing outwards - and refit the two cross-head screws (photo).
3 Tighten the two cross-head screws using a cross-head screwdriver and an open-ended

45.4 Tapping crankshaft rear oil seal into position

46.1 Refitting auxiliary shaft

46.2 Locating auxiliary shaft thrust plate

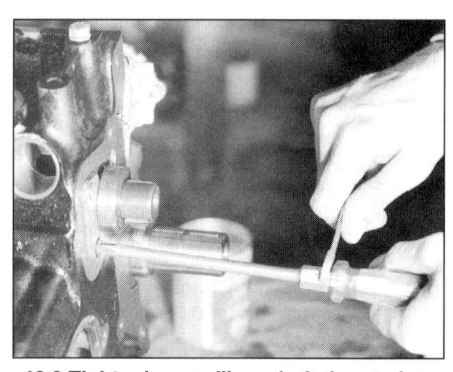

46.3 Tightening auxiliary shaft thrust plate securing screws

46.4 Positioning new gasket on cylinder block front face

46.6a Refitting crankshaft timing cover

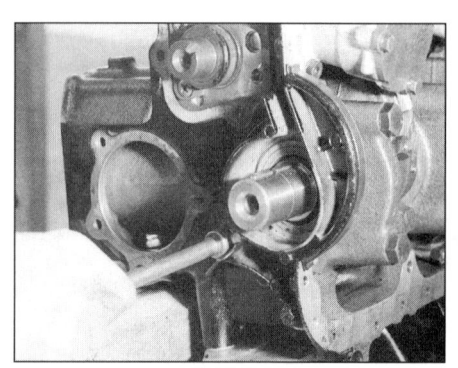

46.6b Tightening crankshaft timing cover securing bolts

Engine 1•27

46.8 Tightening auxiliary shaft timing cover securing bolts

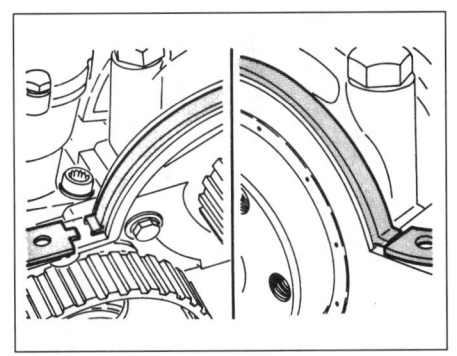

Fig. 1.11 Correct fitment of sump gasket at front and rear main bearing caps (Sec 47)

spanner (photo) or mole wrench to increase turning torque.
4 Smear some grease on to the cylinder block face of a new gasket and carefully fit into position (photo).
5 Apply some gasket sealant to the slot in the underside of the crankshaft timing cover. Insert the shaped seal.
6 Offer up the timing cover and secure with the bolts and spring washers (photos).
7 Smear some grease onto the seal located in the shaft timing cover and carefully ease the cover over the end of the auxiliary shaft.
8 Secure the auxiliary shaft timing cover with the four bolts and spring washers (photo).

47 Sump - refitting

1 Wipe the mating faces of the underside of the crankcase and the sump.
2 Apply jointing compound to the crankcase and sump mating faces.
3 Insert the rubber seal into the groove on the rear seal carrier.
4 Place the sump gasket in position and line up the bolt holes. Ensure that the tabs on the gasket fit under the cut-outs in the rubber seal.
5 Place the sump in position and secure with the bolts (photo).
6 Tighten the sump bolts in a progressive manner to the torque figure given in the Specifications (Fig. 1.12).

47.5 Refitting sump

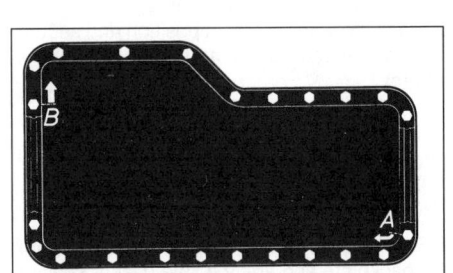

Fig. 1.12 Correct sequence for tightening sump bolts (Sec 47)

1st stage commencing with bolt A
2nd stage commencing with bolt B
3rd stage commencing with bolt A

48 Crankshaft sprocket and pulley and auxiliary shaft Sprocket - refitting

1 Check that the keyways in the end of the crankshaft are clean and the keys are free from burrs. Fit the keys into the keyways (photo).
2 Slide the sprocket into position on the crankshaft. This sprocket is the small diameter one (photo).
3 Ease the drivebelt into mesh with the crankshaft sprocket (photo).
4 Slide the large diameter plain washer onto the crankshaft (photo).
5 Check that the keyway in the end of the

48.1 Refitting Woodruff key to crankshaft

48.2 Sliding on crankshaft sprocket

48.3 Fitting drivebelt to crankshaft sprocket

48.4 Refitting large diameter plain washer

48.6 Fitting sprocket to auxiliary shaft

1•28 Engine

48.7 Refitting crankshaft pulley

48.8 Crankshaft pulley securing bolt and large washer

48.9 Tightening crankshaft pulley securing bolt

auxiliary shaft is clean and the key is free of burrs. Fit the key to the keyway.
6 Slide the sprocket onto the end of the auxiliary shaft (photo).
7 Slide the pulley onto the end of the crankshaft (photo).
8 Refit the bolt and thick plain washer to the end of the crankshaft (photo).
9 Lock the crankshaft pulley with a metal bar and using a torque wrench fully tighten the bolt (photo).

49 Water pump - refitting

1 Make sure that all traces of the old gasket are removed and then apply jointing compound to the gasket face of the cylinder block.

2 Fit a new gasket to the cylinder block.
3 Place the water pump in position and secure it with the four bolts and spring washers (photo).

50 Flywheel and clutch - refitting

1 Remove all traces of the shaped seal from the backplate and apply a little gasket sealant to the backplate. Fit a new seal to the backplate (photo).
2 Wipe the mating faces of the backplate and cylinder block and carefully fit the backplate to the two dowels (photo).
3 Wipe the mating faces of the flywheel and crankshaft and offer up the flywheel to the crankshaft, aligning the previously made

marks unless new parts have been fitted.
4 Fit the six crankshaft securing bolts and lightly tighten.
5 Lock the flywheel using a screwdriver engaged in the starter ring gear and tighten the securing bolts in a diagonal and progressive manner to a final torque wrench setting as given in the Specifications (photo).
6 Refit the clutch disc and pressure plate assembly to the flywheel making sure the disc is the right way round (photo).
7 Secure the pressure plate assembly with the six retaining bolts and spring washers.
8 Centralise the clutch disc using an old input shaft or piece of wooden dowel and fully tighten the retaining bolts (photo).
9 On automatic transmission models, refit the intermediate drive plate and secure with the bolts.

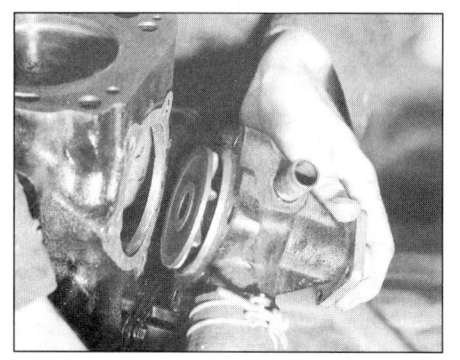

49.3 Refitting the water pump

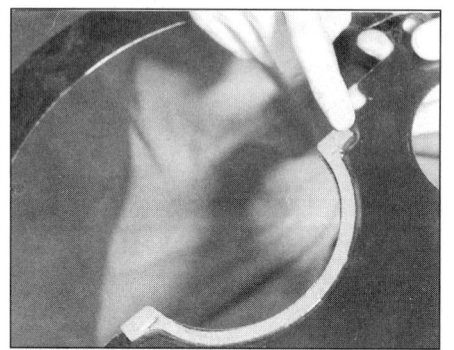

50.1 Fitting new seal to backplate

50.2 Backplate located on dowels in rear of cylinder block

50.5 Tightening flywheel securing bolts

50.6 Refitting clutch

50.8 Tightening clutch securing bolts with disc centralised

Engine 1•29

51.2 Sliding seal down valve stem

51.3 Fit valve spring cap . . .

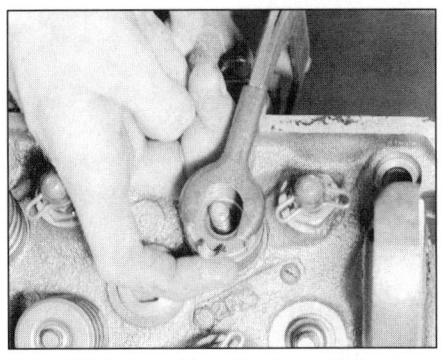

51.4 . . . and insert valve collets

51 Valves - refitting

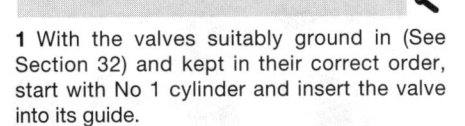

1 With the valves suitably ground in (See Section 32) and kept in their correct order, start with No 1 cylinder and insert the valve into its guide.
2 Lubricate the valve stem with engine oil and slide on a new oil seal. The spring must be uppermost as shown in the photo.
3 Fit the valve spring and cap (photo).
4 Using a special or modified universal valve spring compressor, compress the valve spring until the split collets can be slid into position (photo). Note these collets have serrations which engage in slots in the valve stems. Release the valve spring compressor.
5 Repeat this procedure until all eight valves and valve springs are fitted.

HAYNES HiNT *Tap the head of each valve stem with a soft mallet to settle the components.*

52 Camshaft - refitting

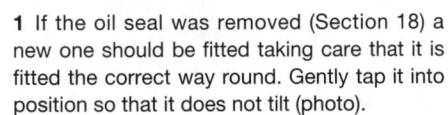

1 If the oil seal was removed (Section 18) a new one should be fitted taking care that it is fitted the correct way round. Gently tap it into position so that it does not tilt (photo).
2 Apply some grease to the lip of the oil seal. Wipe the three bearing surfaces with a clean, non-fluffy rag.
3 Lift the camshaft through the bearings taking care not to damage the bearing surfaces with the sharp edges of the cam lobes. Also take care not to cut the fingers (photo).
4 When the journals are ready to be inserted into the bearings lubricate the bearings with engine oil (photo).
5 Push the camshaft through the bearings until the locating groove in the rear of the camshaft is just rearwards of the bearing carrier.
6 Slide the thrust plate into engagement with the camshaft taking care to fit it the correct way round as previously noted (photo).
7 Secure the thrust plate with the two bolts and spring washers (photo).
8 Check that the keyway in the end of the camshaft is clean and the key is free of burrs. Fit the key into the keyway (photo).
9 Locate the tag on the camshaft sprocket

52.1 Camshaft oil seal correctly fitted

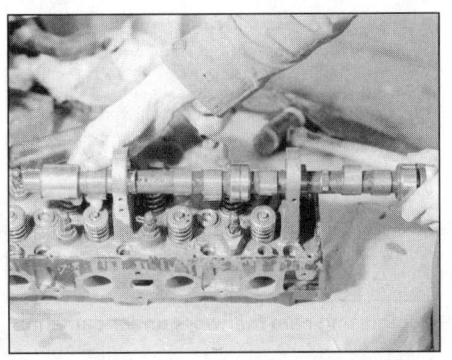

52.3 Refitting camshaft through bearings

52.4 Lubricating camshaft bearings

52.6 Locating camshaft thrust plate

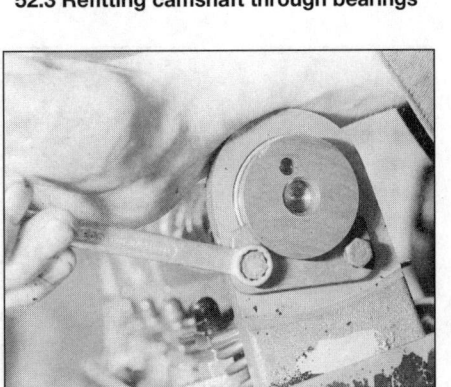

52.7 Tightening thrust plate retaining bolts

52.8 Fitting Woodruff key to camshaft

1•30 Engine

52.9 Camshaft sprocket backplate tag

52.10 Camshaft sprocket backplate refitted

52.11 Refitting camshaft sprocket

backplate and this must locate in the second groove in the camshaft sprocket (photo).
10 Fit the camshaft sprocket backplate, tag facing outwards (photo).
11 Fit the camshaft sprocket to the end of the camshaft and with a soft-faced hammer make sure it is fully home (photo).
12 Refit the sprocket securing bolt and thick plain washer (photo).

53 Cam followers - refitting

1 Undo the ball headed bolt locknut and screw down the bolt fully. This will facilitate refitting the cam followers (photo).
2 Rotate the camshaft until the cam lobe is away from the top of the cylinder head. Pass the cam follower under the back of the cam until the cup is over the ball headed bolt (photo).
3 Engage the cup with the ball headed bolt (photo).
4 Refit the cam follower spring by engaging the ends of the spring with the anchor on the ball headed bolt (photo).
5 Using the fingers pull the spring up and then over the top of the cam follower (photo).
6 Repeat the above sequence for the remaining seven cam followers.
7 Check that the jet holes in the camshaft lubrication pipe are free and offer up to the camshaft bearing pedestals (photo).
8 Refit the pipe securing bolts and spring washers.

52.12 Camshaft sprocket securing bolt and washer

53.1 Slacken ball headed bolt locknut

53.2 Passing cam followers under camshaft

53.3 Cup located over ball headed bolt

53.4 Cam follower spring engaged with the anchor

53.5 Cam follower spring correctly fitted

53.7 Refitting lubrication pipe

Engine 1•31

54.2 Positioning cylinder head gasket onto cylinder block

54.3 Refitting cylinder head

54.4 Refit cylinder head bolts . . .

54.5 . . . and tighten using special tool

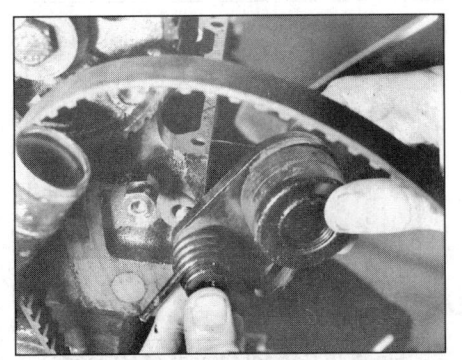

55.1 Refitting drivebelt tensioner

55.3 Using screwdriver to relieve tension of the spring

56.1 Lining up camshaft timing marks

54 Cylinder head - refitting

Before refitting the cylinder head, turn the crankshaft until No 1 piston is roughly 0.8 in (2 cm) before TDC.

1 Wipe the mating faces of the cylinder head and cylinder block.
2 Carefully place a new gasket on the cylinder block and check to ensure that it is the correct way up, and the right way round (photo). If possible, fit a couple of temporary alignment studs to the cylinder block to ensure that the gasket is located exactly.
3 Gently lower the cylinder head being as accurate as possible first time, so that the gasket is not dislodged (photo).
4 Refit the cylinder head bolts taking care not to damage the gasket if it has moved (photo).
5 Using the special tool (21-002) lightly tighten all the bolts (photo).
6 Tighten the cylinder head bolts progressively to a final torque setting as given in Specifications, in the sequence shown in Fig. 1.2.

55 Camshaft drivebelt tensioner and thermostat housing - refitting

1 Thread the shaped bolt through the spring and tensioner plate and screw the bolt into the cylinder head (photo).
2 Tighten the bolt securely using special tool 21-012.
3 Using a screwdriver to overcome the tension of the spring, position the plate so that its securing bolt can be screwed into the cylinder head (photo).
4 Clean the mating faces of the cylinder head and thermostat housing and fit a new gasket.
5 Offer up the thermostat housing and secure in position with the two bolts and spring washers.
6 Tighten the bolts to the specified torque.

56 Camshaft drivebelt - refitting and timing

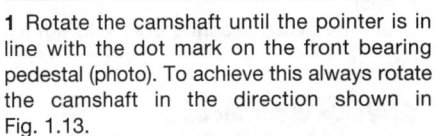

1 Rotate the camshaft until the pointer is in line with the dot mark on the front bearing pedestal (photo). To achieve this always rotate the camshaft in the direction shown in Fig. 1.13.
2 Using a socket wrench on the crankshaft pulley bolt, turn the crankshaft until No. 1 piston is at its TDC position. This is indicated by a mark on the crankshaft sprocket (see Fig. 1.13).

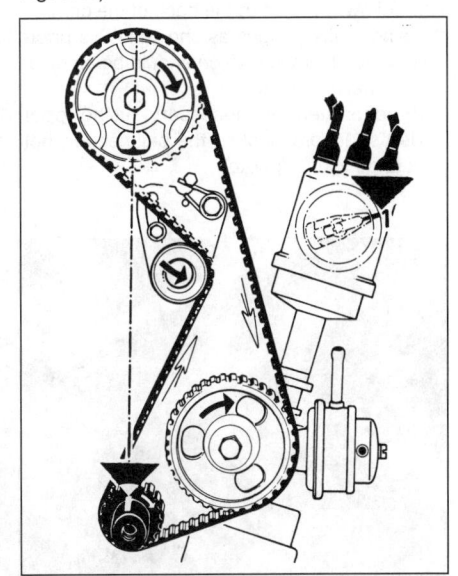

Fig. 1.13 Camshaft, ignition and crankshaft timing marks (Sec 56)

1•32 Engine

56.5 Drivebelt fitted

56.6a Refitting drivebelt guard

56.6b Drivebelt guard securing bolt

3 Engage the drivebelt with the crankshaft sprocket and auxiliary shaft sprocket. Pass the back of the belt over the tensioner jockey wheel and then slide it into mesh with the camshaft sprocket.
4 Slacken the tensioner plate securing bolt and allow the tensioner to settle by rotating the crankshaft twice. Retighten the tensioner plate securing bolt.
5 Line up the timing marks and check that these are correct indicating the belt has been correctly refitted (photo).
6 Refit the drivebelt guard, easing the guard into engagement with the bolt and large plain washer located under the water pump (photos).
7 Refit the guard securing bolts and tighten fully.

57 Valve clearances - checking and adjustment

1 With the engine top cover removed turn the crankshaft until each cam in turn points vertically upwards. This will ensure that the cam follower will be at the back of the cam.
2 Using feeler gauges as shown in this photo check the clearance which should be as given in the Specifications.
3 If adjustment is necessary, using open ended spanners slacken the ball headed bolt securing locknut (photo).

4 Screw the ball headed bolt up or down as necessary until the required clearance is obtained (photo). Retighten the locknut.
5 An alternative method of adjustment is to work to the following table:

Valves open	Valves to adjust
1 ex and 4 in	6 in and 7ex
6 in and 7 ex	1 ex and 4in
2 in and 5ex	3 ex and 8 in
3 ex and 8 in	2 in and 5 ex

Valves are numbered 1 to 8 from the timing belt end of the engine.
6 Refit the top cover and tighten the bolts in the order shown in Fig. 1.15 to the specified torque.

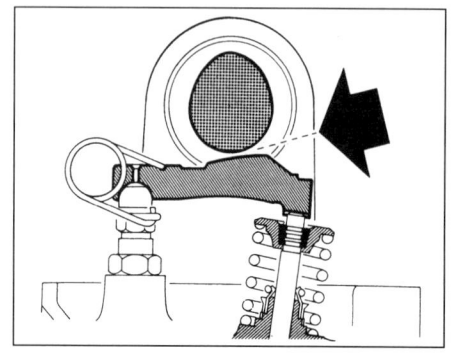

Fig. 1.14 Cam follower and camshaft clearance (Sec 57)

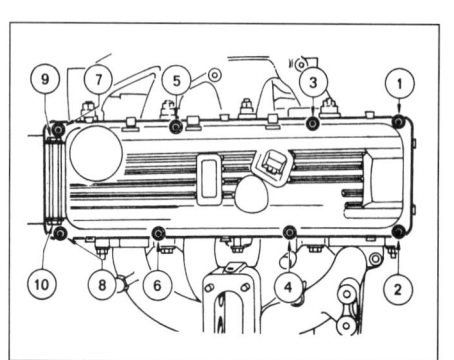

Fig. 1.15 Tightening sequence for top cover securing bolts (Sec 57)

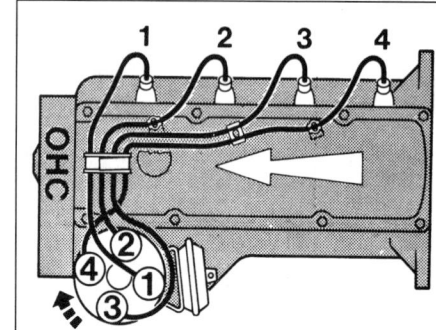

Fig. 1.16 Correct fitment of HT leads (Sec 58)

57.2 Checking cam follower clearance

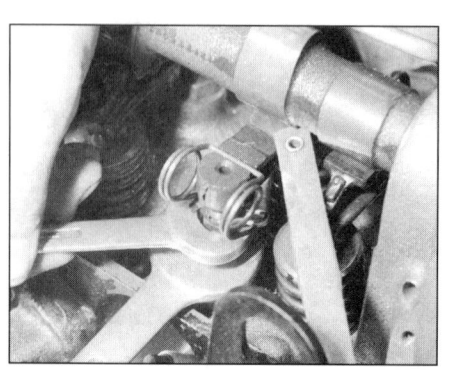

57.3 Slackening ball headed bolt locknut

57.4 Adjusting ball headed bolt

58 Engine - refitting without transmission

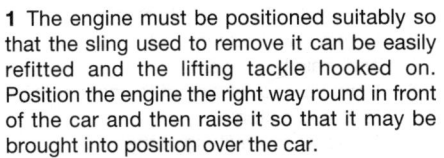

1 The engine must be positioned suitably so that the sling used to remove it can be easily refitted and the lifting tackle hooked on. Position the engine the right way round in front of the car and then raise it so that it may be brought into position over the car.
2 The transmission should be jacked up to its approximately normal position.

Manual gearbox

3 Lower the engine steadily into the engine compartment, keeping all ancillary wires, pipes and cables well clear of the sides. It is best to have a second person guiding the engine while it is being lowered, and to help align the engine with the gearbox.
4 The tricky part is finally mating the engine to the gearbox, which involves locating the gearbox input shaft into the clutch housing and flywheel. Provided that the clutch friction plate has been centred correctly as described in Chapter 5, there should be little difficulty. Grease the splines of the gearbox input shaft first. It may be necessary to rock the engine from side to side in order to get the engine fully home. Under no circumstances let any strain be imparted onto the gearbox input shaft. This could occur if the shaft was not fully located and the engine was raised or lowered more than the amount required for very slight adjustment of position.
5 As soon as the engine is fully up to the gearbox bellhousing refit the bolts holding the two together.
6 The final positioning of the engine brackets onto the mountings requires some attention because the positioning bolts on the mounting are angled inwards and therefore do not exactly line up with the holes in the brackets. However, they are flexibly mounted so, provided two people are doing the work, they may be levered into position whilst the engine is lowered.

Automatic transmission

7 Turn the torque converter in its housing so that the retaining studs are approximately aligned with the corresponding holes in the engine intermediate plate.
8 When the engine is lowered into position the converter can be further adjusted by turning the starter ring accordingly through the starter motor aperture. When the lower studs are aligned with the holes of the intermediate plate the engine can be pushed home to abut against the converter housing. The transmission may have to be raised or lowered further to enable the engine to mate with the transmission.
9 The four special self-locking nuts can now be fitted to secure the intermediate plate to the converter. To gain access to each stud, turn the engine crankshaft pulley retaining nut in a clockwise direction (looking from front to rear) so that each stud in turn is accessible through the starter motor aperture and each nut can be fitted and fully tightened to secure.

All models

10 Insert and tighten the respective bellhousing/engine bolts.
11 Refit the starter motor, distributor, all electrical connections, the fuel lines and carburettor linkages, cooling system hoses and radiator in the reverse order as described in Sections 5 and 6.
12 Reconnect the exhaust pipe to the manifold and refit the plate covering the lower section of the bellhousing. Remove the supporting jack if this has not already been done.
13 Refill the engine with new oil and refill the cooling system. Check for signs of leaks.

59 Engine - refitting with transmission

1 The transmission should be refitted to the engine, taking the same precautions as regards the input shaft as mentioned in Section 58.
2 The general principles of lifting the engine/transmission assembly are the same as for the engine, but the transmission will tilt everything to a much steeper angle. Refitting will certainly require the assistance of a second person.
3 Lift the transmission end of the unit into the engine compartment and then lower and guide the unit down.
4 If a trolley jack is available this is the time to place it under the transmission so that, as the engine is lowered further, the rear end can be supported and raised as necessary - at the same time being able to roll back as required. Without such a jack, support the rear in such a way that it can slide if possible. In any case the transmission will have to be jacked and held up in position when the unit nears its final position.
5 Locate the front mounting brackets on the locating bolts as described in Section 58.
6 Refit the speedometer drive cable with the transmission drive socket and refit the clamping plate and bolt. This MUST be done before the gearbox supporting crossmember is refitted.
7 Fit the transmission crossmember to the transmission and secure it with the bolt which goes through the centre of the crossmember.
8 Using the jack, position the crossmember up to the body frame and refit and tighten the four securing bolts.
9 Refit the transmission remote control change lever and housing as described in Chapter 6.
10 Fit the propeller shaft as described in Chapter 7.
11 *Manual gearbox models:* Refit the clutch cable and adjust as described in Chapter 5. The final connections should then be made to the engine as described in Section 58. In addition to the engine lubricant and coolant, the gearbox should be refilled with fresh oil.
12 *Automatic transmission models:* For refitting the automatic transmission and the engine as a unit, the instructions are similar to those for manual types, but when reconnecting the transmission unit, refer to Chapter 6, Section 16 and reverse the sequence of instructions. Reconnect the respective automatic transmission components and fittings as applicable.

60 Engine - initial start-up after major overhaul or repair

1 Make sure that the battery is fully charged and that all lubricants, coolant and fuel are replenished.
2 Double check all fittings and electrical connections. Ensure that the distributor is correctly fitted and that the ignition timing static setting is correct. If in doubt refer to Chapter 4.
3 Remove the spark plug and the "-" connection from the ignition coil. Turn the engine over on the starter motor until the oil pressure warning light is extinguished or until oil pressure is recorded on the gauge. This will ensure that the engine is not starved of oil during the critical few minutes running after initial start-up. The fuel system will also be primed during this operation.
4 Reconnect the "-" connection on the ignition coil and refit the spark plugs and leads. Start the engine.
5 As soon as the engine fires and runs, keep it going at a fast tickover only (no faster) and bring it up to normal working temperature.
6 As the engine warms up there will be odd smells and some smoke from parts getting hot and burning off oil deposits. The signs to look for are leaks of water or oil, which will be obvious if serious. Check also the exhaust pipe and manifold connections as these do not always find their exact gas tight position until the warmth and vibration have acted on them, and it is almost certain they will need tightening further. This should be done, of course, with the engine stopped.
7 When normal running temperature has been reached, adjust the engine idle speed as described in Chapter 3.
8 Stop the engine and wait a few minutes to see if any lubricant or coolant is dripping out when the engine is stationary.
9 After the engine has run for 20 minutes remove the engine top cover and recheck the tightness of the cylinder head bolts. Also check the tightness of the sump bolts. In both cases use a torque wrench.
10 Road test the car to check that the timing is correct and that the engine is giving the necessary smoothness and power. Do not race the engine - if new bearings and/or pistons have been fitted it should be treated as a new engine and run in at a reduced speed for the first 1000 miles (1600 km).

1•34 Engine

PART B V8 ENGINES

61 General description

As well as the 2.0 litre in-line ohc engine described in Part A of this Chapter, the Granada range of vehicles may also be fitted with the German V6 engines of 2.3 or 2.8 litre capacity. These engines use overhead valves operated by tappets, pushrods and rocker arms from the camshaft which is gear driven by the crankshaft.

The cylinder heads are of the crossflow design with the inlet manifold located on top of the cylinder block between the two cylinder heads and the exhaust manifolds on the outside of the heads.

The combined crankcase and cylinder block Is made of cast iron and houses the pistons, crankshaft and camshaft. Attached to the bottom of the crankcase is a pressed steel sump which acts as a reservoir for the engine oil. The lubrication system is described in Section 83.

Aluminium alloy pistons are connected to the crankshaft by H-section forged steel connecting rods and gudgeon pins. Two compression rings and one oil control ring, all located above the gudgeon pin, are fitted.

The forged crankshaft runs in four main bearings and endfloat is controlled by thrust washers fitted to No 3 main bearing.

The drive gear for the distributor and oil pump is located in front of the rear camshaft bearing and the fuel pump is operated by an eccentric behind the front camshaft bearing.

Refer to Section 98 for modifications made to later models.

62 Major operation with engine in place

The following major operations can be carried out without removing the engine from the car.
a) Removal and refitting of the cylinder heads
b) Removal and refitting of rocker shaft and pushrod
c) Removal and refitting of front engine mountings
d) Removal and refitting of the flywheel (gearbox removed)

63 Major operations requiring engine removal

For the following major operations the engine must be removed from the car.

a) Removal and refitting of the oil sump and oil pump*
b) Removal and refitting of the timing gears
c) Removal of the crankshaft front seal
d) Renewal of the crankshaft rear seal
e) Removal and refitting of the big-end and main bearings
f) Removal and refitting of the connecting rods and pistons
g) Removal and refitting of the crankshaft

*Although it is possible to remove the oil sump if the engine is raised, the steering shaft disconnected and the crossmember lowered, the amount of work involved is not considered worthwhile.

64 Methods of engine removal

1 Although it is possible to remove the engine and gearbox as an assembly it is recommended that the engine is removed as a separate unit. If the engine and gearbox are removed as an assembly it has to be lifted out at a very steep angle and for this a considerable lifting height is required.

2 If the work being undertaken requires the removal of the gearbox as well as the engine it is recommended that the gearbox is removed first, refer to Chapter 6, Section 2 or 8 as appropriate.

65 Engine with carburettor - removal

1 The do-it-yourself owner should be able to remove the engine in about four hours. It is essential to have a good hoist and two axle-stands if an inspection pit is not available.

2 The sequence of operations listed in this Section is not critical as the position of the person undertaking the work, or the tool in his hand, will determine to a certain extent the order in which the necessary operations are performed. Obviously the engine cannot be removed until everything is disconnected from it and the following sequence will ensure that nothing is forgotten.

3 Remove the battery as described in Chapter 10.

4 Open the bonnet and using a soft pencil mark the outline position of both the hinges at the bonnet to act as a datum for refitting.

5 With the help of a second person to take the weight of the bonnet undo and remove the hinge to bonnet securing bolts with plain and spring washers. There are two bolts to each hinge.

6 Lift away the bonnet and put in a safe place where it will not be scratched.

7 Place a container having a capacity of at least 8 Imp pints (4.5 litres) under the engine sump and remove the oil drain plug. Allow the oil to drain out and then refit the plug (photo).

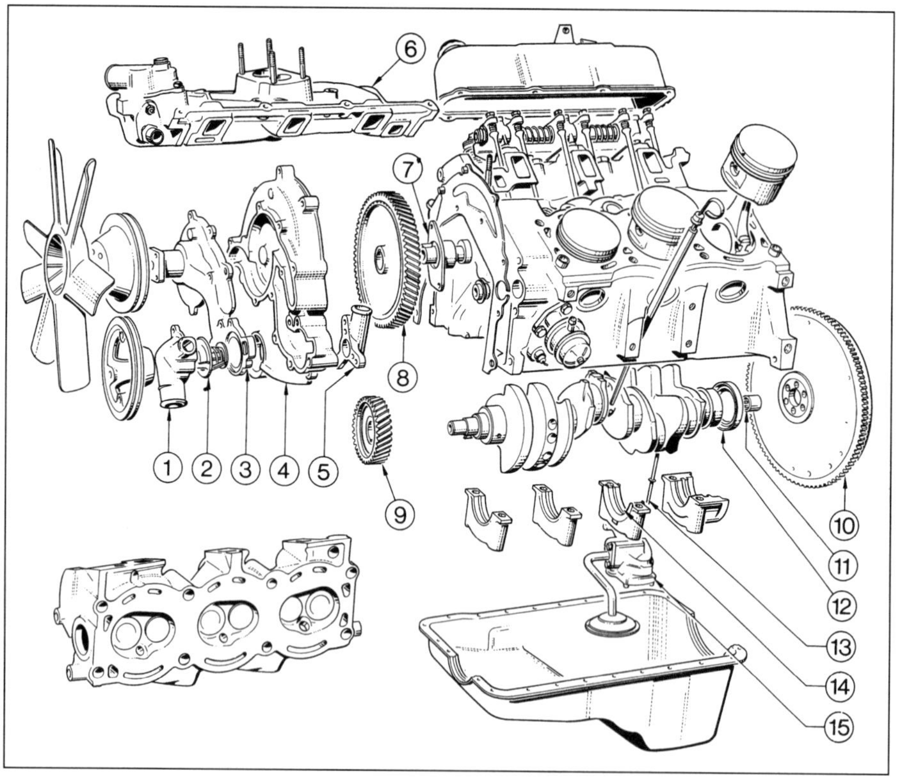

Fig. 1.17 Exploded view of engine main components (Sec 61)

1 Water inlet connection
2 Thermostat
3 Water pump
4 Timing cover
5 By-pass hose flange
6 Inlet manifold
7 Camshaft thrust plate
8 Camshaft gear
9 Crankshaft gear
10 Flywheel
11 Crankshaft pilot bearing
12 Oil seal
13 Oil pump drive shaft
14 Main bearing
15 Oil pump

Engine 1•35

8 Refer to Chapter 3, Section 4 and remove the air cleaner assembly from the top of the carburettor.
9 Mark the HT leads, so that they can be identified for refitting in their original positions, and disconnect them from the spark plugs.
10 Disconnect the HT lead from the centre of the ignition coil. Remove the distributor cap and HT leads from the engine compartment (photo).
11 Unplug the distributor multi-plug connector.
12 Drain the cooling system as described in Chapter 2, Section 2.
13 Disconnect the pressure and return hoses from the power steering reservoir/pump assembly and drain the fluid into a suitable container. Plug the ends of the hoses and the reservoir/pump assembly connectors to prevent the ingress of dirt or water.
14 If the power steering system is not to be dismantled the reservoir/pump assembly can be removed from the engine, without disconnecting the hoses, and placed on one side (photo). To do this slacken the idler pulley bracket securing bolts and remove the power steering pump drivebelt. If the special type wrench is available undo the steering pump to mounting bracket bolts and remove the pump, or alternatively, using an ordinary spanner remove the mounting bracket securing bolts and lift away the steering pump complete with mounting bracket. Note the earth strap attached to one of the bracket mounting bolts.
15 Slacken the clips securing the heater hoses to the water pump, automatic choke or thermostat housing (as applicable), intake manifold and heater connections. Pull off the hoses and remove them from the engine compartment.
16 Remove the radiator as described in Chapter 2.
17 Remove the exhaust heat box from the right-hand exhaust manifold.
18 Detach the multi-connector from the rear of the alternator (photo).
19 Disconnect the wiring from the temperature sender unit and the oil pressure switch. Disconnect the engine to bulkhead earth strap from the rear of the engine.
20 Disconnect the fuel supply pipe from the fuel pump and plug the pipe to prevent the ingress of dirt or fuel loss due to syphoning. Disconnect the fuel return line from the carburettor.
21 Disconnect the throttle control inner cable from the operating rod.
22 Unscrew the throttle control outer cable securing nut, separate the retaining clip and remove the cable from the mounting bracket.
23 Disconnect the servo vacuum hose from the carburettor intermediate flange.
24 Undo and remove the two nuts securing the exhaust downpipe clamp plate to the exhaust manifold (both sides) and lower the exhaust pipes.
25 Working underneath the car disconnect the main starter motor cable and the two wiring connectors from the starter motor solenoid (photo).
26 Remove the two securing bolts and lift out the starter motor (photo).
27 Undo the two securing bolts and remove the clutch housing lower cover plate.
28 Detach the clutch release lever gaiter from the clutch housing and pull it back. Draw back the clutch cable and disconnect it from the clutch operating arm. Remove the clutch cable assembly through the locating hole in the flange on the clutch housing and tuck the cable out of the way where it will not get caught up as the engine is lifted out.
29 On models equipped with automatic

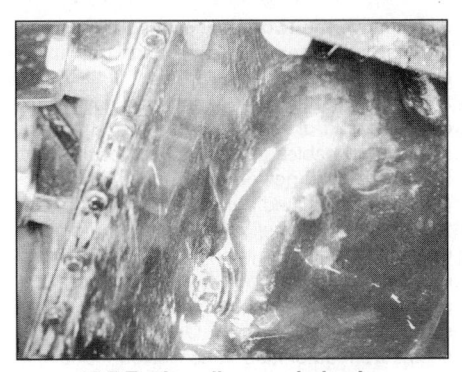

65.7 Engine oil sump drain plug

65.10 Remove the distributor cap

65.14a Remove the power steering pump mounting bolts

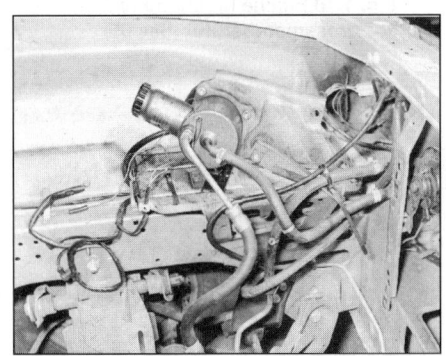

65.14b Place the power steering pump and bracket on the battery tray

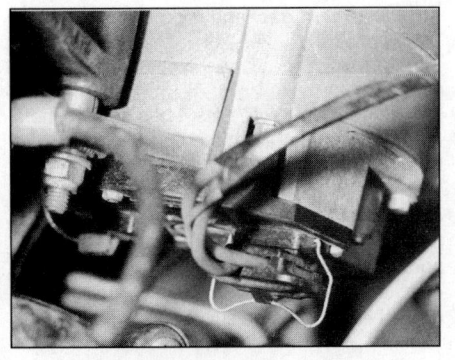

65.18 remove alternator multi-plug

65.25 Disconnect cables from starter solenoid

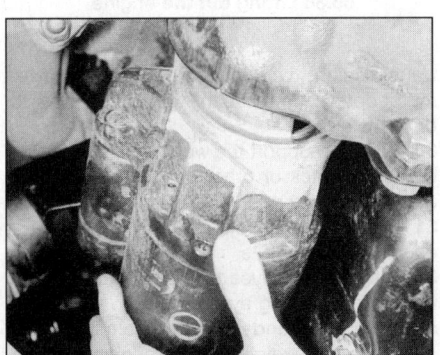

65.26 Removing the starter motor

1•36 Engine

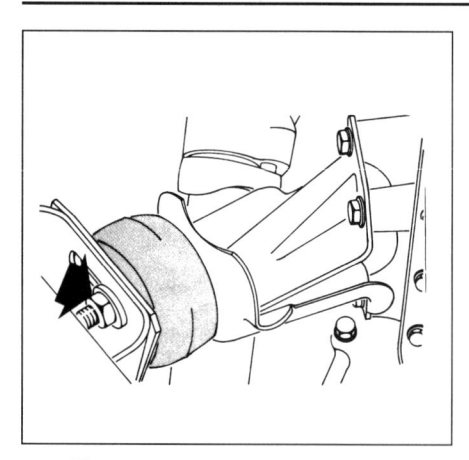

Fig. 1.18 Engine mounting to front crossmember nut (Sec 65)

65.35 Method of retaining torque converter (automatic transmission models)

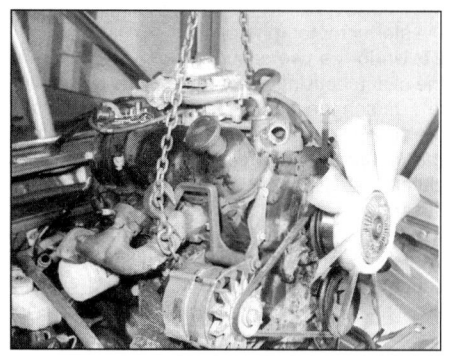

65.36 Lifting out the engine

transmission, working through the starter motor opening, remove the torque converter to drive plate bolts.
30 Suitably support the weight of the gearbox by using a jack or other support such as an axle-stand or wooden blocks.
31 Remove the six clutch housing flange-to-engine securing bolts.
32 Attach lifting eyes to the engine and fit a lifting sling (if lifting eyes are not available use a rope passed under the engine mountings) and take the weight of the engine without lifting it.
33 Remove the two nuts securing the engine mounting rubber insulators to the front crossmember (Fig. 1.18).
34 Check that all the relevant parts have been disconnected or removed and that all wiring and control cables are tucked out of the way where they will not get caught up when the engine is being lifted out.
35 Lift the engine slightly to clear the engine mounting studs from their location in the front crossmember and pull it forwards off the gearbox input shaft. Take care not to put any strain on the input shaft. On removal of the engine from automatic transmission models, it is essential that the torque converter is retained on the transmission. While the engine is being removed press the torque converter rearwards and when the engine is removed secure the converter to the transmission with a piece of wood and metal bar similar to that shown in the photo. The transmission should also be supported by ropes attached to a bar positioned across the engine compartment on the inner wing panels.
36 When the engine is withdrawn from the input shaft, hoist it high enough to clear the front panel, and remove it from the car (photo). Lower the engine and support it so that it will not get damaged.

66 Engine with fuel injection - removal

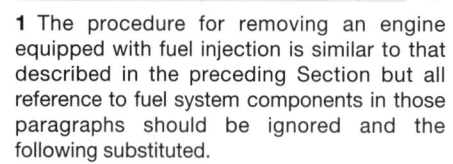

1 The procedure for removing an engine equipped with fuel injection is similar to that described in the preceding Section but all reference to fuel system components in those paragraphs should be ignored and the following substituted.
2 Refer to Chapter 3, Part C for the location and description of the fuel injection system components then disconnect the fuel lines which run between the fuel distributor at the side of the engine compartment and the fuel injector. Observe extreme cleanliness during all work on the fuel injection system and identify each pipe as to position before removing it. Use tape for this with a matching number on the particular injector and the outlet port of the fuel distributor.
3 Disconnect the fuel lines which run between the fuel distributor, the warm up regulator and the start valve.
4 Disconnect the worm drive clips and remove the trunking which runs between the fuel distributor/mixture assembly and the throttle plate housing.
5 Disconnect the throttle linkage from the throttle plate lever and bracket.
6 Identify and then disconnect the electrical plugs from all fuel injection components. These include the thermo time switch, the auxiliary air device, the warm up regulator and the start valve.
7 Disconnect and move away any other pipes or wires which may impede the engine removal operation.
8 Depending upon the type of lifting equipment available (a spreader should be used between the lifting chains or ropes) certain fuel injection components may have to be removed if there is any danger of them being damaged by the lifting gear. The air box is particularly vulnerable and if removed (see Chapter 3) before lifting out the engine will obviate the possibility of strain being put upon it.

67 Engine dismantling - general

1 When the engine is removed from the car, it, and particularly its accessories, are vulnerable to damage. If possible mount the engine on a stand, or failing this, make sure it is supported in such a manner that it will not be damaged whilst undoing tight nuts and bolts.
2 Cleanliness is important when dismantling the engine to prevent exposed parts from contamination. Before starting the dismantling operations, clean the outside of the engine with paraffin, or a good grease solvent if it is very dirty. Carry out this cleaning away from the area in which the dismantling is to take place.
3 If an engine stand is not available carry out the work on a bench or wooden platform. Avoid working with the engine directly on a concrete floor, as grit presents a real source of trouble.
4 As parts are removed, clean them in a paraffin bath. Never immerse parts with internal oilways in paraffin but wipe down carefully with a petrol dampened rag. Clean oilways with wire.
5 It is advisable to have suitable containers to hold small items in their groups as this will help when reassembling the engine and also prevent possible loss.
6 Always obtain complete sets of gaskets when the engine is being dismantled. It is a good policy to always fit new gaskets in view of the relatively small cost involved.

> **HAYNES HINT** *Retain the old gaskets when dismantling the engine with a view to using them as a pattern to make a replacement gasket if a new one is not available.*

7 When possible refit nuts, bolts and washers in their locations as this helps to protect the threads from damage and will also be helpful when the engine is being reassembled, as it establishes their location.
8 Retain unserviceable items until the new parts are obtained, so that the new part can be checked against the old part to ensure that the correct item has been supplied.

Engine 1•37

69.4a Undo the distributor clamp plate bolt . . .

69.4b . . . and lift out the distributor

69.5 Remove the power steering pump drivebelt idler pulley bracket

68 Engine ancillaries - removal

1 Although the items listed may be removed separately with the engine installed (as described in the relevant Chapters) it is more appropriate to take them off after the engine has been removed from the car when extensive dismantling is being carried out. The items comprise:

Carburettor or fuel injection system components (see Chapter 3)
Distributor
Fuel pump
Water pump
Alternator
Power steering pump (if disconnected from the power steering system, see Section 5, paragraph 13)

69 Engine - dismantling

Note. *The following procedure is written for the carburettor engined Granada which was used as a project car for this manual. For models with fuel injection the procedure is similar except that when parts of the system are removed then their exact location should be noted to ensure that they are correctly refitted.*

1 Unscrew the oil filter with a strap or chain wrench. Remove the dipstick.
2 Remove the clutch assembly as described in Chapter 5, (manual gearbox only). Note which way round the clutch disc is fitted. Remove the engine mounting brackets.
3 Remove the power steering pump and drivebelt as described in Chapter 11 (if fitted).
4 Undo the distributor clamp plate securing bolt and withdraw the distributor from the cylinder block (photos).
5 Remove the two bolts attaching the power steering pump drivebelt idler pulley bracket to the front cover and lift away the bracket and pulley (photo).
6 Slacken the alternator attaching bolts and lift off the alternator and fan drivebelt. Undo the three bolts securing the alternator mounting bracket to the cylinder block and remove the alternator and bracket assembly (photo).
7 Refer to Chapter 2 and remove the fan and pulley from the water pump.
8 Disconnect the fuel pipe from the carburettor and withdraw the vent valve from the rocker cover, where fitted.
9 Slacken the securing clips and disconnect the hose from the choke water outlet connection and the bypass hose at the thermostat housing (photo).
10 Remove the four securing bolts, lift off the carburettor and then the intermediate flange from the inlet manifold (photo).
11 Remove the rocker cover securing bolts and lift off the covers.
12 Undo the rocker shaft securing bolts and remove the rocker shafts and oil splash shields. Note which way the splash shields are fitted. Mark the rocker shafts so that they can be refitted in their original positions.

69.6 Removing the alternator and mounting bracket

13 Lift out the pushrods and keep them in their respective positions in relation to the rocker shafts to ensure that they are refitted in their original locations.
14 Remove the inlet and exhaust manifold securing bolts and lift off the manifolds. As sealing compound is used when fitting the inlet manifold it may be necessary to use a screwdriver to lever it free. Do not insert the screwdriver between the mating faces of the inlet manifold and the cylinder block.
15 Remove the cylinder head holding down bolts. **Note:** *If Torx type bolts have been used, these will need to be renewed.* Unscrew the bolts in the reverse order of the tightening sequence as shown in Fig. 1.28.
16 The cylinder head can now be lifted upwards (photo). If the head sticks to the

69.9 Disconnect the choke hoses

69.10 Removing the carburettor intermediate flange

69.16 Lifting off the right-hand cylinder head

1•38 Engine

69.19 The oil pressure switch is located on the left-hand side of the cylinder block

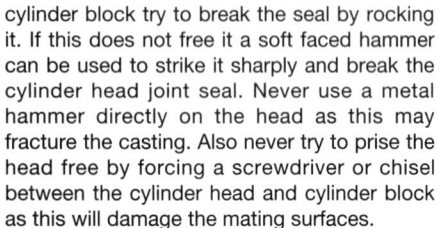

69.21 Removing the flywheel

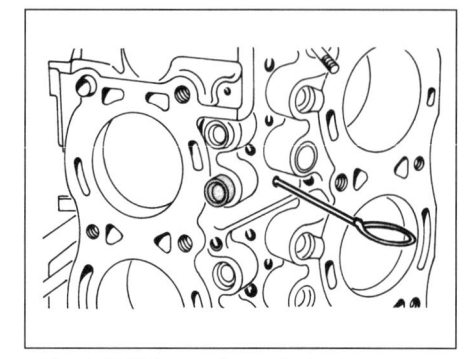

Fig. 1.19 Using a piece of wire to remove the tappets (Sec 69)

cylinder block try to break the seal by rocking it. If this does not free it a soft faced hammer can be used to strike it sharply and break the cylinder head joint seal. Never use a metal hammer directly on the head as this may fracture the casting. Also never try to prise the head free by forcing a screwdriver or chisel between the cylinder head and cylinder block as this will damage the mating surfaces.

17 Undo and remove the two bolts securing the fuel pump to the cylinder block then remove the pump, gasket and operating rod.
18 Unscrew and remove the oil pressure switch (photo).
19 Remove the crankshaft pulley securing bolt and take the pulley off the crankshaft. When the pulley is secured to a vibration damper, also remove the damper central bolt. Restrain the crankshaft from turning by chocking the flywheel.
20 Index mark the position of the flywheel in relation to the crankshaft so that it can be refitted in the same position.
21 Chock the flywheel to restrain it from turning, then remove the six bolts securing the flywheel to the crankshaft and lift off the flywheel (photo).
22 Remove the engine backplate.
23 Position the engine on its side and prise out the valve tappets with a piece of bent brass wire (Fig. 1.19). Place the tappets with their respective pushrods so that they will be refitted in the same tappet bores.
24 Remove the oil sump securing bolts and lift off the sump and sump gasket.

25 Unscrew the two oil securing bolts and remove the oil pump complete with suction pipe and oil strainer. Lift out the oil pump driveshaft, taking note of which way round it is fitted.
26 Remove the bolts securing the water pump to the cylinder block timing cover and lift away the water pump.
27 Undo the thermostat housing securing bolts, remove the housing and lift out the thermostat. Detach the rear water elbow (photo).
28 Remove the bolts securing the timing cover to the front of the cylinder block and lift away the timing cover. Discard the O-ring seals.
29 Unscrew the bolt securing the camshaft gear to the camshaft and pull off the gear (photo).
30 Remove the two camshaft thrust plate securing bolts and withdraw the camshaft together with the spacer and then remove the plate spring and spacer.
31 Unscrew the front intermediate plate attaching bolts and remove the intermediate plate (Fig.1.20).
32 If the crankshaft gear needs to be removed use a standard puller to draw it off the crankshaft.
33 Check that the big-end bearing caps and connecting rods have identification marks. This is to ensure that the correct caps are fitted to the correct connecting rods and at reassembly are fitted in their correct cylinder bores. Note that the pistons have an arrow (or notch) marked on the crown to indicate the forward facing side.
34 Remove the big-end nuts and place to one side in the order in which they are removed.
35 Pull off the big-end caps, taking care to keep them in the right order and the correct way round. If the big-end caps are difficult to remove they can be tapped lightly with a soft faced hammer.

> **HAYNES HiNT** *Ensure that the shell bearings are kept with their respective connecting rods unless they are being renewed.*

36 To remove the shell bearings, press the bearing on the side opposite the groove in both the connecting rod and the cap, and the bearing will slide out.
37 Withdraw the pistons and connecting rods upwards out of the cylinder bores.
38 Make sure that the identification marks are visible on the main bearing caps so that they can be refitted in their original positions at reassembly.
39 Undo the securing bolts and lift off the main bearing caps and the bottom half of each bearing shell, taking care to keep the bearing shell in the right caps (photo).
40 When removing the rear main bearing cap note that this also retains the crankshaft rear oil seal.
41 When removing the No. 3 main bearing cap, note the position of the two half thrust

69.27 Detach the rear water elbow

69.29 Pull off the camshaft gear (note alignment marks)

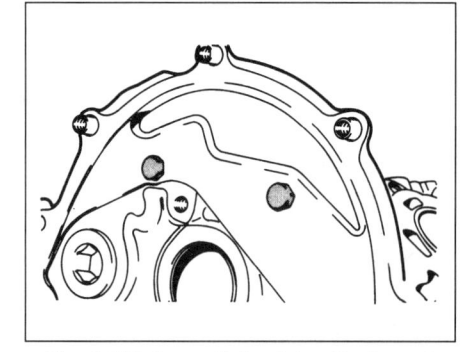

Fig. 1.20 Intermediate plate attachment bolts (Sec 69)

Engine 1•39

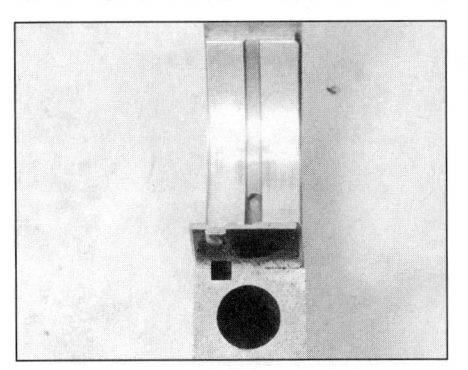

69.39 Remove the bearing shells from the main bearing caps

69.43 No 3 main bearing shell and thrust washers

washers and mark them so that they can be refitted in the same position. It may be necessary to tap the main bearing caps with a soft faced hammer to release them.
42 Lift the crankshaft out of the crankcase and remove the rear oil seal.
43 Remove the upper halves of the main bearing shells and the upper half of the No. 3 bearing thrust washers from the crankcase and place them with the respective main bearing caps (photo).

70 Valve rockers - dismantling, inspection and reassembly

1 Tap out the roll pin from one end of the rocker shaft and remove the spring washer (photo).

2 Slide the rocker arms, rocker supports and springs off the rocker shaft. Keep them in the correct order so that they can be reassembled in the same position. If a rocker support sticks it can be removed by tapping it with a soft faced hammer.
3 Examine the rocker shaft and rocker arms for wear. If the rocker arm surface that contacts the valve stem is considerably worn, renew the rocker arm. If it is worn slightly step-shaped it may be cleaned up with a fine oil stone.
4 Oil the parts and reassemble them on their shafts in the original order. With both rocker shafts fitted the oil holes must face downwards to the cylinder heads. This position is indicated by a notch on one end face of the rocker shaft (Fig. 1.22).

Fig. 1.21 Exploded view of rocker shaft assembly (Sec 70)

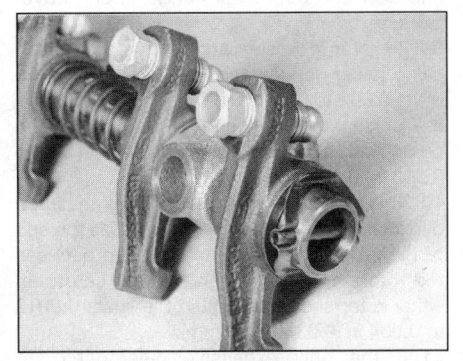

70.1 The rocker shaft has a retaining pin at each end

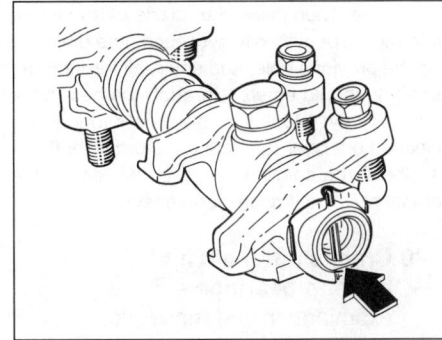

Fig. 1.22 Notch indicating position of oil hole (Sec 70)

71 Valve tappets and pushrods - inspection

1 Examine the valve tappets for wear and damage, renew if suspect.
2 Check the pushrods for signs of bending or wear. Correct or renew as necessary.

72 Camshaft and camshaft bearings - inspection and renovation

1 If there is excessive wear in the camshaft bearings they will have to be renewed. As the fitting of new bearings requires special tools this should be left to your local Ford dealer.
2 The camshaft may show signs of wear on the bearing journals or cam lobes. The main decision to take is what degree of wear necessitates renewing the camshaft, which is expensive. Scoring or damage to the bearing journals cannot be removed by regrinding; renewal of the camshaft is the only solution.
3 The cam lobes may show signs of ridging or pitting on the high points. If ridging is slight then it may be possible to remove it with a fine oil stone or emery cloth. The cam lobes, however, are surface hardened and once the hard skin is penetrated wear will be very rapid.

73 Cylinder heads - dismantling, renovation and reassembly

1 Clean the dirt and oil off the cylinder heads. Remove the carbon deposits from the combustion chambers and valve heads with a scraper or rotary wire brush.
2 Remove the valves by compressing the valve springs with a suitable valve spring compressor and lifting out the collets. Release the valve spring compressor and remove the valve spring retainer, spring and valve (photos). Mark each valve so that they can be fitted in the same location. **Note**: *When removing and refitting the valve spring take care not to damage the valve stem when pressing down the valve spring retainer to remove or refit the collets. If the stem gets damaged the sealing will be ineffective and result in excessive oil consumption and wear of the valve guides. Discard the valve stem seal.*
3 With the valves removed clean out the carbon from the ports.
4 Examine the heads of the valves and the valve seats for pitting and burning. If the pitting on valve and seat is slight it can be removed by grinding the valves and seats together with coarse, and then fine, valve grinding paste. If the pitting is deep the valves will have to be reground on a valve grinding machine and the seats will have to be recut with a valve seat cutter. Both these operations are a job for your local Ford dealer or motor engineering specialist.
5 Check the valves guides for wear by inserting the valve in the guide, the valve stem should move easily in the guide without side

73.2a Compress the valve spring and lift out the collets . . .

73.2b . . . then remove the spring retainer, spring . . .

73.2c . . . and valve

play. Renewal of worn guides requires special tools and should be left to your local Ford dealer.

6 When grinding slightly pitted valves and valve seats with carborundum paste proceed as follows: Apply a little coarse grinding paste to the valve seat and using a suction type valve grinding tool, grind the valve into its seat with a semi-rotary movement, lifting the valve from time to time. A light spring under the valve head will assist in this operation. When a dull matt even surface finish appears on both the valve and the valve seat, clean off the coarse paste and repeat the grinding operation with a fine grinding paste until a continuous ring of light grey matt finish appears on both valve and valve seat. Carefully clean off all traces of grinding paste. Blow through the gas passages with compressed air.

7 Check the valve springs for damage and also check the free length, refer to the Specifications at the beginning of this Chapter. Renew if defective.

8 Lubricate the valve stem with engine oil and insert it in the valve guide. Slide on a new oil seal.

9 Fit the valve spring and valve spring retainer.

10 Use a suitable valve spring compressor to compress the valve spring until the collets can be fitted in position in the slots in the valve stem. Release the valve spring compressor.

> **HAYNES HiNT** *After fitting all the parts, tap the top of the valve springs lightly with a plastic hammer to ensure correct seating of the collets.*

74 Cylinder bores - inspection and renovation

Refer to Part A Section 28 of this Chapter.

75 Piston rings - removal

Refer to Part A Section 21 of this Chapter

76 Pistons and piston rings - examination and renovation

Refer to Part A, Section 29 of this Chapter.

77 Piston ring - refitting

Refer to Part A Section 41 of this Chapter.

78 Connecting rods and gudgeon pin - examination and renovation

Refer to Part A, Section 30 of this Chapter.

79 Crankshaft - examination and renovation

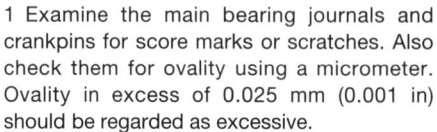

1 Examine the main bearing journals and crankpins for score marks or scratches. Also check them for ovality using a micrometer. Ovality in excess of 0.025 mm (0.001 in) should be regarded as excessive.

2 Contrary to normal practice, Ford stipulate that under no circumstances should the crankshaft be reground; so if damage or excessive wear is evident it will be necessary to obtain a new crankshaft.

3 If the old crankshaft is being refitted, check that the oilways are clear. This can be done by probing with wire or by blowing through with an air line. Then insert the nozzle of an oil gun into the respective oilways, each time blanking off the previous hole, and squirt oil through the shaft. It should emerge through the next hole. Any oilway blockage must obviously be cleaned out prior to refitting the crankshaft.

4 Check the spigot bearing in the rear of the crankshaft and if necessary renew it.

80 Crankshaft main and big-end bearings - examination and renovation

Refer to Part A, Section 27 of this Chapter, and to Fig.1.8.

81 Oil pump - dismantling, inspection and reassembly

1 If oil pump wear is suspected it is possible to obtain a repair kit. Check for wear first as described later in this Section and if confirmed, obtain an overhaul kit or a new pump. The two rotors are a matched pair and form a single replacement unit. Where the rotor assembly is to be re-used the outer rotor, prior to dismantling, must be marked on its front face in order to ensure correct reassembly.

2 Remove the intake pipe and oil strainer.

3 Note the relative position of the oil pump cover and body and then undo and remove the bolts and spring washers. Lift away the cover.

4 Carefully remove the rotors from the housing.

5 Using a centre punch tap a hole in the centre of the pressure relief valve sealing plug, (make a note to obtain a new one).

6 Screw in a self tapping screw and, using an open ended spanner withdraw the sealing plug.

7 Thoroughly clean all parts in petrol or paraffin and wipe dry using a non-fluffy rag. The necessary clearances may now be checked using a machined straight-edge (a good steel rule) and a set of feeler gauges. The critical clearances are between the lobes of the centre rotor and convex faces of the outer rotor; between the rotor and the pump body; and between both rotors and the end cover plate.

8 The rotor lobe clearance may be checked using feeler gauges and should be within the limits 0.002 to 0.008 in (0.05 to 0.20 mm).

9 The clearance between the outer rotor and pump body should be within the limits 0.006 to 0.012 in (0.15 to 0.30 mm).

10 The endfloat clearance may be measured by placing a steel straight-edge across the end of the pump and measuring the gap between the rotors and the straight-edge. The gap in either rotor should be within the limits 0.0012 to 0.004 in (0.03 to 0.10 mm).

11 If the only excessive clearances are endfloat it is possible to reduce them by removing the rotors and lapping the face of the

Engine 1•41

body on a flat bed until the necessary clearances are obtained. It must be emphasised, however, that the face of the body must remain perfectly flat and square to the axis of the rotor spindle, otherwise the clearances will not be equal and the end cover will not be a pressure tight fit to the body. It is worth trying, of course, if the pump is in need of renewal anyway but unless done properly, it could seriously jeopardise the rest of the overhaul. Any variations in the other two clearances should be overcome with a new unit.

12 With all parts scrupulously clean first refit the relief valve and spring and lightly lubricate with engine oil.

13 Using a suitable diameter drift drive in a new sealing plug, flat side outwards until it is flush with the intake pipe mating face.

14 Lubricate both rotors with engine oil and fit them in the body. Fit the oil pump cover and secure with the three bolts tightened in a diagonal and progressive manner to the torque wrench setting given in the Specifications.

15 Fit the driveshaft into the rotor driveshaft and ensure that the rotor turns freely.

16 Fit the intake pipe and oil strainer to the pump body.

82 Flywheel ring gear - inspection and renewal

Refer to Part A, Section 34 of this Chapter.

83 Lubrication system - description

1 The pressed steel oil sump is attached to the underside of the crankcase and acts as a reservoir for the engine oil. The oil pump draws oil from the sump via an oil strainer and intake pipe, then passes it into the full-flow oil filter. The filtered oil flows from the centre of the oil filter element and passes through a short drilling, on the right-hand side, to the oil pressure switch and to the main gallery (on the left-hand side of the crankcase) through a transverse drilling. Refer to Part A, Section 24 for oil filter removal and refitting procedures, but note that the filter is located on the right-hand side of the cylinder block. The engine oil dipstick is on the left-hand side of the engine, and indicates the maximum and minimum oil levels.

2 Four drillings connect the main gallery to the four main bearings and the camshaft bearings in their turn are connected to all the main bearings. The big-end bearings are supplied with oil through diagonal drillings from the nearest main bearing.

3 When the crankshaft is rotating, oil is thrown from the hole in each big-end bearing and ensures splash lubrication of the gudgeon pins and thrust side of the cylinders. The timing gears are also splash lubricated through an oil drilling.

4 The crankshaft third bearing journal intermittently feeds oil under pressure, to the rocker shafts through a drilling in the cylinder block and cylinder head. The oil then passes back to the engine sump, via large drillings in the cylinder block and cylinder head.

84 Crankcase ventilation system - description and maintenance

1 The closed crankcase ventilation system is used to control the emission of crankcase vapour. It is controlled by the amount of air drawn in by the engine when it is running and the throughput of the vent valve.

2 The system is known as the PCV system (positive crankcase ventilation) and the advantage of the system is that should the "blow by" exceed the capacity of the PCV valve, excess fumes are fed into the engine through the air cleaner.

3 At 18 000 miles (30 000 km) pull the valve and hose from the rocker cover on the right-hand side of the engine and wash the valve in a cleaning solvent. Check the valve for freedom of movement. Refit the valve in the rocker cover.

4 Check the security and condition of the system connecting hoses.

85 Engine reassembly - general

1 To ensure maximum life with minimum trouble from an overhauled engine, not only must every part be correctly assembled but everything must be spotlessly clean, all oilways must be clean, locking washers and spring washers must always be fitted where needed and all bearings and other sliding surfaces must be thoroughly lubricated during assembly.

2 Before assembly, renew any bolts, studs and nuts whose threads are in any way damaged and whenever possible use new spring washers. **Note:** *If Torx type cylinder head bolts have been used, these will need to be renewed as well.* Obtain a complete set of new gaskets and all new parts as necessary.

3 When refitting parts ensure that they are refitted in their original positions and directions. Oil seal lips should be smeared with grease before fitting. A liquid gasket sealant should be used where specified to prevent leakage.

4 Apart from your normal tools, a supply of clean rags, an oil can filled with clean engine oil and a torque wrench are essential.

86 Crankshaft - refitting

1 Wipe the bearing shell locations in the crankcase with a clean rag and fit the main bearing upper half shells in position (photo).

2 Clean the main bearing shell locations and fit the half shells in the caps. If the old bearings are being refitted (although this is a false economy unless they are practically new) make sure they are fitted in their original positions.

3 Apply a little grease to each side of the No. 3 main bearing so as to retain the thrust washers.

4 Fit the upper halves of the thrust washers into their grooves each side of the main bearing. The slots must face outwards with the tag located in the groove (photo).

5 Lubricate the crankshaft main bearing

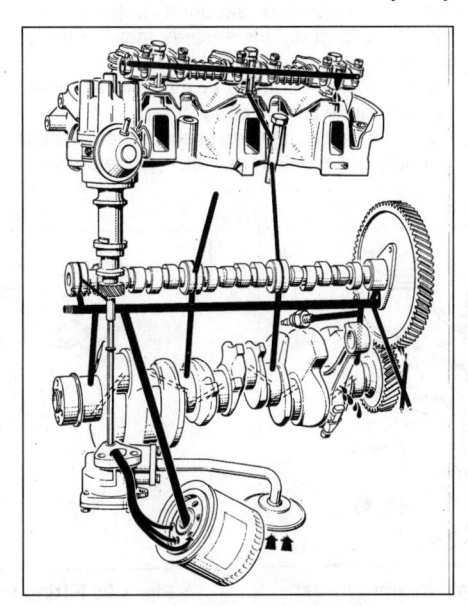

Fig. 1.23 Engine lubrication system (Sec 83)

86.1 Fit the main bearing shells

86.4 Correct position of thrust washers with slots outwards

1•42 Engine

86.5 Liberally lubricate the bearing shells

86.7 Carefully place the crankshaft in position

86.10a Fit the main bearing caps ...

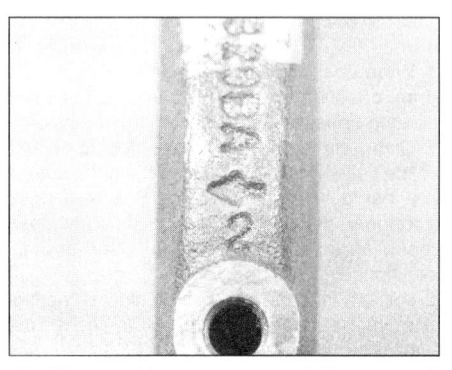

86.10b ... with the arrows pointing toward the front of the engine

86.11 Tightening the main bearing cap bolts

86.13 Checking the crankshaft endfloat with feeler gauges

journals and the main bearing shells with engine oil (photo).
6 Place the rear main bearing oil seal in position on the end of the crankshaft.
7 Carefully lower the crankshaft into the crankcase (photo).
8 Apply a thin coating of sealing compound to the mating faces of the crankcase and the rear main bearing cap.
9 Apply a smear of grease to both sides of the No. 3 main bearing cap so as to retain the thrust washers. Fit the thrust washers with the tag located in the groove and the slots facing outwards.
10 Fit the main bearing caps with the arrows on the caps pointing to the front of the engine (photos).
11 Progressively tighten the main bearing securing bolts to the specified torque (photo), except No. 3 bearing bolts which should only be finger-tight. Now press the crankshaft fully to the rear then slowly press it fully forward, hold it in this position and tighten the No. 3 main bearing cap securing bolts to the specified torque. This ensures that the thrust washers are correctly located.
12 Press the rear main bearing oil seal firmly against the rear main bearing.
13 Using feeler gauges, check the crankshaft endfloat by inserting the feeler gauge between the crankshaft journal side and the thrust washers. The clearance must not exceed the figure given in the Specifications. Different thicknesses of thrust washers are available (photo).
14 Rotate the crankshaft to ensure that it is not binding or sticking. Should it be very stiff to turn or have high spots it must be removed and thoroughly checked.
15 Coat the rear main bearing cap dowel seals with sealing compound and fit in position. Use a blunt screwdriver or similar tool to press them fully in (Fig.1.25). They should be fitted with the rounded face pointing towards the bearing cap.

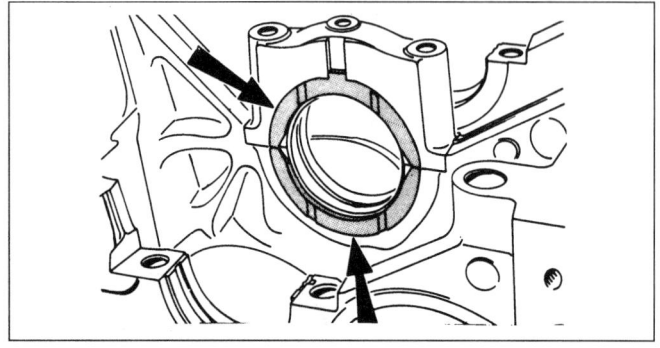

Fig. 1.24 Correct fitted position of No 3 main bearing thrust washers (Sec 86)

Fig. 1.25 Fitting the rear main bearing dowel seals (Sec 86)

Engine 1•43

87.3a Fit the camshaft to the cylinder block ...

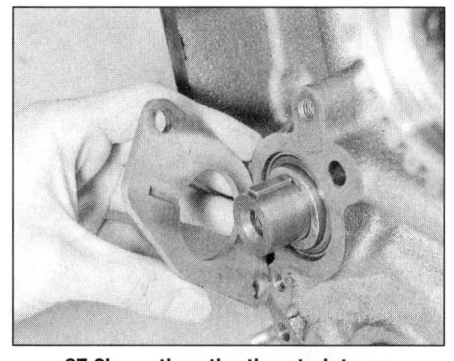

87.3b ... then the thrust plate ...

87.3c ... and self locking bolts

87 Camshaft and front intermediate plate - refitting

1 Slide the spacer onto the camshaft with the chamfered side first and fit the plate spring.
2 Lubricate the camshaft bearings, the camshaft and the thrust plate.
3 Carefully insert the camshaft from the front and fit the thrust plate and self-locking securing bolts. Tighten the bolts to the specified torque (photos).
4 Fit the timing cover guide sleeves and O-ring seals onto the crankcase. The chamfered end of the guide sleeves must face outward towards the timing cover (photo).
5 Ensure the mating faces of the crankcase and the front intermediate plate are clean and then apply sealing compound to both faces. Position the gasket on the crankcase and then fit the intermediate plate.
6 Fit the two centre bolts finger-tight then fit another two bolts temporarily for locating purposes. Tighten the centre securing bolts then remove the temporary fitted locating bolts (photo).

88 Pistons and connecting rods - refitting

1 Wipe clean the connecting rod half of the big-end bearing cap and the underside of the shell bearing and fit the shell bearing in position with its locating tongue engaged with the corresponding cut out in the rod (photo).
2 If the old bearings are nearly new and are being refitted then ensure they are refitted in their correct locations on the correct rods.
3 The pistons, complete with connecting rods, are fitted to their bores from the top of the block.
4 Locate the piston ring gaps in the following manner:
   Top: 150° from one side of the oil control ring helical expander gap
   Centre: 150° from the opposite side of the oil control ring helical expander gap
   Bottom: or control ring Helical expander: opposite the marked piston front side
   Oil control ring intermediate rings: 1 inch (25 mm) each side of the helical expander gap
5 Well lubricate the piston and rings with engine oil.
6 Fit a universal ring compressor and prepare to insert the first piston into the bore. Make sure it is the correct piston-connecting rod

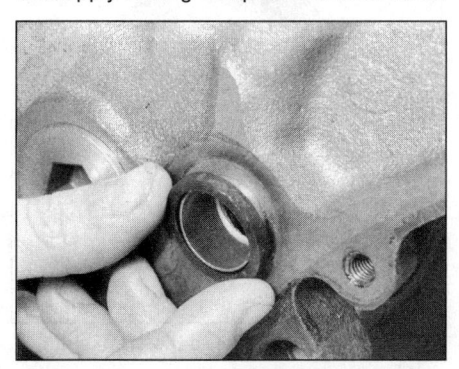

87.4 Chamfered end of guide sleeve facing outward

87.6 Fit the intermediate plate

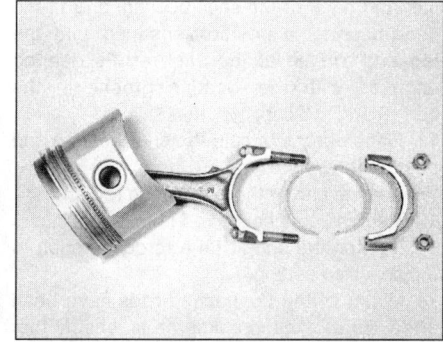

88.1 Piston, piston rings and connecting rod assembly

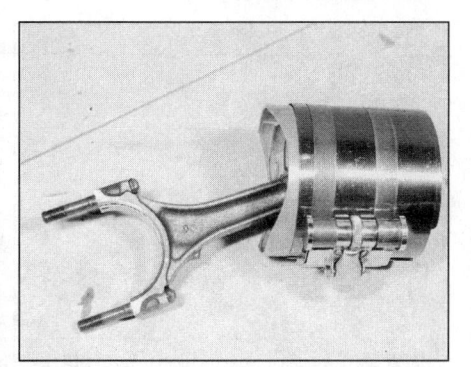

88.6a Piston with piston ring compressor fitted

88.6b Check the numbers on the connecting rod to ensure it is the correct assembly for the cylinder bore

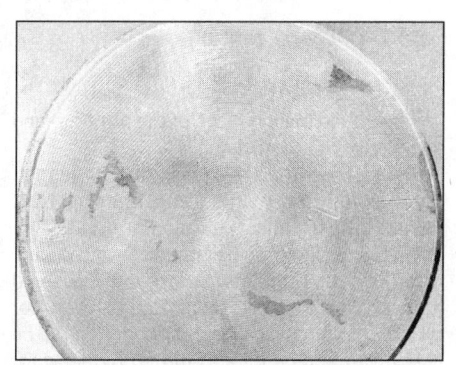

88.6c The arrow on the piston crown must point to the front of the engine

1•44 Engine

88.8 Tap the piston into the cylinder bore

88.11 Fit the big-end bearing cap

89.2 Insert the oil pump drive shaft . . .

assembly for that particular bore, that the connecting rod is the correct way round and that the front of the piston (marked with an arrow or a notch) is to the front of the engine (photos).
7 Again lubricate the piston skirt and insert the connecting rod and piston assembly into the cylinder bore up to the bottom of the piston ring compressor.
8 Gently but firmly tap the piston through the piston ring compressor and into the cylinder bore, using the shaft of a hammer (photo).
9 Generously lubricate the crankpin journals with engine oil and turn the crankshaft so that the crankpin is in the most advantageous position for the connecting rods to be drawn onto it.
10 Wipe clean the connecting rod bearing cap and back of the shell bearing, and fit the shell bearing in position ensuring that the locating tongue at the back of the bearing engages with the locating groove in the connecting rod cap.
11 Generously lubricate the shell bearing and offer up the connecting
rod bearing cap to the connecting rod (photo).
12 Refit the connecting rod nuts.
13 Tighten the bolts with a torque wrench to the specified setting.
14 When all the connecting rods have been fitted, rotate the crankshaft to check that everything is free, and that there are no high spots causing binding.

89 Oil pump - refitting

1 Ensure the mating faces of the oil pump and the crankcase are clean.
2 Insert the hexagonal oil pump driveshaft (photo) with the pointed end, see Fig.1.26, towards the distributor location.
3 Fit the oil pump to the crankcase and secure with two bolts tightened to the specified torque (photos). **Note:** *When a new or overhauled oil pump is being fitted it should be filled with engine oil and turned by hand for one complete revolution before being fitted, this will prime the pump.*

89.3a then fit the oil pump . . .

89.3b and tighten the securing bolts

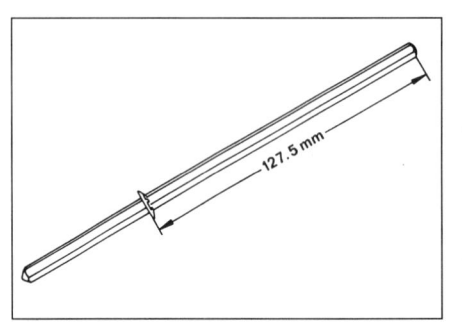

Fig. 1.26 Oil pump hexagonal drive shaft (Sec 89)

90 Flywheel and clutch - refitting

1 Fit the engine backplate on the two locating dowels.
2 Ensure the mating faces of the flywheel and crankshaft are clean and fit the flywheel to the crankshaft, aligning the marks made at dismantling, unless new parts are being fitted (photo).
3 Fit the six flywheel securing bolts and lightly tighten.
4 Chock the flywheel to restrain it from turning, and tighten the securing bolts in a diagonal and progressive manner to the specified torque (photos).
5 Refit the clutch disc and pressure plate

90.2 Fit the flywheel to the crankshaft

90.4a Chock the flywheel to prevent it from turning

Engine 1•45

90.4b Tightening the flywheel bolts with a torque wrench

90.5 Fitting the clutch disc and pressure plate

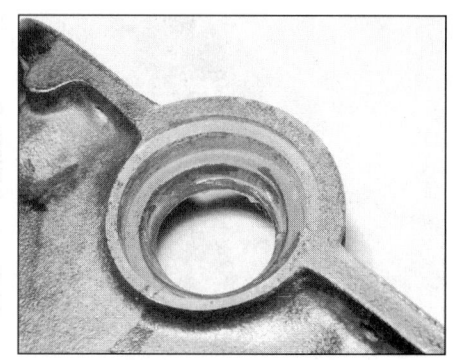

91.1 Renew the seal in the timing cover

assembly to the flywheel making sure the disc is the correct way round (see Chapter 5) (photo).
6 Secure the pressure plate assembly with the retaining bolts, centralise the clutch disc using an old gearbox input shaft or suitable mandrel and then tighten the retaining bolts to the specified torque in a diagonal and progressive manner.

91 Timing gears and timing cover - refitting

1 Fit a new oil seal in the timing cover (photo).
2 Check that the keyways in the end of the crankshaft are clean and the keys are free from burrs. Fit the keys into the keyways.
3 If the crankshaft gear was removed refit it to the crankshaft using a suitable diameter tube to drive it fully home.
4 Fit the camshaft gear on the camshaft with the punch mark in alignment with the mark on the crankshaft gear, as shown in Fig.1.27. Note that there are two punch marks on the crankshaft gear, ensure that the correct mark is aligned (photo). On later models, both 2.3 and 2.8 litre engines are fitted with a nylon camshaft gear. The nylon gear can be fitted to earlier 2.3 engines, but self-locking screws (Ford part No 388379-GS) **must** be used to fasten the camshaft retaining plate.
5 Fit the camshaft gear retaining washer and bolt. Tighten the bolt to the specified torque (photos).
6 Clean the front face of the intermediate plate and the face of the timing cover then coat both mating faces with sealing compound.
7 Position a new gasket on the intermediate plate and fit the timing cover to the cylinder block (photos).
8 Fit the timing cover securing bolts and tighten them to the specified torque.

92 Engine oil sump, water pump, and crankshaft pulley - refitting

1 Clean the mating faces of the crankcase and sump. Ensure that the grooves in the seal carriers are clean.
2 Fit the rubber seals in the grooves of the seal carriers.

Fig. 1.27 Fitting the camshaft gear (Sec 91)

91.4 Fit the camshaft gear on the camshaft...

91.5a ... then the retaining washer and bolt...

91.5b ... and tighten the bolt using a torque wrench

91.7a Stick a new timing cover gasket in position on the intermediate plate...

91.7b ... then fit the timing cover

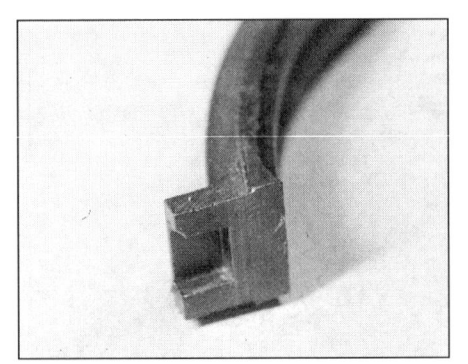

1•46 Engine

92.3a Cutout in the rubber seal

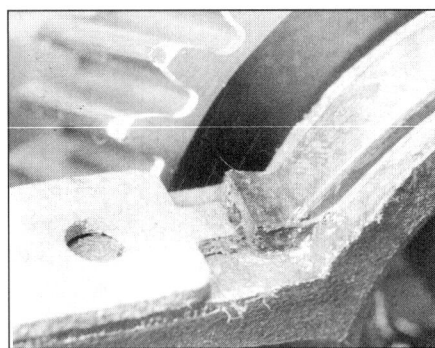

92.3b Slide the sump gasket into the seal

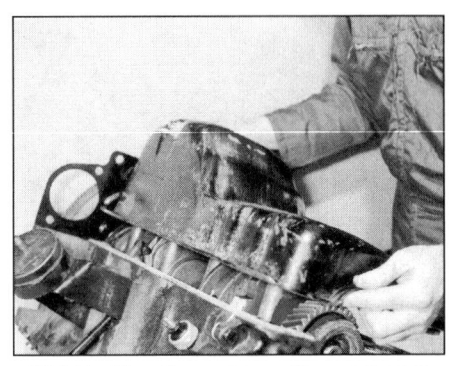

92.4 Position the sump on the crankcase

3 Apply sealing compound on the crankcase and slide the tabs of the gasket under the cut-outs in the rubber seals (photos).
4 Ensure that the gasket hole lines up with the holes in the gasket crankcase and fit the sump. Take care not to dislodge the gasket (photo).
5 Fit the sump securing bolts and tighten them to the correct torque in two stages as specified in Specifications at the beginning of this Chapter.
6 Place a new gasket on the timing cover and fit the water pump (photo). Fit the thermostat and thermostat housing refer to Chapter 2. Fit the rear water elbow.
7 Coat the crankshaft pulley washer with sealing compound. Fit the pulley/damper, washer and securing bolt. Tighten the bolt to the specified torque (photo).

92.6 Fitting the water pump

92.7 Fitting the crankshaft pulley

93 Cylinder heads, rocker shafts and inlet manifold - refitting

Note: *If Torx type cylinder head bolts have been used, these will need to be renewed. Consult your Ford dealer for the correct torque figures for these bolts.*

1 Lubricate the valve tappets with clean engine oil and insert them in the cylinder block. Ensure they are fitted in their original locations (photo).
2 Ensure that the mating faces of the cylinder block and the cylinder heads are clean.
3 Position the new cylinder head gaskets over the guide bushes on the cylinder block. Check that they are correctly located. The right and left-hand gaskets are different. The gaskets are marked FRONT TOP (photo).
4 Carefully lower the cylinder heads onto the cylinder block and fit the holding down bolts (photo).
5 Tighten the bolts in the sequence shown in Fig. 1.28, in three stages as specified in Specifications. (photo).
6 Lubricate the pushrods with engine oil and insert them in the cylinder block (photo).
7 Place the oil splash shields in position on the cylinder heads and fit the rocker shaft

93.1 Insert the tappets in the cylinder block

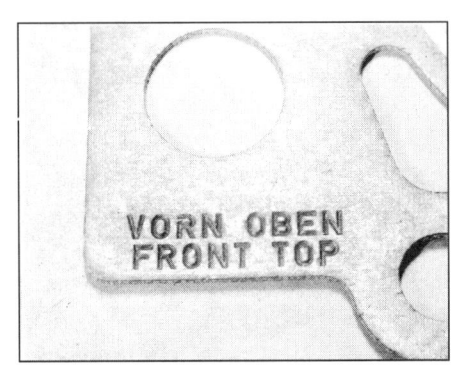

93.3 Cylinder head gasket markings

93.4 Fitting the left-hand cylinder head

93.5 Using a torque wrench to tighten the cylinder head bolts

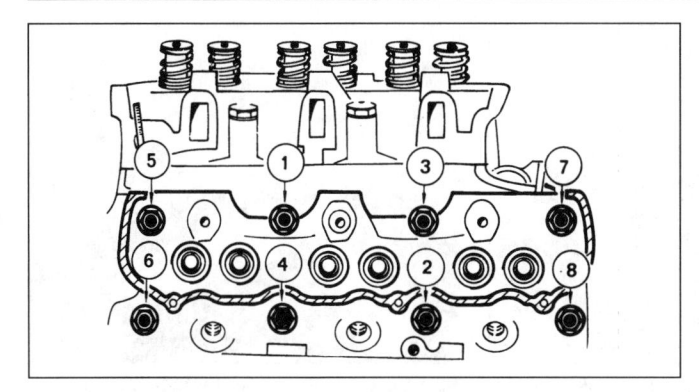

Engine 1•47

Fig. 1.28 Tightening sequence for cylinder head bolts (Sec 93)

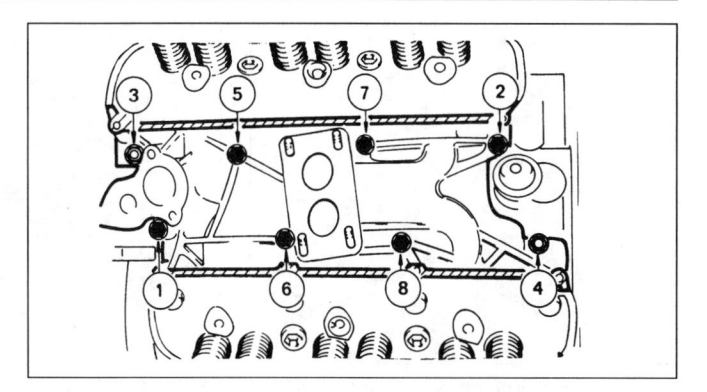

Fig. 1.29 Tightening sequence for inlet manifold bolts (Sec 93)

assemblies. Guide the rocker arm adjusting screws into the pushrod sockets (photos).
8 Tighten the rocker shaft securing bolts progressively to the specified torque.
9 Coat the gasket mating face outer edge of the cylinder heads and inlet manifold with sealing compound.
10 Place a new inlet manifold gasket in position and fit the inlet manifold (photos).
11 Insert the inlet manifold securing bolts and tighten them in the sequence shown in Fig. 1.29. Tighten them to the specified torque in four stages, refer to Specifications (carburettor engine models).
12 Adjust the valve clearances as described in Section 94.

94 Valve clearances - checking and adjustment

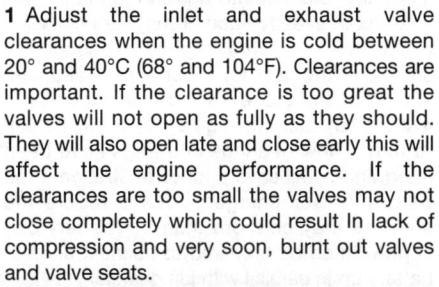

1 Adjust the inlet and exhaust valve clearances when the engine is cold between 20° and 40°C (68° and 104°F). Clearances are important. If the clearance is too great the valves will not open as fully as they should. They will also open late and close early this will affect the engine performance. If the clearances are too small the valves may not close completely which could result In lack of compression and very soon, burnt out valves and valve seats.
2 When turning the engine during valve clearance adjustments always turn the engine in the direction of normal rotation by means of a wrench on the pulley nut.
3 Turn the engine and align the crankshaft pulley mark with the O-mark on the timing cover.
4 If the crankshaft pulley is rotated backwards and forwards slightly, the valves of No. 1 or 5 cylinder will be seen to be rocking (the two rocker arms moving in opposite directions). If the valves of No. 1 cylinder are rocking, rotate the crankshaft through 360° so that those on No. 5 cylinder are rocking.
5 When the valves of No. 5 cylinder are in this position, check the valve clearances of No. 1 cylinder by inserting a feeler gauge of the specified thickness between the rocker arm and the valve stem. Adjust the clearance, if necessary, by turning the rocker arm adjusting

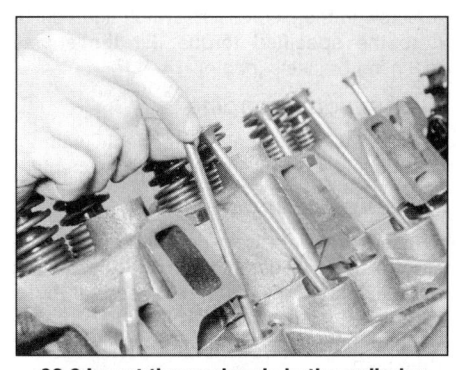

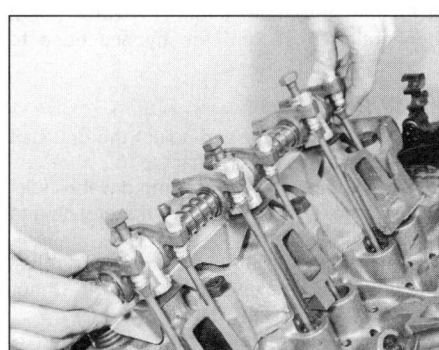

93.6 Insert the pushrods in the cylinder block

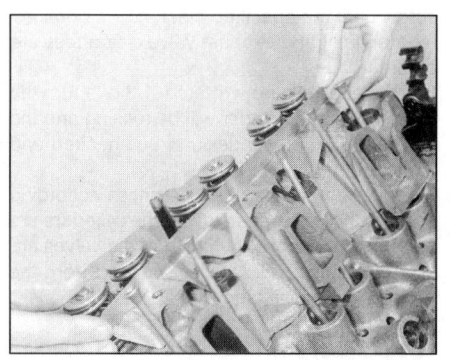

93.7a Fit the oil splash shields in position ..

93.7b ... and then fit the rocker shaft assemblies

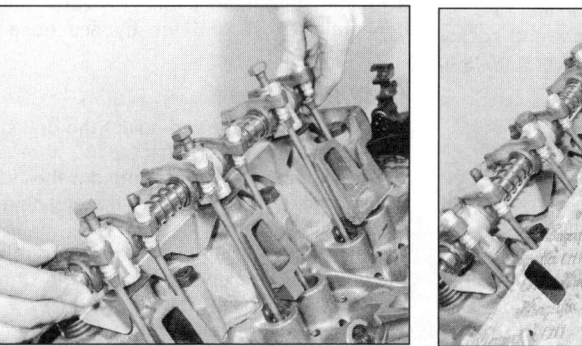

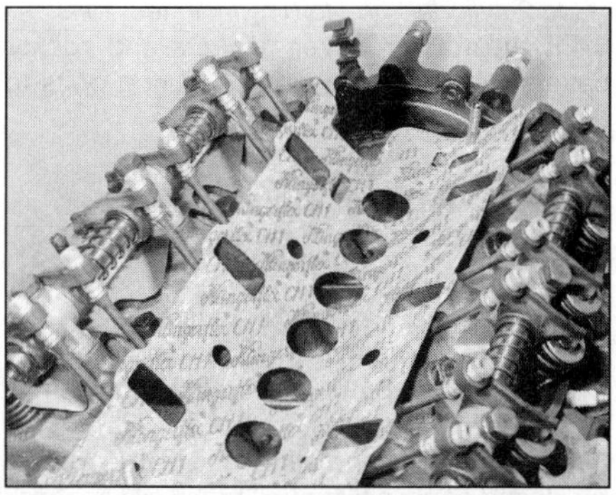

93.10a Place a new inlet manifold gasket in position ...

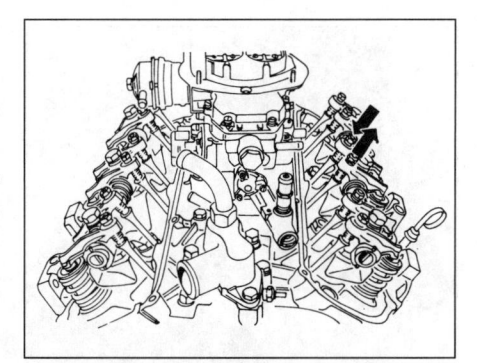

Fig. 1.30 No 5 cylinder valves rocking (Sec 94)

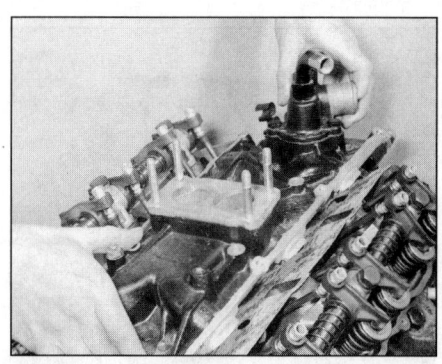

93.10b ... and then fit the inlet manifold

1•48 Engine

94.5 Adjusting the valve clearances

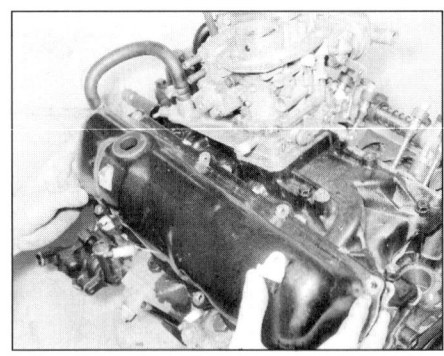

94.8 Fitting the left-hand valve rocker cover

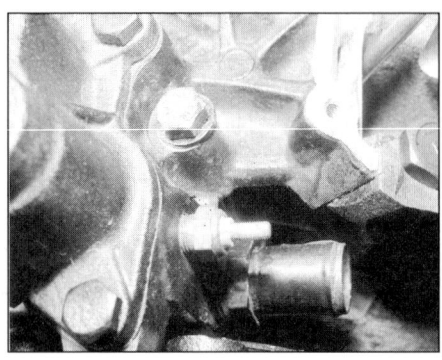

95.1 Screw in the water temperature switch

screw until the specified clearance is obtained (photo). Inlet and exhaust valve clearances are different.
6 If the engine is now rotated 1/3 of a turn the valves of No. 3 cylinder will be rocking and the valves of No. 4 cylinder can be checked and adjusted,
7 Proceed to adjust the clearances according to the firing order as follows. The cylinders are numbered as shown in Fig.1.31: the valves are listed in their correct order, working from the front of the engine (see Fig. 1.32):

Valves rocking	Valves to adjust
No 5 cylinder	No 1 cylinder (in, ex)
No 3 cylinder	No 4 cylinder (in, ex)
No 6 cylinder	No 2 cylinder (in, ex)
No 1 cylinder	No 5 cylinder (ex, in)
No 4 cylinder	No 3 cylinder (ex, in)
No 2 cylinder	No 6 cylinder (ex, in)

8 Fit the rocker cover gaskets and rocker covers (photo). Tighten the securing bolts to the specified torque.

96 Ancillary components - refitting

1 Screw in the water temperature switch and tighten it to the specified torque (photo).
2 Refit the distributor and time the ignition as described in Chapter 4.
3 Clean the oil filter mating face of the cylinder block and apply a film of oil to the oil filter seal. Screw the new oil filter cartridge in until the rubber seal makes contact with the housing, then tighten a further ¾ turn.

4 Screw in the oil pressure switch and tighten it to the specified torque. Fit the engine mounting brackets (photo).

Fuel injection engine

5 Refit the individual components without altering their original settings or adjustments.

Carburettor engines

6 Insert the fuel pump operating rod in the side of the cylinder block, fit the insulation washer, fuel pump and securing bolts (photos).
7 Refit the carburettor as described in Chapter 3.
8 Connect the engine breather hose to rocker cover and the fuel line between the fuel pump and carburettor.
9 Fit the automatic choke to water outlet connecting hose and the bypass hose to thermostat housing.

All engines

10 Fit the alternator and adjust the drivebelt tension as described in Chapter 10.
11 Fit the power steering pump drivebelt idler pulley bracket to the front of the engine and secure with two bolts.
12 Fit the power steering pump and adjust the drivebelt tension as described in Chapter 11.
13 Refit the exhaust manifolds. On models built after February 1978 note the following:
a) The retaining studs must protrude from the cylinder head by 1.14±0.06 in (29.0±1.5 mm)

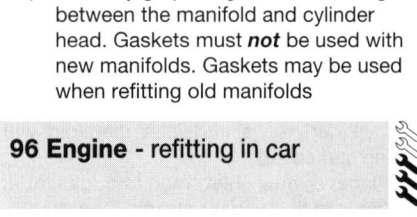

b) Use only graphite grease for sealing between the manifold and cylinder head. Gaskets must *not* be used with new manifolds. Gaskets may be used when refitting old manifolds

96 Engine - refitting in car

1 Refitting the engine in the car is a reversal of the removal procedure. A little trouble in getting the engine properly slung so that it takes up a suspended attitude similar to its final position will pay off when it comes to locating it on the front engine mountings.
2 Ensure that all loose leads, cables, hoses etc, are tucked out of the way. If not, it is easy to trap one and so cause additional work after the unit is refitted in the car.
3 Carefully lower the engine whilst an assistant guides it into position (photo). If the gearbox is already fitted, it may be necessary to turn the crankshaft slightly to engage the splines of the gearbox input shaft. Take care not to put any undue strain on the input shaft.
4 It is likely that the engine will initially be stiff to turn if new bearings and rings have been fitted and it will save a lot of frustration if the battery is well charged. After a rebore the stiffness may be more than the battery can cope with so be prepared to connect another battery up in parallel with jump leads.
5 The following check list should ensure that the engine starts safely and with the minimum of delay:

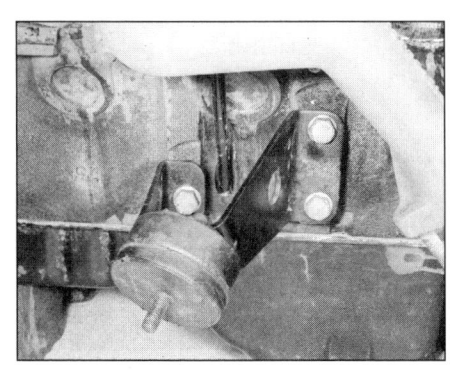

95.4 Left-hand engine mounting and bracket

95.6a Insert the fuel pump operating rod . . .

95.6b . . . and then fit the fuel pump

Engine 1•49

96.3 Lowering the engine into the engine compartment

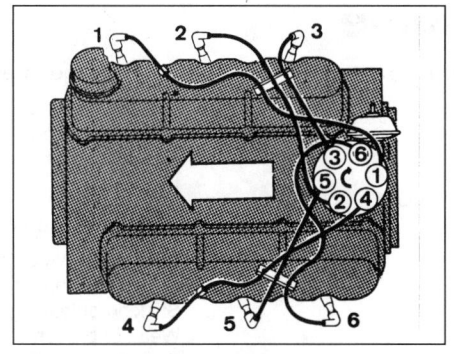

Fig. 1.31 Cylinder numbering and HT lead connections (Sec 96)

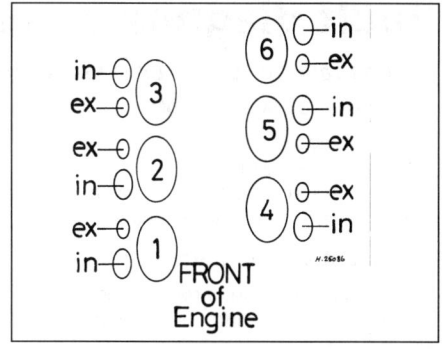

Fig. 1.32 Location of inlet and exhaust valves

a) Fuel lines connected and tightened
b) Water hoses connected and secured with clips
c) Coolant drain plugs fitted and tightened
d) Cooling system replenished
e) Sump drain plug fitted and tight
f) Oil in engine
g) LT wiring connected to distributor and coil
h) Spark plugs tight
j) Valve clearances set correctly
k) Rotor arm fitted in distributor
l) HT leads connected correctly to distributor, spark plugs and coil (Fig. 1.31)
m) Throttle cable connected
n) Earth straps reconnected
o) Starter motor leads connected
p) Alternator leads connected
q) Battery fully charged and leads connected to clean terminals

6 On fuel injection engines, refer to Section 66 and reverse the operation described in paragraphs 1 to 7.

97 Engine - initial start-up after major overhaul or repair

Refer to Part A Section 60 of this Chapter.

98 Engine modifications - later models

1 A number of modifications were made to the V6 engines introduced from June 1979, primarily to improve performance and fuel economy. The main differences are as follows.
2 On 2.3 litre engines, the piston height was increased to give a higher compression ratio. At the same time, the combustion chambers were redesigned to accommodate larger valves, and the valve springs and valve stem seals were uprated. The head gaskets differ from earlier models as a result.
3 On all engines, modified oil scraper piston rings were introduced, having a lower coefficient of friction.
4 Bronze bushes are now used to support the camshaft. The valve timing has been changed on all but fuel injection engines - see Specifications.
5 Modifications have also been made to the inlet and exhaust manifolds to improve gas flow.
6 Ancillary components such as the carburettor and distributor have been modified to make full advantage of the mechanical changes in the engine. Additionally, the thermostatically controlled fan coupling, previously fitted only to the 2.8 litre engine, is now fitted to all Granada engines. Refer to the respective Chapters for details.
7 In view of the above, particular care must be taken when ordering replacement engine parts. Note that early and late model components are not necessarily interchangeable. Consult your Ford dealer if in doubt and always quote your car's model year and engine number when ordering parts.

Fault diagnosis - engine

Engine fails to turn over when starter operated

No current at starter motor
Flat or defective battery
Loose battery leads
Defective starter solenoid or switch or broken wiring
Engine earth strap disconnected

Current at starter motor
Jammed starter motor drive pinion
Defective starter motor or solenoid

Engine turns over but will not start

No spark at spark plug
Ignition damp or wet
Ignition leads to spark plugs loose
Shorted or disconnected low tension leads
Dirty, incorrectly set or pitted contact breaker points
Faulty condenser
Defective ignition switch
Ignition leads connected wrong way round
Faulty coil
Contact breaker point spring earthed or broken

Excess of petrol in cylinder or carburettor
Too much choke allowing too rich a mixture to wet plugs flooding
Float damaged or leaking or needle not seating
Float lever incorrectly adjusted

Engine stalls and will not start

No spark at spark plug
Ignition failure

No fuel getting to engine
No petrol in petrol tank
Petrol tank breather choked
Obstruction in carburettor(s)
Water in fuel system

Engine misfires or idles unevenly

Intermittent spark at spark plugs
Ignition leads loose
Battery leads loose on terminals
Battery earth strap loose on body attachment point
Engine earth lead loose
Low tension leads to SW (+) and CB (-) terminals on coil loose
Low tension lead from CB (-) terminal side to distributor loose
Dirty, or incorrectly gapped plugs
Dirty, incorrectly set, or pitted contact breaker points
Tracking across inside of distributor cover
Ignition too retarded
Faulty coil

No fuel at carburettor or fuel injection system
No petrol in petrol tank
Vapour lock in fuel line (In hot conditions or at high altitude)
Blocked float chamber needle valve
Fuel pump filter blocked
Choked or blocked carburettor jets
Faulty fuel pump

Fuel shortage at engine
Mixture too weak
Air leak in carburettor
Air leak at inlet manifold to cylinder head, or inlet manifold to carburettor

Mechanical wear
Incorrect valve clearances
Burnt out exhaust valves
Sticking or leaking valves
Weak or broken valve springs
Worn valve guides or stems
Worn pistons and piston rings

Lack of power and poor compression

Fuel/air mixture leaking from cylinder
Burnt out exhaust valves
Sticking or leaking valves
Worn valve guides and stems
Weak or broken valve springs
Blown cylinder head gasket (accompanied by increase in noise)
Worn pistons and piston rings
Worn or scored cylinder bores

Incorrect adjustments
Ignition timing wrongly set. Too advanced or retarded
Contact breaker points incorrectly gapped
Incorrect valve clearances
Incorrectly set spark plugs

Carburation and ignition faults
Dirty contact breaker points
Distributor automatic balance weights or vacuum advance and retard mechanisms not functioning correctly
Faulty fuel pump giving top end fuel starvation

Excessive oil consumption

Oil being burnt by engine
Badly worn, perished or missing valve stem oil seals
Excessively worn valve stems and valve guides
Worn piston rings
Worn pistons and cylinder bores
Excessive piston ring gap allowing blow-by
Piston oil return holes choked

Oil being lost due to leaks
Leaking oil filter gasket
Leaking sump gasket
Loose sump plug

Unusual noises from engine

Excessive noise and/or clearances due to mechanical wear
Camshaft fibre gear stripped (V6 engines only)
Worn valve gear (noisy tapping from rocker box)
Worn big-end bearing (regular heavy knocking)
Worn timing belt and gears (rattling from front of engine)
Worn main bearings (rumbling and vibration)
Worn crankshaft (knocking rumbling and vibration)

Chapter 2 Cooling system

For modifications, and information applicable to later models, see Supplement at end of manual

Contents

Antifreeze precautions 15
Cooling system - draining 2
Cooling system - filling 4
Cooling system - flushing 3
Fan belt - adjustment 11
Fan belt - removal and refitting 10
Fan - removal and refitting 12
Fault diagnosis - cooling system ....See end of Chapter
General description 1
Radiator - refitting 6
Radiator - removal, inspection and cleaning 5
Temperature gauge and sender unit - removal and refitting 14
Temperature gauge - fault diagnosis 13
Thermostat - removal, testing and refitting 7
Water pump - dismantling and overhaul 9
Water pump - removal and refitting 8

Degrees of difficulty

Easy, suitable for novice with little experience	**Fairly easy,** suitable for beginner with some experience	**Fairly difficult,** suitable for competent DIY mechanic	**Difficult,** suitable for experienced DIY mechanic	**Very difficult,** suitable for expert DIY or professional

Specifications

System type .... Pressurised assisted by pump and fan. Sealed on later models

Water pump type .... Centrifugal

Radiator
Type .... Corrugated fin
Pressure cap setting .... 13 lbf/in^2 (0.9 kgf/cm^2)

Thermostat
	ohc engine	V6 engines
Type	Wax	Wax
Location	Front of cylinder head	Bottom of LH side of front cover
Starts to open*	89°C to 93°C	89°C to 93°C
Fully open*	103°C to 106°C	107°C to 108°C

For used thermostats allow ± 3°C (6°F)

Coolant type/specification .... Antifreeze to Ford spec SSM-97B9103-A

Cooling system capacity including heater .... 12.85 pints (7.3 litres) 18 pints (10.22 litres)

Torque wrench settings
	lbf ft	Nm
OHC engine		
Water pump:		
M8 bolts	15	20
M10 bolt	30	40
Thermostat housing	12	17
Fan blades (fixed fan)	7	10
Fan clutch-to-water pump hub	26	35
Fan-to-fan clutch	7	10
V6 engine		
Water pump	7	10
Thermostat housing	7	10
Pulley-to-water pump hub	9	12
Fan-to-fan clutch	7	10

2•2 Cooling system

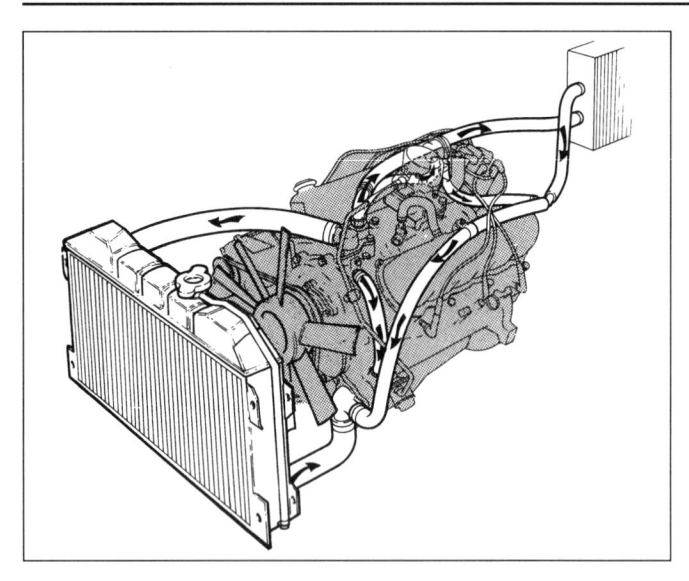

Fig. 2.1 Cooling system layout - V6 engines without expansion tank (arrows show direction of flow) (Sec 1)

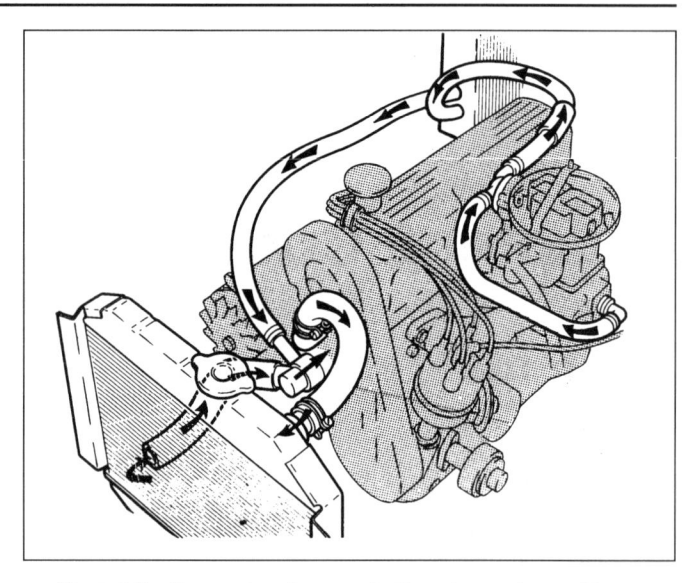

Fig. 2.2 Cooling system layout - in-line ohc engines without expansion tank (arrows show direction of flow) (Sec 1)

1 General description

The engine coolant is circulated by a pressurised, water pump assisted, thermosyphon system. Pressure is maintained at 13 lbf/in^2 (0.9 kgf/cm^2) by a spring loaded radiator cap. This has the effect of increasing the boiling point of the coolant. If the coolant temperature rises above the increased boiling point, the extra pressure in the system forces the internal, spring loaded, part of the cap off its seat. This exposes the overflow pipe down which steam from the boiling water may escape, thus relieving the pressure in the system. When the coolant cools after the engine has been switched off, a valve opens in the cap to admit air to prevent the creation of a vacuum within the radiator header tank. It is therefore important to ensure that the radiator cap is in good condition and that the spring behind the sealing washer has not weakened.

Later models (from 1979/1980 onwards) were fitted with a slim top tank radiator and a separate coolant expansion container. Although carburettor-equipped models built between July 1979 and February 1980 retained the early radiator, they were fitted with a smaller (1 litre) coolant expansion tank.

The cooling system comprises the radiator, top and bottom water hoses, heater hoses, thermostat, drain tap and impeller water pump. The impeller water pump is mounted on the front of the engine and carries the fan blades. Drive is by vee belt and pulley.

On carburettor models the inlet manifold is water heated and an automatic choke is also operated by the coolant.

On automatic transmission models a transmission oil cooler is located in the radiator bottom tank.

Depending on engine variant three types of cooling fan may be fitted. These are the standard fixed centre fan, a viscous drive fan and a temperature sensitive viscous drive fan. The fixed centre fan revolves continuously with engine speed. The viscous drive fan utilises a torque limiting device and allows the blades to revolve at varying speeds relative to engine speed. This feature reduces power absorption and noise at high engine speeds. The temperature sensitive viscous drive fan works in the same way but has the additional ability of idling the fan when cooling is not required.

The system functions in the following fashion: Cold water in the bottom of the radiator circulates up the lower radiator hose to the water pump where it is pushed round the water passages in the cylinder block, helping to keep the cylinder bores and pistons cool. The water then travels down the radiator where it is rapidly cooled by the in-rush of cold air through the radiator core, which is created by both the fan and the motion of the car. The water, now much cooler, reaches the bottom of the radiator when the cycle is repeated.

When the engine is cold the thermostat (which is a valve that opens and closes according to the temperature of the water) maintains the circulation of the same water in the engine.

Only when the correct minimum operating temperature has been reached, as shown in the Specifications, does the thermostat begin to open, allowing water to return to the radiator.

2 Cooling system - draining

1 If the engine is cold, remove the filler cap from the radiator by turning the cap anti-clockwise. If the engine is hot, then turn the filler cap to the first stop, and leave until the pressure in the system has had time to be released. If with the engine very hot the cap is released suddenly the drop in pressure can result in the water boiling. With the pressure released the cap can be removed.

HAYNES HiNT *Use a rag over the cap to protect your hand from escaping steam*

2 If antifreeze is used in the cooling system, drain it into a bowl having a capacity of at least 12 Imp pints (7 litres) for re-use.

3 Disconnect the lower radiator hose and allow to drain. Also remove the engine drain plug which is located at the rear left-hand side of the cylinder block (in-line ohc engines) and two plugs either side of the cylinder block (V6 engines) (photo).

4 When the water has finished draining, probe the drain plug orifice with a short piece of wire to dislodge any particles of rust or sediment which may be causing a blockage.

5 It is important to note that the heater cannot be drained completely during the cold weather so an antifreeze solution must be used. Always use an antifreeze with an ethylene glycol base.

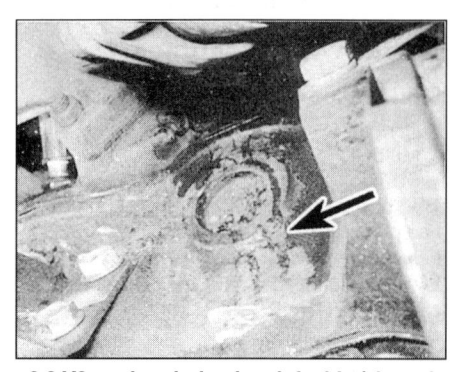

2.3 V6 engine drain plug, left side (viewed from below)

3 Cooling system - flushing

1 In time the cooling system will gradually lose its efficiency as the radiator becomes choked with rust, scale deposits from the water, and other sediment. To clean the system out, remove the radiator filler cap and engine drain plug and leave a hose running in the filler cap neck for ten to fifteen minutes. It is possible to use a good proprietary flushing agent.

2 In very bad cases the radiator should be reverse flushed. This can be done with the radiator in position. The cylinder block plug is removed and a hose with a suitable tapered adaptor placed in the drain plug hole. Water under pressure is then forced through the radiator and out of the header tank filler cap neck.

3 It is recommended that some polythene sheeting is placed over the engine to stop water finding its way into the electrical system.

4 The hose should now be removed and placed in the radiator cap filler neck, and the radiator washed out in the usual manner.

4 Cooling system - filling

1 Refit the cylinder block drain plugs and reconnect the lower radiator hose.

2 Fill the system slowly to ensure that no air locks develop. Ensure that the valve in the heater is open (control at HOT), otherwise an air lock may form in the heater.

HAYNES HiNT *The best type of water to use in the cooling system is rainwater; use this whenever possible.*

3 Do not fill the system higher than within a ½ inch (12.7 mm) of the filler neck. Overfilling will merely result in wastage, which is especially to be avoided when antifreeze is in use.

4 It is usually found that air locks develop in the heater radiator so the system should be vented during refilling by detaching the heater supply hose from the elbow connection on the water outlet housing.

5 Pour coolant in to the radiator filler neck whilst the end of the heater supply hose is held at the elbow connection height. When a constant stream of water flows from the supply hose quickly refit the hose. If venting is not carried out it is possible for the engine to overheat. Should the engine overheat for no apparent reason then the system should be vented before seeking other causes.

6 Only use antifreeze mixture with ethylene glycol base.

7 Refit the filler cap and turn it firmly clockwise to lock it in position.

5 Radiator - removal, inspection and cleaning

1 Drain the cooling system, as described in Section 2 of this Chapter.

2 Slacken the two clips which hold the top and bottom radiator hoses on the radiator and carefully pull off the two hoses (photo). On later models, disconnect and detach the coolant expansion pipe connecting hose at the radiator filler neck overflow pipe.

3 Undo and remove the four bolts that secure the radiator shroud to the radiator side panels and move the shroud over the fan blades. This is only applicable when a shroud is fitted (photos).

4 Vehicles with automatic transmission. Position a drain tray beneath the radiator, disconnect both oil cooler pipes and pull clear of radiator (photo).

5 Undo and remove the four bolts that secure the radiator to the front panel. The radiator may now be lifted upwards and away from the engine compartment. The fragile matrix must not be touched by the fan blades as it easily punctures (photos).

6 Lift the radiator shroud from over the fan blades and remove from the engine compartment.

7 With the radiator away from the car any leaks can be soldered or repaired with a suitable radiator sealant. Clean out the inside of the radiator by flushing as described earlier in this Chapter. When the radiator is out of the car it is advantageous to turn it upside down and reverse flush. Clean the exterior of the radiator by carefully using a compressed air jet or a strong jet of water to clear away any road dirt, flies etc.

8 If you are renewing an early radiator and you intend to fit a later slim top tank type, you will need to fit an expansion container. Consult your Ford dealer for the correct type and fitting

5.2 Radiator top hose connections (V6 engines)

5.3a Remove the fan radiator shroud securing bolts . . .

5.3b . . . and position shroud over fan blades

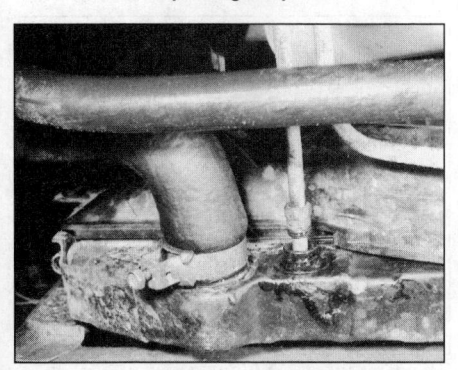

5.4 Remove the automatic transmission oil cooler pipes from the radiator

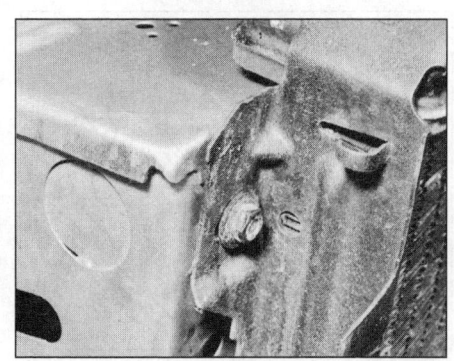

5.5a Remove the radiator mounting bolts . . .

5.5b . . . and lift out the radiator

2•4 Cooling system

details should this be the case, as a special service kit will also be required depending on model and type.

9 Inspect the radiator hoses for cracks, internal or external perishing and damage by overtightening of the securing clips. Also inspect the overflow pipe. Renew the hoses if suspect. Examine the radiator hose clips and renew them if they are rusted or distorted.

6 Radiator - refitting

1 Refitting the radiator and shroud is the reverse sequence to removal.

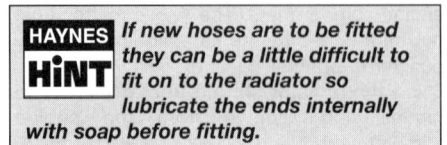

HAYNES HINT: *If new hoses are to be fitted they can be a little difficult to fit on to the radiator so lubricate the ends internally with soap before fitting.*

2 Refill the cooling system as described in Section 4.
3 *Vehicles with automatic transmission*. Refer to Chapter 6, Section 15, and check and top-up the automatic transmission fluid.
4 Ensure all hoses are secure, free from kinks and correctly positioned.

7 Thermostat - removal, testing and refitting

1 Partially drain the cooling system as described in Section 2.

In-line ohc engines

2 Slacken the top radiator hose to the thermostat housing and remove the hose.
3 Undo and remove the two bolts and spring washers securing the thermostat housing to the cylinder head.
4 Carefully lift the thermostat housing away from the cylinder head. Recover the joint washer adhering to either the housing or cylinder head.
5 Using a screwdriver ease the clip securing the thermostat to the housing. Note which way round the thermostat is fitted in the housing and also that the bridge is 90° to the outlet.

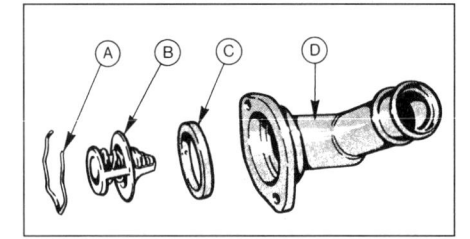

Fig. 2.3 Thermostat components - in-line ohc engines (Sec 7)

A Retaining clip   B Thermostat
C Sealing ring   D Housing

6 The thermostat may now be withdrawn from the housing. Recover the seal from inside the housing.

V6 engines

7 Disconnect the radiator bottom hose at the thermostat housing and remove the hose.
8 Slacken the heater hose clip at the thermostat housing and pull off the hose.
9 Remove the three retaining bolts and lift away the thermostat housing, sealing ring and thermostat (photos).

All engines

10 Test the thermostat for correct functioning by suspending it on a string in a saucepan of cold water together with a thermometer. Heat the water and note the temperature at which the thermostat begins to open.
11 Compare the noted temperatures with those given in the Specifications at the beginning of this Chapter. If they differ considerably the thermostat must be renewed.
12 If the thermostat does not fully open in boiling water, or does not close down as the water cools, then it must be discarded and a new one fitted. Should the thermostat be stuck open when cold this will usually be apparent when removing it from the housing.
13 Refitting the thermostat is the reverse sequence to removal. Always ensure that the mating faces are clean and flat. If the thermostat housing is badly corroded fit a new housing. Always use a new gasket. Tighten the securing bolts to the specified torque wrench setting.

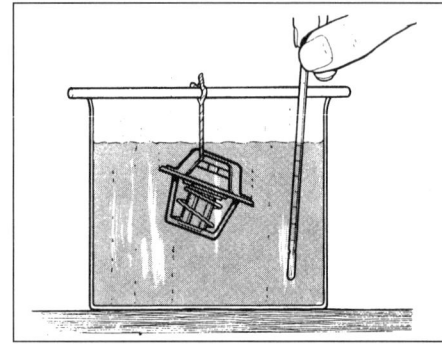

Fig. 2.4 Testing of the thermostat (Sec 7)

7.9a Remove the thermostat housing . . .

7.9b . . . and then the thermostat and sealing ring

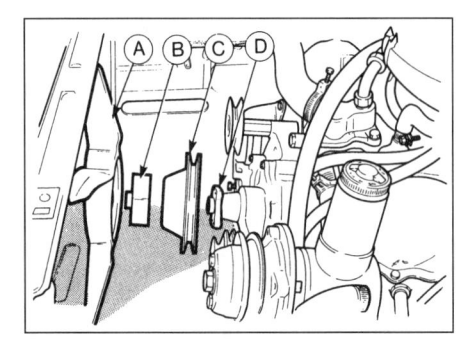

Fig. 2.5 Fan assembly - standard fans (Sec 8)

A Fan   B Spacer   C Pulley   D Hub

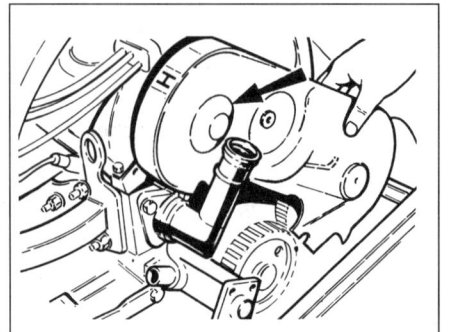

Fig. 2.6 Removing the timing belt cover - in-line ohc engine (arrowed) (Sec 8)

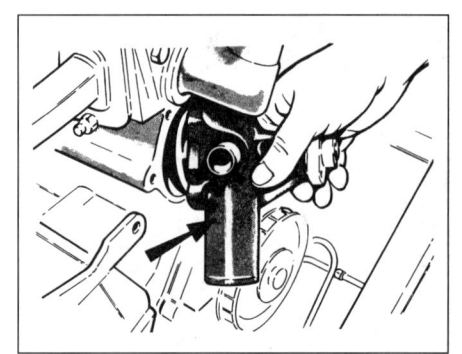

Fig. 2.7 Lifting out the water pump - in-line ohc engines (arrowed) (Sec 8)

Cooling system 2•5

8 Water pump - removal and refitting

1 Drain the cooling system as described in Section 2.

In-line ohc engines

2 Refer to Section 5 and remove the radiator (and shroud if fitted).
3 Slacken the alternator mounting bolts and push the alternator towards the cylinder block. Lift away the fan belt.
4 Undo and remove the four bolts and washers that secure the fan assembly to the water pump spindle hub. Lift away the fan and pulley (Fig. 2.5). On later models with a viscous fan coupling, note that the fan clutch is secured to the water pump hub by a bolt having a *left-hand thread*. See Section 12 for details.
5 Slacken the clip that secures the heater hose to the water pump. Pull the hose from its union on the water pump.
6 Slacken the clamp that secures the lower hose to the water pump. Pull the hose from the pump.
7 Remove three bolts and take off timing belt cover (Fig. 2.6).
8 Undo and remove the four bolts and spring washers that secure the water pump to the cylinder block. Lift away the water pump and recover the gasket (Fig. 2.7).

V6 engines

9 Remove the radiator and shroud as described in Section 5
10 Slacken the alternator mounting bolts and remove the fan belt.
11 Remove the fan as described in Section 12.
12 Remove the thermostat housing and thermostat as described in Section 7.
13 Undo the securing bolts and remove the water pump assembly (Fig. 2.8).

All engines

14 Refitting the water pump is the reverse sequence to removal. The following additional points should, however, be noted:
 a) Make sure the mating faces of the cylinder block and water pump are clean. Always use a new gasket

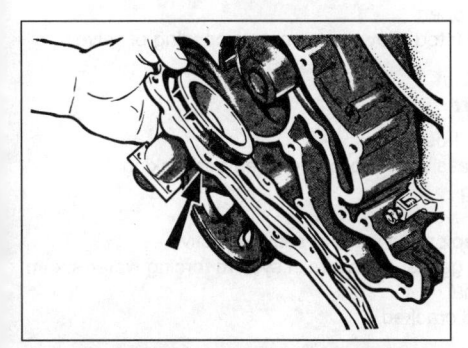

Fig. 2.8 Removing the water pump assembly - V6 engines (arrowed) (Sec 8)

 b) Tighten the water pump securing bolts to the specified torque wrench setting
 c) Tighten the water pump fan and pulley bolts to the specified torque wrench setting
15 Refill the cooling system as described in Section 4.

9 Water pump - dismantling and overhaul

1 Water pump failure is indicated by leaks, noisy operation and/or excessive slackness of the pump spindle.
2 Although new parts are available, the dismantling of the pump and the fitting of the seals and bearings requires a press and correctly sized mandrels. In the event of pump failure, it is recommended that a new pump is fitted rather than trying to repair the old one.

10 Fan belt - removal and refitting

1 If the fan belt has worn or has stretched unduly it should be renewed. The most common reason for renewal is that the belt has broken in service. For this reason it is recommended that a spare belt should always be carried in the car.
2 On in-line ohc engines with power assisted steering two belts are fitted. These belts must always be renewed in pairs never individually.
3 On V6 engines with power assisted steering the power steering drivebelt must be removed first. Refer to Chapter 11 for removal of this belt.
4 Loosen the alternator mounting bolts and move the alternator towards the engine.
5 Slip the old belt over the pulley wheels and lift it over the fan blades.
6 Fit a new belt over the pulleys and adjust it as described in Section 11. Note: *After fitting a new belt it will require adjustment after approximately 250 miles (400 km).*

11 Fan belt - adjustment

1 It is important to keep the fan belt correctly adjusted and it is considered that this should be a regular maintenance task every 6000 miles (10000 km). If the belt is loose it will slip, wear rapidly and cause the alternator and water pump to malfunction. If the belt is too tight the alternator and water pump bearings will wear rapidly causing premature failure of these components.
2 The fan belt tension is correct when there is 0.5 in (12.7 mm) of lateral movement at the mid-point position of the belt run between the alternator pulley and the water pump.
3 Adjust the fan belt, slacken the alternator securing bolts and move the alternator in or out until the correct tension is obtained.

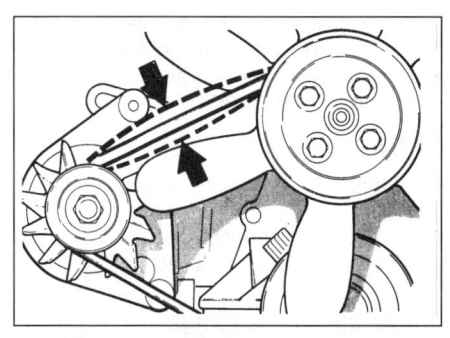

Fig. 2.9 Fan belt tension measuring point (Sec 11)

> **HAYNES HiNT** *It is easier if the alternator bolts are only slackened a little so it requires some effort to move the alternator. In this way the tension of the belt can be arrived at more quickly than by making frequent adjustment.*

4 When the correct adjustment has been obtained fully tighten the alternator mounting bolts.

12 Fan - removal and refitting

Before removing the fan determine the type of fan assembly fitted. Standard fixed centre fans have seven blades, viscous drive fans have eight blades.
1 Remove the fan shroud (where fitted) as described in Section 5.
2 Slacken the alternator mounting bolts and move the alternator towards the engine.
3 Remove the fan belt from the fan pulley.

Fixed centre fans

4 Remove the four bolts and spring washers securing the fan to the hub (Fig. 2.10).
5 Lift off the fan and pulley from the hub.

Viscous drive fans

6 Hold the fan blades and remove the centre bolt securing the fan assembly to the extension shaft. (In some cases there are four bolts behind the blades) (photo). Note that on 2 litre ohc engines, the centre bolt has a *left-hand thread*.

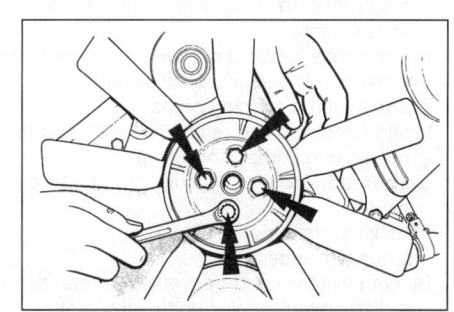

Fig. 2.10 Removing the fan retaining bolts - standard fans (arrowed) (Sec 12)

2•6 Cooling system

7 Lift the fan assembly off the extension shaft.
8 Remove the four nuts and bolts securing the viscous clutch to the fan and separate the clutch and fan (Fig. 2.13).

All fans

9 Refitting is a reverse of the removal procedure.
10 Ensure that the fan belt is correctly adjusted as described in Section 11.

13 Temperature gauge - fault diagnosis

1 If the temperature gauge fails to work either the gauge, the sender unit, the wiring or the connections are at fault.
2 It is not possible to repair the gauge or the sender unit if found to be at fault and they must be replaced by new units.
3 First check the wiring connections are sound. Check the wiring for breaks using an ohmmeter. The sender unit and gauge should be tested by substitution.

14 Temperature gauge and sender unit - removal and refitting

1 The temperature gauge is removed as described in Chapter 10.
2 Ensure that the cooling system is not pressurised by removing and refitting the radiator cap. This will minimise the amount of coolant loss.
3 Disconnect the wiring from the sender unit and unscrew the unit. The sender unit is located in the cylinder head left-hand side just below the manifold on in-line ohc engines and on V6 engines just below the coolant top hose connection on the front of the left-hand cylinder head.
4 Refitting is the reverse of the removal procedure. Smear sealant on the threads of the sender unit. Check the coolant level and top-up as necessary.

12.6 Viscous drive fan mounting bolts

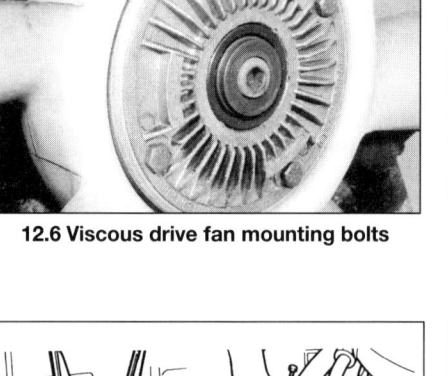

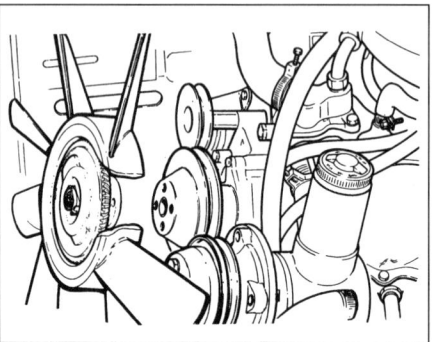

Fig. 2.12 Alternative method of attaching viscous fans (Sec 12)

15 Antifreeze precautions

1 Apart from the protection against freezing conditions which the use of antifreeze provides, its use is essential to minimise corrosion of the cooling system.
2 The cooling system is initially filled with a solution of 45% antifreeze and it is recommended that this percentage is maintained.
3 With long-life types of antifreeze mixtures, renew the coolant every two years. With other types, drain and refill the system every twelve months. Whichever type is used, it must be of the ethylene glycol base.
4 The following table gives a guide to protection against frost but a mixture of less than 30% concentration will not give protection against corrosion:

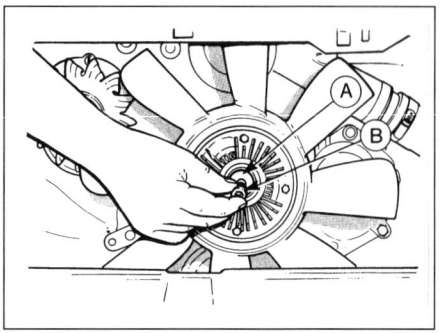

Fig. 2.11 Removing the centre securing bolt - viscous drive fans (Sec 12)

A Interlocking coupling   B Centre bolt

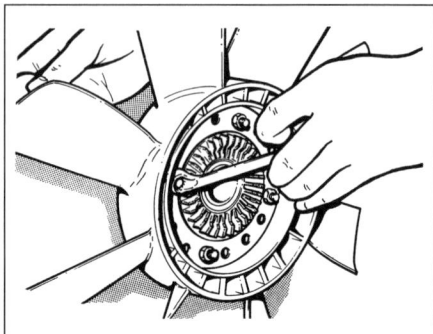

Fig. 2.13 Separating viscous clutch and fan (Sec 12)

Amount of antifreeze	Protection to
45%	-32°C (-26°F)
40%	-25°C (-13°F)
30%	-16°C (+3°F)
25%	-13°C (+9°F)
20%	-9°C (+15°F)
15%	-7°C (+20°F)

Fault diagnosis - cooling system

Overheating

Insufficient water in cooling system
Fan belt slipping (accompanied by a shrieking noise on rapid engine acceleration)
Radiator core blocked or radiator grille restricted
Bottom water hose collapsed, impeding flow
Thermostat not opening properly
Ignition advance and retard incorrectly set (accompanied by loss of power, and perhaps misfiring)
Carburettor incorrectly adjusted (mixture too weak, carburettor models)
Exhaust system partially blocked
Viscous fan inoperative
Blown cylinder head gasket (water/steam being forced down the radiator overflow pipe under pressure)
Engine not yet run-in
Brakes binding

Cool running

Thermostat jammed open
Incorrect thermostat fitted allowing premature opening of valve
Thermostat missing

Loss of cooling water

Loose clips on water hose
Top, bottom or by-pass water hoses perished and leaking
Radiator core leaking
Thermostat gasket leaking
Radiator pressure cap spring worn or seal ineffective
Blown cylinder head gasket (pressure in system forcing water/steam down overflow pipe)
Cylinder wall or head cracked

Chapter 3 Fuel and exhaust systems

For modifications, and information applicable to later models, see Supplement at end of manual

Contents

General description	1

Part A In-line ohc engines

Accelerator cable - removal and refitting	12
Accelerator linkage - adjustment	13
Accelerator pedal assembly - removal and refitting	11
Air cleaner - removal and refitting	2
Exhaust system - general description	21
Exhaust system - removal and refitting	22
Fuel filler neck - removal and refitting	8
Fuel gauge sender unit - removal, testing and refitting	10
Fuel pump - description	3
Fuel pump - removal and refitting	6
Fuel pump - routine servicing	4
Fuel pump - testing	5
Fuel tank - cleaning	9
Fuel tank - removal and refitting	7
Weber carburettor - adjustment	15
Weber carburettor automatic choke - adjustment	20
Weber carburettor automatic choke - removal, overhaul and refitting	19
Weber carburettor - dismantling and reassembly (general)	17
Weber carburettor - dismantling, cleaning, inspection and reassembly	18
Weber carburettor - general description	14
Weber carburettor - removal and refitting	16

Part B V6 Engines

Accelerator cable - removal and refitting	30
Accelerator linkage - adjustment	31
Accelerator pedal assembly - removal and refitting	29
Air cleaner - general description	23
Air cleaner - removal and refitting	24
Air cleaner thermostatically controlled valve - testing	25
Exhaust system - general description	40
Exhaust system - removal and refitting	41
Fuel pump - description	26
Fuel tank - general	27
Fuel gauge sender unit - removal and refitting	28
Solex carburettor - adjustment	33
Solex carburettor automatic choke - adjustment	38
Solex carburettor automatic choke - dismantling, inspection and reassembly	39
Solex carburettor - dismantling and reassembly (general)	35
Solex carburettor - dismantling cleaning and inspection	36
Solex carburettor fast idle speed - adjustment	37
Solex carburettor - removal and refitting	34
Solex carburettor - general description	32

Part C Fuel injection system

Air cleaner assembly - removal and refitting	43
Fuel/air mixture - adjustment	46
General description and principle of operation	42
Main components - removal and refitting	44
Maintenance and adjustment - general	45

Fault diagnosis – fuel and exhaust systems

Fault diagnosis - all models . . . . See end of Chapter

Degrees of difficulty

Easy, suitable for novice with little experience	Fairly easy, suitable for beginner with some experience	Fairly difficult, suitable for competent DIY mechanic	Difficult, suitable for experienced DIY mechanic	Very difficult, suitable for expert DIY or professional

Specifications

In-line ohc engines

Fuel pump

Type	Mechanical driven by pushrod from auxiliary shaft
Delivery pressure	4 to 5 lbf/in^2 (0.28 to 0.36 kgf/cm^2)

Air cleaner element . . . . Champion W110

Weber carburettor

Venturi diameter:
- Primary . . . . 1.02 in (26 mm)
- Secondary . . . . 1.06 in (27 mm)

Main jet - primary:
- Manual transmission 135
- Automatic transmission 132

Main jet - secondary:
- Manual transmission 130
- Automatic transmission 140

Air correction jet - primary:
- Manual transmission 170
- Automatic transmission 175

3•2 Fuel and exhaust systems

Weber carburettor (cont.)

Air correction jet - secondary
   Manual transmission 125
   Automatic transmission 125
Diffuser tube:
   Primary F66
   Secondary F66
Idling jet:
   Primary 45
   Secondary 45
Mixture % CO 1.5
Idling speed .... 775 to 825 rpm
Fast idle speed .... 1900 to 2100 rpm
Float level:
   Brass float .... 1.59 to 1.63 in (40.5 to 41.5 mm)
   Plastic float .... 1.37 to 1.41 in (34.8 to 35.5 mm)
Vacuum pull down setting .... 0.27 to 0.29 in (6.75 to 7.25 mm)
Choke phasing setting .... 0.05 to 0.07 in (1.25 to 1.75 mm)

V6 engines

Fuel pump

Type .... Mechanical driven by pushrod from camshaft
Delivery pressure .... 4 to 5 lbf/in^2 (0.28 to 0.36 kgf/cm^2)

Air cleaner

Type .... Thermostatically controlled
Element:
   2.3 litre .... Champion W119
   2.8 litre (carburettor) .... Champion W167
   2.8 litre (fuel injection) .... Champion U507

Solex carburettor

Venturi diameter:
   2.3 litre .... 0.98 in (25 mm)
   2.8 litre .... 1.10 in (28 mm)
Main jet:
   2.3 litre 130
   2.8 litre manual 147.5
   2.8 litre automatic 145
Idle jet:
   2.3 litre 42.5
   2.8 litre 45
Mixture % CO 1.5
Idle speed:
   Early models (to mid-1979) .... 800 ± 25 rpm
   Later models .... 850 ± 50 rpm
Basic idle % CO 2.75
Basic idle speed .... 600 rpm
Fast idle speed .... 1700 to 1900 rpm
Float level setting .... 0.55 to 0.59 in (14 to 15 mm)
Pump stroke direction setting (refer to Section 36) .... 0.14 to 0.26 in (3.5 to 6.5 mm)
Choke plate pull down:
   2.3 litre .... 0.12 in (3.1 mm)
   2.8 litre .... 0.16 in (4.0 mm)
Choke phasing setting .... 0.012 to 0.024 in (0.3 to 0.6 mm)
Modulation spring gap:
   2.3 litre .... 0.07 in (1.9 mm)
   2.8 litre .... 0.08 in (2.1 mm)

Fuel injection system

Type .... Ford continuous injection

Fuel filter .... Champion L203

Idle speed

Manual transmission .... 875 to 925 rpm
Automatic transmission .... 825 to 875 rpm
Mixture % CO .... 1.25 ± 2

Torque wrench settings

	lbf ft	Nm
Fuel pump to cylinder block	13	18
Exhaust manifold clamp nuts	20	27
Exhaust pipe U-bolts	20	27

Fuel and exhaust systems

1 General description

The fuel system on all models (except 2.8 litre fuel injection, see Sec 42) consists of a rear mounted fuel tank, a mechanically operated fuel pump and either a Weber or Solex twin venturi carburettor with automatic choke.

The fuel tank is positioned below the luggage compartment and is held in position by two metal straps. A combined fuel outlet pipe and sender unit is located in the front face of the tank and can be removed with the tank in position. Fuel tank ventilation is via the filler cap located in the right-hand rear quarter panel.

The mechanical fuel pump is located on the left-hand side of the engine and is connected to the tank by a nylon pipe. Located in the fuel pump is a nylon filter and access is gained via a sediment cap.

The air cleaner fitted to all models is of the renewable paper element type. On in-line ohc engines an adjustable spout is fitted to the air cleaner body allowing manual setting for winter or summer operation. On V6 engines a thermostatically controlled valve automatically adjusts air intake temperature to a predetermined figure.

PART A
IN-LINE OHC ENGINES

2 Air cleaner - removal and refitting

The renewable paper element type air cleaner is fitted onto the top of the carburettor installation and is retained in position by securing nuts and washers, on a flange at the top of the carburettor. Additional support brackets are used. To remove the air cleaner assembly proceed as follows:

1 Note the direction in which the air intake is pointing.
2 Undo and remove the bolt and spring washer that secures the support bracket to the top cover.
3 Undo and remove the bolt and spring washer that secures the air cleaner long support bracket located next to the distributor.
4 Undo and remove the four self-tapping screws on the top cover.
5 There is one additional screw that should be removed and this is located above the air intake.
6 Carefully lift away the top cover. At this stage the element may be lifted out.
7 If it is necessary to remove the lower body, first bend back the lock tabs and then undo and remove the four securing nuts.
8 Lift away the two tab washers and the reinforcement plate.
9 The lower body together with the support brackets may now be removed from the top of the carburettor.
10 Refitting the air cleaner is the reverse sequence to removal.

3 Fuel pump - description

1 The mechanical fuel pump is mounted on the left-hand side of the engine and is driven by an auxiliary shaft. This type of pump cannot be dismantled for repair other than cleaning the filter and sediment cap. Should a fault appear in the pump it may be tested and if confirmed, it must be discarded and a new one obtained. One of two designs may be fitted, this depending on the availability at the time of production of the car.

4 Fuel pump - routine servicing

1 At intervals of 3000 miles (5000 km), undo and pull the fuel pipe from the pump inlet tube (photo).
2 Undo and remove the centre screw and "O"-ring and lift off the sediment cap, filter and seal.
3 Thoroughly clean the sediment cap, filter and pumping chamber using a paintbrush and clean petrol to remove any sediment.
4 Reassembly is the reverse sequence to dismantling. Do not over tighten the centre screw as it could distort the sediment cap.

5 Fuel pump - testing

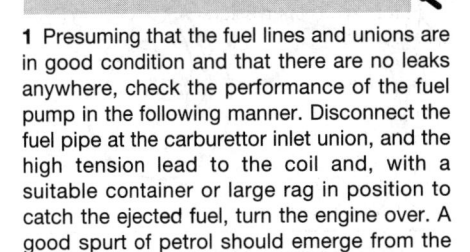

1 Presuming that the fuel lines and unions are in good condition and that there are no leaks anywhere, check the performance of the fuel pump in the following manner. Disconnect the fuel pipe at the carburettor inlet union, and the high tension lead to the coil and, with a suitable container or large rag in position to catch the ejected fuel, turn the engine over. A good spurt of petrol should emerge from the end of the pipe every second revolution.

6 Fuel pump - removal and refitting

1 Remove the inlet and outlet pipes at the pump and plug the ends to stop petrol loss or dirt finding its way into the fuel system.
2 Undo and remove two bolts and spring washers that secure the pump to the cylinder block.
3 Lift away the fuel pump and gasket and recover the pushrod.
4 Refitting the fuel pump is the reverse sequence to removal but there are several additional points that should be noted:
  a) Do not forget to refit the pushrod
  b) Tighten the pump securing bolts to the specified torque wrench setting
  c) If a crimped type hose clamp was fitted, it will have been damaged on removal, and should be replaced by a suitable screw type clamp
  d) Before reconnecting the pipe from the fuel tank to the pump inlet, move the end to a position lower than the fuel tank so that fuel can syphon out. Quickly connect the pipe to the pump inlet
  e) Disconnect the pipe at the carburettor and turn the engine over (with the HT lead to the coil disconnected) until petrol issues from the open end. Quickly connect the pipe to the carburettor union

7 Fuel tank - removal end refitting

1 The fuel tank is positioned at the rear of the car and is supported on two straps.
2 Remove the filler cap and, using a length of rubber hose or plastic pipe approximately 0.25 in (6.35 mm) bore, syphon as much petrol out as possible until the level is below the level of the sender unit.
3 Disconnect the battery earth terminal, release the fuel gauge sender unit wire and the fuel feed pipe from the sender unit.

HAYNES HINT *Certain vehicles have a fuel return pipe fitted. in this case, take a careful note of the correct pipe connections.*

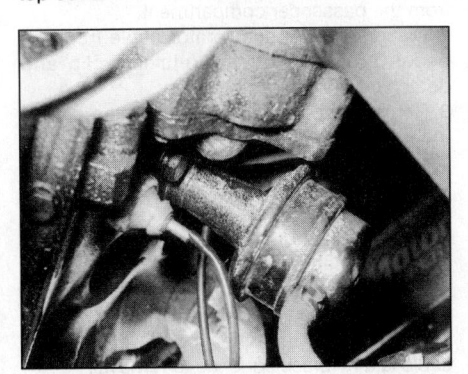

4.1 The fuel pump and fuel feed pipe

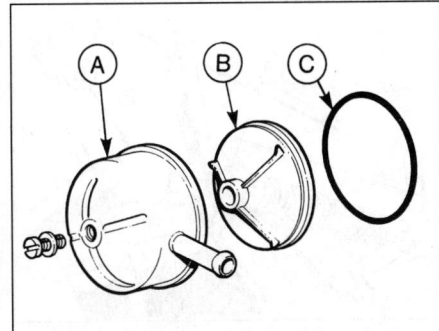

Fig. 3.1 Cleaning the fuel pump filter (Sec 4)
A Cover B Filter C Seal

3•4 Fuel and exhaust systems

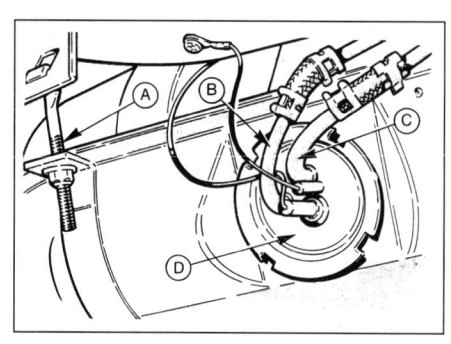

Fig. 3.2 Fuel gauge sender unit connections (Sec 7)

A Fuel tank securing strap
B Fuel supply pipe
C Fuel return pipe
D Sender unit

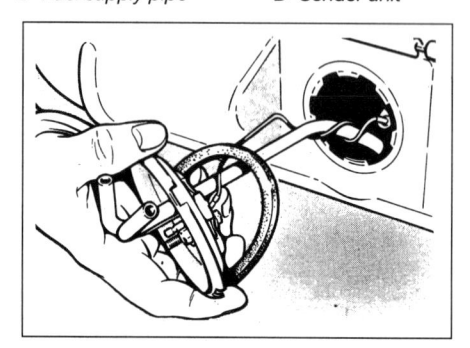

Fig. 3.3 Removing the fuel gauge sender unit (Sec 7)

4 Plug the ends of the pipes to stop petrol loss or dirt ingress.
5 Using two screwdrivers in the slots in the sender unit retaining ring unscrew the sender unit from the fuel tank. Lift away the sealing ring and the sender unit noting that the float must hang downwards.
6 Undo and remove the tank strap retaining nuts and lower the tank. As the tank is removed detach the fuel filler and vent pipes.
7 Refitting is the reverse sequence to removal. Tighten the support strap securing nuts until 1.6 to 1.8 in (40 to 45 mm) of thread is protruding through the nut.
8 Refill the fuel tank and reconnect the battery earth terminal. Test the operation of the fuel gauge sender unit by switching on the ignition. Wait 30 seconds and observe the gauge reading.

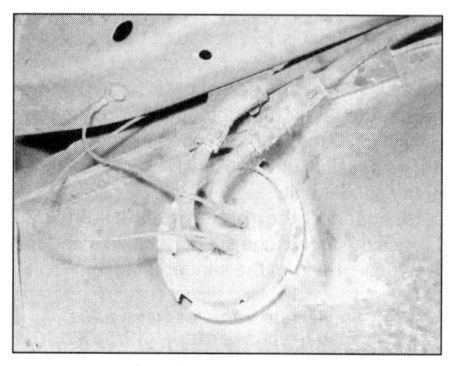

10.1 The fuel gauge sender unit located in the front of the petrol tank

8 Fuel filler neck - removal and refitting

1 Remove the fuel tank as described in Section 7.
2 Open the fuel filler flap and detach the single screw securing the filler neck to the body (photo).
3 From under the car disconnect the vent pipe from the tank and lift out the filler neck from its location in the body.
4 Refitting is a reverse of the removal sequence.

9 Fuel tank - cleaning

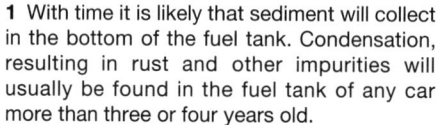

1 With time it is likely that sediment will collect in the bottom of the fuel tank. Condensation, resulting in rust and other impurities will usually be found in the fuel tank of any car more than three or four years old.
2 When the tank is removed it should be vigorously flushed out with paraffin and turned upside down to drain. If facilities are available at the local garage the tank may be steam cleaned and the exterior repainted with a lead based paint.
3 Never weld or bring a naked light close to an empty fuel tank until it has been steam cleaned out for at least two hours or, washed internally with boiling water and detergent and allowed to stand for at least three hours.

10 Fuel gauge sender unit - removal, testing and refitting

1 The fuel gauge sender unit can be removed with the fuel tank in position. (Refer to Section 7 paragraphs 2, 3, 4 and 5 (photo).
2 If the operation of the sender unit is suspect, check that the rheostat is not damaged and that the wiper contact is bearing against the coil.
3 Replacement is a straightforward reversal of the removal sequence. Always fit a new seal to the recess in the tank to ensure no leaks develop.

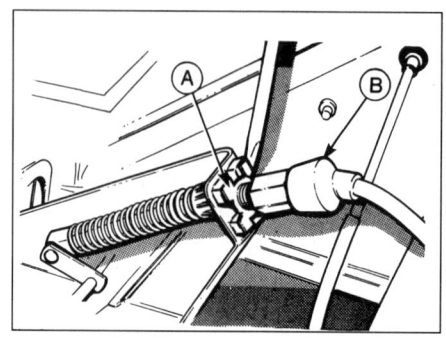

Fig. 3.4 Removing the accelerator cable retaining clip (Sec 12)

A Retaining clip   B Cable

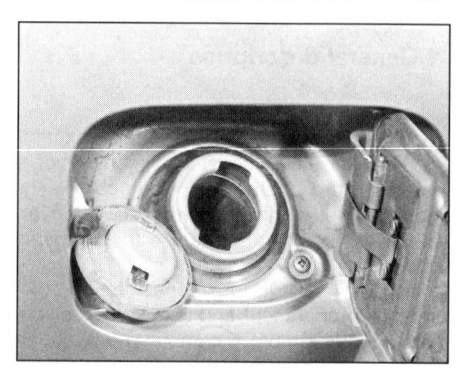

8.2 Fuel filler cap, neck and securing screw

4 The float arm should hang downwards. Test the operation of the fuel gauge sender unit by switching on the ignition. Wait 30 seconds and observe the gauge reading.

11 Accelerator pedal assembly - removal and refitting

1 Remove the right-hand lower trim panel from the passenger compartment.
2 Disconnect the accelerator cable from the pedal by prising off the retaining clip and detaching the cable.
3 Remove the single screw and nut securing the pedal assembly to the bulkhead. The screw is accessible from inside the car and the nut is accessible from the engine compartment.
4 Lift away the accelerator pedal assembly.
5 Refitting is a reverse of the removal procedure.
6 If necessary adjust the cable as described in Section 13.

12 Accelerator cable - removal and refitting

1 Remove the right-hand lower trim panel from the passenger compartment.
2 Disconnect the accelerator cable from the pedal by prising off the retaining clip and detaching the cable.
3 Remove the retaining clip from the end of the inner cable and disconnect the cable end from the ball socket on the accelerator linkage.
4 Prise out the cable retaining clip from its bracket on the inlet manifold. Depress the lugs individually with a screwdriver and twist out the throttle cable retainer.
5 Pull the cable through into the engine compartment and remove from the car.
6 Refitting and reconnecting the accelerator cable is a reverse sequence to removal.
7 It will now be necessary to adjust the linkage as described in Section 13.

13 Accelerator linkage - adjustment

1 Refer to Section 2 and remove the air cleaner.
2 Fully depress the accelerator pedal and wedge it with a length of wood between the pedal and the front seat.
3 Wind back the cable adjuster until the carburettor linkage is just in the fully open position, with no slack in the cable and no strain on the linkage.
4 *Automatic transmission models:* refer to Chapter 6, Section 19 and check kickdown cable adjustment to ensure this does not prevent the carburettor linkage from fully opening.
5 Release the accelerator pedal and refit the air cleaner.

14 Weber carburettor - general description

This carburettor is of the sequential type incorporating a fully automatic strangler type choke and internally vented float chamber.

The carburettor body comprises two castings which form the upper and lower bodies. The upper incorporates the float chamber cover, float pivot brackets, fuel inlet and return unions, gauze filter, spring-loaded needle valve, twin air intakes, choke plates and the section of the power valve controlled by vacuum.

Incorporated in the lower body is the float chamber, accelerator pump, two throttle barrels and integral main venturies, throttle plates, spindles, levers, jets and the petrol power valve.

The throttle plate opening is in a preset sequence so that the primary starts to open first and is then followed by the secondary, in such a manner that both plates reach full throttle position at the same time. The primary barrel, throttle plate and venturi are smaller than the secondary, whereas the auxiliary venturi size is identical in both the primary and secondary barrels.

All the carburation systems are located in the lower body and the main progression systems operate in both barrels, whilst the idling and the power valve systems operate in the primary barrel only and the full load enrichment system in the secondary barrel.

The accelerator pump discharges fuel into the primary barrel.

A connection for the vacuum required to control the distributor advance/retard vacuum unit is located on the lower body.

In addition, the idle mixture adjustment and basic idle adjustment screws are "tamper-proofed".

As from May 1981, a redesigned accelerator pump discharge nozzle and Low Vacuum Enrichment (LOVE) device are fitted to this carburettor to improve drive when the engine is cold. The LOVE device is also known as an anti-stall device.

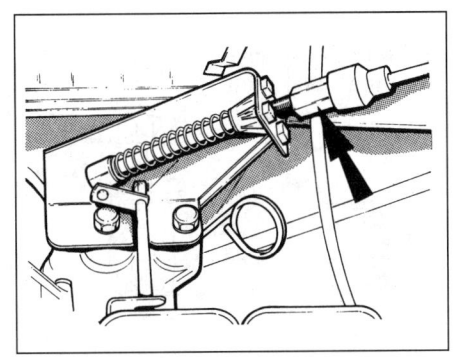

Fig. 3.5 Accelerator cable adjusting nut, arrowed (Sec 13)
Cable retaining clip removed

15 Weber carburettor - adjustment

In view of the increasing awareness of the dangers of exhaust pollution and the very low levels of carbon monoxide (CO) emission for which this carburettor is designed, the slow running mixture setting and the basic idle setting on Weber carburettors, should not be adjusted without the use of a proper CO meter (exhaust gas analyser).

1 Warm up the engine to its normal operating temperature.
2 Connect a CO meter and a tachometer, if the latter is not already fitted to the car, according to manufacturer's instructions.
3 Clear the engine exhaust gases by running the engine at 3000 rpm for approximately 30 seconds and allow the engine to idle.
4 Wait for the meter to stabilise and compare the CO and idle speed readings against those given in the Specifications at the beginning of this Chapter.
5 Adjust the idle speed screw to give the correct rpm,
6 During normal routine maintenance servicing, normally no adjustment of the mixture (CO level) will be required. If however the CO level is found to be incorrect the following procedure should be adopted.
7 Remove the air cleaner assembly as described in Section 2.
8 Using a small electricians screwdriver prise out the tamper proof plug covering the mixture adjusting screw.
9 Loosely refit the air cleaner, it is not necessary to bolt it in position.
10 Clear the engine exhaust gases by running the engine at 3000 rpm for approximately 30 seconds and allow the engine to idle.
11 Adjust the mixture screw and the idle screw until the correct idle speed and CO reading are obtained. If the correct readings are not obtained within 10 to 30 seconds clear the engine exhaust gases as described in paragraph 10 and repeat the adjustment procedure until correct readings are obtained.
12 Refit the air cleaner and a new tamper proof plug

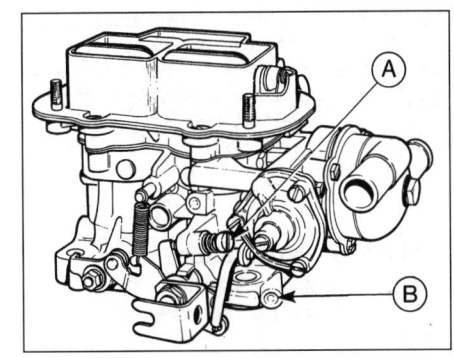

Fig. 3.6 Weber dual venturi carburettor (Sec 15)

A Idle speed adjusting screw
B Mixture (CO) adjusting screw

16 Weber carburettor - removal and refitting

1 Disconnect the battery earth lead.
2 Remove the air cleaner as described in Section 2.
3 Depressurize the cooling system by removing and refitting the radiator cap. Disconnect the water hoses from the choke housing and plug the hose ends to prevent loss of coolant.
4 Disconnect the accelerator linkage rod from the carburettor.
5 Disconnect the fuel supply and return pipes from the carburettor noting their positions. If crimped hose clips are fitted they must be cut off and renewed with screw type clips on reassembly.
6 Remove the carburettor to distributor vacuum pipe.
7 Undo the four retaining nuts and lift off the carburettor. Remove the gasket (and spacer if fitted).
8 Refitting is the reverse of the removal procedure but the following points must be noted:
   a) Ensure that all mating surfaces are clean and new gaskets are fitted
   b) Where a spacer is fitted position a gasket on each side of it.

17 Weber carburettor - dismantling and reassembly (general)

1 With time, the component parts of the carburettor will wear and petrol consumption increase. The diameter of drillings and jets may alter, and air and fuel leaks may develop around spindles and other moving parts. Because of the high degree of precision involved it is best to purchase an exchange carburettor. This is one of the few instances where it is better to take the latter course rather than to rebuild the component oneself.
2 It may be necessary to dismantle the carburettor to clear a blocked jet. The accelerator pump itself may need attention

3•6 Fuel and exhaust systems

and gaskets may need renewal and providing care is taken, there is no reason why the carburettor may not be completely reconditioned at home, but ensure a full repair kit can be obtained before you strip the carburettor down. *Never* clean out jets with wire or similar but blow them out with compressed air or air from a car tyre pump.

18 Weber carburettor - dismantling, cleaning, inspection and reassembly

1 Initially remove the carburettor from the car as described in Section 16, then clean the exterior with a water soluble cleaner.
2 Carefully prise out the U-clip with a screwdriver, and disconnect the choke plate operating link.
3 Remove the six screws and detach the carburettor upper body.
4 Unscrew the brass nut located at the fuel intake and detach the fuel filter.
5 Tap out the float retaining pin, and detach the float and needle valve.
6 Remove the three screws and detach the power valve diaphragm assembly.
7 Unscrew the needle valve housing,
8 Unscrew the jets and jet plugs from the carburettor body, noting the positions in which they are fitted.
9 From beneath the carburettor, remove the two primary diffuser tubes.
10 Remove four screws and detach the accelerator pump diaphragm, taking care that the spring is not lost (Fig.3.10).
11 Using a small electricians screwdriver prise out the tamper proof plug covering the mixture adjusting screw and remove the screw. Note the number of turns taken to remove it so it may be refitted in approximately the same position.
12 Remove four screws and detach the anti-stall diaphragm, taking care that the spring is not lost. Note: *This is only fitted on certain variants.*
13 Clean the jets and passageways using clean, dry compressed air. Check the float assembly for signs of damage or leaking. Inspect the power valve and pump diaphragms and gaskets for splits or deterioration. Examine the mixture screw,

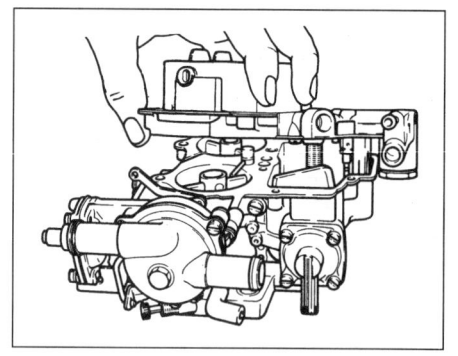

Fig. 3.7 Removing the carburettor upper body - Weber carburettor (Sec 18)

needle valve seat and throttle spindle for signs of wear. Renew parts as necessary.
14 When reassembling, refit the accelerator pump diaphragm assembly.
15 Fit the mixture screw and spring in the same position as originally fitted.
16 Slide the two diffuser tubes into position, then refit the jets and jet plugs.
17 Refit the anti-stall diaphragm assembly, if applicable.
18 Loosely fit the three screws to retain the power valve diaphragm assembly, then compress the return spring so that the diaphragm is not twisted or distorted. Lock the retaining screws and release the return spring.
19 Hold the diaphragm down, block the air bleed with a finger then release the diaphragm. If the diaphragm stays down it has correctly sealed to the housing.
20 Refit the needle valve housing, needle valve and float assembly to the upper body.
21 *Float level adjustment:* Hold the upper body vertically so that the needle valve is closed by the float, then measure the dimension from the face of the upper body to the base of the float. Adjust to the specified figure by bending the tag.
22 Refit the fuel inlet filter and brass nut.
23 Position a new gasket and refit the carburettor upper body to the main body. Ensure that the choke link locates correctly through the upper body.
24 Reconnect the choke link and refit the U-clip.
25 Adjust the carburettor as described in Section 15.

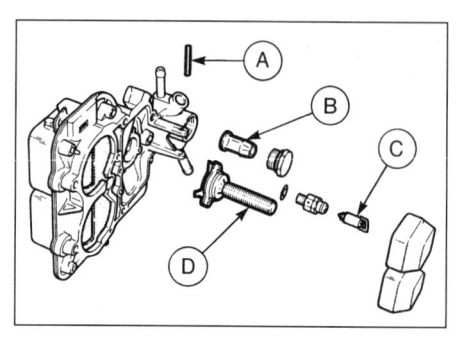

Fig. 3.8 Upper body components - Weber carburettor (Sec 18)

A Float retaining pin   B Filter
C Needle valve   D Power valve

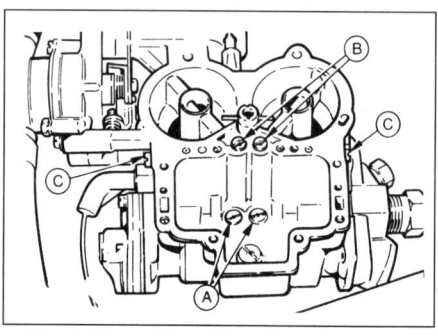

Fig. 3.9 Location of jets in main body - Weber carburettor (Sec 18)

A Main jets   B Air correction jets   C Idle jets

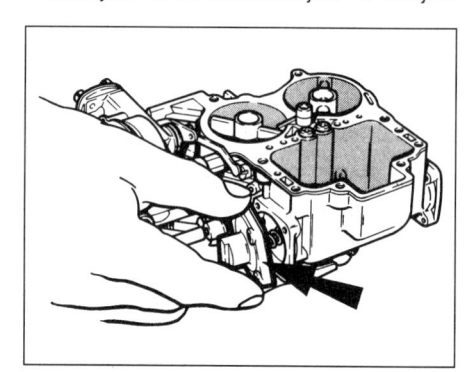

Fig. 3.10 Removing accelerator pump diaphragm, arrowed - Weber carburettors (Sec 18)

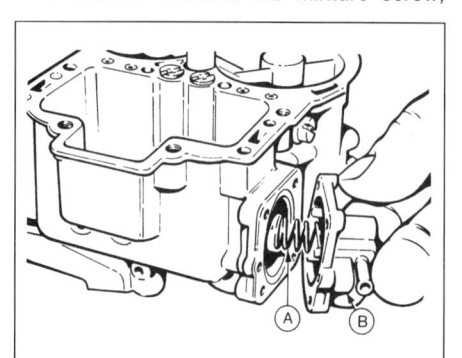

Fig. 3.11 (A) Anti stall diaphragm spring. (B) Diaphragm housing

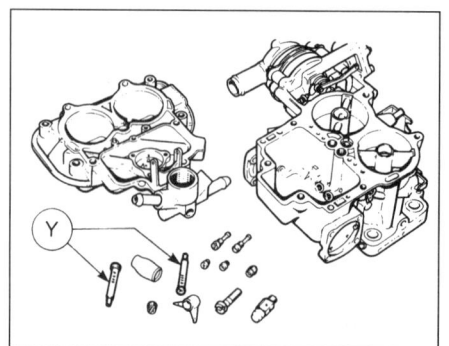

Fig. 3.12 Items to be cleaned (Y-primary diffuser tubes) - Weber carburettor (Sec 18)

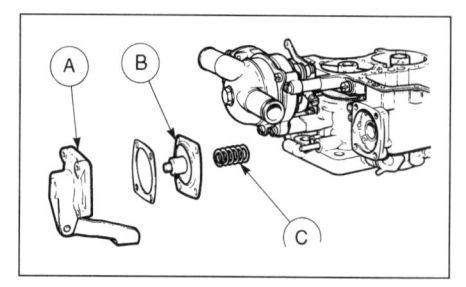

Fig. 3.13 Accelerator pump components - Weber carburettor (Sec 18)

A Pump housing   B Diaphragm
C Return spring

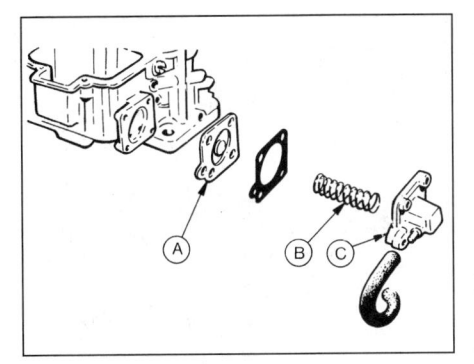

Fuel and exhaust systems 3•7

Fig. 3.14 Anti-stall device components - Weber carburettor (Sec 18)

A Diaphragm B Return spring C Housing

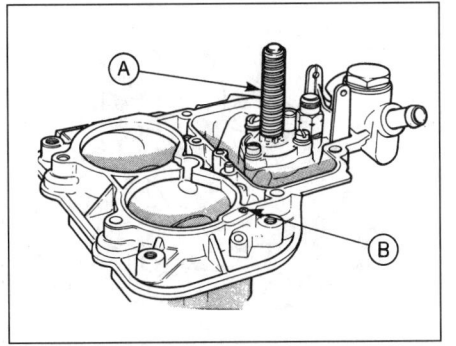

Fig. 3.15 Power valve diaphragm assembly - Weber carburettor (Sec 18)

A Power valve B Air bleed

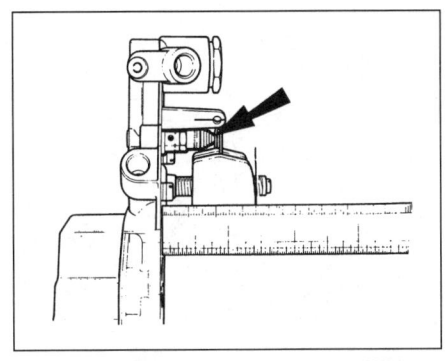

Fig. 3.16 Float level adjusting tag - Weber carburettor (Sec 18)

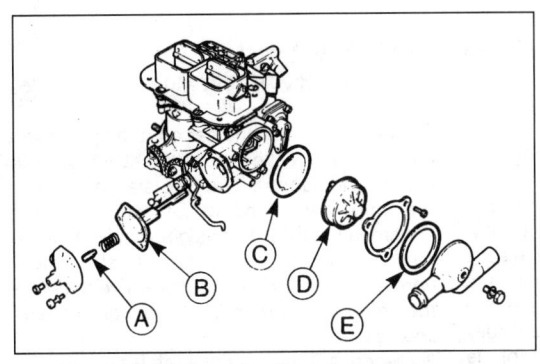

Fig. 3.17 Choke outer housing and vacuum diaphragm - Weber carburettor (Sec 19)

A Adjusting screw
B Vacuum diaphragm
C Heat shield
D Housing assembly
E Gasket

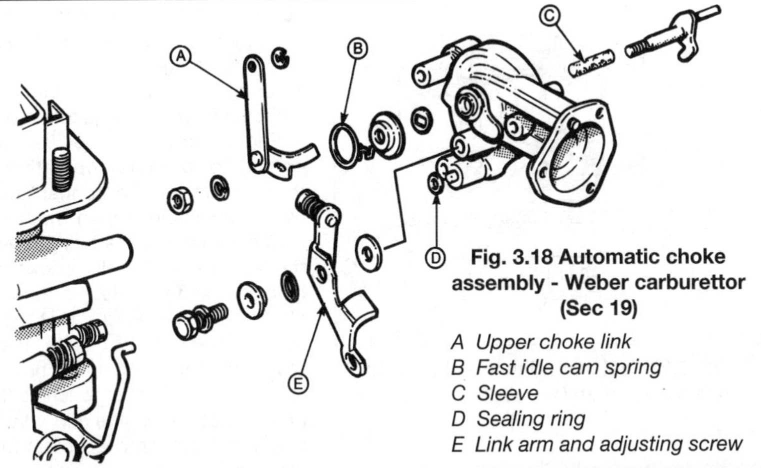

Fig. 3.18 Automatic choke assembly - Weber carburettor (Sec 19)

A Upper choke link
B Fast idle cam spring
C Sleeve
D Sealing ring
E Link arm and adjusting screw

19 Weber carburettor automatic choke - removal, overhaul and refitting

1 Disconnect the battery earth lead.
2 Remove the air cleaner, as described in Section 2.
3 Remove the three screws, detach the cover and move it clear of the carburettor. For access to the lower screw it will be necessary to make up a suitably cranked screwdriver.
4 Detach the internal heat shield.
5 Remove the single U-clip and disconnect the choke plate operating link.
6 Remove the three screws, disconnect the choke link at the operating lever and detach the choke assembly. For access to the lower screw the cranked screwdriver will again be required.
7 Remove the three screws and detach the vacuum diaphragm assembly.
8 Dismantle the remaining parts of the choke mechanism.
9 Clean all the components, inspect them for wear and damage and wipe them dry with a lint-free cloth. Do not use any lubricants during reassembly.
10 Reassemble the choke mechanism, after checking the diaphragm and sealing ring for splits.

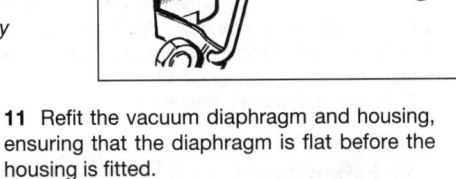

11 Refit the vacuum diaphragm and housing, ensuring that the diaphragm is flat before the housing is fitted.
12 Ensure that the O-ring is correctly located in the choke housing then reconnect the lower choke link. Position the assembly and secure it with the three screws; ensure that the upper choke link locates correctly through the carburettor body.
13 Reconnect the upper choke link to the choke spindle.
14 Check the vacuum pull-down and choke phasing, as described in Section 20.
15 Refit the internal heat shield ensuring that the hole in the cover locates correctly onto the peg cast in the housing.
16 Connect the bi-metal spring to the choke lever, position the choke cover and loosely fit the three retaining screws.
17 Rotate the cover until the marks are aligned, then tighten the three screws.
18 Reconnect the battery, run the engine and adjust the fast idle speed, as described in Section 11.
19 Refit the air cleaner.

20 Weber carburettor automatic choke - adjustment

Note: *The procedure is described for a carburettor which is fitted in the car, but with the exception of fast idle speed adjustment, can be carried out on the bench if required where the carburettor has been removed.*

1 Disconnect the battery earth lead.
2 Remove the air cleaner, as described in Section 2.
3 Remove the three screws, detach the choke cover and move it clear of the carburettor. For access to the lower screw it will be necessary to make up a suitable cranked screwdriver.
4 Detach the internal heat shield.
5 Vacuum pull-down: Fit an elastic band to the choke plate lever and position it so that the

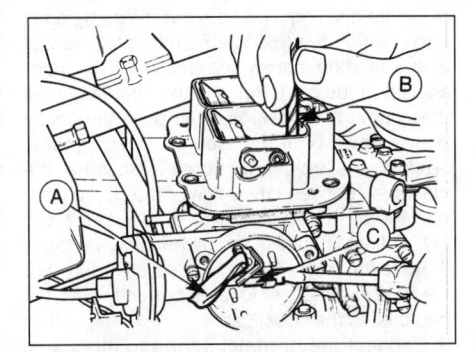

Fig. 3.19 Checking vacuum pull down - Weber carburettor (Sec 20)

A Elastic band B Twist drill (gap gauge)
C Operating rod

3•8 Fuel and exhaust systems

Fig. 3.20 Adjusting vacuum pull down - Weber carburettor (Sec 20)

A Screwdriver   B Housing

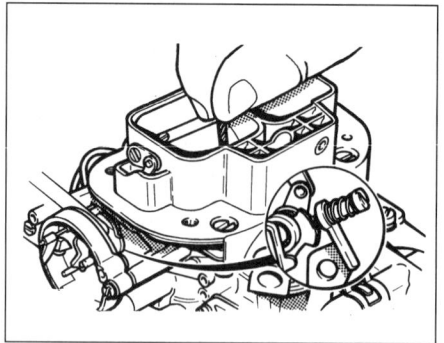

Fig. 3.21 Checking choke phasing gap using a twist drill. Inset shows fast idle adjusting screw on upper step of cam (Sec 20)

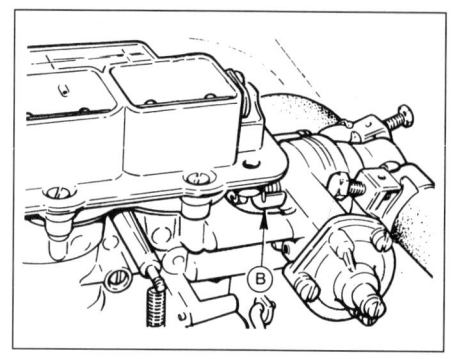

Fig. 3.22 Adjust choke phasing gap if necessary by bending tag (B) (Sec 20)

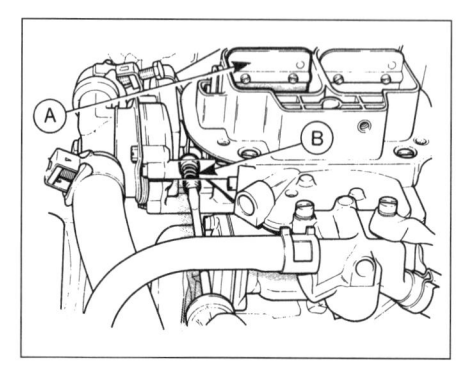

Fig. 3.23 Fast idle adjustment (B) with choke plates open (A) - Weber carburettor (Sec 20)

choke plates are held closed. Open, then release, the throttle to ensure that the choke plates close fully. Unscrew the plug from the diaphragm unit then manually push open the diaphragm up to its stop from inside the choke housing. Do not push on the rod as it is spring loaded but push on the diaphragm plug body. The choke plate pull-down should now be measured, using an unmarked twist drill shank between the edge of the choke plate and the air horn wall, and compared with the specified figure. Adjust, if necessary, by screwing the adjusting screw in or out, using a short bladed screwdriver. Refit the end plug and detach the elastic band on completion.

6 Choke phasing: Hold the throttle partly open and position the fast idle cam so that the fast idle adjusting screw locates on the upper section of the cam. Release the throttle to hold the cam in this position, then push the choke plates down until the step on the cam jams against the adjusting screw. Measure the clearance between the edge of the choke plate and the air horn wall using a specified sized drill. Adjust, if necessary, by bending the tag.

7 Refit the internal heat shield ensuring that the hole in the cover locates correctly onto the peg cast in the housing.

8 Connect the bi-metal spring to the choke lever, position the choke cover and loosely fit the three retaining screws.

9 Rotate the cover until the marks are aligned then tighten the three screws.

10 Reconnect the battery, run the engine and adjust the fast idle speed as described in the following paragraph.

11 Fast idle speed adjustment: Ideally a tachometer will be required in order to set the fast idle rpm to the specified value. Run the engine up to normal operating temperature, then switch off and connect the tachometer (where available). Open the throttle partially, hold the choke plates fully closed then release the throttle so that the choke mechanism is held in the fast idle position. Release the choke plates, checking that they remain fully open (if they are not open, the assembly is faulty or the engine is not at operating temperature). Without touching the accelerator pedal, start the engine and adjust the fast idle screw as necessary to obtain the correct fast idle rpm.

12 Finally refit the air cleaner.

21 Exhaust system - general description

The exhaust system consists of a cast iron manifold, a down pipe, a front resonator and rear silencer assembly.

The system is flexibly attached to the floor pan by two circular rubber mountings.

At regular intervals the system should be checked for corrosion, joint leakage, the condition and security of the flexible mountings and the tightness of the joints.

22 Exhaust system - removal and refitting

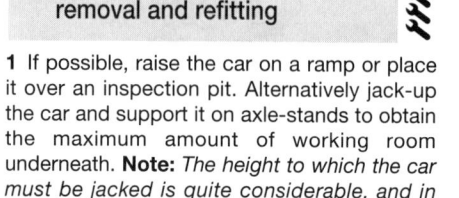

1 If possible, raise the car on a ramp or place it over an inspection pit. Alternatively jack-up the car and support it on axle-stands to obtain the maximum amount of working room underneath. **Note:** *The height to which the car must be jacked is quite considerable, and in normal circumstances it is preferable to entrust the removal and refitting to an exhaust specialist*

2 Disconnect the battery earth lead.

3 Slacken and remove the exhaust U-bolt clamp located just in front of the rear axle.

4 Apply some easing oil to assist in separating the joint. Twist the exhaust rear section back and forth to free and separate the joint.

5 Unhook the rear section from the rear rubber mounting and guide the section clear of the vehicle over the rear axle.

6 Undo the two nuts and separate the exhaust clamp at the manifold.

7 Unhook the front exhaust section from the rubber mounting and lift it off the vehicle.

8 Remove the U-bolt clamp that secures the front down pipe to the resonator section and separate the two pipes.

9 Remove the manifold clamp sealing ring from the down pipe.

10 Clean the mating surfaces of the clamp, sealing ring and manifold with emery cloth to remove any carbon build up. Examine the

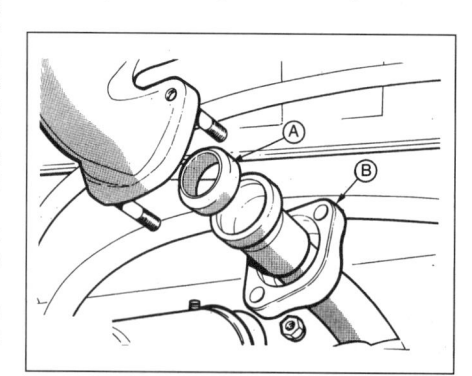

Fig. 3.24 Manifold clamp (B) and sealing ring (A) (Sec 22)

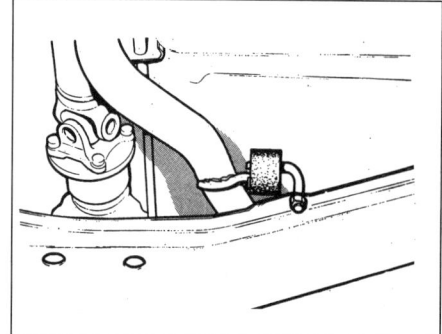

Fig. 3.25 Exhaust rubber mounting on rear crossmember (Sec 22)

Fuel and exhaust systems 3•9

rubber mountings for deterioration and renew if necessary.
11 Apply exhaust sealer to the joint between down pipe and front section and assemble the two joints.
12 Fit the sealing ring to the down pipe and offer up the front assembly to the car. Suspend the rear end from the rubber mounting and loosely fit the manifold clamp. Do not fully tighten the clamp at this stage.
13 Position the rear section of the exhaust system over the rear axle and suspend it from the rear rubber mounting.
14 Align the front and rear exhaust sections and connect by pushing together. Use an exhaust sealer to ensure a gas tight seal and ensure the two pipes are pushed fully together.
15 Correctly align the system. Ensure there is no strain on the rubber mountings and that a clearance of at least 0.8 in (20 mm) exists between the exhaust system and vehicle body, or axle.
16 Securely tighten all U-bolt clamps and the exhaust manifold clamp.
17 Reconnect the battery and run the engine to ensure there are no exhaust leaks.

PART B
V6 ENGINES

23 Air cleaner - general description

The air cleaner fitted to V6 engines is thermostatically controlled and has the ability to automatically regulate air intake temperature at a constant pre-determined figure during normal driving conditions.

The air cleaner has two sources of air supply, cool air entering from a standard air intake spout and hot air entering from a heat box mounted on the exhaust manifold. The air supply through the air cleaner is controlled by a flap valve mounted in the spout. The flap valve mixes cool air from the standard intake with hot air from the heat box to achieve the required air intake temperature.

The flap valve is controlled by a vacuum diaphragm unit which holds it fully open as long as the vacuum is maintained above 4.0 in (100 mm) of mercury. Under these conditions only hot air from the heat box enters the air cleaner.

As the vacuum applied to the diaphragm unit falls, the flap valve progressively closes allowing cool air to enter the air cleaner. The vacuum feed for the diaphragm unit is supplied from the inlet manifold via a heat sensor.

The heat sensor is located inside the air cleaner and senses the temperature of the air actually entering the carburettor. The unit consists of a bi-metal strip, a ball valve and two vacuum take off points. Air temperature causes the bi-metal strip to move opening or closing the ball valve. Movement of the ball valve opens or closes a port, which is open to the atmosphere, thus maintaining or reducing the strength of the vacuum.

The combined effect of this system is to control air intake temperature under closed or part throttle conditions and to provide cold air at or near full throttle to achieve good performance.

24 Air cleaner - removal and refitting

The renewable paper element type air cleaner is fitted onto the top of the carburettor and is retained in position by two screws. To remove the air cleaner proceed as follows:

24.3 Air cleaner paper element

1 Remove the two screws securing the air cleaner to the carburettor.
2 Lift the air cleaner off the carburettor and unplug the vacuum supply hose at the carburettor.
3 To change the paper element, unclip the clasps around the circumference of the air cleaner body and lift off the top cover. The element can now be removed (photo).
4 Refitting is a reverse of the removal sequence. Ensure that the gasket is positioned correctly on the carburettor flange when refitting the air cleaner.

25 Air cleaner thermostatically controlled valve - testing

1 By looking through the main air cleaner spout, note the position of the flap valve. With the engine stationary the flap should be fully closed.
2 Start the engine and check the operation of the flap valve. With the engine idling the flap should open and stay open to allow hot air into the air cleaner. **Note:** *It is essential that this test is carried out with the engine cold.*
3 If the flap remains closed when the engine is started then there is a fault in either the diaphragm unit or heat sensor and they should be tested as follows.
4 Check the vacuum supply pipe for leaks and renew it if suspect.
5 Disconnect the vacuum supply pipe from the heat sensor. Observe the operation of the flap valve while at the same time apply suction to the end of the supply pipe. If the flap valve remains closed then either the diaphragm or the flap valve is faulty. If the flap now opens, the fault lies in the heat sensor.

26 Fuel pump - description

The V6 engine fuel pump is identical to the unit fitted to in-line ohc engine models except that it is actuated from the engine camshaft not the auxiliary shaft. A general description, removal, refitting and testing procedures are given in Part A of this Chapter.

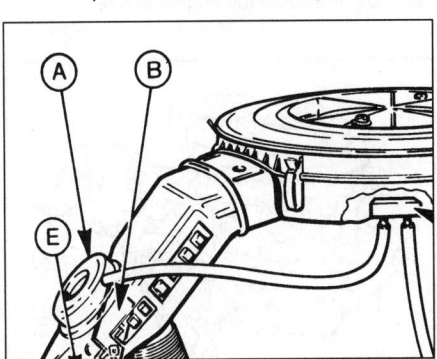

Fig. 3.26 Thermostatic air cleaner (Sec 23)

A Vacuum unit B Flap valve
C Heat sensor D Hot air inlet E Cool air inlet

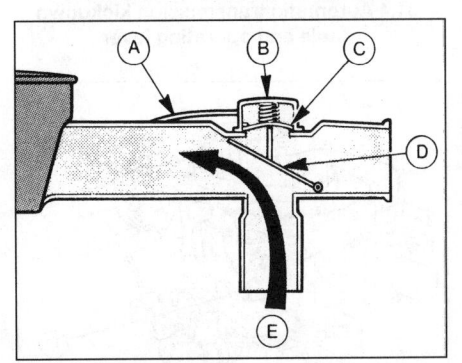

Fig. 3.27 Flap valve open - hot air through cleaner (Sec 23)

A Vacuum hose B Vacuum unit
C Diaphragm D Flap valve E Hot air inlet

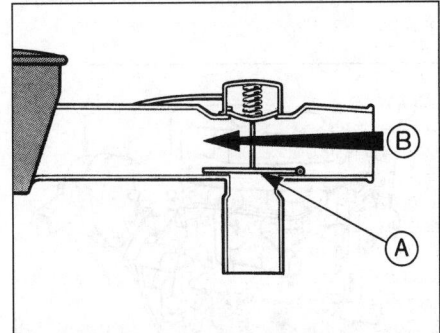

Fig. 3.28 Flap valve closed - cool air through cleaner (Sec 23)

A Flap valve B Cool air inlet

3•10 Fuel and exhaust systems

27 Fuel tank - general

Fuel tank cleaning, removal and refitting procedures are the same as those shown in Part A of this Chapter.

28 Fuel gauge sender unit - removal and refitting

The procedure is the same as that for the in-line ohc engine, in Part A of this Chapter.

29 Accelerator pedal assembly - removal and refitting

The procedure is the same as that for the in-line ohc engine, in Part A, of this Chapter.

30 Accelerator cable - removal and refitting

The procedure is the same as that for the in-line ohc engine, in Part A of this Chapter.

31 Accelerator linkage - adjustment

1 Refer to Section 24 and remove the air cleaner.
2 Fully depress the accelerator pedal and wedge it with a length of wood between the pedal and the front seat.
3 Wind back the cable adjuster until the carburettor linkage is just in the fully open position, with no slack in the cable and no strain on the linkage (photo)
4 *Automatic transmission models:* refer to Chapter 6 Section 19 and check the kickdown cable adjustment to ensure this does not prevent the carburettor linkage from fully opening (photo).
5 Release the accelerator pedal and refit the air cleaner.

32 Solex dual venturi carburettor - general description

The Solex dual venturi carburettor incorporates a fully automatic strangler type choke to ensure easy starting whilst the engine is cold. The float chamber is internally vented,
The carburettor body consists of two castings which form the upper and lower bodies. The upper incorporates the float chamber cover, twin air intakes, choke plates, and the automatic choke housing.
Incorporated in the lower body is the float chamber, accelerator pump, power valve, float and needle valve, fuel inlet and return unions, two throttle barrels and integral venturis, throttle plates, splines and jets. An anti-stall device and fuel return line system are also incorporated.
The carburettor idle system is of the bypass type and the idle mixture adjustment and basic idle adjustment screws are "tamper proofed".
Fuel is drawn from the float chamber and past an air bleed. This air bleed only allows enough air to give a rich fuel/air mixture which passes to the bypass idle mixture screw. Turning the screw in weakens the mixture, turning it out gives a richer mixture. The mixture is then drawn past the idle speed screw and passes into the main air flow. The idle speed screw regulates the amount of mixture passing through the idle system and therefore controls the idle speed, it does not affect the mixture ratio at idle. On later models this screw too is tamper proofed.
The anti-stall device works in the following way. During normal operating conditions, vacuum from the inlet manifold pulls a diaphragm back against spring tension and this allows fuel to be drawn, through internal drillings, from the float chamber, and into the anti-stall device. When the engine is about to stall the vacuum drops and the spring operates the diaphragm which pumps the reserve of fuel back through the internal drillings to the discharge tubes to enrich the fuel/air mixture and prevent the engine from stalling. Two non-return valves are fitted, one at the float chamber which allows fuel to be drawn from the chamber but not to return, and the other at the discharge tubes.

33 Solex carburettor - adjustment

In view of the increasing awareness of the dangers of exhaust pollution and the very lower levels of carbon monoxide (CO) emission for which this carburettor is designed, the by-pass idle mixture setting and the basic idle mixture setting on Solex carburettors, should not be adjusted without the use of a proper CO meter (exhaust gas analyser).

Early models (coolant-heated choke)

1 Warm up the engine to its normal operating temperature.
2 Connect a CO meter and a tachometer according to the manufacturers' instructions.
3 Clear the engine exhaust gases by running the engine at 3000 rpm for approximately 30 seconds and allow the engine to idle.

31.3 Solex carburettor accelerator linkage

31.4 Automatic transmission kickdown cable and operating lever

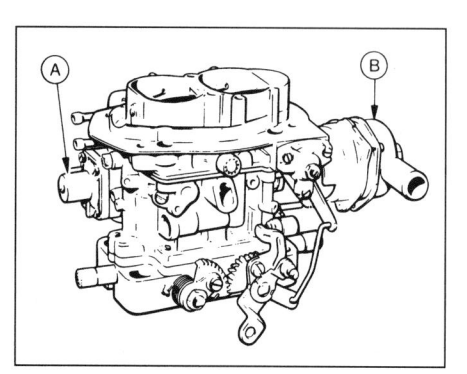

Fig. 3.29 Solex dual venturi carburettor (Sec 32)

A Anti-stall device   B Choke housing

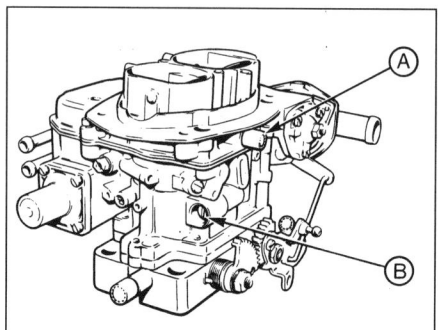

Fig. 3.30 By-pass idle adjusting screw location - Solex carburettor (Sec 33)

A Mixture screw   B Idle speed screw

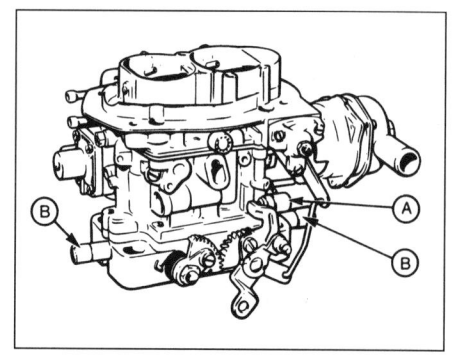

Fig. 3.31 Basic idle adjusting screw location - Solex carburettor (Sec 33)

A Basic idle speed screw
B Basic idle mixture screws

Fuel and exhaust systems 3•11

4 Wait for the meters to stabilise and compare the CO and idle speed readings against those given in the Specifications at the beginning of this Chapter.
5 Adjust the by-pass idle speed screw which is located in the main body of the carburettor to achieve the correct idle rpm (Fig. 3.30).
6 If the CO figure is not within the specified figure the by-pass idle mixture screw will have to be adjusted, and this will necessitate removal of the tamper proof cap.
7 The tamper proof cap can be removed from the idle mixture adjusting screw by compressing it with a pair of pliers to break it and then prising it off.
8 Clear the engine exhaust gases as described in paragraph 3 and adjust the by-pass idle mixture and speed screws to achieve the correct CO reading at the specified idle speed. **Note:** *During normal routine servicing no further carburettor adjustments normally need to be made. If however it is not possible to achieve the specified CO refining using the above procedure, or if the carburettor has been overhauled adjust the basic idle system as follows.*
9 Close the by-pass idle system by screwing in the by-pass idle speed control screw.
10 Break off and remove the tamper proof caps from the throttle stop screw and the two mixture adjustment screws.
11 Screw in the two mixture adjusting screws fully home and then unscrew them evenly five turns each.
12 Start the engine and adjust the throttle stop screw to achieve the specified basic idle speed.
13 Adjust the two basic mixture adjusting screws to achieve the specified basic idle CO.
14 When the basic settings have been achieved, re-open the by-pass system and adjust the idle speed and mixture settings as described previously in this Section.
15 Once all settings are correct, fit new tamper proof caps.

Later models (electrically-heated choke)

16 The only adjustment possible without removing tamperproof caps is to the bypass idle mixture. The adjusting screw is located next to the bypass idle speed screw (B in

34.3 Disconnect automatic choke coolant hoses (early models)

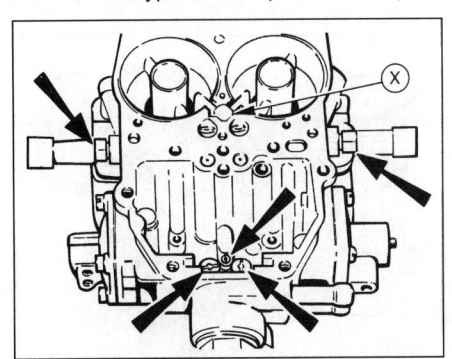

Fig. 3.30). The bypass idle speed screw itself is fitted with a tamperproof cap.
17 Basic idle speed is set by means of a shear-head screw, no further adjustment being possible.

34 Solex carburettor - removal and refitting

1 Disconnect the battery earth lead.
2 Remove the air cleaner as described in Section 24.
3 On models with a coolant-heated choke, ensure that the cooling system is not pressurised by removing the radiator cap, then disconnect and plug the coolant hoses from the choke housing (photo).
4 On models with an electrically-heated choke, disconnect the electrical lead from the choke housing.
5 Disconnect the throttle cable from the carburettor.
6 Disconnect the fuel feed and fuel return pipes. If crimped type clamps are used on the hoses they must be prised open (photo).
7 Disconnect the vacuum hose from the carburettor.
8 Undo and remove the four securing nuts and lift off the carburettor and gasket.
9 Refitting the carburettor is the reverse of the removal procedure, but the following points must be noted:
 a) Ensure that the mating surfaces are clean and always use a new gasket

34.6 Fuel feed and return pipe carburettor connections (Note crimped type clamps fitted)

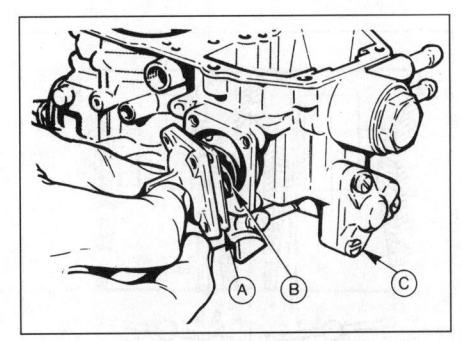

 b) When connecting the fuel feed and return pipes ensure that they are connected the right way round. Two arrows on the casting indicate the direction of flow. Fit screw type hose clips as replacements for crimped type clamps
 c) Top-up the cooling system before running the engine (early models)
 d) Check and adjust the engine by-pass idle settings as described in Section 33

35 Solex carburettor - dismantling and reassembly (general)

1 With time, the component parts of the carburettor will wear and petrol consumption increase. The diameter of drillings and jets may alter, and air and fuel leaks may develop around spindles and other moving parts. Because of the high degree of precision involved it is best to purchase an exchange carburettor. This is one of the few instances where it is better to take the latter course rather than to rebuild the component oneself.
2 It may be necessary to partially dismantle the carburettor to clear a blocked jet. The accelerator pump itself may need attention and gaskets may need renewal and providing care is taken, there is no reason why the carburettor may not be completely reconditioned at home, but ensure a full repair kit can be obtained before you strip the carburettor down. *Never* clean out jets with wire or similar but blow them out with compressed air or air from a car tyre pump.

36 Solex carburettor - dismantling, cleaning, inspection and reassembly

1 Remove the carburettor from the engine as described in Section 24 and thoroughly clean the exterior.
2 Remove the screws securing the upper body to the lower body then disconnect the choke link and lift off the carburettor upper body.
3 Lever out the nylon float locking tab with a small screwdriver, lift out the float retaining pin, then the float and needle valve.
4 Unscrew the five jets, arrowed in Fig. 3.32,

Fig. 3.32 Location of carburettor jets, arrowed - Solex carburettor (Sec 36)

X Accelerator pump discharge tubes

Fig. 3.33 Removing the accelerator pump - Solex carburettor (Sec 36)

*A Pump diaphragm   B Return spring
C Power valve assembly*

3•12 Fuel and exhaust systems

Fig. 3.34 Dismantling the anti-stall device - Solex carburettor (Sec 36)

A Diaphragm B Return spring

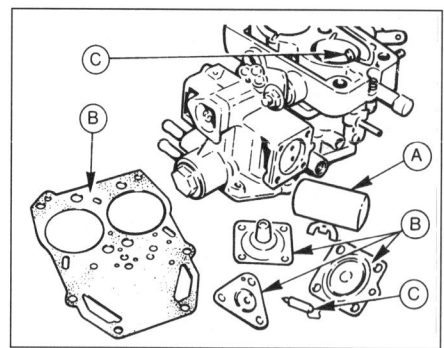

Fig. 3.35 Inspecting dismantled components - Solex carburettor (Sec 36)

A Check for leaks B Check for splits
C Check for wear

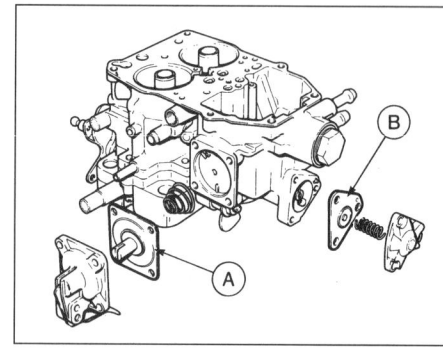

Fig. 3.36 Refitting the accelerator pump and power valve - Solex carburettor (Sec 36)

A Pump diaphragm
B Power valve diaphragm

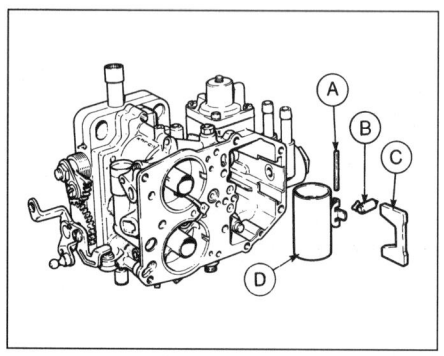

Fig. 3.37 Refitting the needle valve and float assembly - Solex carburettor (Sec 36)

A Retaining pin B Needle valve
C Locking tab D Float

noting the positions in which they are fitted, and then lever out the accelerator supply tubes.

5 Remove the four securing screws and detach the accelerator pump diaphragm assembly, taking care not to lose the spring.
6 Undo the three securing screws and remove the power valve diaphragm assembly.
7 Break off the tamper proof caps and remove the mixture adjusting screws.

8 Remove the four securing screws and remove the anti-stall diaphragm and spring.
9 Clean out the float chamber and the upper body. Clean all the jets and passageways using clean, dry compressed air. Check the float assembly for signs of damage or leaking. Inspect the power valve and pump diaphragms for splits or deterioration. Examine the mixture screws, needle valve and throttle spindle for signs of wear and damage. Renew defective parts as necessary.
10 Refit the two mixture adjusting screws.
11 Fit the accelerator and power valve assemblies with the parts assembled in the order shown in Fig. 3.36.
12 Fit the anti-stall diaphragm assembly, taking care that the diaphragm does not get kinked or twisted.
13 Refit the five jets. Place the accelerator supply tubes in their location and tap them into place in the body.
14 Fit the needle valve, make sure the spring clip is on the valve and then fit the retaining pin to the float. Position the assembly with the float tag behind the needle valve spring and fit the nylon locking tag (Fig. 3.37).
15 To adjust the float level setting slowly fill the float chamber with petrol until the float fully closes the needle valve. Measure the distance between the gasket face and the float. Adjust

to the specified setting by bending the float level adjusting tag.
16 To check the pump stroke fuel direction partly fill the float chamber with petrol. Operate the accelerator pump and check the fuel direction in relation to the throttle plates, see Fig. 3.39. Adjust the position of the pump discharge tubes to obtain the specified dimension at A.
17 Reconnect the choke link, position a new gasket on the lower body then fit the upper body and the securing screws.
18 On early models to synchronise the throttle plates unscrew the basic idle speed adjusting screw until it is clear of the throttle mechanism. Loosen the synchronisation adjusting screw. Hold the choke plates open and flick the throttle to close both throttle plates. Press down both plates to ensure that they are fully closed and tighten the synchronising screw. Note that the synchronisation screw has a left-hand thread.
19 On later models, the basic idle adjusting screw is preset during manufacture and its end sheared off. Adjustment by this means is therefore no longer possible.
20 Refit the carburettor and adjust the idle settings as described in Section 33.

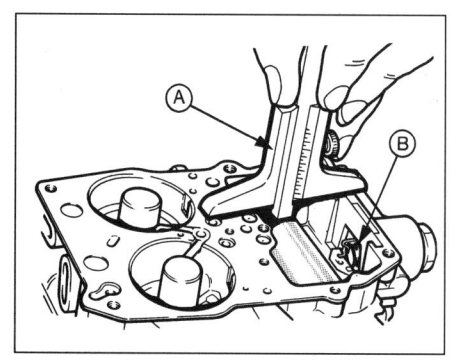

Fig. 3.38 Measuring the float level - Solex carburettor (Sec 36)

A Depth gauge B Adjusting tag

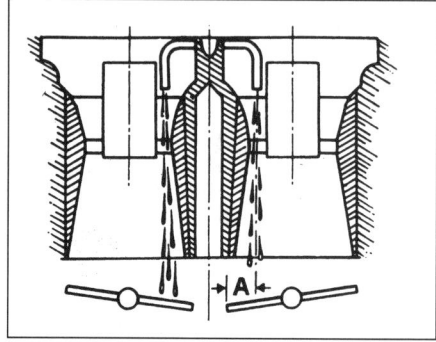

Fig. 3.39 Checking the pump stroke fuel direction - Solex carburettor (Sec 36)

Dimension A = 0.14 to 0.26 in (3.5 to 6.5 mm)

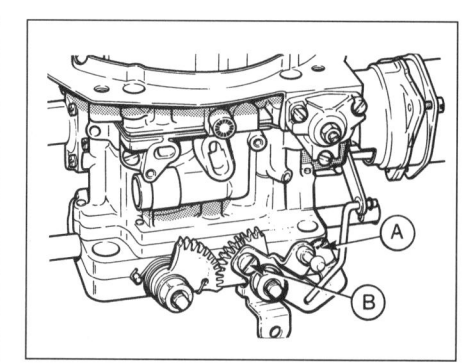

Fig. 32.40 Synchronising the throttle plates - Solex carburettor (Sec 36)

A Basic idle adjusting screw
B Throttle synchronising screw

Fuel and exhaust systems 3•13

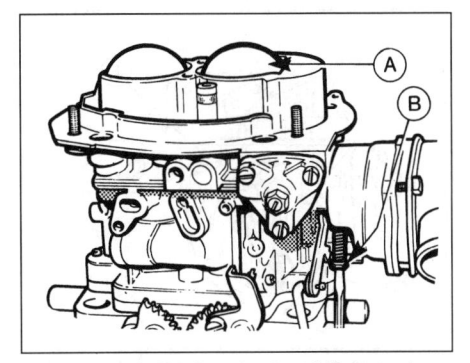

Fig. 3.41 Adjusting the fast idle speed - Solex carburettor (Sec 37)

A Choke plates B Fast idle adjusting screw

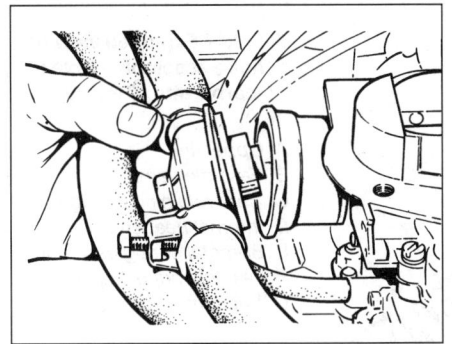

Fig. 3.42 Removing the choke bi-metal spring housing - Solex carburettor (Sec 38)

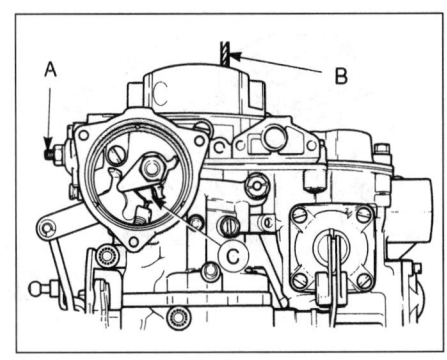

Fig. 3.43 Adjusting the choke plate pull down - Solex carburettor (Sec 38)

A Adjusting screw B Twist drill
C "High cam" position (fast idle)

37 Solex carburettor fast idle speed - adjustment

1 Check and if necessary, reset, the by-pass idle settings as described in Section 33.
2 Remove the air cleaner and without disconnecting the vacuum supply pipe, position it clear of the carburettor.
3 Open the throttle partially, hold the choke plates fully closed, then release the throttle so that the choke mechanism is held in the fast idle position.
4 Release the choke plates and check that they return to the fully open position. (If not, the assembly is faulty or the engine is not at operating temperature.)
5 Without touching the accelerator pedal, start the engine and check the fast idle speed. Adjust by screwing the fast idle screw in or out as necessary. To gain access to the screw, stop the engine and partly open the throttle. A ½ turn of the screw alters the engine speed by approximately 100 rpm.
6 Refit the air cleaner.

38 Solex carburettor automatic choke - adjustment

1 Check and if necessary reset the by-pass idle settings as described in Section 33.
2 Remove the air cleaner and without disconnecting the vacuum supply pipe, position it clear of the carburettor.
3 Remove the securing screws and pull the choke housing and bi-metal spring assembly clear of the carburettor.
4 Detach the internal heat shield.

Vacuum pull-down

5 With the engine at normal operating temperature, partially open the throttle then hold the choke plates fully closed and release the throttle. The choke mechanism will now be held in the high cam position (fast idle).
6 Release the choke plates and start the engine without touching the accelerator pedal. Carefully close the choke plates until resistance is felt and then hold in this position.
7 Measure the clearance using an unmarked twist drill shank between the edge of the choke plate and the air horn wall, and compare it with the specified figure. Adjust, if necessary, by screwing the diaphragm adjusting screw in or out as required.

Choke phasing

8 With the engine at the normal operating temperature and the choke mechanism at the fast idle position start the engine without touching the accelerator pedal then close the choke plates to the pull-down position. Hold the choke plates, partly open the throttle and allow the fast idle cam to return to its normal position.
9 Release the throttle and stop the engine. With the choke plates held in the pull-down position the fast idle screw should locate on the cam next to the high cam stop leaving a small operating clearance. If necessary adjust by bending the phasing adjusting tag.

Modulator spring gap

10 Remove the carburettor upper body as described in Section 36 paragraph 2. Measure the clearance between the modulator spring and the choke lever with a twist drill shank. If the clearance is not as specified adjust by bending the spring to obtain the correct clearance (Fig. 3.45).

Reassembly

11 Refit the internal heat shield.
12 Connect the bi-metal spring to the choke lever, position the choke cover and loosely fit the three retaining screws. Rotate the cover until the marks are aligned then tighten the three screws (Fig. 3.46).
13 Adjust the fast idle as described in Section 37.
14 Refit the air cleaner assembly.

Fig. 3.44 Checking the choke phasing adjustment - Solex carburettor (Sec 38)

A Fast idle cam B Adjusting tag
X Operating clearance (see Specifications)

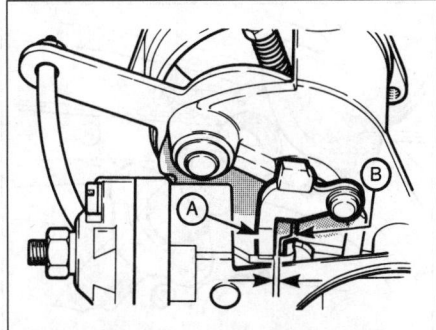

Fig. 3.45 Adjusting the modulator spring gap - Solex carburettor (Sec 38)

A Choke lever B Modulator spring

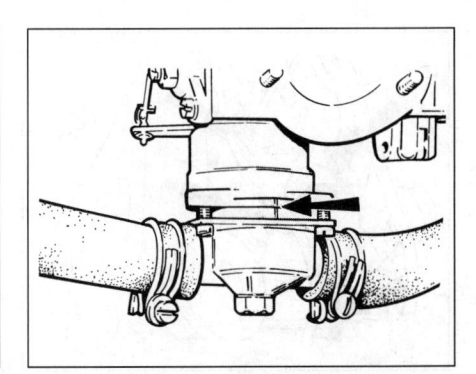

Fig. 3.46 Alignment marks on choke housing - Solex carburettor (Sec 38)

3•14 Fuel and exhaust systems

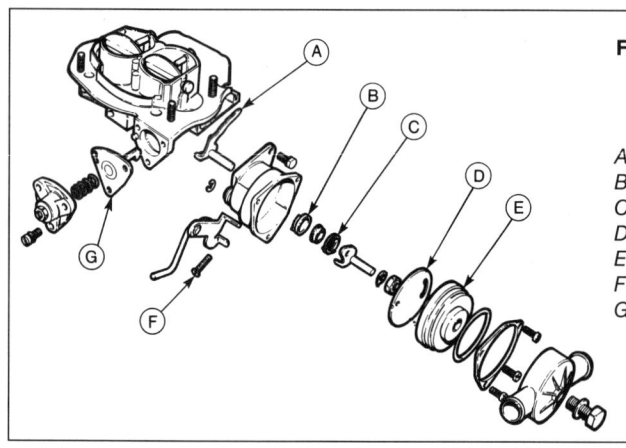

Fig. 3.47 Exploded view of choke assembly - Solex carburettor (Sec 39)

A Upper link
B Fast idle cam
C Cam retaining spring
D Internal heat shield
E Bi-metal spring assembly
F Fast idle adjusting screw
G Vacuum diaphragm

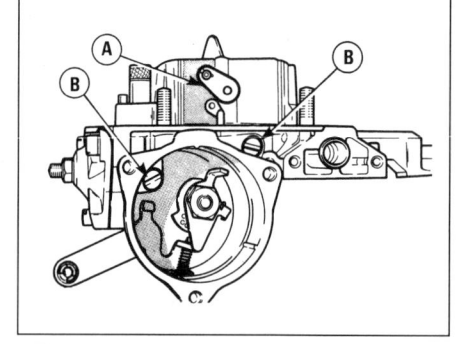

Fig. 3.48 Removing choke housing from upper body - Solex carburettor (Sec 39)

A Operating link   B Securing screws

39 Solex carburettor automatic choke - dismantling inspection and reassembly

1 Remove the air cleaner as described in Section 24.
2 Remove the retaining screws and pull the choke housing and bi-metal assembly away from the carburettor.
3 Detach the internal heat shield.
4 Remove the U-clip and disconnect the choke operating link from the choke plate spindle.
5 Remove the upper body as described in Section 36 paragraph 2.
6 Remove the securing screws and detach the choke assembly from the upper body.
7 Release the locking tab, remove the retaining nut and dismantle the choke assembly.
8 Remove the securing screws and remove the vacuum diaphragm from the upper body, take care not to lose the return spring.
9 Clean all the parts, inspect them for wear and damage and wipe them dry with a clean cloth. Do not use any lubricants during reassembly.
10 Refit the vacuum diaphragm and housing, ensuring that the diaphragm is flat before tightening the securing screws.
11 Reassemble the choke assembly in the order shown in Fig. 3.47 then fit the assembly to the upper body.
12 Reconnect the choke link to the choke spindle, fit the U-clip and the dust cover.
13 Check and, if necessary, adjust the modulator spring gap as described in Section 38.
14 Fit the upper body and connect the choke link.
15 Check and adjust the vacuum pull-down and choke phasing as described in Section 38.
16 Refit the internal heat shield.
17 Connect the bi-metal spring to the choke lever, position the choke and bi-metal housing assembly on the upper body and loosely fit the retaining screws. Rotate the housing until the marks are aligned, then tighten the screws.
18 Adjust the fast idle speed as described in Section 37.

40 Exhaust system - general description

1 The exhaust system consists of two cast iron manifolds located on either side of the engine, two down pipes, two front silencers and one rear silencer assembly.
2 The system is flexibly attached to the floor pan by circular rubber mountings.
3 At regular intervals the system should be checked for corrosion, joint leakage, the condition and security of the flexible mountings and the tightness of the joints.

41 Exhaust system - removal and refitting

1 If possible raise the car on a ramp or place it over an inspection pit. Alternatively jack-up the car and support it on axle-stands to obtain the maximum amount of working room underneath.
2 Disconnect the battery earth lead.
3 Slacken and remove the exhaust U-bolt clamp located just in front of the rear axle.
4 Apply some easing oil to assist in separating the joint. Twist the exhaust rear section back and forth to free and separate the joint.
5 Unhook the rear section from the rear rubber mounting and guide the section clear of the vehicle over the rear axle (photo).
6 Remove the flexible tube located between the heat box on the right-hand manifold and the air cleaner spout.
7 Undo the four screws securing the two halves of the heat box to the manifold and remove the heat box.
8 Remove the two nuts from each manifold and separate the two clamps. Remove the clamp securing the right-hand exhaust down

Fig. 3.49 Location of heat box securing screws (Sec 41)

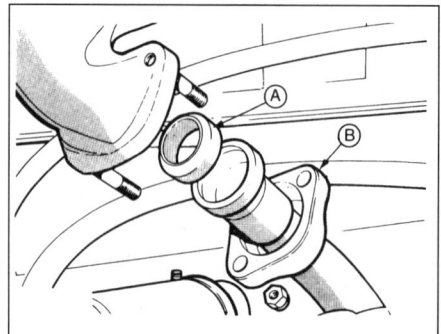

Fig. 3.50 Exhaust down pipe to manifold joint (Sec 41)

A Sealing ring   B Manifold clamp

41.5 Exhaust rear rubber mounting

Fuel and exhaust systems 3•15

pipe to the bracket on the clutch housing (where fitted) (photos).

9 Detach the front section from the rubber mounting and guide the section clear of the car.

10 Undo the two U-bolt clamps located just forward of the front silencers. Apply some easing oil to the joint and remove the down pipes from the front silencer section. Remove the manifold sealing rings from the down pipes.

11 Clean the mating surfaces of the sealing ring, clamp and manifold with emery cloth to remove any carbon build up. Examine the rubber mountings for deterioration and renew if necessary.

12 Apply exhaust sealer to the joint between the down pipes and the front section and assemble the joints.

13 Fit the sealing rings to the down pipes and offer up the assembly to the car.

14 Suspend the rear end from the rubber mounting and loosely fit the manifold clamps. Do not fully tighten the clamps at this stage.

15 Position the rear section of the exhaust system over the rear axle and suspend it over the rear rubber mounting.

16 Align the front and rear exhaust sections and connect by pushing together. Use an exhaust sealer to ensure a gas tight seal and ensure the two pipes are pushed fully together.

17 Correctly align the system. Ensure there is no strain on the rubber mountings and that a clearance of at least 0.8 in (20 mm) exists between the exhaust system and the vehicle body or axle.

18 Refit and securely tighten all U-bolt clamps and the exhaust manifold clamps.

19 Reconnect the battery and run the engine to ensure there are no exhaust leaks.

PART C
FUEL INJECTION SYSTEM

42 General description and principle of operation

The fuel injection system fitted to the 2.8 litre V6 Granada is of the continuous injection type and supplies a precisely controlled quantity of atomized fuel to each cylinder under all operating conditions.

This system when compared with conventional carburettor arrangements, achieves a more accurate control of the air/fuel mixture resulting in reduced emission levels and improved performance.

The main components of the fuel injection system fall into three groups:

   A Fuel tank
       Fuel pump
       Fuel accumulator
       Fuel filter
       Fuel distributor and mixture control
       Throttle and injector valves
       Air box

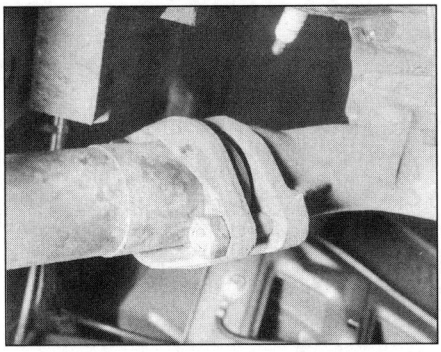

41.8a Exhaust down pipe to manifold clamp securing nut

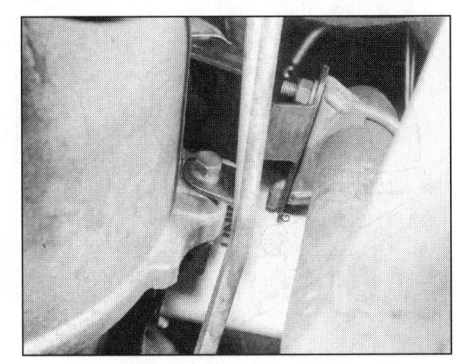

41.8b Right-hand exhaust down pipe steady clamp (viewed from below)

   B Warm up regulator
       Auxiliary air device
       Starter valve
       Thermo time switch
   C Wiring
       Relays
       Safety switch

The fuel pump is of electrically operated, roller cell type. A pressure relief valve is incorporated in the pump to prevent excessive pressure build up in the event of a restriction in the pipelines.

The fuel accumulator has two functions, (i) to dampen the pulsation of the fuel flow, generated by the pump and (ii) to maintain fuel pressure after the engine has been switched off. This prevents a vapour lock developing with consequent hot starting problems.

The fuel filter incorporates two paper filter elements to ensure that the fuel reaching the injection system components is completely free from dirt.

The fuel distributor/mixture control assembly. The fuel distributor controls the quantity of fuel being delivered to the engine, ensuring that each cylinder receives the same amount. The mixture control assembly incorporates an air sensor plate and control plunger. The air sensor plate is located in the main air stream between the air cleaner and the throttle butterfly. During idling, the air flow lifts the sensor plate which in turn raises a control plunger which allows fuel to flow past the plunger and out of the injector valves. Increases in engine speed cause increased air flow which raises the control plunger and so admits more fuel.

The throttle valve assembly is mounted in the main air intake between the mixture control assembly and the air box. The throttle valve plate is controlled by a cable connected to the accelerator pedal.

The injector valves are located in the inlet

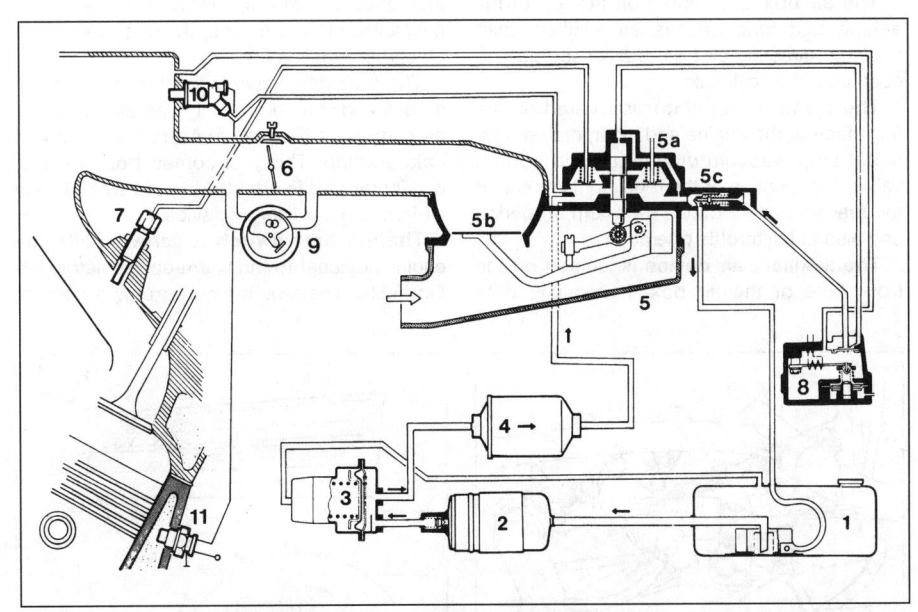

Fig. 3.51 Layout of fuel injection system (Sec 42)

1 Fuel tank	5 Mixture control unit	6 Throttle plate assembly
2 Fuel pump	5a Fuel distributor	7 Injector valve
3 Fuel accumulator	5b Air sensor	8 Warm up regulator
4 Fuel filter	5c Pressure regulator	9 Auxiliary air device
		10 Start valve
		11 Thermo time switch

3•16 Fuel and exhaust systems

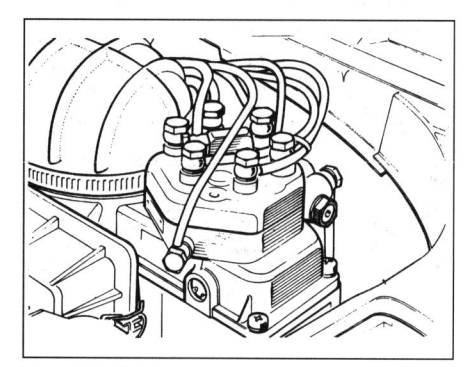

Fig. 3.52 Mixture control and fuel distributor assembly - Fuel injection (Sec 42)

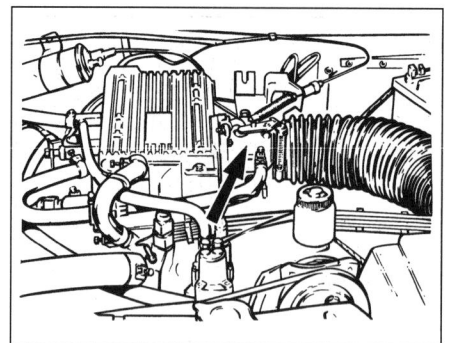

Fig. 3.53 Throttle valve plate assembly - Fuel injection (Sec 42)

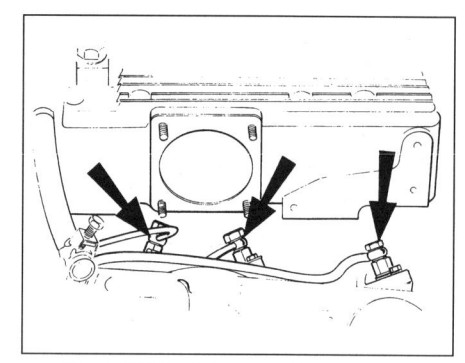

Fig. 3.54 The injector valves, air box removed - Fuel injection (Sec 42)

Fig. 3.55 The air box, arrowed - Fuel injection (Sec 42)

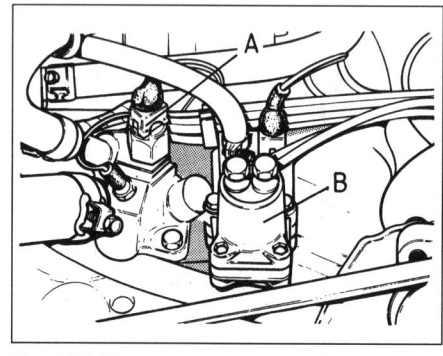

Fig. 3.56 Warm-up regulator - Fuel injection (Sec 42)

A Thermo time switch   B Warm up regulator

Fig. 3.57 Auxiliary air device, arrowed - Fuel injection (Sec 42)

manifold and are designed to open at a fuel pressure of 46.9 lbf/in^2 (3.3 kgf/cm^2).

The air box is mounted on the top of the engine and functions as an auxiliary inlet manifold directing air from the sensor plate to each individual cylinder.

The warm up regulator is located on the front face of the engine and incorporates a bi-metal strip, vacuum diaphragm and control valve. The function of the regulator is to enrich the fuel/air mixture during the warm up period and also at full throttle operation.

The auxiliary air device is located on the front face of the air box. It consists of a pivoted plate, bi-metal strip and heater coil. The purpose of this device is to supply an increased volume of fuel/air mixture during cold idling rather similar to the fast idle system on carburettor layouts.

The fuel start valve is located on the main air box and is an electrically operated injector. Its purpose is to spray fuel into the air box at cold starting. This fuel comes from the fuel distributor and is atomized in the air box with air from the auxiliary air device.

Thermo time switch is screwed into the engine adjacent to the warm up regulator (see Fig. 3.56). The switch incorporates a contact set, bi-metal strip and two heated elements. The switch controls the start valve by limiting the period of fuel injection from the valve and also preventing the valve from injecting any fuel at all when starting a hot engine.

One of two thermo time switches available will be fitted, either an 80°C or 35°C switch. If a long cranking time is necessary to restart the engine when hot, you probably have the 80°C switch fitted and this can be readily changed for the 35°C switch to reduce the cranking time required. Consult your Ford dealer to confirm which switch type you have.

Electrical relays and fuses. These are

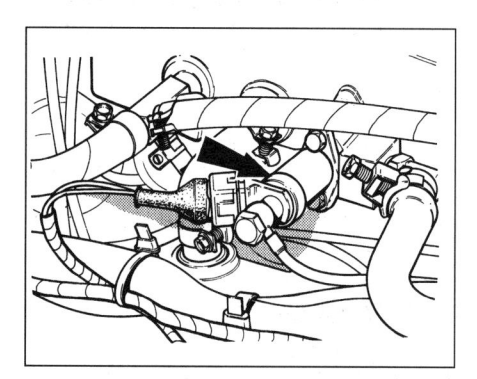

Fig. 3.58 Fuel start valve, arrowed - Fuel injection (Sec 42)

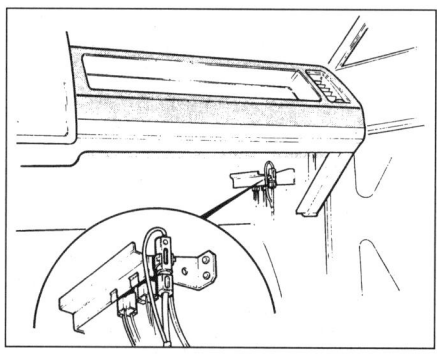

Fig. 3.59 Location of electrical relays and fuse - Fuel injection (Sec 42)

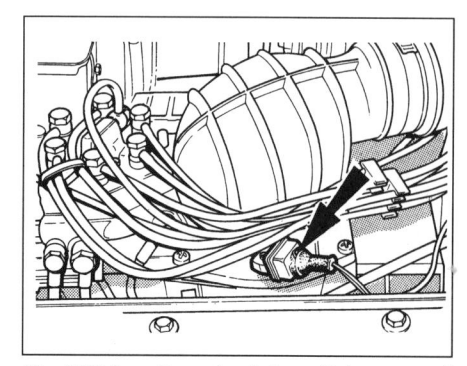

Fig. 3.60 Location of safety switch, arrowed - Fuel injection (Sec 42)

Fuel and exhaust systems

located under the facia panel on the right-hand side in the interior of the car. The relays comprise the main control relay and the power supply relay.

The safety switch, is provided to cut off all power to the fuel injection system if the air sensor plate (mixture control assembly) is in the rest position even if the ignition switch is on. The safety switch would cut off the power to the fuel pump for example if a fuel line was damaged in an accident or burst through deterioration. The switch is located close to the mixture control unit.

The fuel tank is similar to the one described in Section 7 for carburettor type vehicles except that before removing it, the pipes to the fuel pump and accumulator and the pipe from pressure regulator must be disconnected and plugged.

43 Air cleaner assembly - removal and refitting

1 Disconnect the battery earth terminal.
2 Unclip the four clasps securing the air cleaner assembly to its housing.
3 Partially remove the air cleaner, detach the intake pipe and remove the air cleaner and filter assembly.
4 Refitting is a reverse of the removal sequence.

44 Main components - removal and refitting

Fuel pump

1 The fuel pump is bolted to the lower face of the floor pan just ahead of the fuel tank.
2 Disconnect the battery.
3 With the rear of the car over an inspection pit or suitably raised, clamp the fuel inlet hose between the fuel tank and the pump using a brake hose clamp or similar. If there is only very little fuel in the tank, an alternative to clamping the hose is to drain the tank once the inlet hose is disconnected.
4 Disconnect the fuel inlet and outlet pipes from the pump.
5 Carefully identify the electrical connections to the pump and disconnect them.
6 Remove the two U-bolt nuts and lift away the pump.
7 Installation is a reversal of removal.

Fuel accumulator

8 The fuel accumulator is mounted adjacent to the fuel pump.
9 Disconnect the battery.
10 Place the rear of the car over an inspection pit or raise it on ramps or axle-stands.
11 Disconnect the fuel pipes from the fuel accumulator and catch the small quantity of fuel which will be released. Disconnect the vent pipe also.
12 Remove the two mounting nuts and remove the accumulator.

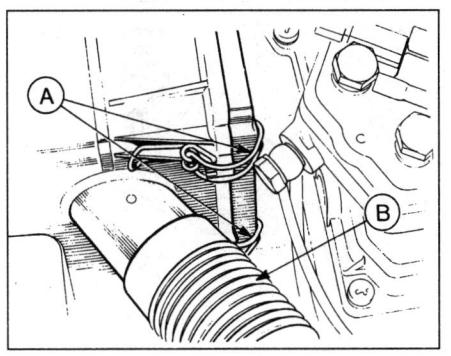

Fig. 3.61 Air cleaner assembly - Fuel injection (Sec 43)

A Securing clip   B Intake pipe

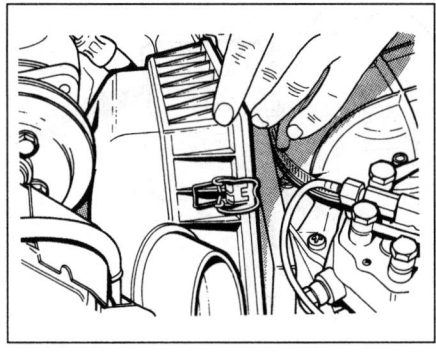

Fig. 3.62 removing the air cleaner - Fuel injection (Sec 43)

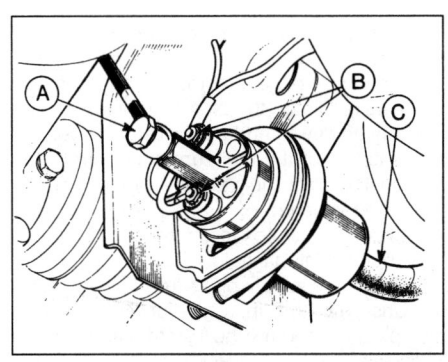

Fig. 3.63 The fuel pump - Fuel injection (Sec 44)

A Fuel outlet   B Terminals   C Fuel inlet

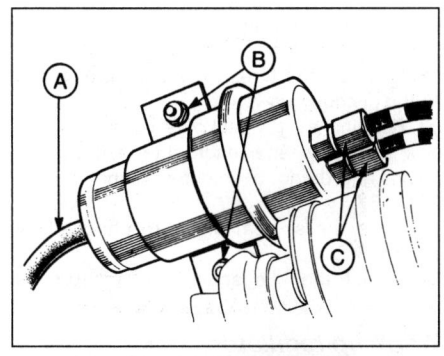

Fig. 3.64 The fuel accumulator - Fuel injection (Sec 44)

A Vent pipe   B Mounting nuts
C Fuel pipe connections

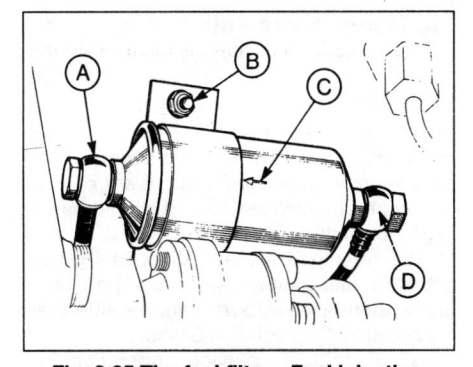

Fig. 3.65 The fuel filter - Fuel injection (Sec 44)

A Fuel outlet   B Mounting nut
C Direction of flow   D Fuel inlet

13 Refitting is a reversal of removal.

Fuel filter

14 The fuel filter is located close to the fuel pump and accumulator.
15 Disconnect the battery and place the rear of the car over an inspection pit or raise it on ramps or axle-stands.
16 Disconnect the fuel pipes from the filter and catch the small quantity of fuel which will be released, in a container.
17 Unbolt the filter and install the new one but

Fig. 3.66 The fuel distributor, assembly securing screws arrowed - Fuel injection (Sec 44)

make sure that the fuel flow directional arrow marked on the filter casing is pointing the correct way (fuel flow *from* tank).
18 Reconnect the fuel pipes and reconnect the battery.

Fuel distributor

19 Disconnect the battery.
20 Disconnect the fuel pipes from the distributor. Catch the small quantity of fuel which will leak out of the disconnected warm

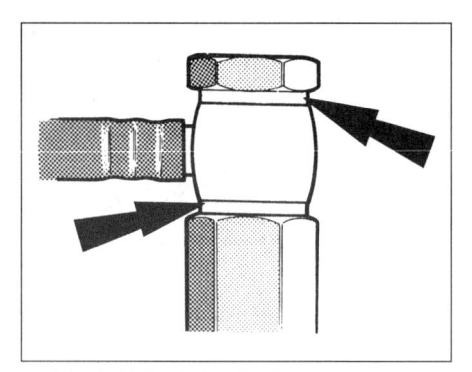

Fig. 3.67 Fuel pipe banjo type unions, copper washers arrowed - Fuel injection (Sec 44)

Fig. 3.68 Air intake trunking from mixture control unit - Fuel injection (Sec 44)

A Throttle bracket   B Main air supply hose

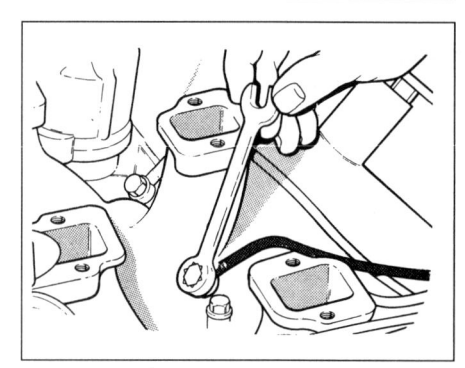

Fig. 3.69 Removing fuel pipe from injector - Fuel injection (Sec 44)

up regulator feed pipe as it is disconnected from the distributor.
21 Unscrew the three mounting screws and lift the distributor away.
22 Refitting is a reversal of removal but use a new copper washer each side of the banjo unions and do not overtighten the union bolts (Fig. 3.67).
23 Have the system pressure checked by your dealer.
24 Adjust the idle speed and mixture as described in Section 46.

Warm up regulator

25 Disconnect the battery.
26 Disconnect the plug and vacuum hose from the regulator.
27 Disconnect the fuel lines from the regulator and catch the small quantity of fuel which will be released.
28 Unscrew and remove the two mounting bolts and lift the regulator away.
29 Refitting is a reversal of removal but use a new copper washer each side of the banjo unions. Do not overtighten the union centre bolts.

Fuel start valve

30 Disconnect the battery.
31 Disconnect the electric plug and fuel supply pipe from the valve.
32 Unscrew the two socket headed mounting bolts using an Allen key.
33 Refitting is a reversal of removal but use a new mounting flange gasket.
34 Use a new copper washer one each side of the banjo union and do not overtighten the centre bolt.

Auxiliary air device

35 Disconnect the battery.
36 Disconnect the electric plug and two air hoses from the device.
37 Unscrew the two mounting bolts and lift the assembly away.
38 Refitting is a reversal of removal.

Fuel injectors

39 Disconnect the battery.
40 Disconnect the accelerator cable and bracket from the throttle assembly.

41 Disconnect the main air supply trunking from the mixture control unit.
42 Disconnect the electric plugs from the warm up regulator, the thermo time switch, the auxiliary air device and the fuel start valve.
43 Disconnect the fuel supply pipe from the start valve also the vacuum pipe from the warm up regulator.
44 Unbolt the air box (eight bolts) and move it clear of the inlet manifold.
45 Unbolt the supply pipes from the injectors.
46 Unscrew the injector forked securing clamps and withdraw the injectors and their O-ring seals.
47 Observe extreme cleanliness when fitting the injectors and use new O-ring seals, otherwise refitting is a reversal of removal.

45 Maintenance and adjustment - general

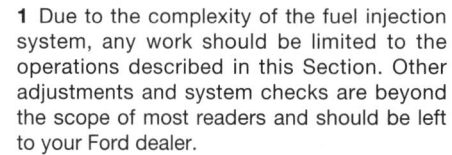

1 Due to the complexity of the fuel injection system, any work should be limited to the operations described in this Section. Other adjustments and system checks are beyond the scope of most readers and should be left to your Ford dealer.
2 After reference to Fault diagnosis if it is definitely established that a component is faulty then a new part can be fitted as described in the preceding Section.
3 As a routine service operation or

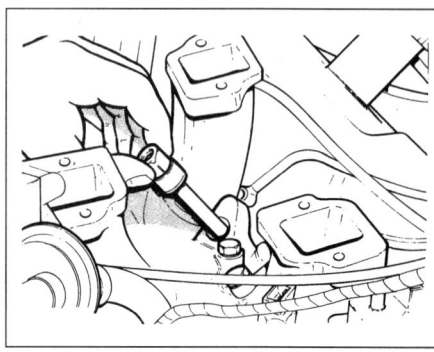

Fig. 3.70 Withdrawing a fuel injector - Fuel injection (Sec 44)

immediately upon malfunction of the system, check the security of all connecting plugs, fuel and vacuum hoses. Check the system fuses and relays.
4 The mixture setting is pre-set during production of the car and should not normally require adjustment. If new components of the system have been fitted however, the mixture can be adjusted after reference to Section 46.
5 The only adjustment which can be carried out without special equipment is to vary the engine idle speed by means of the screw mounted in the throttle housing. Use the screw to set the engine speed to that specified when the engine is at the normal operating temperature.

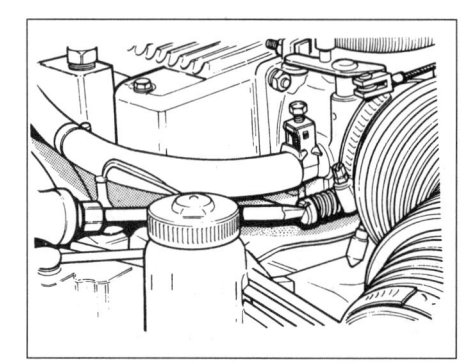

Fig. 3.71 Idle speed screw - Fuel injection (Sec 45)

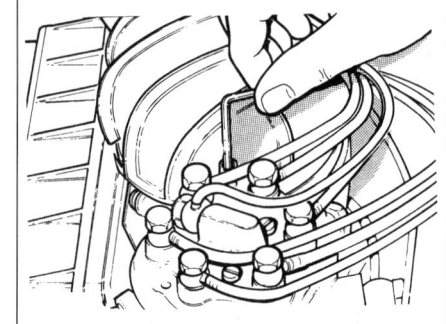

Fig. 3.72 Adjusting the mixture control screws - Fuel injection (Sec 46)

46 Fuel/air mixture and adjustment

1 As already explained, this work will normally only be required after installation of new components to the fuel injection system.
2 Unless an exhaust gas analyser is available, leave the work to your Ford dealer. If a tachometer is not fitted to your particular model, an externally mounted accurate instrument will have to be obtained.
3 Run the engine until it is at the normal operating temperature.
4 Connect an exhaust gas analyser and a tachometer (if not fitted as standard equipment) in accordance with the manufacturers' instructions.
5 Increase the engine speed to 3000 rpm and hold it there for 30 seconds to stabilise the exhaust gases and then allow the engine to return to idling.
6 Check the readings on the test instruments with those specified. If adjustment is required first turn the idle speed screw to give the correct idle speed.
7 Break off the tamper proof cap from the mixture control screw on top of the fuel distributor.
8 Stabilise the exhaust gases as described in paragraph 5.
9 Insert a 3 mm Allen key into the head of the mixture screw and turn the screw until the correct CO reading is obtained. Readjust the idle speed screw.
10 If the mixture adjustment cannot be finalised within 30 seconds from the moment of stabilising the exhaust gases, repeat the operations described in paragraph 5 before continuing the adjustment procedure.

Fault diagnosis – fuel and exhaust systems

Carburettor models

Fuel consumption excessive

Air cleaner choked and dirty giving rich mixture
Fuel leaking from carburettor, fuel pumps, or fuel lines
Float chamber flooding
Generally worn carburettor
Distributor condenser faulty
Balance weights or vacuum advance mechanism in distributor faulty
Carburettor incorrectly adjusted, mixture too rich
Idling speed too high
Contact breaker gap incorrect
Valve clearances incorrect
Incorrectly set spark plugs
Tyres under-inflated
Wrong spark plugs fitted
Brakes dragging

Insufficient fuel delivery or weak mixture due to air leaks

Petrol tank air vent restricted
Partially clogged filter in pump
Fuel pump diaphragm leaking or damaged
Seal in fuel pump damaged
Fuel pump valves sticking due to petrol gumming
Too little fuel in fuel tank (prevalent when climbing steep hills)
Union joints on pipe connections loose
Split in fuel pipe on suction side of fuel pump
Inlet manifold to block or inlet manifold to carburettor gasket leaking

Fuel injection models

Engine will not start (cold or hot)

Air flow sensor plate incorrectly set
Faulty fuel pump
Fuel tank empty

Engine will not start (HOT)

Auxiliary air device not closing

Engine will not start (COLD)

Auxiliary air device not opening
Start valve faulty
Thermo time switch not closing

Engine misfires on road

Loose fuel pump electrical connections

Unsatisfactory road performance

System fuel pressure incorrect

Rough idling (always)

Mixture adjustment incorrect

Rough idling (during warm-up)

Auxiliary air device not operating correctly

Excessive fuel consumption

Mixture adjustment incorrect
Leak in system fuel lines

Engine runs on

Air flow sensor plate or control plunger not moving freely
Injection valves leaking or their opening pressure too low

Notes

Chapter 4 Ignition system

For modifications, and information applicable to later models, see Supplement at end of manual

Contents

Bosch distributor (in-line ohc engine) - dismantling 6
Bosch distributor (in-line ohc engine) - reassembly 9
Condenser (in-line ohc engine) - removal, testing and refitting 4
Contact breaker points (in-line ohc engine) - adjustment 2
Contact breaker points (in-line ohc engine) - removal and refitting 3
Distributor (in-line ohc engine) - inspection and repair 8
Distributor (in-line ohc engine) - lubrication 11
Distributor (in-line ohc engine) - removal and refitting 5
Distributor (V6 engines) - dismantling 14
Distributor (V6 engines) - removal and refitting 13
Fault diagnosis - ignition system ....See end of Chapter
General description 1
Ignition system - fault finding 17
Ignition timing (in-line ohc engine) 12
Ignition timing (V6 engines) 15
Motorcraft distributor (in-line ohc engine) - dismantling 7
Motorcraft distributor (in-line ohc engine) - reassembly 10
Spark plugs and HT leads 16

Degrees of difficulty

| Easy, suitable for novice with little experience | Fairly easy, suitable for beginner with some experience | Fairly difficult, suitable for competent DIY mechanic | Difficult, suitable for experienced DIY mechanic | Very difficult, suitable for expert DIY or professional |

Specifications

In-line ohc engine

Spark plugs

Type .... Champion F7YCC or F7YC
Electrode gap (up to 1983):
  F7YCC spark plugs .... 0.032 in (0.8 mm)
  F7YC spark plugs .... 0.025 in (0.6 mm)

HT leads .... Champion CLS 4

Firing order .... 1-3-4-2 (No 1 at timing belt end)

Coil

Type .... 7 volt used in conjunction with 1.5 ohm ballast resistor
Primary resistance .... 0.95 to 1.60 ohms
Secondary resistance .... 5000 to 9300 ohms

Distributor

Type .... Bosch or Motorcraft
Contact breaker gap:
  Bosch .... 0.016 to 0.020 in (0.4 to 0.5 mm)
  Motorcraft .... 0.025 in (0.64 mm)
Rotation of rotor .... Clockwise viewed from top
Condenser capacity:
  Bosch .... 0.18 to 0.26 mfd
  Motorcraft .... 0.21 to 0.25 mfd
Dwell angle .... 48° to 52°
Ignition timing:
  To 1981 .... 8° BTDC
  1982 on .... 12° BTDC

Torque wrench setting	lbf ft	Nm
Spark plugs	18	25

V6 engines

Spark plugs

Type .... Champion RN7YCC or RN7YC
Electrode gap:
  RN7YCC spark plugs .... 0.032 in (0.8 mm)
  RN7YC spark plugs .... 0.025 in (0.6 mm)

4•2 Ignition system

HT leads	Champion CLS 6
Firing order	1-4-2-5-3-6
Coil	
Type	8 volt used in conjunction with 1.1 ohm ballast resistor
Primary resistance	1.1 to 1.3 ohms
Secondary resistance	7500 to 9500 ohms
Distributor	
Type	Motorcraft breakerless
Rotation of rotor	Clockwise viewed from top
Ignition timing:	
Carburettor models	9° BTDC
Fuel injection models	12° BTDC
Torque wrench setting	**lbf ft** **Nm**
Spark plugs	25 34

1 General description

To achieve optimum performance from an engine and to meet stringent exhaust emission requirements, it is essential that the fuel/air mixture in the combustion chamber is ignited at exactly the right time relative to engine speed and load. The ignition system provides the spark necessary to start the mixture burning and the instant at which ignition occurs is varied automatically as engine operating conditions change.

The ignition system consists of a primary (low tension) circuit which comprises the battery, ignition switch, contact breaker points (in-line ohc engine) condenser (in-line ohc engine), magnetic armature (V6 engines), magnetic pick-up (V6 engine), amplifier module (V6 engine) and primary windings of the ignition coil. The ignition secondary (high tension) circuit is made up of the secondary coil windings, the distributor rotor, cap, spark plugs and associated cables.

The in-line ohc engine system functions in the following manner: Battery voltage is fed to the primary circuit of the ignition coil via the ignition switch and contact breaker points. As the contact breaker points open the primary

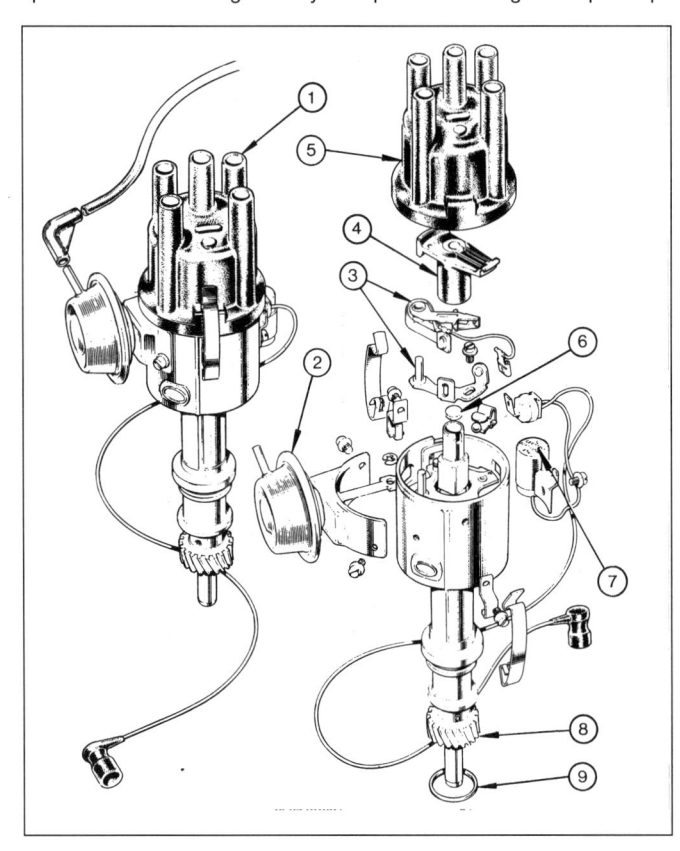

Fig. 4.1 Bosch distributor as fitted on in-line ohc engines (Sec 1)

1 Distributor assembly
2 Vacuum advance unit
3 Contact breaker points
4 Rotor arm
5 Distributor cap
6 Felt pad
7 Condenser
8 Drive gear
9 Seal

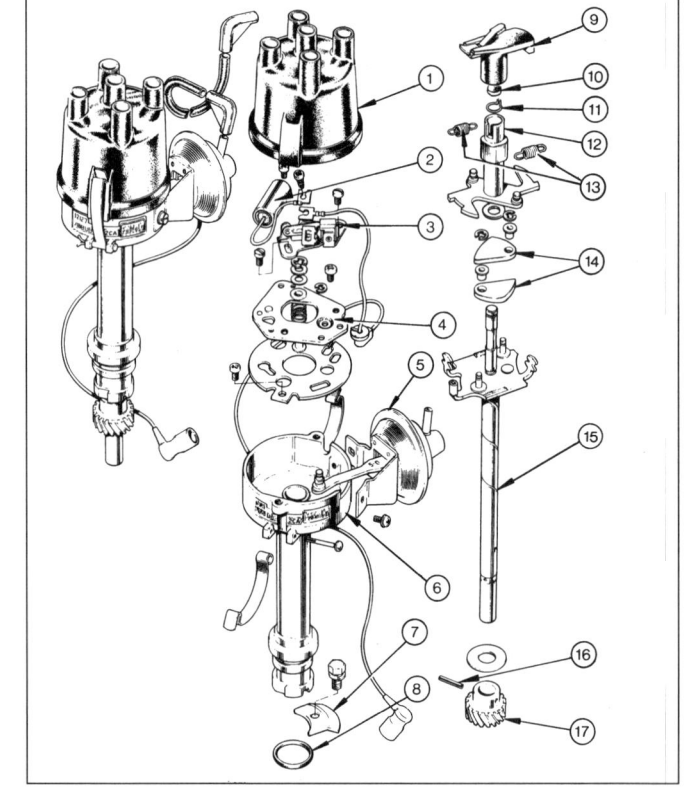

Fig. 4.2 Motorcraft distributor as fitted on in-line ohc engines (Sec 1)

1 Distributor cap
2 Condenser
3 Contact breaker points
4 Base plate
5 Vacuum advance unit
6 Distributor body
7 Clamp plate
8 Seal
9 Rotor arm
10 Felt pad
11 Circlip
12 Cam
13 Advance springs
14 Advance weights
15 Shaft
16 Pin
17 Drive gear

Ignition system 4•3

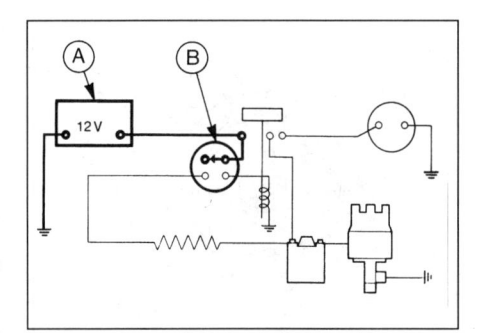

Fig. 4.3 Ignition switch in "off" position (Sec 1)

A Battery B Ignition switch

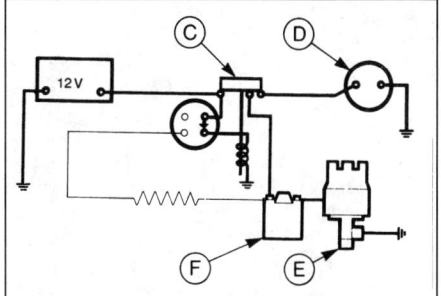

Fig. 4.4 Ignition switch in "start" position (Sec 1)

C Starter solenoid D Starter motor
E Distributor F Coil

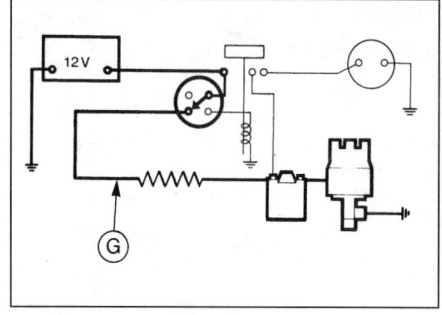

Fig. 4.5 Ignition switch in "on" position (Sec 1)

G Ballast resistor wire

ignition circuit is switched off and a very high voltage is induced in the secondary circuit of the ignition coil. This voltage is directed to the appropriate cylinder by the rotor arm and segments in the distributor cap.

The way in which the primary circuit is switched is different on V6 engines. These engines have an armature with magnetic spokes in place of the usual contact breaker cam in the distributor. As the armature rotates, the spokes pass in front of a magnetic pick-up and each time the pick-up senses the presence of a spoke it sends a signal to an amplifier module. The amplifier then switches the primary circuit off and the ignition system then functions in the same manner as for the in-line ohc engine. After allowing sufficient time for the primary circuit to collapse, the amplifier switches the primary circuit on again so that the cycle can be repeated when the next magnetic spoke passes the pick-up coil.

With both types of system the ignition is advanced and retarded automatically to ensure that the spark occurs at just the right instant for the particular load at the prevailing engine speed.

The ignition advance is controlled both mechanically and by a vacuum operated system. The mechanical governor mechanism comprises two weights, which move out from the distributor shaft as the engine speed rises due to centrifugal force. As they move outwards they rotate the cam relative to the distributor shaft and so advance the spark. The weights are held in position by two light springs and it is the tension of the springs which is largely responsible for correct spark advancement.

The vacuum control consists of a diaphragm, one side of which is connected via a small bore tube to the carburettor, and the other side to the contact breaker plate. Depression in the inlet manifold and carburettor, which varies with engine speed and throttle opening, causes the diaphragm to move so moving the contact breaker plate, and advancing or retarding the spark. A fine degree of control is achieved by a spring in the vacuum assembly.

To facilitate cold engine starting a high output ignition coil is used incorporating a high resistance wire in the ignition coil feed circuit.

During starting this ballast resistor wire is bypassed allowing the full available battery voltage to be passed to the coil. This ensures that during cold starting, when the starter motor current draw is high, sufficient voltage is still available at the coil to produce a powerful spark. Under normal running the 12 volt supply is directed through the ballast resistor before reaching the coil.

Two makes of distributor are used on the in-line ohc engine. These are Motorcraft (black cap) and Bosch (red cap). They are similar in design with the exception of the condenser location, which is external on the Bosch unit.

On V6 engines with breakerless system, a Motorcraft distributor is used. Both the distributor cap and ignition coil on these systems are coloured blue.

2 Contact breaker points (in-line ohc engine) - adjustment

1 Release the two clips securing the distributor cap to the distributor body and lift away the cap. Clean the cap inside and out with a dry cloth. Closely inspect the inside of the cap and the four segments. If there are any signs of cracking or if the segments are burned or scored the cap will have to be renewed.

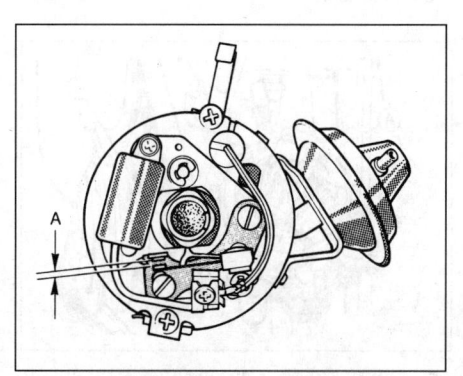

Fig. 4.6 Contact breaker points gap "A" - Motorcraft distributor illustrated (Sec 2)

2 Inspect the carbon brush contact located in the top of the cap to ensure that it is not broken and stands proud of the plastic surface.

3 Lift off the rotor arm and on Motorcraft units check the contact spring on top of the rotor arm. It must be clean and have adequate tension to ensure a good contact.

4 Gently prise the contact breaker points open and examine the condition of their faces. If they are rough, pitted or dirty it will be necessary for new points to be fitted.

5 If the points are satisfactory, or have been renewed, measure the gap between the points with feeler gauges, by turning the crankshaft until the heel of the breaker arm is on a high point of the cam. Consult the Specifications at the beginning of this Chapter for the correct setting.

6 If the gap is incorrect, slacken the contact plate securing screw/s. Bosch distributor 1 screw, Motorcraft distributors 2 screws.

7 Insert a screwdriver into the slot provided at the edge of the contact plate and move the plate until the gap is correct.

8 Tighten the screw/s and recheck the gap.

9 Refit the rotor arm ensuring that it is correctly located in the slot at the top of the cam. Refit the distributor cap and retain in position with the two clips.

10 On modern engines a more accurate method of setting the points gap is by means of a dwell meter. Not only does this method give a precise points gap but it also evens out any variations in gap caused by wear in the distributor shaft or bushes or differences in any of the heights of the four cam peaks.

11 The dwell angle is the number of degrees through which the distributor cam turns during the period between the instants of closure and opening of the contact breaker points. It can only be checked with a dwell meter connected in accordance with the maker's instructions.

12 If the angle is not as given in the Specifications, adjust the points gap. If the dwell angle is too large increase the points gap, if the dwell angle is too small, reduce the points gap.

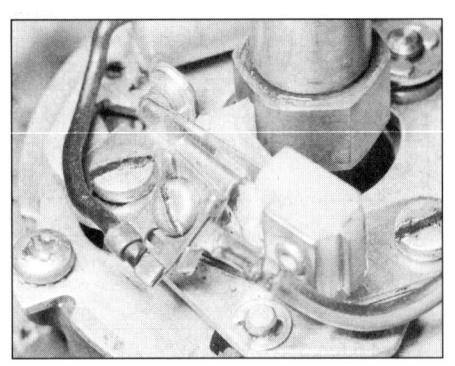

4•4 Ignition system

3.3 Removing the contact breaker low tension leads (Motorcraft distributor illustrated)

3.4 Location of contact breaker securing screws (Motorcraft distributor illustrated)

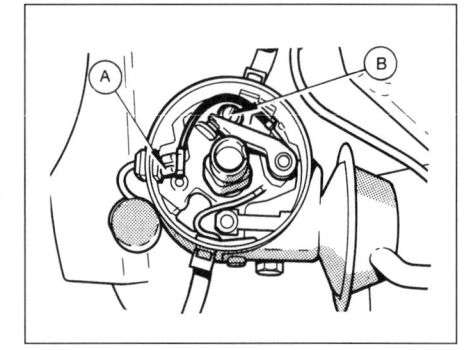

Fig. 4.7 Low tension lead connection (A) and contact breaker screw (B) in Bosch distributor (Sec 3)

3 Contact breaker points (in-line ohc engine) - removal and refitting

1 Release the two clips securing the distributor cap to the distributor body and lift away the cap.
2 Lift the rotor arm off the distributor cam.
3 Remove the low tension lead/s to the contact breaker assembly. Bosch: pull off the lead from the terminal. Motorcraft: slacken the screw and slide out the forked ends (photo).
4 Remove the securing screws and lift out the contact breaker points. (Bosch 1 screw, Motorcraft 2 screws) (photo).
5 Smear a trace of grease onto the cam and refit the contact breaker points and securing screw/s.
6 Refit the low tension leads and adjust the contact breaker points as described in Section 2.
7 Refit the rotor arm and distributor cap.

4 Condenser (in-line ohc engine) - removal, testing and refitting

1 The purpose of the condenser (sometimes known as a capacitor) is to ensure that when the contact breaker points open there is no sparking across them which would cause rapid wear of their faces and lead to engine misfire.
2 The condenser is fitted in parallel with the contact breaker points. If it becomes faulty, it will cause ignition failure as the contact breaker points will be prevented from correctly interrupting the low tension circuit.
3 If the engine becomes very difficult to start or begins to miss after several miles of running and the breaker points show signs of excessive burning, then the condition of the condenser must be suspect.

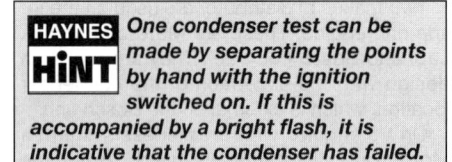

> **HAYNES HINT** *One condenser test can be made by separating the points by hand with the ignition switched on. If this is accompanied by a bright flash, it is indicative that the condenser has failed.*

4 Without special test equipment the only safe way to diagnose condenser trouble is to replace a suspect unit with a new one and note if there is any improvement.
5 To remove the condenser from the distributor take off the distributor cap and rotor arm.
6 Bosch: Disconnect the low tension leads from the coil and to the contact breaker points. Release the condenser cable from the side of the distributor body and then undo and remove the screw that secures the condenser to the side of the distributor body. Lift away the condenser.

7 Motorcraft: Slacken the self tapping screw holding the condenser lead and low tension lead to the contact breaker points. Slide out the forked terminal on the end of the condenser low tension lead. Undo and remove the condenser retaining screw and remove the condenser from the breaker plate.
8 To refit the condenser, simply reverse the order of removal.

5 Distributor (in-line ohc engine) - removal and refitting

1 Unclip and remove the distributor cap and position it clear of the distributor body.
2 Remove the distributor low tension lead from the (-) terminal of the ignition coil.
3 Pull off the rubber union holding the vacuum pipe to the distributor vacuum advance housing.
4 Rotate the engine by hand until the TDC mark (large notch) on the crankshaft pulley is in line with the pointer on the timing cover and the rotor arm is pointing to the No. 1 cylinder segment in the distributor cap. Mark the position of the distributor body in relation to the cylinder head (Figs. 4.8 and 4.9).
5 Remove the single bolt at the base of the distributor and slowly remove the distributor.
6 As the distributor is removed, mark the position of the rotor arm relative to the distributor body. This will be of help when refitting the distributor.
7 To refit the distributor ensure that the timing marks are still aligned as detailed in paragraph 4 of this Section.
8 Hold the distributor over the cylinder head so that the body to cylinder head marks are in alignment.
9 Position the rotor arm towards the mark on the distributor body and slide the assembly into position on the engine. As the gears mesh, the rotor will turn and align with No.1 segment in the distributor cap.
10 Refit the securing bolt to the base of the distributor but do not fully tighten at this stage.
11 Refit the cap and vacuum advance pipe to the distributor and low tension lead to the coil.
12 Refer to Section 12 and accurately set the ignition timing.

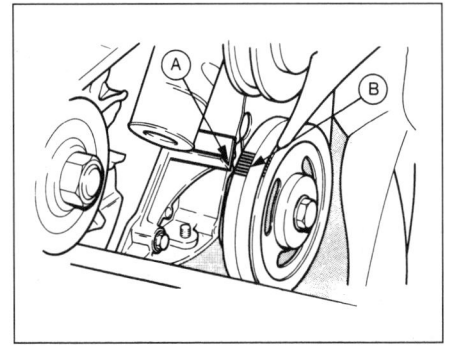

Fig. 4.8 Timing marks on in-line ohc engine (Sec 5)

A Timing pointer   B Notches on pulley

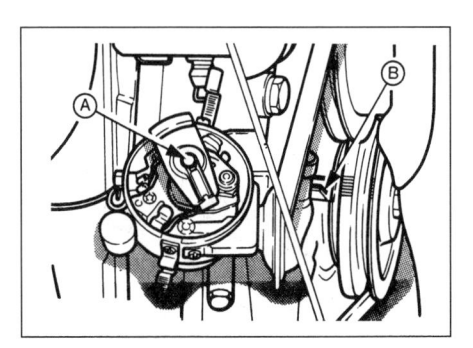

Fig. 4.9 Position of distributor and pulley, No 1 cylinder at TDC - in-line ohc engine, Bosch distributor illustrated (Sec 5)

A Rotor arm   B Timing pointer

Ignition system 4•5

6 Bosch distributor (in-line ohc engine) - dismantling

1 With the distributor on the bench, pull the rotor arm off the distributor cam spindle.
2 Remove the contact breaker points as described in Section 3.
3 Unscrew and remove the condenser securing screw and lift away the condenser and connector.
4 Carefully remove the small circlip from the pull rod of the vacuum advance unit.
5 Undo and remove the two screws that secure the vacuum advance unit to the side of the distributor body. Lift away the unit.
6 The distributor cap spring clip retainers may be removed by undoing and removing the screws securing them to the side of the distributor body. **Note:** *This is the limit of the dismantling which should be attempted since none of the parts beneath the breaker plate, including the drive gear, can be renewed separately on the Bosch distributor. If they are worn, renew the distributor complete.*

7 Motorcraft distributor (in-line ohc engine) - dismantling

1 With the distributor on the bench, pull the rotor arm off the distributor cam spindle.

7.8 Mechanical advance mechanism. Note different size spring (typical)

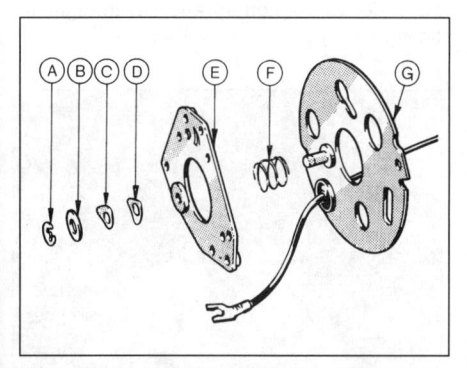

Fig. 4.10 Motorcraft distributor base plate assembly (Sec 7)

A Circlip B Washer C Wave washer
D Wave washer E Upper plate
F Spring G Base plate

2 Remove the contact breaker points as described in Section 3.
3 Next prise off the small circlip from the vacuum unit pivot post.
4 Take out the two screws that hold the breaker plate to the distributor body and lift away.
5 Undo and remove the condenser retaining screw and lift away the condenser.
6 Take off the circlip, flat washer and two wave washers from the pivot post. Separate the two plates. Be careful not to lose the spring now left on the pivot post.
7 Pull the low tension wire and grommet from the lower plate.
8 Undo the two screws holding the vacuum unit to the body. Take off the unit.
9 Make a sketch of the position of the cam plate assembly in relation to the bumpstop, noting the identification letters. Also note which spring - thick or thin - is fitted to which post (photo).
10 Dismantle the spindle by taking out the felt pad in the top. Remove the spring clip using small electrical pliers.
11 Prise off the bumpstop and lift out the cam plate assembly. Remove the thrust washer.
12 It is only necessary to remove the spindle and lower plate if it is excessively worn. If this is the case, with a suitable diameter parallel pin punch tap out the gear lock pin.
13 The gear may now be drawn off the shaft with a universal puller. If there are no means of holding the legs these must be bound together with wire to stop them springing apart during removal.
14 Finally withdraw the shaft from the distributor body.

8 Distributor (in-line ohc engine) - inspection and repair

1 Check the contact breaker points for wear, burning or pitting. Check the distributor cap for signs of tracking indicated by a thin black line between the segments. Renew the cap if any signs of tracking are found.
2 If the metal portion of the rotor arm is badly burned or loose, renew the arm. If only slightly

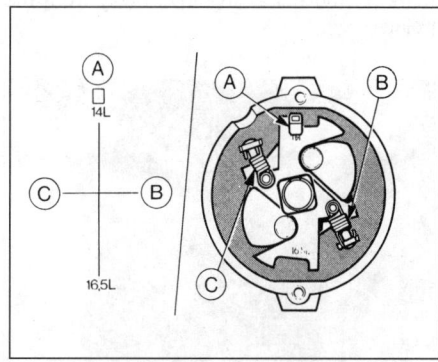

Fig. 4.11 Cam plate and advance spring location - typical (Sec 7)

A Bump stop B Thick advance spring
C Thin advance srping

burned clean the end with a fine file. Check that the contact spring has adequate pressure and the bearing surface is clean and in good condition.
3 Check that the carbon brush in the distributor cap is unbroken and stands proud of its holder.
4 Examine the centrifugal weights and pivots for wear and the advance springs for slackness. They can be checked by comparing with new parts. If they are slack they must be renewed.
5 Check the points assembly for fit on the breaker plate, and the cam follower for wear.
6 Examine the fit of the spindle in the distributor body. If there is excessive side movement it will be necessary either to fit a new bush or obtain a new body.
7 Check the operation of the vacuum advance unit by sucking the outlet pipe connection. A strong resistance should be felt and the pull rod should move toward the body during suction. If no resistance is felt or if the pull rod does not move the unit must be renewed.

9 Bosch distributor (in-line ohc engine) - reassembly

1 Place the distributor cap retaining spring clip and retainers on the outside of the distributor body and secure the retainers with the two screws.
2 Position the contact breaker point assembly in the breaker plate in such a manner that the entire lower surface of the assembly contacts the plate. Refit the contact breaker point assembly securing screw but do not fully tighten yet.
3 Hook the diaphragm assembly pull rod into contact with the pivot pin.
4 Secure the diaphragm to the distributor body with the two screws. Also refit the condenser to the terminal side of the diaphragm bracket securing screw. The condenser must firmly contact its lower stop on the housing.
5 Apply a little grease or petroleum jelly to the cam and also to the heel of the breaker lever.
6 Reset the contact breaker points, as described in Section 2 and then refit the rotor arm and distributor cap.

10 Motorcraft distributor (in-line ohc engine) - reassembly

1 Reassembly is a straightforward reversal of the dismantling process but there are several points which must be noted.
2 Check that the drive gear is not 180° out of position, as the pin bores may be slightly misaligned. Secure it with a new pin.
3 Coat the upper shaft with a lithium base grease, ensuring that the undercut is filled.
4 When fitting the cam spindle assembly, first refit the thrust washer, then refer to the sketch made in Section 7 item 8. Check that the assembly moves freely without binding.

4•6 Ignition system

5 Position the spring clip legs opposite the rotor arm slot.
6 Before assembling the breaker plate make sure that the nylon bearing studs are correctly located in their holes in the upper breaker plate, and the small earth spring is fitted on the pivot post.
7 When all is assembled reset the contact breaker points, as described in Section 2.

11 Distributor (in-line ohc engine) - lubrication

1 It is important that the distributor cam is lubricated with vaseline petroleum jelly or grease at 6000 miles (10 000 km) or 6 monthly intervals. Also the automatic timing control weights and cam spindle are lubricated with engine oil.
2 Great care should be taken not to use too much lubricant as any excess that finds its way onto the contact breaker points could cause burning and misfiring.
3 To gain access to the cam spindle, lift away the distributor cap and rotor arm. Apply no more than two drops of engine oil onto the felt pad. This will run down the spindle when the engine is hot and lubricate the bearings.
4 To lubricate the automatic timing control, allow a few drops of oil to pass through the holes in the contact breaker base plate through which the four sided cam emerges. Apply not more than one drop of oil to the pivot post of the moving contact breaker point. Wipe away excess oil and refit the rotor arm and distributor cap.

12 Ignition timing (in-line ohc engine)

1 Turn the crankshaft until No. 1 piston is coming up to TDC on the compression stroke. This can be checked by removing No. 1 spark plug and feeling the pressure being developed in the cylinder as the crankshaft is rotated. If this check is not made it is all too easy to set the timing 180° out. The crankshaft can most easily be turned by placing a suitably sized socket and ratchet on the crankshaft pulley bolt.
2 Continue turning the crankshaft until the pointer on the timing cover is in line with the appropriate groove on the crankshaft pulley. The large groove is TDC; each groove to the side of TDC equals 2°. Consult the Specifications at the front of this Chapter for the correct setting.
3 Slacken the clamp bolt at the base of the distributor until it is just possible to turn the distributor body.
4 Gently turn the distributor until the contact breaker points are just opening when the rotor is pointing towards the segment in the distributor cap which is connected to No. 1 spark plug lead. The best way to do this is to turn the distributor body clockwise until the heel of the contact arm is off the cam and then slowly turn the distributor anti-clockwise until the points just open.
5 Tighten the distributor clamp bolt enough to hold the distributor, but do not overtighten.
6 Set in this way the timing should be approximately correct but a more accurate method is by using a stroboscopic timing light.
7 Clean the timing scale on the crankshaft pulley and mark the specified timing degree line and also the pointer on the timing cover with white paint or chalk.
8 Connect a timing light between No. 1 spark plug and No. 1 spark plug lead.
9 Slacken the distributor clamp bolt until it is just possible to turn the distributor body.
10 Run the engine until the normal operating temperature is reached. Disconnect and plug the vacuum advance pipe at the carburettor.
11 With the engine idling at the speed given in the Specifications, point the timing light at the crankshaft pulley. The white painted marks will appear stationary and if the timing is correct they will be in alignment.
12 If they are not directly opposite each other, turn the distributor slightly one way or the other until the marks appear in line.
13 Increase the engine speed and note whether the white mark on the pulley moves away from the pointer. If it does the distributor mechanical advance is functioning.
14 With the engine idling the vacuum advance can be checked by sucking the advance pipe that was removed from the carburettor. This should also cause the white mark on the pulley to move away from the pointer.
15 Remove the timing light, reconnect the HT lead to the spark plug and tighten the distributor clamp bolt.

13 Distributor (V6 engines) - removal and refitting

1 Remove the air cleaner as described in Chapter 3.
2 Unscrew the two screws securing the distributor cap to the body. Lift off the cap and position it out of the way (photo).
3 Remove the vacuum advance pipe from the distributor.
4 Disconnect the distributor multi-plug connector. To disconnect the plug, the wires must be held and pulled, not the plug itself (photo).
5 Turn the engine until the rotor arm is pointing to the No. 1 cylinder segment in the distributor cap and the notch in the crankshaft pulley wheel is in line with the correct mark on the timing cover. (The timing marks are 0° to 15° in 3° intervals, see the Specifications for correct setting. Mark the position of the distributor body in relation to the cylinder head.
6 Remove the single bolt at the base of the distributor and slowly lift out the distributor (photo).
7 As the distributor is removed, mark the position of the rotor arm relative to the distributor body. This will be of help when refitting the distributor.
8 To refit the distributor ensure that the timing marks are still aligned as detailed in paragraph 4 of this Section. Hold the distributor over the cylinder head so that the body to the cylinder head marks are in alignment.
9 Position the rotor arm towards the mark on the distributor body and slide the assembly into position on the engine. As the gears mesh, the rotor will turn and align with No. 1 segment in the distributor cap.
10 Refit the securing bolt to the base of the distributor but do not fully tighten at this stage.
11 Refit the vacuum advance pipe, distributor cap and multi-plug connector.
12 Refer to Section 15 and set the ignition timing.

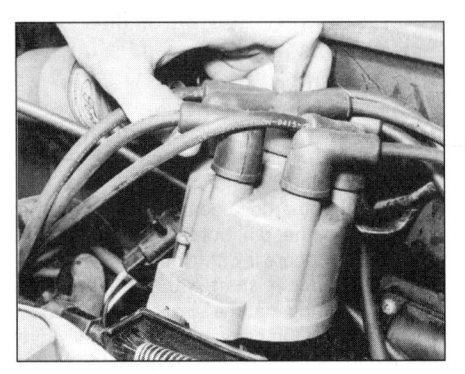

13.2 Lifting off the distributor cap

13.4 Disconnect the distributor multi-plug by pulling the wires

13.6 Distributor clamp plate securing bolt

Ignition system 4•7

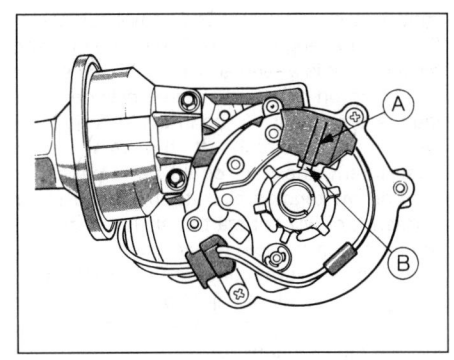

Fig. 4.12 No 1 rotor segment lined up with marks on trigger housing (Sec 13)

A Trigger coil housing   B Rotor

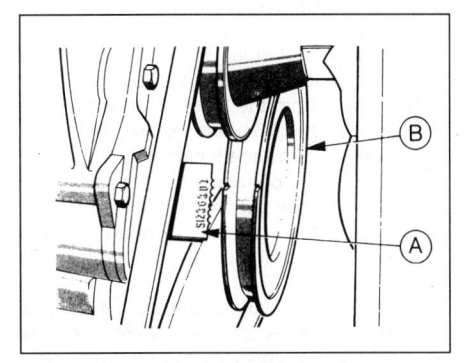

Fig. 4.13 Location of ignition timing marks (Sec 13)

A Timing plate   B Crankshaft pulley

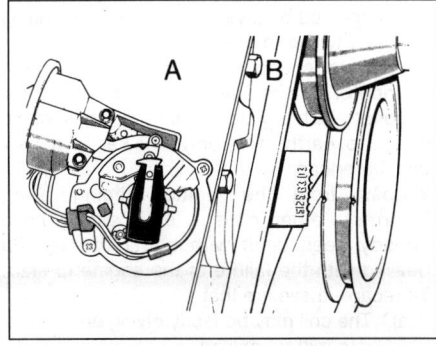

Fig. 4.14 Position of pulley and distributor prior to removal (Sec 13)

A Rotor arm adjacent No 1 cylinder segment
B Pulley in line with timing marks

14 Distributor (V6 engines) - dismantling

1 The Motorcraft distributor fitted to these engines is a sealed unit and only the rotor arm and distributor cap can be serviced separately. Any problems with the distributor itself can only be cured by fitting a new unit.

15 Ignition timing (V6 engines)

1 Remove the air cleaner as described in Chapter 3.
2 Remove the two screws, lift off the distributor cap and position it out of the way.
3 Turn the engine until the rotor arm is pointing towards the No. 1 cylinder segment in the distributor cap and the notch in the crankshaft pulley wheel is in line with the correct mark on the timing cover. (The timing marks are 0° to 15° in 3° intervals, see Specifications for correct setting.)
4 Loosen the clamp bolt at the base of the distributor and turn the complete distributor until the No. 1 rotor segment lines up with the two lines on the trigger coil housing (see Fig. 4.12).
5 Tighten the clamp bolt and refit the rotor arm, distributor cap and air cleaner.
6 Set in this way the timing will be approximately correct but a more accurate method is by using a stroboscopic timing light.
7 Highlight the specified timing marks on the timing cover and crankshaft pulley with white paint or chalk.
8 Connect a timing light between No. 1 spark plug and No. 1 spark plug lead.
9 Disconnect the vacuum advance pipe at the carburettor or air box (fuel injection) and plug the pipe then run the engine until normal operating temperatures are reached.
10 With the engine idling at the speed given in the Specifications, point the timing light at the crankshaft pulley. The white painted marks will appear stationary and if the timing is correct they will be in alignment.
11 To adjust the timing, stop the engine, slacken the clamp bolt and turn the distributor slightly. Tighten the clamp bolt, start the engine and recheck the timing. Repeat this procedure until the marks are in line.
12 To check the mechanical advance, increase the engine speed and note whether the white mark on the pulley moves away from the mark on the pointer. If it does the mechanical advance is functioning.
13 With the engine idling the vacuum advance can be checked by sucking the advance pipe that was removed from the carburettor. This should also cause the white mark on the pulley to move away from the mark on the pointer.
14 Remove the timing light, reconnect the HT lead to the spark plug and refit the air cleaner. Refit the vacuum pipe.

16 Spark plugs and HT leads

1 The correct functioning of the spark plugs is vital for the correct running and efficiency of the engine. It is essential that the plugs fitted are appropriate for the engine, and the suitable type is specified at the beginning of this chapter. If this type is used and the engine is in good condition, the spark plugs should not need attention between scheduled replacement intervals. Spark plug cleaning is rarely necessary and should not be attempted unless specialist equipment is available as damage can easily be caused to the firing ends.
2 Examination of the spark plugs will give a good indication of the overall condition of the engine.
3 If the insulator nose of the spark plug is clean and white, with no deposits, this is indicative of a weak mixture, or too hot a plug (a hot plug transfers heat away from the electrode slowly - a cold plug transfers it away quickly).
4 The plugs fitted as standard are as listed in the Specifications at the beginning of this Chapter. If the top and insulator noses are covered with hard black-looking deposits, then this is indicative that the mixture is too rich. Should the plug be black and oily, then it is likely that the engine is fairly worn, as well as the mixture being too rich.
5 If the insulator nose is covered with light tan to greyish brown deposits, then the mixture is correct and it is likely that the engine is in good condition.
6 The spark plug gap is of considerable importance, as, if it is too large or too small, the size of the spark and its efficiency will be seriously impaired. The spark plug should be set to the figure given in the Specifications at the beginning of this Chapter.
7 To set it, measure the gap with a feeler

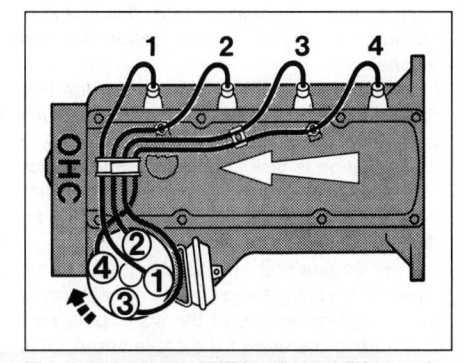

Fig. 4.15 Layout of HT leads and firing order - in-line ohc engine (Sec 16)

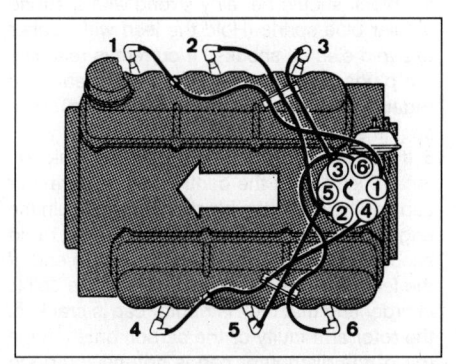

Fig. 4.16 Layout of HT leads and firing order - V6 engines (Sec 16)

gauge, and then bend open, or close, the outer plug electrode until the correct gap is achieved. The centre electrode should never be bent as this may crack the insulation and cause plug failure if nothing worse.

8 Refit the plugs, and refit the leads from the distributor in the correct firing order, which is given in the Specifications.

9 The plug leads require no routine attention other than being kept clean and wiped over regularly.

10 At intervals of 6000 miles (10 000 km) or 6 months, pull the leads off the plugs and distributor one at a time and make sure no water has found its way onto the connections. Remove any corrosion from the brass ends, wipe the collars on top of the distributor, and refit the leads.

17 Ignition system - fault finding

By far the majority of breakdown and running troubles are caused by faults in the ignition system either in the low tension or high tension circuits.

There are two main symptoms indicating ignition faults. Either the engine will not start or fire, or the engine is difficult to start and misfires. If it is a regular misfire, ie. the engine is running on only two or three cylinders, the fault is almost sure to be in the secondary or high tension circuit. If the misfiring is intermittent, the fault could be in either the high or low tension circuits. If the car stops suddenly, or will not start at all, it is likely that the fault is in the low tension circuit. Loss of power and overheating, apart from faulty carburation or fuel injection settings, are normally due to faults in the distributor or to incorrect ignition timing.

Fault diagnosis – ignition system

Engine fails to start

1 If the engine fails to start and the car was running normally when it was last used, first check there is fuel in the petrol tank. If the engine turns over normally on the starter motor and the battery is evidently well charged, then the fault may be in either the high or low tension circuits. First check the HT circuit.

Note: *If the battery is known to be fully charged, the ignition light comes on, and the starter motor fails to turn the engine* **check the tightness of the leads on the battery terminals** *and also the secureness of the earth lead to its connection to the body. It is quite common for the leads to have worked loose, even if they look and feel secure. If one of the battery terminal posts gets very hot when trying to work the starter motor this is a sure indication of a faulty connection to that terminal.*

2 One of the commonest reasons for bad starting is wet or damp spark plug leads and distributor. Remove the distributor cap. If condensation is visible internally dry the cap with a rag and also wipe over the leads. Refit the cap. A moisture dispersant can be very effective.

3 If the engine still fails to start, check that current is reaching the plugs, by disconnecting each plug lead in turn at the spark plug end, and holding the end of the cable about 3/16 inch (5 mm) away from the cylinder block. Spin the engine on the starter motor.

4 Sparking between the end of the cable and the block should be fairly strong with a strong regular blue spark. (Hold the lead with rubber to avoid electric shocks.) If current is reaching the plugs, then remove them and clean and regap them to 0.025 inch (0.60 mm). The engine should now start.

5 If there is no spark at the plug leads take off the HT lead from the centre of the distributor cap and hold it to the block as before. Spin the engine on the starter once more. A rapid succession of blue sparks between the end of the lead and the block indicate that the coil is in order and that the distributor cap is cracked, the rotor arm faulty or the carbon brush in the top of the distributor cap is not making good contact with the spring on the rotor arm. Possibly, the points are in bad condition (mechanical contact breaker). Renew them as described in this Chapter, Section 3.

6 If there are no sparks from the end of the lead from the coil check the connections at the coil end of the lead. If it is in order start checking the low tension circuit.

7 Use a 12V voltmeter or a 12V bulb and two lengths of wire. On conventional distributors switch on the ignition and ensure that the points are open. On breakerless distributors ensure that the segments on the rotor are not adjacent the trigger coil. Make a test between the low tension wire to the coil (+) terminal and earth. A reading of 7 to 8 volts should be obtained. No reading indicates a break in the supply from the ignition switch or a fault in the ballast resistance wire. A correct reading indicates a faulty coil or condenser, or a broken lead between the coil and the distributor. On breakerless ignition systems the trigger coil and amplifier module are suspect. Have them checked by a Ford dealer.

8 Take the condenser wire off the points assembly and with the points open test between the moving point and earth. If there is now a reading then the fault is in the condenser. Fit a new one as described in this Chapter, Section 4.

9 With no reading from the moving point to earth, take a reading between earth and the CB or negative (-) terminal of the coil. A reading here shows a broken wire which will need to be renewed between the coil and distributor. No reading confirms that the coil has failed and must be renewed, after which the engine will run once more. Remember to refit the condenser wire to the points assembly.

Engine misfires

1 If the engine misfires regularly run it at a fast idling speed. Pull off each of the plug caps in turn and listen to the note of the engine. Hold the plug cap in a dry cloth or with a rubber glove as additional protection against a shock from the HT supply.

2 No difference in engine running will be noticed when the lead from the defective coil is removed. Removing the lead from one of the good cylinders will accentuate the misfire.

3 Remove the plug lead from the end of the defective plug and hold it about 3/16 inch (5 mm) away from the block. Start the engine. If the sparking is fairly strong and regular the fault must lie in the spark plug.

4 The plug may be loose, the insulation may be cracked, or the points may have burnt away giving too wide a gap for the spark to jump. Worse still, one of the points may have broken off.

5 If there is no spark at the end of the plug lead, or if it is weak and intermittent, check the ignition lead from the distributor to the plug. If the insulation is cracked or perished, renew the lead. Check the connections at the distributor cap.

6 If there is still no spark, examine the distributor cap carefully for tracking. This can be recognised by a very thin black line running between two or more electrodes, or between an electrode and some other part of the distributor. These lines are paths which now conduct electricity across the cap thus letting it run to earth. The only answer is a new distributor cap.

7 Apart from the ignition timing being incorrect; other causes of misfiring have already been dealt with under the section dealing with the failure of the engine to start. To recap - these are that:
   a) The coil may be faulty giving an intermittent misfire
   b) There may be a damaged wire or loose connection in the low tension circuit
   c) The condenser may be short circuiting (if fitted)
   d) There may be a mechanical fault in the distributor (broken driving spindle or contact breaker spring, if fitted)

8 If the ignition timing is too far retarded, it should be noted that the engine will tend to overheat, and there will be a quite noticeable drop in power. If the engine is overheating and the power is down, and the ignition timing is correct, then the carburettor or fuel injection should be checked, as it is likely that this is where the fault lies.

Chapter 5 Clutch

For modifications, and information applicable to later models, see Supplement at end of manual

Contents

Clutch assembly - refitting 4
Clutch assembly - removal 2
Clutch cable - removal and refitting 5
Clutch - dismantling and inspection 3
Clutch pedal and self adjusting mechanism - removal and refitting 7
Clutch release bearing - removal and refitting 6
Fault diagnosis - clutch .... See end of Chapter
General description 1

Degrees of difficulty

Easy, suitable for novice with little experience	Fairly easy, suitable for beginner with some experience	Fairly difficult, suitable for competent DIY mechanic	Difficult, suitable for experienced DIY mechanic	Very difficult, suitable for expert DIY or professional

Specifications

Type .... Single dry plate, diaphragm spring, cable operated

Lining diameter | **2.0 and 2.3 litre** | **2.8 litre**
Inner .... 6.02 in (153 mm) | 6.12 in (155.6 mm)
Outer .... 8.5 in (216 mm) | 9.13 in (232 mm) 9.35 in (242 mm)
Lining thickness .... 0.151 in (3.85 mm) | 0.150 in (3.81 mm)

Number of torsion springs 6

Total pedal travel .... 7.3 in (185 mm)

Torque wrench settings | **lbf ft** | **Nm**
Pressure plate to flywheel bolts .... 13 | 18
Bellhousing to engine
 2.0 and 2.3 litre .... 32 | 44
 2.8 litre .... 25 | 34

1 General Description

All models covered by this manual are fitted with a single diaphragm spring clutch. The unit comprises a steel cover which is dowelled and bolted to the rear face of the flywheel and contains the pressure plate, diaphragm spring and fulcrum rings.

The clutch disc is free to slide along the splined first motion shaft and is held in position between the flywheel and the pressure plate by the pressure of the pressure plate spring. Friction lining material is riveted to the clutch disc and it has a spring cushioned hub to absorb transmission shocks and to help ensure a smooth take off.

The circular diaphragm spring is mounted on shoulder pins and held in place in the cover by two fulcrum rings. The spring is also held to the pressure plate by three spring steel clips which are riveted in position.

The clutch is self adjusting and is actuated by a cable controlled by the clutch pedal. The clutch release mechanism consists of a

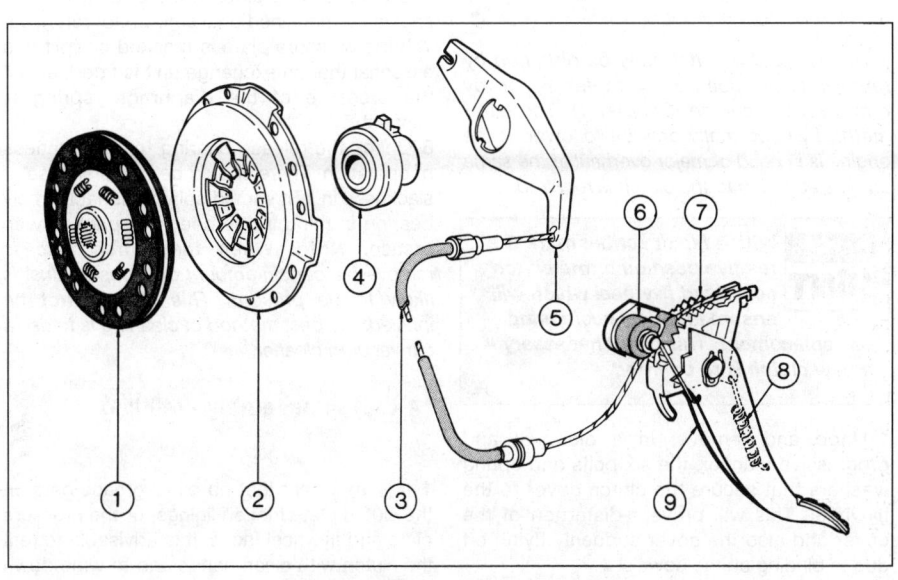

Fig. 5.1 Clutch components (Sec 1)

1 Clutch disc (driven plate)
2 Pressure plate
3 Clutch cable
4 Release bearing
5 Release lever
6 Pawl
7 Toothed segment
8 Tension spring
9 Clutch pedal

release fork and bearing which are in permanent contact with the release fingers on the pressure plate assembly. There should therefore never be any free play at the release fork. Wear of the friction material in the clutch is adjusted out automatically by means of the self adjuster mounted on the clutch pedal.

Depressing the clutch pedal actuates the clutch release arm by means of the cable. The release arm pushes the release bearing forwards to bear against the release fingers so moving the centre of the diaphragm spring inwards. The spring is sandwiched between two annular rings which act as fulcrum points. As the centre of the spring is pushed in, the outside of the spring is pushed out, so moving the pressure plate backwards and disengaging the pressure plate from the clutch disc.

When the clutch pedal is released the diaphragm spring forces the pressure plate into contact with the high friction linings on the clutch disc and at the same time pushes the clutch disc a fraction of an inch forwards on its splines so engaging the clutch disc with the flywheel. The clutch disc is now firmly sandwiched between the pressure plate and the flywheel so the drive is taken up.

The self adjusting mechanism consists of a toothed segment, pawl and tension springs, mounted on the clutch pedal, with the clutch operating cable attached to the toothed segment. As free play develops on the operating cable due to wear of the friction linings or stretching of the cable, the tension spring causes the toothed segment to rotate slightly thus taking up any free play as it develops. The toothed segment is prevented from rotating in the opposite direction by the action of the pawl.

2 Clutch assembly - removal

Access to the clutch may be obtained by removing the gearbox (Chapter 6) or by removing the engine (Chapter 1). The latter method will normally only be followed if the engine is in need of major overhaul at the same time as attention to the clutch is required.

HAYNES HiNT *With a file or scriber mark the relative position of the clutch cover and flywheel which will ensure identical positioning on replacement. This is not necessary if a new clutch is to be fitted.*

1 Undo and remove, in a diagonal and progressive manner, the six bolts and spring washers that secure the clutch cover to the flywheel. This will prevent distortion of the cover and also the cover suddenly flying off due to binding on the dowels.
2 With all the bolts removed lift the clutch assembly from the locating dowels. Note which way round the friction plate is fitted and lift it from the clutch cover.

3 Clutch - dismantling and inspection

1 It is not practical to dismantle the pressure plate assembly and the term "dismantling" is usually used for simply fitting a new clutch friction plate.
2 If a new clutch disc is being fitted it is a false economy not to renew the release bearing at the same time. This will preclude having to replace it at a later date when wear on the clutch linings is still very small.
3 If the pressure plate assembly requires renewal an exchange unit must be purchased. This will have been accurately set up and balanced to very fine limits.
4 Examine the clutch disc friction linings for wear and loose rivets and the disc for rim distortion, cracks, broken hub springs, and worn splines. The surface of the friction linings may be highly glazed, but as long as the clutch material pattern can be clearly seen this is satisfactory. Compare the amount of lining wear with a new clutch disc at the stores in your local garage. If worn the friction plate must be renewed.
5 It is always best to renew the clutch driven plate as an assembly to preclude further trouble, but, if it is wished to merely renew the linings, the rivets should be drilled out and not knocked out with a punch. The manufacturers do not advise that only the linings be renewed.

HAYNES HiNT *It is far more satisfactory to renew the driven plate complete rather than to try and economise by only fitting new friction linings.*

6 Check the machined faces of the flywheel and the pressure plate. If either is grooved it should be machined until smooth, or renewed.
7 If the pressure plate is cracked or split it is essential that an exchange unit is fitted, also if the pressure of the diaphragm spring is suspect.
8 Check the release bearing for smoothness of operation. There should be no harshness or slackness in it. It should spin reasonably freely bearing in mind it has been pre-packed with grease. **Note:** *When the clutch disc is removed, a certain amount of asbestos dust is likely to be present: This* **should not** *be inhaled: the best method of cleaning is to use a car vacuum cleaner.*

4 Clutch assembly - refitting

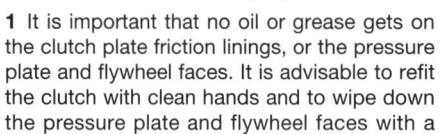

1 It is important that no oil or grease gets on the clutch plate friction linings, or the pressure plate and flywheel faces. It is advisable to refit the clutch with clean hands and to wipe down the pressure plate and flywheel faces with a clean rag before assembly begins.
2 Place the clutch plate against the flywheel, ensuring that it is the correct way round. The projecting torsion spring plate should be furthest from the flywheel.
3 Refit the clutch cover assembly loosely on the dowels. Refit the six bolts and spring washers and tighten them finger-tight so that the clutch plate is gripped but can still be moved.
4 The clutch disc must now be centralised so that when the engine and gearbox are mated, the gearbox first motion shaft splines will pass through the splines in the centre of the driven plate.
5 Centralisation can be carried out quite easily by inserting a round bar or long screwdriver through the hole in the centre of the clutch, so that the end of the bar rests in the small hole in the end of the crankshaft containing the spigot bush. Ideally an old Ford first motion shaft should be used (Fig. 5.2).

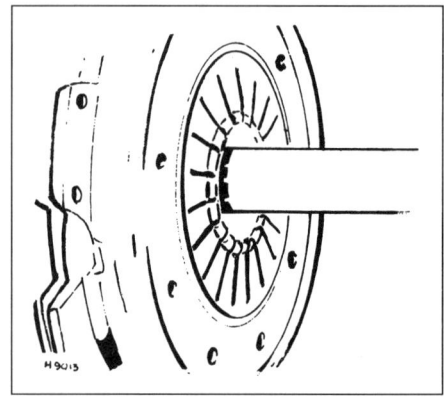

Fig. 5.2 Centralising the clutch disc (Sec 4)

6 Using the first motion shaft spigot bush as a fulcrum, moving the bar sideways or up and down will move the clutch disc in whichever direction is necessary to achieve centralisation.
7 Centralisation is easily judged by removing the bar and viewing the driven plate hub in relation to the hole in the centre of the clutch cover plate diaphragm spring. When the hub appears exactly in the centre of the hole all is correct. Alternatively the first motion shaft will fit the bush and centre of the clutch hub exactly, obviating the need for visual alignment.
8 Tighten the clutch bolts firmly in a diagonal sequence to ensure that the cover plate is pulled down evenly and without distortion of the flange. Finally tighten the bolts to the specified torque, see Specifications at the beginning of this Chapter.

5 Clutch cable - removal and refitting

1 Open the bonnet and for safety reasons disconnect the battery.
2 Apply the handbrake, chock the rear wheels and jack-up the front of the car supporting it firmly on axle-stands.
3 Ease off the release lever rubber gaiter from the side of the clutch housing.

Clutch 5•3

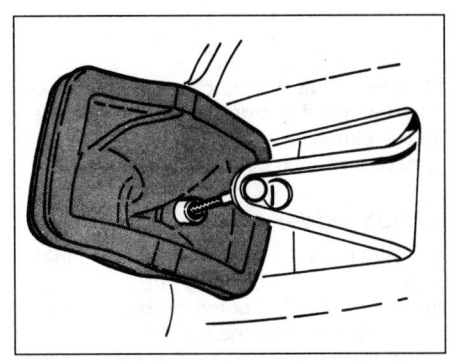

Fig. 5.3 Clutch release lever, gaiter and cable (Sec 5)

4 Hold the release lever arm to prevent it disengaging internally, pull the exposed portion of cable downwards and hold it with a pair of pliers to prevent it slipping back. (Fig. 5.5).
5 It will now be possible to lift the cable ball end from the slotted end of the release lever.
6 Remove the rubber gaiter and withdraw the cable from its fitting in the clutch housing.
7 Remove the driver's side lower trim panel from the facia.
8 Disconnect the other end of the cable from the toothed segment of the self adjusting mechanism on the pedal.
9 Withdraw the cable from its mounting and into the engine compartment.
10 Refitting is a straightforward reversal of the removal sequence.

6 Clutch release bearing - removal and refitting

1 With the gearbox and engine separated to provide access to the clutch, attention can be given to the release bearing located in the bellhousing, over the input shaft.
2 The release bearing is a relatively inexpensive but important component and unless it is nearly new it is a mistake not to renew it during an overhaul of the clutch.
3 The release bearing and arm can be withdrawn from the clutch housing.
4 To free the bearing from the release arm, simply rotate the bearing through 90° and remove. Note which way round the bearing is fitted.
5 Refitting is a straightforward reversal of removal.

7 Clutch pedal and self adjusting mechanism - removal and refitting

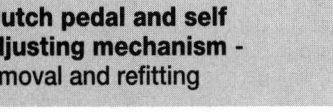

1 Release the clutch operating cable from the pedal, refer to Section 5.
2 Remove the spring clip and washer from the clutch pedal spindle.
3 Slide the clutch pedal and self adjusting mechanism complete, sideways off the spindle.
4 Remove the two spindle bushes from the pedal and lift out the toothed segment and tensioning spring.
5 Remove one clip from the pin locating the pawl to the pedal.
6 Pull the pin out of the pedal holding the pawl and spring at the same time.
7 Refitting is a reverse of the removal sequence bearing in mind the following points.
8 Lubricate the pivot pin before reassembly.
9 When refitting the toothed segment to the pedal, rotate the segment until the pawl is resting on the untoothed end. (Fig. 5.8) The first application of the clutch pedal will then automatically adjust the clutch free movement.

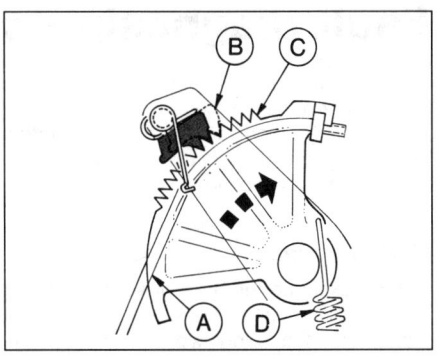

Fig. 5.4 Clutch self adjusting mechanism components (Sec 7)

A Clutch cable   B Pawl
C Toothed segment   D Tension spring

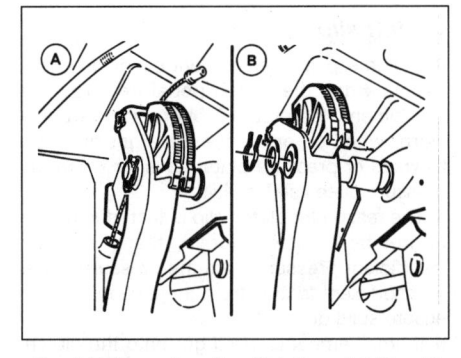

Fig. 5.5 Removing the clutch pedal (Sec 7)

A Disconnect clutch cable
B Remove clutch pedal

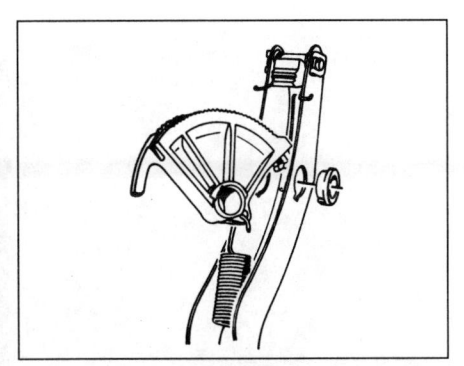

Fig. 5.6 Removing the toothed segment (Sec 7)

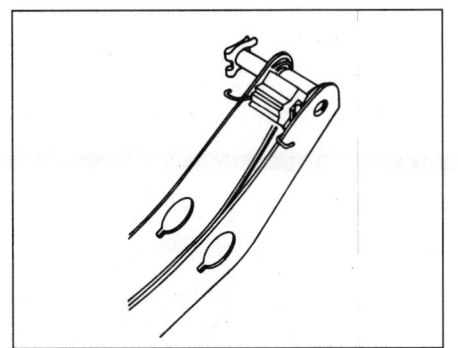

Fig. 5.7 Removing the pawl assembly (Sec 7)

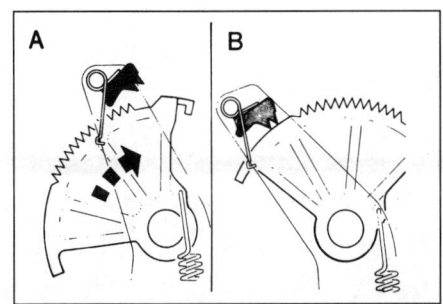

Fig. 5.8 Setting the toothed segment prior to refitting (Sec 7)

A Release pawl and swing segment
B Rest pawl on flat section of segment

Fault diagnosis – clutch

There are four main faults to which the clutch and release mechanism are prone. They may occur by themselves or in conjunction with any of the other faults. They are clutch squeal, slip, spin and judder.

Clutch squeal

1 If on taking up the drive or when changing gear, the clutch squeals, this is a good indication of a badly worn clutch release bearing.

2 As well as regular wear due to normal use, wear of the clutch release bearing is much accentuated if the clutch is ridden, or held down for long periods in gear, with the engine running.

To minimise wear of the release bearing the car should always be taken out of gear at traffic lights and for similar holdups.

Clutch slip

3 Clutch slip is a self-evident condition which occurs when the clutch pedal free-travel is insufficient, the clutch friction plate is badly worn, when oil or grease have got onto the flywheel or pressure plate faces, or when the pressure plate itself is faulty.

4 The reason for clutch slip is that, due to one of the faults listed above, there is either insufficient pressure from the pressure plate, or insufficient friction from the friction plate to ensure solid drive.

5 If small amounts of oil get onto the clutch, they will be burnt off under the heat of clutch engagement, and in the process, gradually darken the linings. Excessive oil on the clutch will burn off leaving a carbon deposit which can cause quite bad slip, or fierceness, spin and judder.

6 If clutch slip is suspected, and conformation of this condition is required, there are several tests which can be made.

7 With the engine in second or third gear and pulling lightly up a moderate incline sudden depression of the accelerator pedal may cause the engine to increase its speed without any increase in road speed. Easing off on the accelerator will then give a definite drop in engine speed without the car slowing.

8 In extreme cases of clutch slip the engine will race under normal acceleration conditions.

9 If slip is due to oil or grease contamination of the linings, the cause of the leak must be traced and cured and the clutch driven plate renewed.

Clutch spin

10 Clutch spin is a condition which occurs when the release arm travel is excessive, there is an obstruction in the clutch either on the primary grease splines or in the operating lever itself, or the oil may have partially burnt off the clutch linings and have left a resinous deposit which is causing the clutch disc to stick to the pressure plate or flywheel.

11 The reason for clutch spin is that due to any, or a combination of, the faults just listed, the clutch pressure plate is not completely freeing from the centre plate even with the clutch pedal fully depressed.

12 If clutch spin is suspected, the condition can be confirmed by extreme difficulty in engaging first gear from rest, difficulty in changing gear, and very sudden take-up of the clutch drive at the fully depressed end of the clutch pedal travel as the clutch is released.

13 Check that the clutch cable is correctly adjusted and if in order the fault lies internally in the clutch. It will then be necessary to remove the clutch for examination.

Clutch judder

14 Clutch judder is a self evident condition which occurs when the gearbox or engine mountings are loose or too flexible, when there is oil on the faces of the clutch friction plate or when the clutch pressure plate has been incorrectly adjusted during assembly.

15 The reason for clutch judder is that due to one of the faults just listed, the clutch pressure plate is not freeing smoothly from the friction disc, and is snatching.

16 Clutch judder normally occurs when the clutch pedal is released in first gear or reverse gear, and the whole car shudders as it moves backwards or forwards.

Variation of clutch pedal operation

17 This fault may be noticeable on models produced previous to May 1981 which were fitted with the original clutch pedal adjuster pawl spring. Models produced from this date are fitted with a stronger spring to prevent this fault, which is caused by the adjuster quadrant rotating by up to one-third of a tooth when the engine speed is increased. What happens is that the clutch diaphragm release fingers are moved back at higher engine speeds, causing release bearing movement, and this is then transmitted through the cable to the pedal adjuster mechanism. Fitment of an additional pawl spring (Ford part number 77FB-7L585-AB), or the later strengthened type spring, will prevent this action.

Chapter 6
Manual and automatic transmission

For modifications, and information applicable to later models, see Supplement at end of manual

Contents

Automatic transmission - fluid level checking 6
Automatic transmission - general description 5
Automatic transmission - removal and refitting 7
Automatic transmission selector mechanism - adjustment 9
Automatic transmission selector mechanism - removal,
  overhaul and refitting 8
Automatic transmission vacuum diaphragm unit - removal
  and refitting 12
Downshift cable - removal, refitting and adjustment 10
Fault diagnosis - automatic transmission ....See end of Chapter
Fault diagnosis - manual gearbox ....See end of Chapter
Gearbox (B type) - removal and refitting 2
Gearbox (E type) - removal and refitting 3
Manual gearbox - general description and maintenance 1
Starter inhibitor/reverse lamp switch - removal and refitting 11

Degrees of difficulty

| **Easy,** suitable for novice with little experience | **Fairly easy,** suitable for beginner with some experience | **Fairly difficult,** suitable for competent DIY mechanic | **Difficult,** suitable for experienced DIY mechanic | **Very difficult,** suitable for expert DIY or professional |

Specifications

Manual gearbox (B type)

Type .... Four forward (all synchromesh) and one reverse. Floor mounted gear lever

Gear ratios

	Standard	Heavy duty
First	3.65: 1	3.36: 1
Second	1.97: 1	1.81: 1
Third	1.37: 1	1.26: 1
Fourth	1.00: 1	1.00 :1
Reverse	3.66: 1	3.36: 1

Layshaft

Laygear cluster endfloat .... 0.006 to 0.018 in (0.15 to 0.45 mm)
Diameter of layshaft:
  Standard .... 0.68 in (17.32 mm)
  Heavy duty .... 0.76 in (19.31 mm)

Lubricant

Lubricant capacity .... 3.0 pint (1.7 litre)
Lubricant type/specification .... Gear oil, viscosity SAE 80EP, to Ford spec SQM-2C-9008-A

Manual gearbox (E type)

Type .... Four forward (all synchromesh) and one reverse. Floor mounted gear lever

Gear ratios

First 3.16: 1
Second 1.94: 1
Third 1.41: 1
Fourth 1.00: 1
Reverse 3.35: 1

Layshaft

Laygear cluster endfloat .... 0.006 to 0.018 in (0.15 to 0.45 mm)
Diameter of layshaft .... 0.76 in (19.33 mm)

6•2 Manual and automatic transmission

Lubricant
Lubricant capacity	3.52 pint (2.0 litre)
Lubricant type/specification	Gear oil, viscosity SAE 80EP, to Ford spec SQM-2C-9008-A

Automatic transmission
Type
Ford C3

Torque converter
Type	Trilock (hydraulic)
Converter ratio:	
2.0 litre in-line ohc engine	2.35: 1
2.3 litre V6 engine	2.15: 1
2.8 litre V6 engine	2.22/2.35: 1

Transmission ratios
First	2.47: 1
Second	1.47: 1
Third	1.00: 1
Reverse	2.11: 1

Lubricant
Lubricant capacity (approx):	
2.0 litre in-line ohc engine	11 pint (6.3 litre)
2.3 litre V6 engine	11 pint (6.3 litre)
2.8 litre V6 engine	13 pint (7.4 litre)
Lubricant type:	
Early models (black dipstick)	ATF to Ford spec SQM-2C-9007-AA
Later models (red dipstick)	ATF to Ford spec SQM-2C-9010-A

Torque wrench settings

	lbf ft	Nm

Manual gearbox (B type)
	lbf ft	Nm
Clutch housing to gearbox	38	53
Clutch housing to engine	32	44
Input shaft bearing retainer	7	10
Extension housing to gearbox	32	44
Gearbox top cover	7	10
Rear crossmember to body	16	22
Rear mounting to crossmember	13	18
Rear mounting to gearbox	38	53

Manual gearbox (E type)
	lbf ft	Nm
Clutch housing to gearbox	32	44
Clutch housing to engine	25	34
Input shaft bearing retainer	14	19
Mainshaft nut	27	38
Extension housing to gearbox	42	58
Selector shaft bracket to extension housing	13	18
Selector housing cover	7	10
Gearbox top cover	7	10
Rear crossmember to body	16	22
Rear mounting to crossmember	13	18
Rear mounting to gearbox	38	53
Gear lever cap (fitted with O-ring modification)	15	20

Automatic transmission
	lbf ft	Nm
Converter housing to engine	22 to 27	30 to 37
Converter drain plug	20 to 29	27 to 40
Drive plate to converter	26 to 30	36 to 41
Oil sump bolts	12 to 17	16 to 24
Fluid line to connector	7 to 10	9 to 14
Connector to transmission	10 to 15	14 to 20
Oil cooler line to connector	12 to 15	16 to 20
Downshift cable bracket	12 to 17	16 to 24
Downshift lever nut:		
Outer	7 to 11	10 to 15
Inner	30 to 40	41 to 54
Inhibitor switch	12 to15	16 to 20

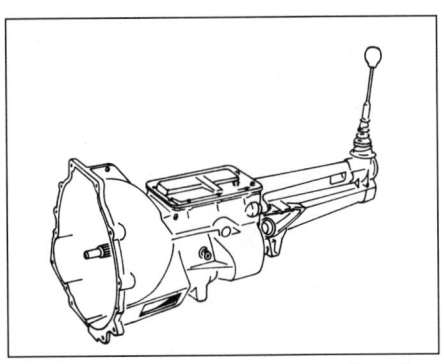

Manual and automatic transmission 6•3

Fig. 6.1 B type manual gearbox (Sec 1)

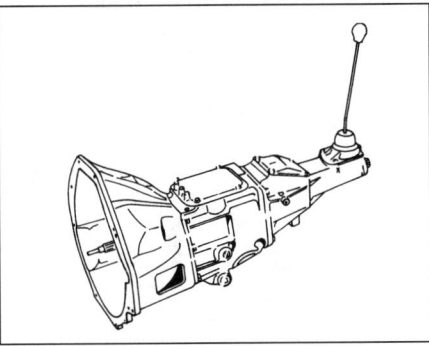

Fig. 6.2 E type manual gearbox (Sec 1)

1 Manual gearbox - general description and maintenance

One of two types of manual gearbox are fitted to the Granada models covered by this manual. These are the B type gearbox as fitted to 2.0 litre in-line ohc engine models and 2.3 litre V6 engine models and the E type gearbox used exclusively on the 2.8 litre V6 engine models.

Both gearbox types are equipped with four forward and one reverse gear. All forward gears are engaged through baulk ring synchromesh units to obtain smooth silent gear changes. All forward gears on the mainshaft are in constant mesh with their corresponding gears on the layshaft and are helically cut to achieve quiet running.

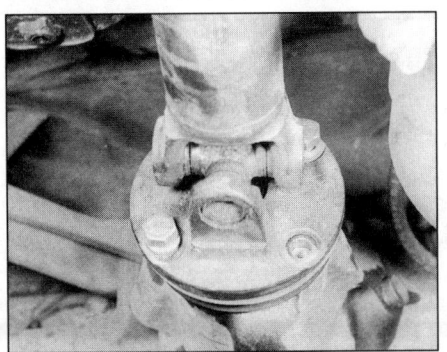

The layshaft reverse gear has straight cut spur teeth and drives the 1st/2nd gear synchroniser hub on the mainshaft through an interposed sliding idler gear.

Gears are engaged either by a single selector rail and forks on the B type gearbox or by a three rail system and forks on the E type gearbox.

No routine oil changes are required. At the specified intervals wipe around the plug on the left side of the gearbox, remove it then, if necessary, add oil to bring the level up to the filler hole. Refit the plug securely on completion. Note: *Foaming of gearbox oil can occur if the vehicle has been running prior to the oil level being checked. If this has occurred, allow the vehicle to stand for a few hours before any checking or topping-up is done.*

2 Gearbox (B type) - removal and refitting

The gearbox can be removed by separating it from the rear of the engine at the bellhousing, then lowering and removing it from under the car.

1 If a hoist or an inspection pit is not available, jack-up the car and support it on axle-stands or solid blocks of wood. Make sure it is raised high enough to provide a good working height under the car.
2 Disconnect the battery earth terminal.
3 Refer to Chapter 12 and remove the centre console.
4 Pull up the gear lever inner gaiter and using a screwdriver bend back the lock tabs on the gear lever locking ball cap. Lift the gear lever from the gearbox.
5 From under the car mark the mating flanges of the propeller shaft and final drive so that they may be reconnected in their original positions, then undo and remove the four attaching bolts (photo).
6 Remove the two bolts securing the propeller shaft centre bearing assembly to the body then lower the propeller shaft at the rear and draw it rearwards to detach it from the gearbox.

HAYNES HiNT *Secure a polythene bag over the end of the gearbox extension housing to prevent the oil draining out and also prevent the ingress of dirt.*

7 Make a note of the connections to the starter motor terminals and then disconnect the wiring.
8 Remove the bolts securing the starter motor to the housing and lift away the starter motor (photo).
9 Undo and remove the nuts securing the exhaust pipe flanges to the manifolds and allow the exhaust pipes to hang on their mountings.
10 On certain V6 models there is an additional clamp on the right-hand exhaust pipe which must be removed.
11 Disconnect the wiring from the reversing light switch located on the side of the gearbox extension housing (photo).
12 Using a pair of circlip pliers, remove the circlip retaining the speedometer drive cable end to the gearbox extension housing (Fig. 6.3). Withdraw the cable from the extension housing.
13 Pull back the rubber gaiter over the clutch operating arm. Firmly pull the clutch cable downwards by hand in front of the operating arm and hold it from slipping back with a pair of pliers. Unhook the cable end from the slot in the operating arm and withdraw the complete cable from the locating hole in the clutch housing.

2.5 Disconnect the propeller shaft from the final drive flange

2.8 Remove the starter motor from the bellhousing

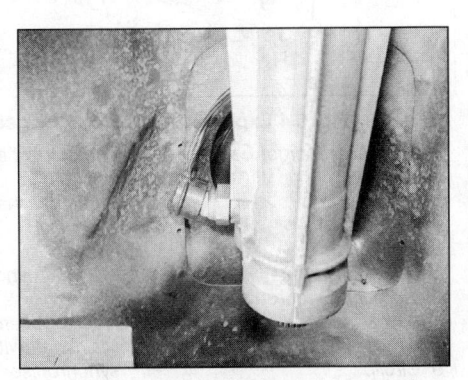

2.11 Pull the connector from the reversing light switch

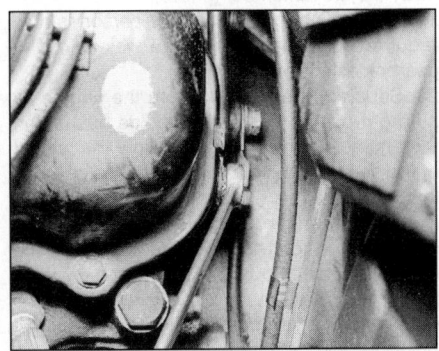

2.16 Location of earth strap on top bellhousing bolt (V6 engine)

6•4 Manual and automatic transmission

2.17 Remove the rear mounting to gearbox housing securing bolts . . .

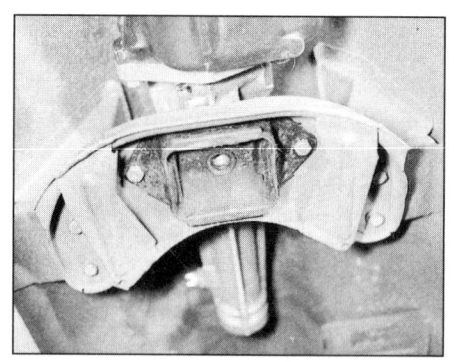

2.18 . . . and then the crossmember to body bolts (arrowed)

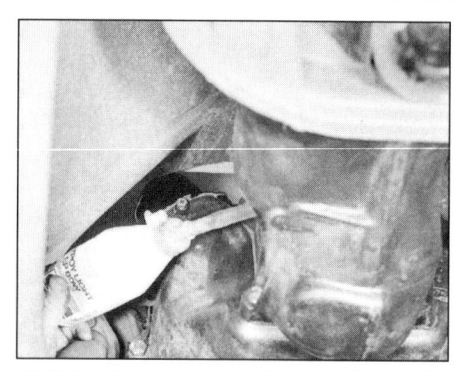

2.21 Don't forget to refill the gearbox with oil

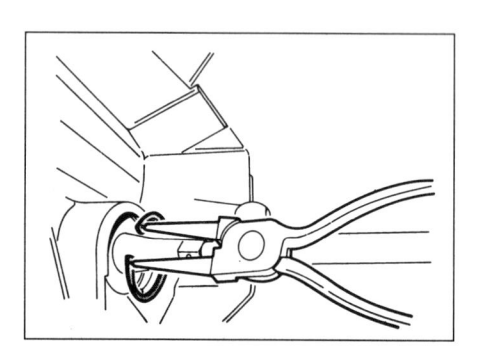

Fig. 6.3 Removing the speedometer cable retaining circlip from manual gearbox (Sec 2)

14 Remove the two bolts securing the clutch housing lower cover plate. On in-line ohc engines also remove the bolt securing the steady bar and swing the bar clear of the gearbox.
15 Suitably support the weight of the gearbox with a jack or axle-stand. Insert a block of wood between the rear of the engine and the bulkhead. This will prevent the rear of the engine dropping when the gearbox extension housing mounting is removed.
16 Undo and remove the bolts securing the bellhousing to the rear of the engine. Note the earth strap on V6 engines (photo).
17 Remove the bolt securing the rubber mounting to the gearbox extension housing (photo).
18 Remove the bolts securing the gearbox support crossmember to the body and lift away the crossmember (photo).
19 The assistance of a second person is now required to help in taking the weight of the gearbox as it is removed.
20 Separate the gearbox from the engine by pulling it rearwards from the guide bushes on the engine. Take care not to allow the weight of the gearbox to put any strain on the gearbox input shaft as it is easily bent. As the gearbox is moved rearwards it will be necessary to lower the rear end to give clearance from the underside of the body.
21 Refitting the gearbox is the reverse of the removal procedure but the following additional points should be noted:
 a) Lightly grease the gearbox input shaft splines, but take care not to over-grease or the clutch disc could be contaminated
 b) Ensure that the engine rear plate is correctly located on the bellhousing guide bushes
 c) Tighten the bellhousing to engine securing bolts evenly to the specified torque
 d) Fill the gearbox with the specified gear oil (photo)

3 Gearbox (E type) - removal and refitting

1 The procedure is the same as detailed in Section 2 for removal and refitting of the B type gearbox but on vehicles equipped with a fuel injection system, the following variations must be noted.

2 Lowering of the gearbox prior to and during its removal will obviously mean that the rear of the engine tilts downward. To avoid strain on the connecting components of the fuel system, disconnect the following:
 Fuel lines from fuel distributor to injectors
 Trunking from throttle plate housing
 Fuel line from fuel distributor to start valve
 Fuel line from fuel distributor to warm-up regulator

4 Gearbox - overhaul

Overhauling a manual transmission unit is a difficult and involved job for the DIY home mechanic. In addition to dismantling and reassembling many small parts, clearances must be precisely measured and, if necessary,

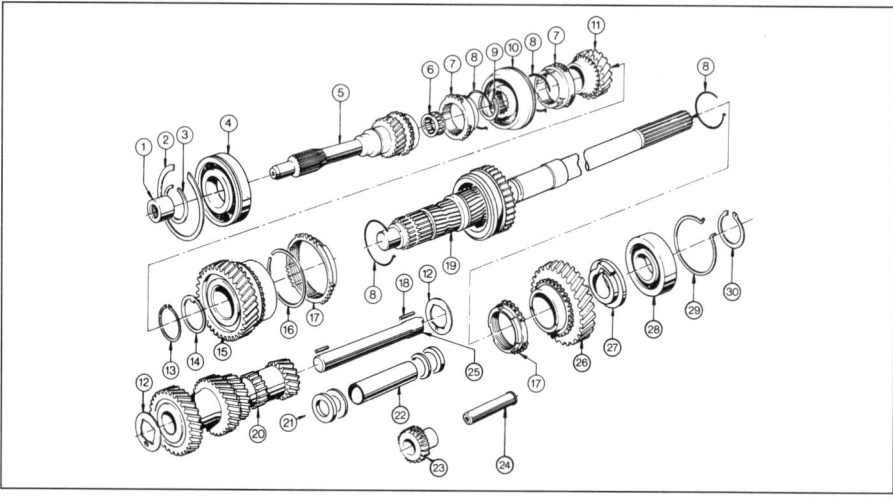

Fig. 6.4 Exploded view of B type gearbox internal components (Sec 4)

1 Input shaft spigot bearing
2 Circlip
3 Circlip
4 Bearing
5 Input shaft
6 Needle bearing
7 3/4 gear synchroniser baulk ring
8 Synchroniser spring clip
9 Circlip
10 3/4 gear synchroniser
11 3rd gear
12 Thrust washer
13 Circlip
14 Thrust washer
15 2nd gear
16 Circlip
17 1/2 gear synchroniser baulk ring
18 Needle rollers (19 or 21)
19 Mainshaft with 1/2 gear synchroniser and reverse
20 Laygear
21 Spacer shims
22 Spacer tube (standard gearbox)
23 Reverse idler gear
24 Idler shaft
25 Layshaft
26 1st gear
27 Oil scoop ring
28 Bearing
29 Circlip
30 Circlip

Manual and automatic transmission 6•5

1 Main drive gear bearing retainer
2 O-ring
3 Oil seal
4 Blanking plug
5 Gearbox top cover
6 Gasket
7 3/4 gear selector fork
8 Reverse relay lever
9 Reverse selector boss
10 Spring pin
11 Lock plate
12 1/2 gear selector fork
13 Selector rail
14 Blanking plug
15 Oil seal
16 Gasket
17 Extension housing
18 Gear knob
19 Locknut
20 Gear lever assembly
21 Gear case
22 Side plug
23 Spring
24 Locking plunger, selector rail
25 Oil filler plug
26 Relay lever pin
27 Speedometer driver cover
28 Speedometer drive gear
29 Oil seal
30 Circlip
31 Bush, extension housing
32 Oil seal
33 Extension housing, rear cover

A Rubber strap
B Damping bush
C Tab washer
D Metal cup

Fig. 6.5 Exploded view of B type gearbox casing and selector mechanism. Inset shows gear lever assembly fitted to later models to prevent rattles. Some gearboxes have a detachable bellhousing (Sec 4)

1 Main drive gear bearing
2 Gasket
3 Detent balls with springs
4 Transmission case gasket
5 Transmission case cover
6 Speedometer pinion
7 Dowel
8 Extension housing cover gasket
9 Extension housing top cover
10 Selector housing gasket
11 Selector housing cover
12 Cap bolt with lock plate
13 Spring
14 Detent ball
15 Plunger
16 Spring
17 Reversing light switch
18 Gear lever assembly
19 Oil seal
20 Extension housing bush
21 Selector shaft bearing support
22 Reverse relay arm roll pin
23 Bearing support
24 Extension housing gasket
25 Selector rail with 1st/2nd gear selector fork
26 Filler plug
27 Drive gear, bearing retainer oil seal
28 Circlip
29 Plunger
30 Selector rail with 3rd/4th gear selector fork
31 Plunger
32 Selector rail with reverse gear selector fork
33 Selector shaft
34 Selector finger
35 Reverse gear relay lever

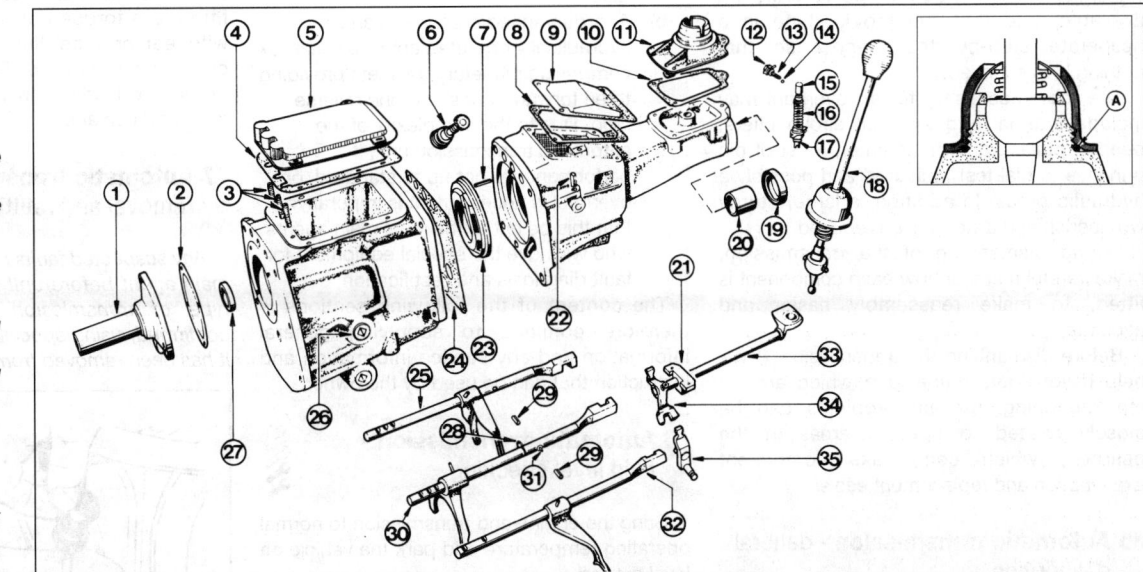

Fig. 6.6 Exploded view of E type gearbox casing and selector mechanism. Inset shows O-ring (A) which can be fitted to prevent gear lever vibration (Sec 4)

6•6 Manual and automatic transmission

1 Circlip
2 Circlip
3 Grooved ball bearing
4 Input shaft
5 Needle roller bearing
6 Circlip
7 3rd/4th gear synchroniser baulk ring
8 3rd/4th gear synchroniser hub/sleeve assembly
9 3rd gear
10 Transmission mainshaft
11 Speedometer drive locking ball
12 2nd gear
13 1st/2nd gear synchroniser baulk ring
14 1st/2nd gear synchroniser with reverse
15 Circlip
16 1st gear
17 Oil scoop ring
18 Intermediate bearing support
19 Bearing
20 Spacer
21 Speedometer worm drive gear
22 Lock plate
23 Mainshaft nut
24 Layshaft thrust washer
25 Shim
26 Needle rollers (22)
27 Laygear
28 Layshaft thrust washer
29 Layshaft
30 Reverse idler gear
31 Idler shaft

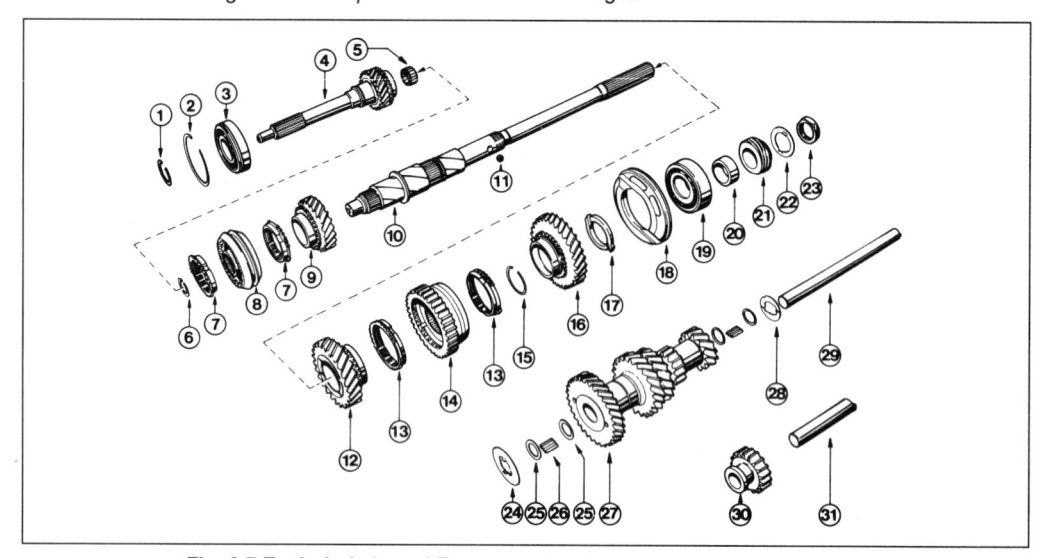

Fig. 6.7 Exploded view of E type gearbox internal components (Sec 4)

changed by selecting shims and spacers. Internal transmission components are also often difficult to obtain, and in many instances, are extremely expensive. Because of this, if the transmission develops a fault or becomes noisy, the best course of action is to have the unit overhauled by a specialist repairer, or to obtain an exchange reconditioned unit.

Nevertheless, it is not impossible for the more experienced mechanic to overhaul the transmission, provided the special tools are available, and that the job is done in a deliberate step-by-step manner so that nothing is overlooked.

The tools necessary for an overhaul may include internal and external circlip pliers, bearing pullers, a slide hammer, a set of pin punches, a dial test indicator, and possibly a hydraulic press. In addition, a large, sturdy workbench and a vice will be required.

During dismantling of the transmission, make careful notes of how each component is fitted, to make reassembly easier and accurate.

Before dismantling the transmission, it will help if you have some idea which area is malfunctioning. Certain problems can be closely related to specific areas in the gearbox, which can make component examination and replacement easier.

5 Automatic transmission - general description

The automatic transmission takes the place of the clutch and gearbox, which are, of course, mounted behind the engine.

The unit has a large aluminium content which helps to reduce its overall weight and it is of compact dimensions. A transmission oil cooler is fitted as standard and ensure cooler operation of the transmission under trailer towing conditions. A vacuum connection to the inlet manifold provides smoother and more consistent downshifts under load than is the case with units not incorporating this facility.

The system comprises two main components:

a) A three element hydrokinetic torque converter coupling capable of torque multiplication at an infinitely variable ratio
b) A torque/speed responsive and hydraulically operated epicyclic gearbox comprising planetary gearsets providing three forward ratios and one reverse ratio. Due to the complexity of the automatic transmission unit, if performance is not up to standard, or overhaul is necessary, it is imperative that this be left to the local main agents who will have the special equipment for fault diagnosis and rectification

The content of the following sections is therefore confined to supplying general information and any service information and instruction that can be used by the owner.

6 Automatic transmission - fluid level checking

1 Bring the engine and transmission to normal operating temperature and park the vehicle on level ground.
2 With the engine idling, move the transmission selector lever through all positions three times, then engage P and wait for at least one minute.
3 With the engine still idling, withdraw the transmission dipstick. Wipe it clean, reinsert it, withdraw it for the second time and read off the level.
4 If necessary top-up with the specified oil, through the dipstick tube, to the MAX mark.
5 Always keep the exterior of the transmission unit clean and free from mud and oil and the air intake grilles must not be obstructed.
6 Two types of automatic transmission fluid are available (see Specifications for type numbers) and it is important that they are not mixed. The colour of the dipstick for fluid level checking denotes which fluid type to use.
7 If a transmission requiring later type fluid is fitted to a torque converter previously used with earlier type fluid, or vice versa, the components should be flushed to avoid mixing of the two fluids. Consult your Ford dealer if this condition arises.

7 Automatic transmission - removal and refitting

Any suspected faults must be referred to the main agent before unit removal, as with this type of transmission the fault must be confirmed, using specialist equipment, before it has been removed from the car.

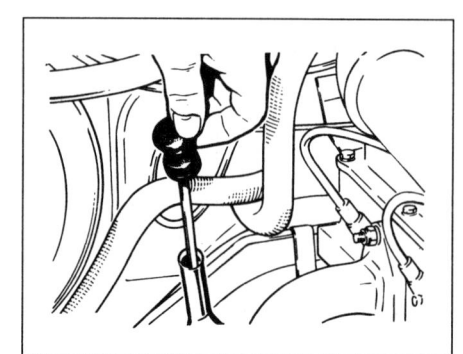

Fig. 6.8 Checking the automatic transmission fluid level (Sec 6)

Manual and automatic transmission 6•7

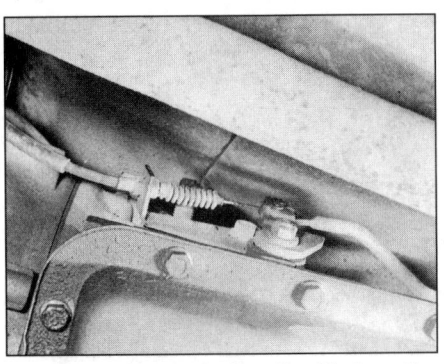

7.6 Remove the "kick-down" cable from the downshift lever

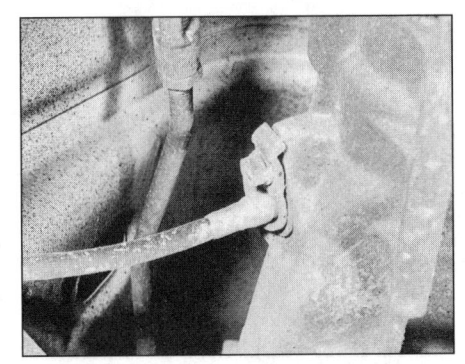

7.8 Remove the speedometer cable

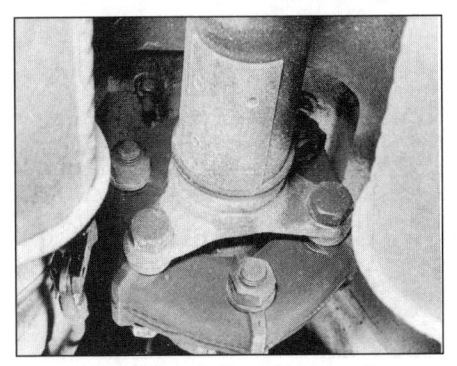

7.9 Propeller shaft rubber coupling

1 Open the engine compartment lid and place old blankets over the wings to prevent accidental scratching of the paintwork.
2 Undo and remove the battery earth connection nut and bolt from the battery terminal.
3 Remove the starter motor, refer to Chapter 10.
4 Position the vehicle on a ramp or over an inspection pit or, if these are not available, jack it up and support securely under the bodyframe.

> **HAYNES HiNT** *Make sure that there is sufficient clearance under the vehicle to permit withdrawal of the transmission.*

5 Unscrew and remove the upper bolts which secure the torque converter housing to the engine. One of these bolts secures the dipstick tube support bracket.
6 Disconnect the "kick-down" cable from the transmission downshift lever and bracket (photo).
7 Disconnect the plug from the reverse inhibitor switch.
8 Unscrew the lock plate bolt and remove the speedometer cable (photo).
9 Refer to Chapter 7, and remove the propeller shaft (photo).
10 Undo and remove the nuts that secure the exhaust downpipe(s) to the manifold(s). Ease the clamps from the studs, lower the

> **HAYNES HiNT** *To stop accidental dirt ingress, wrap some polythene around the end of the automatic transmission unit and secure with string or wire.*

downpipe(s) and collect the sealing ring(s). On certain V6 engines a further support clamp must be removed from the right-hand downpipe.
11 Remove the lower bolts from the torque converter cover plate.
12 Disconnect the oil cooler pipes from the transmission. Plug the pipe ends to prevent dirt ingress.
13 Remove the two selector rod spring clips and take out the selector rod (photos).
14 Disconnect the vacuum pipe from the vacuum diaphragm.
15 The torque converter should next be disconnected from the crankshaft driving plate. Rotate the crankshaft until each bolt may be seen through the starter motor aperture. Undo each bolt and turn one at a time until all the bolts are free (photo).
16 Place an additional jack under the automatic transmission unit and remove the five bolts securing the transmission support crossmember to the transmission and the underside of the body (photo).
17 On vehicles equipped with fuel injection, before the transmission is lowered, disconnect the following components to avoid distortion or strain.

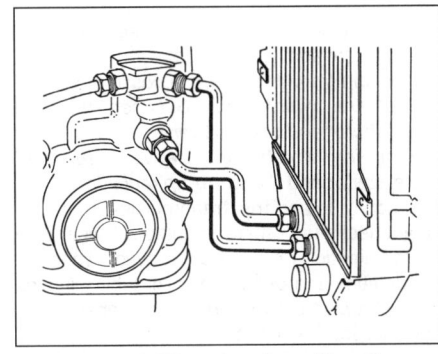

Fig. 6.9 Oil cooler pipes (Sec 7)

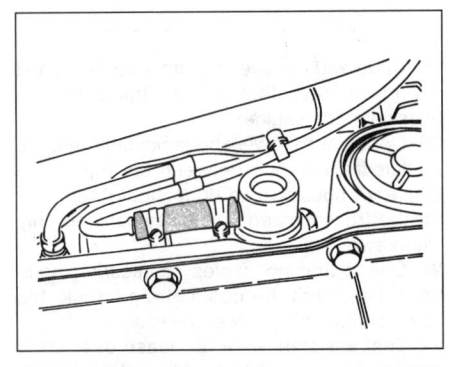

Fig. 6.10 Disconnect the vacuum pipe (Sec 7)

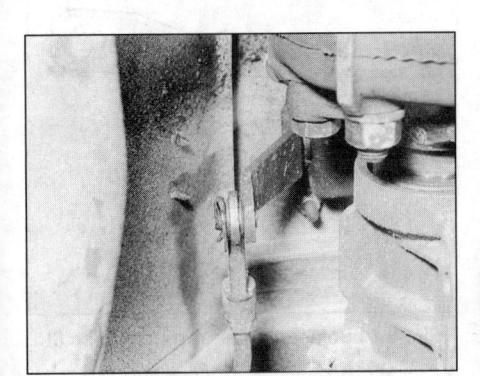

7.13a Selector rod, rear . . .

7.13b . . . and front securing spring clips

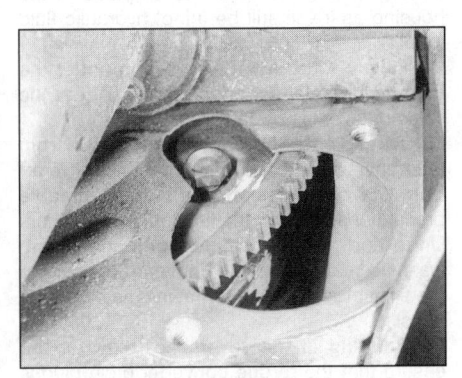

7.15 Torque converter to drive plate securing bolts

6•8 Manual and automatic transmission

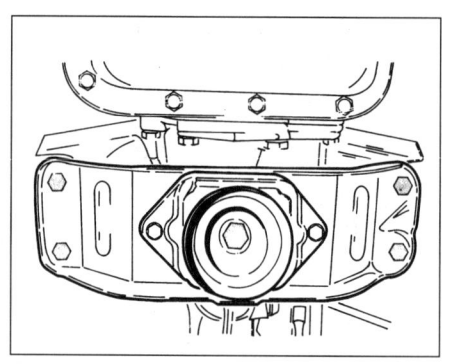

Fig. 6.11 Remove the support crossmember (Sec 7)

7.16 Remove the transmission support crossmember

7.23 Converter retained in place in transmission housing. Drain plug (arrowed)

Fuel pipes which run between the fuel distributor and the injectors
Trunking from throttle plate housing
Fuel line which runs between the fuel distributor and the start valve
Fuel line which runs between the fuel distributor and the warm-up regulator

18 Slowly lower the transmission unit and engine jacks until there is sufficient clearance for the dipstick tube to be removed. Withdraw the dipstick and pull the oil filler tube (dipstick tube) sharply from the side of the transmission unit. Collect the O-ring.

HAYNES HiNT *Plug the oil filler opening to prevent the ingress of dirt.*

19 Undo and remove the remaining bolts and spring washers that secure the converter housing to the engine.
20 Continue to lower the jacks until there is sufficient clearance between the top of the converter housing and underside of the floor for the transmission unit to be satisfactorily withdrawn.
21 Check that no cables or securing bolts have been left in position and tuck the speedometer cable out of the way.
22 The assistance of at least one other person is now required because of the weight of the complete unit.
23 Carefully pull the unit rearwards and, when possible, hold the converter in place in the housing as it will still be full of hydraulic fluid (photo).
24 Finally withdraw the unit from under the car and place on wooden blocks so that the selector lever is not damaged or bent.
25 To separate the converter housing from the transmission case first lift off the converter from the transmission unit, taking suitable precautions to catch the fluid upon separation. Undo and remove the six bolts and spring washers that secure the converter housing to the transmission case. Lift away the converter housing.
26 Refitting is the reverse of removal but ensure that the torque converter drain plug is in line with the hole in the driveplate. To check that the torque converter is positively

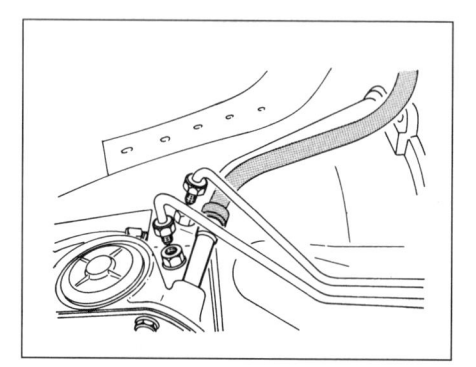

Fig. 6.12 Pull out the oil filler tube (Sec 7)
Oil cooler pipes disconnected

engaged, measure the distance A between the converter housing to engine mating face and the end of the stub shaft. This should be at least 0.4 in (10 mm) (Fig. 6.13).
27 Adjust the selector cable and inhibitor switch as described later in this Chapter.
28 Refill the transmission unit with the specified fluid before starting the engine and check the oil level as described in Section 6.

8 Automatic transmission selector mechanism - removal, overhaul and refitting

1 Prise off the selector lever escutcheon, and withdraw the illumination mounting from the selector lever.
2 Remove the spring clip and detach the selector rod from the selector lever.
3 If required, remove the spring clip and detach the selector rod from the transmission selector lever.
4 Remove the four bolts retaining the selector lever housing to the transmission tunnel and remove the housing.
5 Remove the rubber plug from the side of the selector lever housing, unscrew the nut and press the lower lever out of the housing.
6 Unscrew the locknut on the inhibitor cable and withdraw the pawl, spring and guide bush.
7 Remove the Allen screw from the T-handle

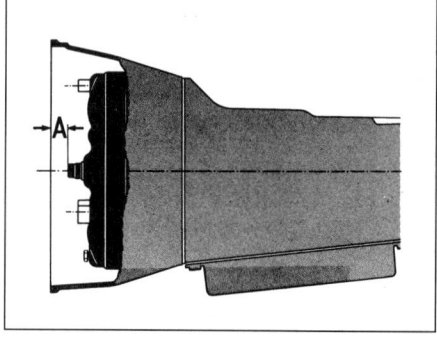

Fig. 6.13 Checking torque converter engagement with transmission oil pump (Sec 7)

A At least 0.4 in (10 mm)

and remove the handle. Remove the push button and spring.
8 Using a pin punch, drive out the retaining pin and remove the inhibitor mechanism and inhibitor cable.
9 Refitting is a reversal of the above procedure noting that the T-handle Allen screw is inserted from the front, so that the push button is nearest the driver.
10 Adjust the mechanism, refer to Section 9.

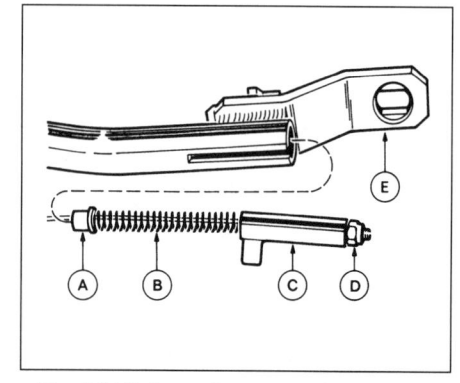

Fig. 6.14 Selector lever assembly (Sec 8)

A Guide bush
B Spring
C Selector pawl
D Lock nut
E Selector lever

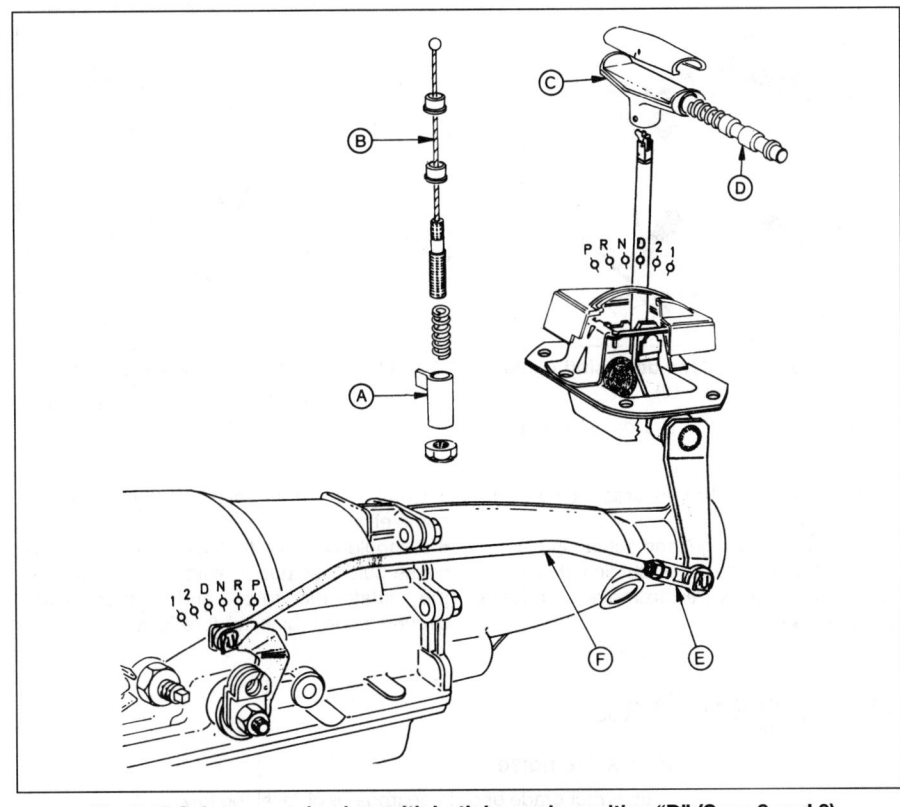

Manual and automatic transmission 6•9

Fig. 6.15 Selector mechanism with both levers in position "D" (Secs 8 and 9)

A Selector pawl
B Selector cable
C Selector lever handle
D Push button
E Adjusting link
F Selector rod

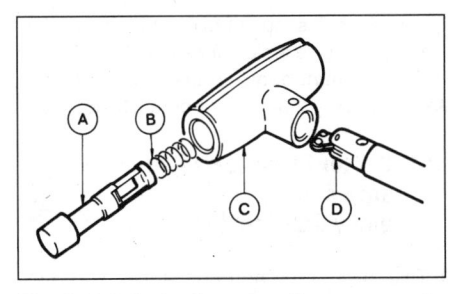

Fig. 6.16 Selector lever handle components (Sec 8)

A Inhibitor push button
B Spring
C Handle
D Selector lever

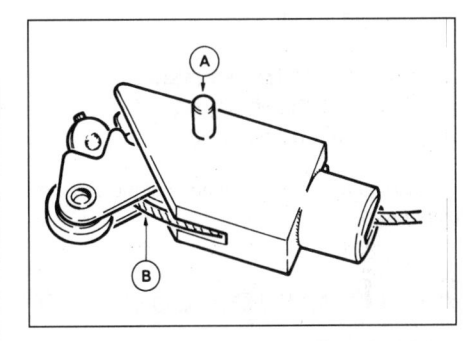

Fig. 6.17 Removing the inhibitor cable (Sec 8)

A Straight pin   B Inhibitor cable

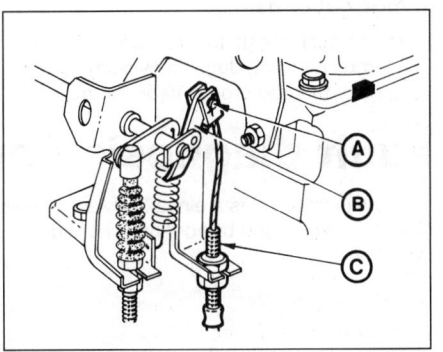

Fig. 6.19 Downshift cable removal (Sec 10)

A Pin   B Connecting lever
C Threaded sleeve

9 Automatic transmission selector mechanism - adjustment

1 Prise off the selector lever escutcheon, and from under the car, remove the plug in the side of the selector housing.
2 Adjust the inhibitor cable locknut X to give a dimension of 0.004 to 0.008 in (0.1 to 0.2 mm) (Fig. 6.18).
3 Refit the plug and escutcheon.
4 With the transmission shift lever and the manual selector lever in D, adjust the selector rod link until it can be reconnected without strain.

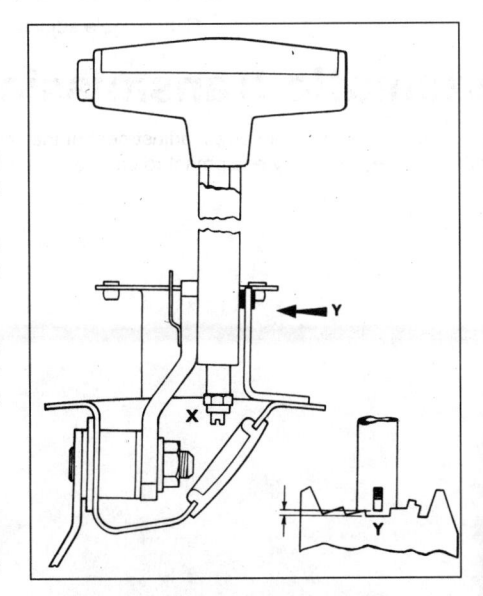

10 Downshift cable - removal, refitting and adjustment

1 Remove the split pin and disconnect the cable from the carburettor linkage (photo).
2 Slacken the adjusting nut from the mounting bracket, pull back the outer sheath and unhook the cable from the slot.
3 Unhook the cable from the transmission lever and bracket.
4 Refitting is a reversal of the above, but the inner nut at the upper end should be screwed on completely, and the outer nut only a few turns.

Fig. 6.18 Selector mechanism adjustment (Sec 9)

X Locknut   Y 0.04 to 0.08 in (0.1 to 0.2 mm)

5 Depress the accelerator pedal fully, and check that the throttle plate is fully open.
6 Using a screwdriver, lever the downshift cable lever upwards, pulling the inner cable fully upwards.

10.1 Downshift cable carburettor attachment

6•10 Manual and automatic transmission

7 Turn the adjusting nut to lengthen or shorten the cable to give a clearance of 0.02 to 0.05 in (0.5 to 1.3 mm) between the downshift lever and the accelerator shaft. Tighten the locknut.

11 Starter inhibitor/reverse lamp switch - removal and refitting

1 This switch is non-adjustable and any malfunction must be due to a wiring fault, a faulty switch or wear in the internal actuating cam.
2 When removing and installing the switch, always use a new O-ring seal and tighten to the specified torque.

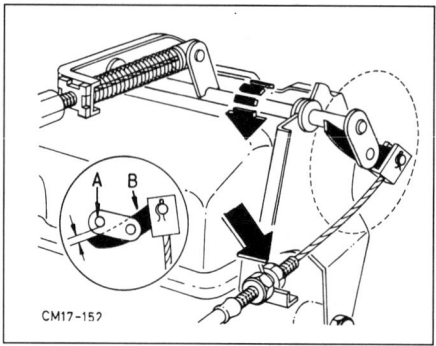

12 Automatic transmission vacuum diaphragm unit - removal and refitting

1 Disconnect the vacuum hose at the vacuum diaphragm.

Fig. 6.20 Adjusting the downshift cable (Sec 10)

A Accelerator shaft   B Downshift lever

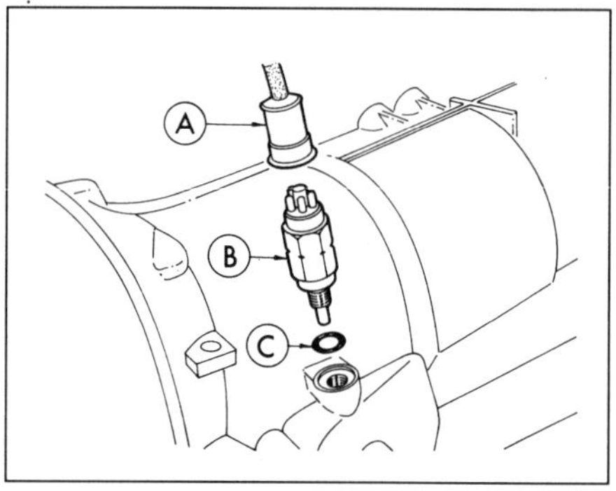

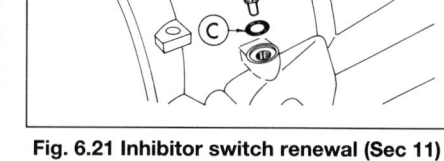

2 Remove the diaphragm bracket bolt and then the bracket.
3 Lift out the diaphragm and actuating pin.
4 Refit in the reverse order, but first check that the throttle valve is free to move and use a new O-ring seal.

Fig. 6.21 Inhibitor switch renewal (Sec 11)

A Wiring plug   B Switch   C O-ring seal

5 Check the transmission fluid level on completion.
6 It should be noted that failure of this unit can cause fluid loss due to fluid being drawn into the inlet manifold, and under some circumstances "pinking" (pre-ignition).

Fault diagnosis - manual gearbox

Weak or ineffective synchromesh
Synchronising cones worn, split or damaged
Baulk ring synchromesh dogs worn, or damaged

Jumps out of gear
Broken gearchange fork rod spring
Gearbox coupling dogs badly worn
Selector fork rod groove badly worn

Excessive noise
Incorrect grade of oil in gearbox or oil level too low
Bush or needle roller bearings worn or damaged
Gear teeth excessively worn or damaged
Countershaft thrust washers worn allowing excessive endplay

Excessive difficulty in engaging gear
Clutch cable adjustment incorrect

Fault diagnosis - automatic transmission

Faults in these units are nearly always the result of low fluid level or incorrect adjustment of the selector linkage or downshift cable. Internal faults should be diagnosed by your main Ford dealer who has the necessary equipment to carry out the work.

Chapter 7 Propeller shaft

For modifications, and information applicable to later models, see Supplement at end of manual

Contents

Fault diagnosis - propeller shaft . . . . .See end of Chapter
General description 1
Propeller shaft centre bearing - removal and refitting 3
Propeller shaft - removal and refitting 2
Propeller shaft rubber coupling - removal and refitting 4
Universal joints - tests for wear 5

Degrees of difficulty

Easy, suitable for novice with little experience	Fairly easy, suitable for beginner with some experience	Fairly difficult, suitable for competent DIY mechanic	Difficult, suitable for experienced DIY mechanic	Very difficult, suitable for expert DIY or professional

Specifications

Type . . . . Two piece tubular steel with rubber mounted centre bearing. Hardy-Spicer universal joints with alternative rubber coupling on some models

Yoke outer diameter . . . . 1.37 in (34.9 mm)

Torque wrench settings   lbf ft   Nm
Propeller shaft to drive pinion flange . . . . 42 to 48   57 to 67
Centre bearing mounting bolts . . . . 13 to 17   18 to 23
Rubber coupling to propeller shaft flange bolts . . . . 55 to 59   76 to 82

1 General description

Drive is transmitted from the gearbox to the rear axle by a finely balanced tubular propeller shaft, split into two halves and supported at the centre by a rubber mounted bearing.

Fitted to the front, centre and rear of the propeller shaft assembly are universal joints which allow slight movement of the complete power unit and the rear axle on their rubber mountings. Each universal joint comprises a four legged centre spider, four needle roller bearing sets and two yokes. On all automatic transmission models and Ghia variants the front universal joint is replaced by a rubber coupling.

Fore and aft movement of the power unit is absorbed by a sliding spline located at the gearbox end. The yoke flange of the rear universal joint is fitted to the rear axle and is secured to the pinion flange by four bolts and lockwashers.

The propeller shaft universal joints cannot be renewed without special equipment because the joint spiders are staked into the yokes in a position determined during electronic balancing. When joint wear is detected either a new propeller shaft should be obtained, or the complete unit should be passed to a suitably equipped engineering workshop for repair.

2 Propeller shaft - removal and refitting

1 Jack-up the rear of the car, or position the rear of the car over a pit or on a ramp.
2 If the rear of the car is jacked-up, supplement the jack with support blocks, so that danger is minimised should the jack collapse.
3 If the rear wheels are off the ground, place the car in gear (automatic transmission in "P") or put the handbrake on, to ensure that the propeller shaft does not turn when loosening the four bolts securing the propeller shaft to the rear axle flange.
4 The propeller shaft is carefully balanced to fine limits and it is important that it is refitted in exactly the same position it was in prior to removal.

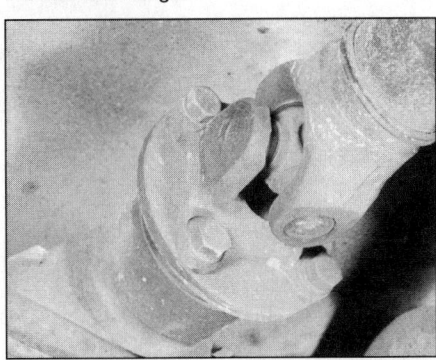

> **HAYNES HiNT** *Scratch a mark on the propeller shaft and rear axle flanges, to ensure accurate mating during reassembly.*

5 Unscrew and remove the four lockbolts and securing washers which hold the flange on the propeller shaft to the flange on the rear axle (photo).
6 Undo and remove the two bolts holding the

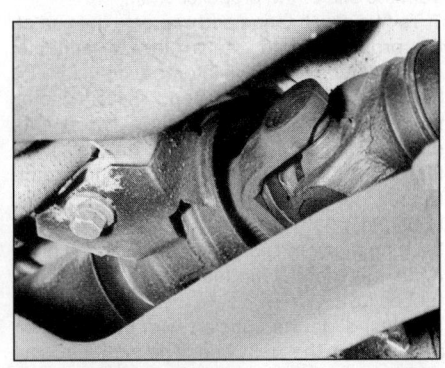

2.5 Propeller shaft rear axle flange mounting bolts

2.6 Propeller shaft centre bearing mounting bolts

7•2 Propeller shaft

centre bearing housing to the underframe. Note the position and number of any shims that may be fitted (photo).

7 Lower the centre of the shaft and separate the two flanges at the rear. Pull the complete shaft rearwards to disengage it from the gearbox mainshaft splines.

8 Place a large can or tray under the rear of the gearbox extension to catch any oil which is likely to leak through the spline lubricating holes when the propeller shaft is removed.

9 Refitting is a reverse of the above procedure. Ensure that the mating marks scratched on the propeller shaft and rear axle flanges line up and that any shims at the centre bearing are refitted. Tighten the rear flange securing bolts before finally tightening the centre support bearing mounting bolts. This will ensure that the centre bearing finds its own position and does not cause distortion of the rubber.

10 Check the level of the gearbox oil and make good any loss.

3 Propeller shaft centre bearing - removal end refitting

Depending on vehicle type a propeller shaft with a non-sealed centre bearing or a sealed centre bearing may be fitted (Fig. 7.1). In the case of bearing failure on the sealed type, a complete housing and bearing assembly must be fitted.

1 Refer to Section 2 and remove the complete propeller shaft assembly.

2 Using a blunt chisel carefully prise open the centre bearing support retaining bolt locking tab.

3 Slacken the bolt in the end of the yoke and with a screwdriver ease out the U-shaped retainer through the side of the yoke.

4 Mark the propeller shaft and yoke for correct refitting and separate the two units.

Non-sealed bearing type

5 Lift off the insulator rubber together with the collar from the ball race. Remove the insulating rubber from the collar.

6 Refer to Fig. 7.3 and using a universal two leg puller draw the bearing together with cap from the end of the propeller shaft.

7 To fit a new bearing and cap onto the end of the propeller shaft use a piece of suitable diameter tube and drive into position.

8 With a pair of pliers bend the six metal tongues of the collar slightly outward and carefully insert the insulator rubber. It is important that the flange of the insulator rubber, when fitted into the support, is uppermost.

9 Using a pair of "parrot jaw" pliers or a chisel bend the metal tongues rearward over the rubber lip.

10 Next slide the support with the insulator rubber over the ball race. The semi circular recess in the support periphery must be positioned toward the front end of the car when fitted.

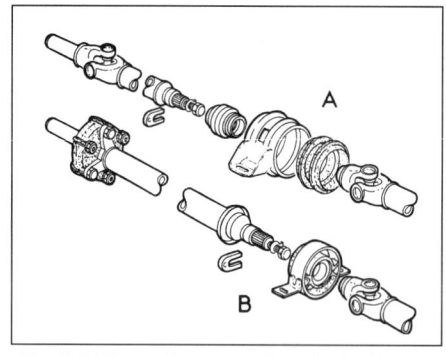

Fig. 7.1 Alternative propeller shafts (Sec 3)

A Non-sealed centre bearing
B Sealed centre bearing

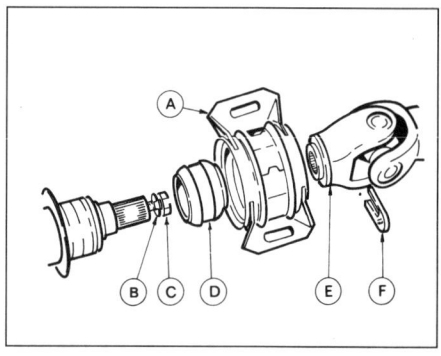

11 Screw in the bolt together with locking tab into the propeller shaft forward end bearing leaving just sufficient space for the U-shaped retainer to be inserted.

12 Assemble the two propeller shaft halves in their original positions as denoted by the two previously made marks or by the double tooth.

13 Refit the U-shaped retainer with the tagged end towards the splines.

14 Finally tighten the retainer securing bolt and bend over the lockwasher.

Sealed bearing type

15 Support the inner sleeve of the centre bearing assembly in a vice and knock the propeller shaft out of the bearing with a hammer.

16 With the propeller shaft supported in a vice drive the new bearing assembly onto the shaft using a suitable diameter tube.

17 Assemble the two propeller shaft halves in their original positions as denoted by the previously made marks or by the double tooth.

18 Refit the U-shaped retainer with the tagged end towards the splines.

19 Tighten the retainer securing bolt and bend over the lockwasher.

Fig. 7.2 Centre bearing components (Sec 3)

A Collar and insulator   D Bearing and cap
B Locking tab   E Yoke
C Bolt   F U-retainer

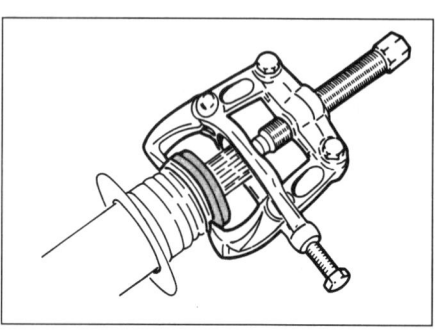

Fig. 7.3 Pulling off the bearing and caps (Sec 3)

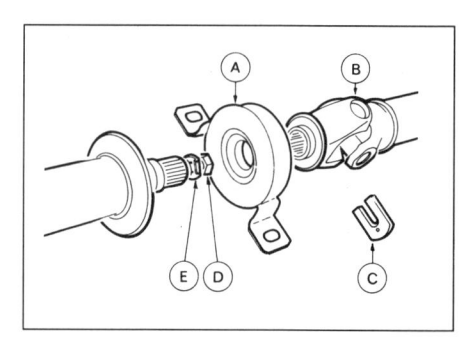

Fig. 7.4 Sealed centre bearing components (Sec 3)

A Centre bearing B Yoke C U-retainer
D Bolt E Locking tab

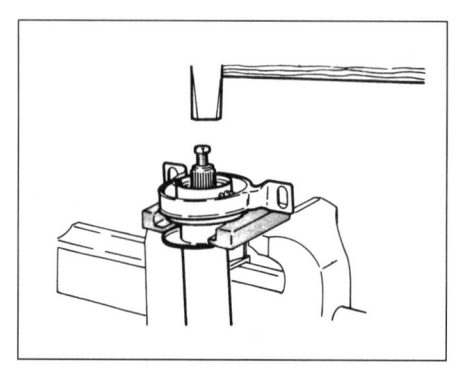

Fig. 7.5 Separating propeller shaft from centre bearing (Sec 3)

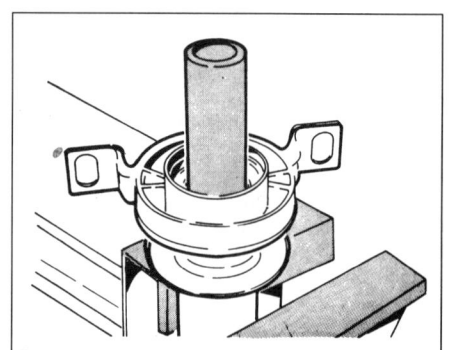

Fig. 7.6 Using a suitable tube to refit centre bearing (Sec 3)

Propeller shaft 7•3

4.3 Note position of rubber coupling securing nuts and bolts before removal

4 Propeller shaft rubber coupling - removal and refitting

1 Refer to Section 2 and remove the propeller shaft.
2 Fit a compressor around the circumference of the coupling and tighten it until it just begins to compress the coupling. If a compressor is not available, two large worm drive hose clips joined end to end will serve the same purpose.
3 Make a note of the relative positions of the flanges and then progressively slacken and remove the six nuts, spring washers and bolts. Note which way round each bolt is fitted (photo).
4 Remove the coupling from the flanges. If the coupling is to be refitted, leave the compressor in position.
5 Refitting is a reverse sequence to removal. Tighten the securing bolts to the figure given in the Specifications and remove the compressor.

5 Universal joints - tests for wear

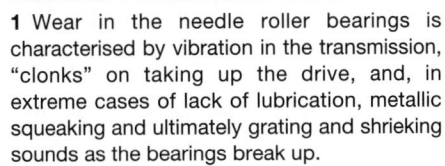

1 Wear in the needle roller bearings is characterised by vibration in the transmission, "clonks" on taking up the drive, and, in extreme cases of lack of lubrication, metallic squeaking and ultimately grating and shrieking sounds as the bearings break up.
2 It is easy to check if the needle roller bearings are worn with the propeller shaft in position, by trying to turn the shaft with one hand, the other hand holding the rear axle flange when the rear universal joint is being checked, and the front half coupling when the front universal joint is being checked. Any movement between the propeller shaft and the front and the rear half coupling is indicative of considerable wear. If worn, a new assembly will have to be obtained. Check by trying to lift the shaft and noticing any movement of the joints.
3 The centre bearing is a little more difficult to test for wear when mounted on the car. Undo and remove the two support securing bolts, spring and plain washers and allow the propeller shaft centre to hang down. Test the centre bearing for wear by gripping the support and rocking it. If movement is evident the bearing is probably worn and should be renewed as described in Section 3.
4 When a rubber coupling is fitted to the front of the propeller shaft, carefully inspect for signs of oil contamination or the rubber breaking up. If this condition is evident the coupling must be renewed as described in Section 4.

Fault diagnosis - propeller shaft

Vibration

Wear in sliding sleeve splines
Worn universal joint bearings
Propeller shaft out of balance
Propeller shaft distorted

Knock, or clunk when taking up drive

Worn universal joint bearings
Worn rear axle drive pinion splines
Loose rear drive flange bolts
Excessive backlash in rear axle gears

Notes

Chapter 8 Rear axle

For modifications, and information applicable to later models, see Supplement at end of manual

Contents

Drive shafts - removal, overhaul and refitting 2
Fault diagnosis - rear axle .... See end of Chapter
Final drive - removal and refitting 5
General description and maintenance 1
Rear stub axle bearing and seals - renewal 4
Rear stub axle shaft - removal and refitting 3

Degrees of difficulty

Easy, suitable for novice with little experience	Fairly easy, suitable for beginner with some experience	Fairly difficult, suitable for competent DIY mechanic	Difficult, suitable for experienced DIY mechanic	Very difficult, suitable for expert DIY or professional

Specifications

Axle type .... Hypoid, semi-trailing

Ratio
In-line ohc engine 3.89:1
2.3 litre engine 3.64:1
2.8 litre engine 3.45:1

Lubricant
Lubricant capacity .... 3.43 pints (1.95 litres)
Lubricant type/specification .... Gear oil, viscosity SAE 90EP, to API GL5

Torque wrench settings   lbf ft   Nm
Rear axle to crossmember bolts ....   53 to 66   71 to 92
Rear axle rear mounting (castellated nut) ...   123 to 151   172 to 216
Bearing hub to suspension arm ....   30 to 36   41 to 51
Rear hub centre nut ....   185 to 198   255 to 275
Drive shaft socket head bolts ....   28 to 31   38 to 43

1 General description and maintenance

The rear axle and suspension assembly is of independent design utilising a semi-trailing axle with coil spring suspension and double acting shock absorbers.

The differential unit is of the two pinion design and is driven by a hypoid crown wheel and pinion.

Engine torque is transmitted from the propeller shaft to the drive pinion via an extension shaft which is supported by a conventional ball race. A splined coupling connects the extension shaft to the drive pinion. This has two roller bearings which are pre-loaded with a collapsible spacer. The correct drive pinion/crownwheel mesh is adjusted using a selective shim between the drive pinion and rear taper roller bearing.

The crownwheel is bolted to the differential carrier and the whole assembly is fitted with taper roller bearings which are pre-loaded with selective shims between the bearings and differential housing. Engine torque is transmitted from the differential assembly through two pinion gears to the side gears. These side gears are secured with circlips as well as being splined to the side drive shafts. Each drive shaft is fitted with two constant velocity joints to allow for suspension travel.

The contents of this Chapter are confined to the removal and refitting of the final drive assembly, the drive shafts and associated components. Details of the suspension assembly are given in Chapter 11.

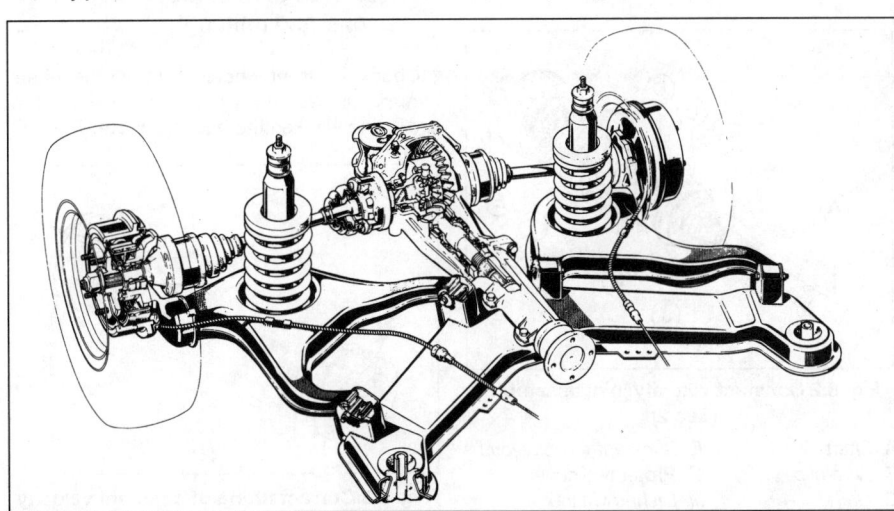

Fig. 8.1 Final drive unit mounted on rear suspension crossmember (Sec 1)

8•2 Rear axle

Overhauling the rear axle requires a variety of special tools and it is not recommended that any attempt should be made to do this. If a rear axle is defective it should be replaced by an exchange unit.

No routine oil changes are required. At the specified intervals wipe around the plug on the left side of the differential housing, just forward of the driveshaft flange. Remove the plug then, if necessary, add oil to bring the level up to the filler hole. Refit the plug securely on completion.

2.2a Drive shaft inner constant velocity joint

2.2b Drive shaft outer constant velocity joint

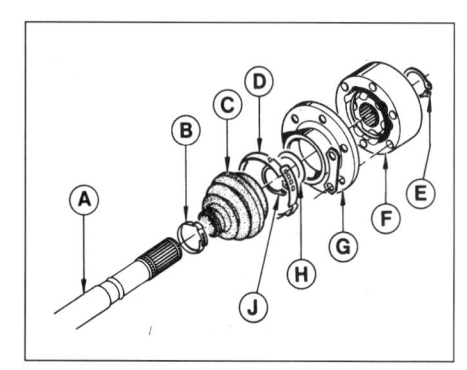

Fig. 8.2 Constant velocity joint assembly (Sec 2)

A Shaft
B Wire clip
C Gaiter
D Clip
E Outer circlip
F Constant velocity joint
G Protective cover
H Dished washer
J Inner circlip

2 Drive shafts - removal, overhaul and refitting

1 Chock the front wheels, jack-up the rear of the car and support it on firmly based axle-stands.
2 Mark the drive shafts left and right as they are not interchangeable. From both ends of the drive shaft remove the six socket headed screws and three washers per flange and lower the shafts to the ground (photos).
3 It should be noted that two types of half shaft are fitted to vehicles covered by this manual. They are identical in appearance but the difference lies in the diameter of the constant velocity joint.
4 To dismantle the constant velocity joint first remove the outer circlip and withdraw the constant velocity joint from the shaft.
5 Remove the dished washers noting which way round they are fitted and then remove the inner circlips.
6 Slacken the gaiter clamps and then remove the cover and rubber sleeves.
7 Thoroughly clean all parts and inspect for wear or damage to the gaiter. Obtain new parts as necessary.
8 To reassemble first fit the gaiters and covers and secure with the gaiter clamps ensuring that the ends of the outer clamps are between the bolt holes.
9 Insert the inner circlips and then fit the dished washers with the small diameter towards the half shaft and constant velocity joint. The identification groove must be away from the half shaft.
10 Refit the outer circlip.
11 Apply 2 oz (55 g) of grease to each constant velocity joint. The grease should be to Ford specification S-MIC-75A or equivalent.
12 Refitting the half shaft is the reverse sequence to removal. Make sure that the mating faces of the half shaft, final drive and stub axle are clean.

3 Rear stub axle shaft - removal and refitting

1 Chock the front wheels, remove the wheel trim and slacken the wheel nuts. Remove the circlip and slacken the hub nut. **Note:** *The hub nut on the left-hand side of the vehicle has a left-hand thread and the hub nut on the right has a right-hand thread.*
2 Jack-up the rear of the car and support it on axle-stands. Remove the road wheel.
3 Remove the speed nut from the brake drum and with the handbrake released remove the drum. If it is tight, gently tap its circumference with a copper mallet.
4 Undo and remove the six screws securing the half shaft flange to the stub axle. Support the weight of the half shaft with wire or string.
5 It is now necessary to draw off the hub flange. If the flange is an easy fit it may be removed with a large three leg universal puller. Otherwise special Ford tool P1039-A will have to be borrowed from the local Ford garage and used for this operation.
6 Remove the hub nut and washer and with the hub flange released the stub axle may now be withdrawn from the rear of the backplate and lower arm.
7 Refitting the stub axle is the reverse sequence to removal. It will be found beneficial if the hub nut is finally tightened when the road wheel has been refitted and the rear of the car lowered to the ground.
8 Tighten the hub nut to the correct torque figure given in the Specifications at the front of this Chapter.

4 Rear stub axle bearing and seals - renewal

1 Refer to Section 3 and remove the stub axle.
2 Clean the area around the rear of the brake wheel cylinder and remove the brake pipe union. Plug the end of the brake pipe and wheel cylinder to prevent ingress of dirt and loss of brake fluid.
3 Undo and remove the four bolts retaining the brake back plate and bearing hub. Lower the back plate to the ground and lift away the hub assembly.
4 Using a screwdriver carefully remove the outer grease seal.
5 The bearing assembly and second seal may now be pressed out using a tube of suitable diameter and a large bench vice.

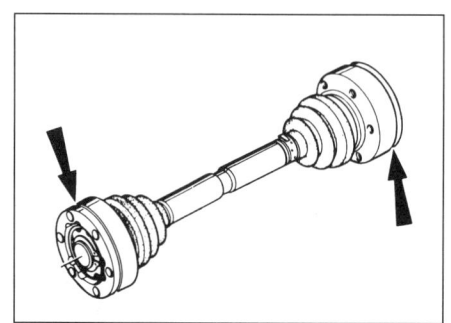

Fig. 8.3 Correct fitting of constant velocity joints with groove on joint away from drive shaft - arrowed (Sec 2)

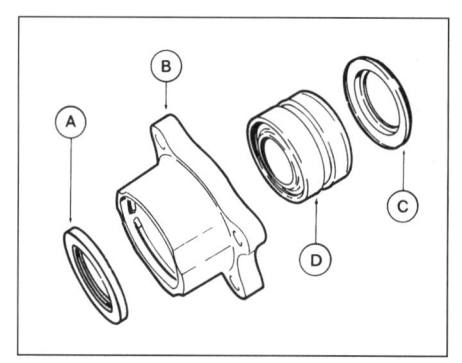

Fig. 8.4 Rear stub axle bearing assembly (Sec 4)

A Outer oil seal B Hub C Inner oil seal
D Bearing

Rear axle 8•3

6 Carefully clean the bearing and grease seal seats to remove any burrs.
7 Fit a new bearing using the same tube used to remove the old bearing until it abuts against the stop.
8 Fit the two grease seals with the lips facing inward using a suitable diameter tube and lightly lubricate the lip.
9 Refit the hub assembly and stub axle, this is the reverse sequence to removal.
10 Bleed the braking system as described in Chapter 9.

5 Final drive - removal and refitting

1 Chock the front wheels, jack-up the rear of the car and support it on firmly based axle-stands.
2 Detach the exhaust pipe from the rear rubber mounting. Unscrew and remove the nuts securing the exhaust pipe mountings to the crossmember and remove the mountings.
3 Undo and remove the single bolt securing the brake 3-way union to the body and allow it to hang free.
4 Refer to Chapter 7 and disconnect the propeller shaft from the rear axle flange.
5 Undo and remove the six socket headed screws each side, securing the inner ends of the drive shafts to the two rear axle flanges. Lift the drive shafts off the flanges and make a note of any shims that may be fitted. Tie the

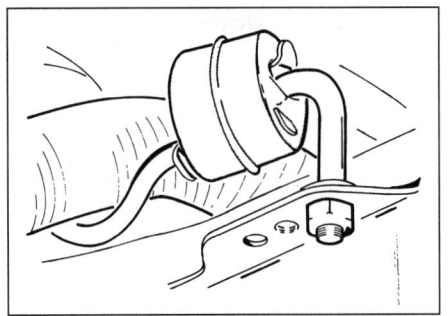

Fig. 8.5 Exhaust mounting crossmember attachment (Sec 5)

drive shafts to a suitable place on the body and allow the shafts to hang free.
6 Position a jack or stand under the rear axle casing and take the weight of the unit.
7 Extract the split pin from the castellated nut securing the rear axle mounting to the body. Undo and remove the castellated mounting nut and thrust pad. It may be necessary to hold the bolt from inside the luggage compartment during this operation.
8 Using a second jack suitably positioned, take the weight of the rear suspension crossmember.
9 Bend back the lock tabs and undo and remove the two crossmember mounting bolts from the centre of the guide plates.
10 Undo and remove the four guide plate securing bolts and lift off the bolts and guide plates.

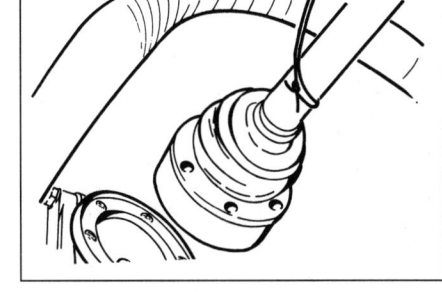

Fig. 8.6 Tie the drive shafts to a suitable place on the body (Sec 5)

11 The rear axle assembly is secured to the crossmember with four through bolts and nuts. These may now be removed using a socket and extension bar on the bolt heads below the cross-member and a spanner on the nuts above.
12 Carefully lower the two jacks until sufficient clearance exists to withdraw the rear axle unit rearwards between the petrol tank and exhaust pipe. Take care not to stretch the handbrake cables and ensure that no undue strain is placed on any components as the two assemblies are lowered.
13 Refitting the rear axle assembly is the reverse sequence to removal. Tighten all nuts and bolts to the torque figure given in the Specifications and secure the rear mounting castellated nut with a new split pin.

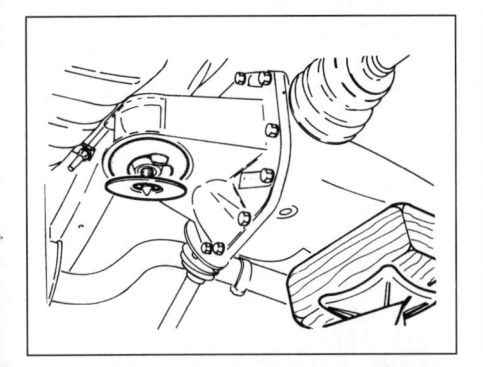

Fig. 8.7 Final drive rear mounting (Sec 5)

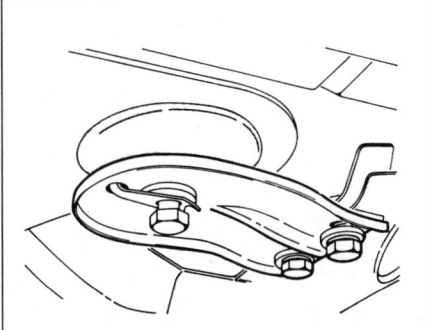

Fig. 8.8 Crossmember guide plate and mounting bolts (Sec 5)

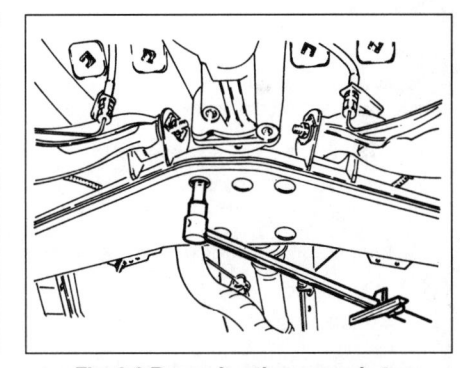

Fig. 8.9 Removing the rear axle to crossmember mounting bolts (Sec 5)

Fault diagnosis - rear axle

Vibration
Worn axleshaft bearings
Loose drive flange bolts
Out of balance propeller shaft
Wheels out of balance
Defective tyre

Noise
Insufficient lubricant
Excessive wear of gears and bearings

Clunk on acceleration or deceleration
Incorrect mesh of crownwheel and pinion
Excessive backlash due to wear of crownwheel and pinion teeth
Worn axleshaft or differential side gear splines
Loose drive flange bolts
Worn drive pinion flange splines
Worn propeller shaft universal joints

Oil leakage
Defective or worn pinion or axleshaft oil seal
Blocked axle housing breather

Notes

Chapter 9 Braking system

For modifications, and information applicable to later models, see Supplement at end of manual

Contents

Bleeding the hydraulic system 2	Front brake caliper - removal and refitting 5
Brake master cylinder - dismantling, overhaul and refitting 15	Front brake discs - removal, inspection and refitting 7
Brake master cylinder - removal and refitting 14	Front brake pads - inspection, removal and refitting 4
Brake pedal - removal and refitting 16	General description 1
Brake pressure control valve - removal and refitting 19	Handbrake - adjustment 11
Drum brake backplate - removal and refitting 10	Handbrake cable - removal and refitting 12
Drum brake shoes - inspection, removal and refitting 8	Handbrake lever - removal and refitting 13
Drum brake wheel cylinder - removal, overhaul and refitting 9	Stop light switch - removal and refitting 20
Fault diagnosis - braking system ....See end of Chapter	Vacuum servo unit - description 17
Flexible hoses - inspection, removal and refitting 3	Vacuum servo unit - removal and refitting 18
Front brake caliper - dismantling and reassembly 6	

Degrees of difficulty

Easy, suitable for novice with little experience	Fairly easy, suitable for beginner with some experience	Fairly difficult, suitable for competent DIY mechanic	Difficult, suitable for experienced DIY mechanic	Very difficult, suitable for expert DIY or professional

Specifications

General
System type	Dual line hydraulic, servo assisted on all four wheels
Front brake	Disc with twin piston caliper
Rear brake	Self adjusting, drum
Handbrake	Mechanical, on rear wheels only
Brake fluid type/specification	Hydraulic fluid to Ford spec SAM-C9103-A

Front brakes
Disc diameter	10.3 in (262 mm)
Maximum disc run-out	0.002 in (0.05 mm)
Caliper cylinder diameter	2.13 in (54.0 mm)
Minimum pad thickness	0.06 in (1.5 mm)

Rear brakes
Drum diameter:
Saloon	9.0 in (228.6 mm)
Estate	10.0 in (254.0 mm)

Shoe width:
Saloon	1.75 in (44.5 mm)
Estate	2.25 in (57.2 mm)

Wheel cylinder diameter:
Saloon	0.75 in (19.05 mm)
Estate	0.81 in (20.64 mm)
Minimum lining thickness	0.04 in (1.00 mm)

Torque wrench settings
	lbf ft	Nm
Caliper to stub axle	45 to 55	65 to 75
Brake disc to hub	40 to 50	54 to 67.5
Rear brake backplate to axle housing	15 to 18	20.6 to 24.6
Hydraulic unions	5 to 7	6.8 to 9.5
Bleed valve	8	10.2
Master cylinder to servo	17	23

9•2 Braking system

1 General description

The Granada utilises a servo assisted, hydraulic, dual line braking system, with disc brakes fitted to the front wheels and drum brakes fitted to the rear. The front disc brake calipers have a separate hydraulic system to that of the rear drum brake wheel cylinders. In the event of failure of the hydraulic pipes to the front or rear brakes, half the system will still operate. Servo assistance in this condition is still available.

The front calipers are a twin piston design and are mounted on the rear of the suspension hub carrier. The brake disc is secured to the wheel hub and rotates between the two halves of the caliper. Inside each half of the caliper is a hydraulic cylinder this being interconnected by a drilling which allows hydraulic fluid pressure to be transmitted to both halves. A piston operates in each cylinder and is in contact with the outer face of the brake pad. By depressing the brake pedal, hydraulic fluid pressure is increased by the servo unit and transmitted to the caliper by a system of metal and flexible pipes, whereupon the pistons are moved outwards so pushing the pads onto the face of the disc and slowing down the rotational speed of the disc.

The rear drum brakes have one double acting cylinder operating two shoes per wheel.

When the brake pedal is depressed, hydraulic fluid pressure, increased by the servo unit is transmitted to the rear brake wheel cylinders by a system of hydraulic pipes. The pressure moves the pistons in the cylinders outwards so pushing the shoe linings into contact with the inside circumference of the brake drum and slowing down the rotational speed of the drum.

Certain models have a brake pressure control valve connected in the hydraulic circuit to the rear wheels. This valve allows the hydraulic pressure to the rear brakes to rise at a lower rate than the pressure to the front brakes. This prevents the rear wheels from locking before the front wheel irrespective of whether the vehicle is lightly or heavily loaded.

The handbrake is cable operated and provides an independent and mechanical means of rear brake application.

The entire system is self adjusting. Wear of the front disc pads is automatically taken up by the movement of the caliper pistons and a self adjusting mechanism actuated by the operation of the footbrake is incorporated on the rear brakes. A manual method of handbrake adjustment is provided but its use should only be necessary when a new cable is being fitted or after rear brake overhaul.

A handbrake warning light is mounted on the facia panel of all models. This light is illuminated whenever the handbrake is applied. On certain models a float is fitted in the master cylinder and is wired to the warning light. In this case the light doubles as a low brake fluid warning light.

Note: *Brake system components must be kept clean of any foreign material. Where necessary hydraulic components can be washed in clean brake fluid or methylated spirit and dried with a lint free rag. Do not use petrol or paraffin. Never re-use brake fluid. Fluid that has been exposed to atmosphere for a short period of time must not be used. Any brake fluid allowed to come into contact with vehicle paintwork must be cleaned off immediately using cold water. Never inhale dust from brake linings. A vacuum cleaner, brush or rag should be used to clean the linings. Any dust should be contained in a sealed bag for disposal.*

2 Bleeding the hydraulic system

1 Removal of all the air from the hydraulic system is essential to the correct working of the braking system, and before undertaking this, examine the fluid reservoir cap to ensure that the vent hole is clear. Check the level of fluid in the reservoir and top-up if required.

2 Check all brake line unions and connections for possible seepage, and at the same time check the condition of any rubber hoses.

3 If the condition of the caliper or wheel cylinders is in doubt, check for possible sign of fluid leakage.

4 If there is any possibility that incorrect fluid has been used in the system, drain all the fluid out and flush through with methylated spirit.

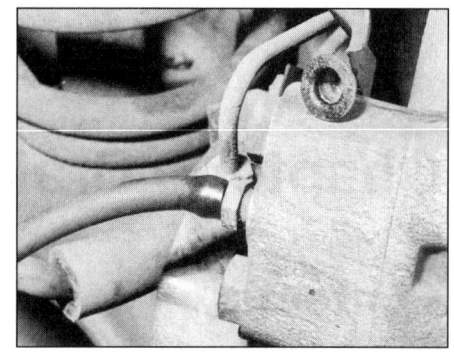

2.6 Front caliper bleed screw with dust cap removed and bleed tube fitted

Renew all piston seals and cups since they will be affected and could possibly fail under pressure.

5 Gather together a clean jam jar, a 12 inch (300 mm) length of tubing which fits tightly over the bleed screws and a tin of the correct brake fluid.

6 To bleed the system, clean the area around the bleed valves and start on the front right-hand bleed screw by first removing the rubber cup over the end of the screw (photo).

7 Place the end of the tube in the clean jar which should contain sufficient fluid to keep the end of the tube below the surface during the operation.

8 Open the bleed screw ½ turn with a spanner and depress the brake pedal. After slowly releasing the pedal, pause for a moment to allow the fluid to recoup in the master cylinder and then depress it again. This will force air from the system. Continue until no more air bubbles can be seen coming from the tube. At intervals make certain that the reservoir is kept topped up, otherwise air will enter at this point again.

9 Finally press the pedal down fully and hold it there whilst the bleed screw is tightened.

10 Repeat this operation on the second front brake, and then the rear brakes, starting with the right-hand brake unit.

11 When completed check the level of the fluid in the reservoir and then check the feel of the brake pedal, which should be firm and free from any "spongy" action, which is normally associated with air in the system,

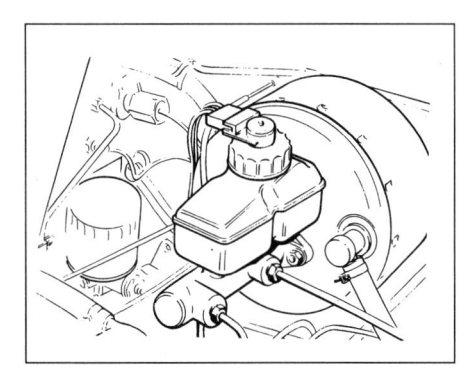

Fig. 9.1 Brake fluid reservoir (Sec 2)

Fig. 9.2 Front caliper bleed valve - A (Sec 2)

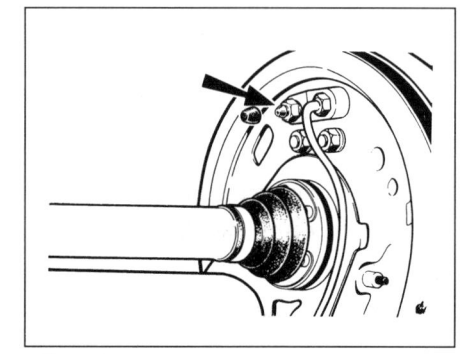

Fig. 9.3 Rear wheel cylinder bleed valve - arrowed (Sec 2)

Braking system 9•3

3.2a Front flexible hose upper union connection

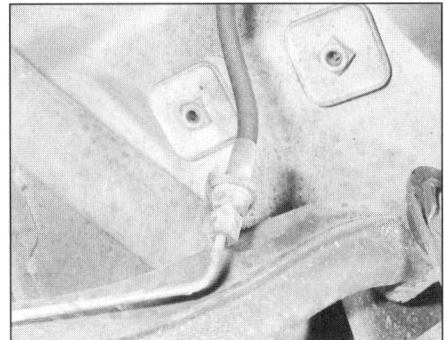

3.2b Rear brake hose union connection

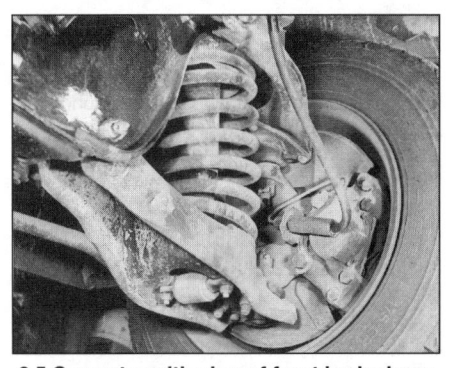

3.5 Correct positioning of front brake hose

3 Flexible hoses - inspection, removal and refitting

1 Inspect the condition of the flexible hydraulic hoses leading to each of the front disc brake calipers and the two at the front of the rear suspension arms. If they are swollen, damaged chafed or have any signs of deterioration they must be renewed.
2 To remove a front flexible hose, wipe the union and brackets free from dust and undo the union nuts from the metal pipe ends (photos). Plug the ends of the metal pipes to prevent loss of fluid and ingress of dirt.
3 Undo and remove the locknuts and washers securing each flexible hose end to its bracket and lift away the hose.

4 Refitting is the reverse sequence to removal. Ensure that the hose is not kinked and does not chafe on the tyre when the wheels are turned from lock to lock (photo).
5 Bleed the hydraulic system as described in Section 2. It will only be necessary to bleed the particular circuit on which the hose has been changed.
6 To remove the rear flexible hoses follow the instructions for the front hoses.

4 Front brake pads - inspection, removal and refitting

1 Jack-up the front of the car and place on firmly based axle-stands. Remove the front wheel.
2 Inspect the amount of friction material left on the pads. The pads must be removed when the thickness has been reduced to a minimum of 0.12 in (3.0 mm).
3 When the pistons are moved into their respective bores to accommodate new pads, the level could rise sufficiently for the fluid to overflow. Place absorbent cloth around the reservoir or syphon a little fluid out so preventing paintwork damage caused by being in contact with the hydraulic fluid.
4 Using a pair of long nosed pliers extract the two small spring clips that hold the main retaining pins in place (photo).
5 Remove the main retaining pins that run through the caliper and the metal backing of the pads and shims (photo).

6 Lift away the wire anti-rattle spring (photo).
7 The friction pads can now be removed from the caliper. If they prove difficult to remove by hand a pair of long nosed pliers can be used. Lift away the shims (photo).
8 Carefully clean the recesses in the caliper in which the friction pads and shims lie, and the exposed faces of each piston from all traces of dirt or rust.
9 Using a suitable flat iron bar levering against the disc, carefully push the pistons back into their bores until there is sufficient room to accommodate the new pads (Fig. 9.4).
10 Fit the new friction pads into the caliper. Place the shims in position behind the pads with the arrows pointing upwards. Insert the antirattle wire spring and refit the retaining pins and spring clips.

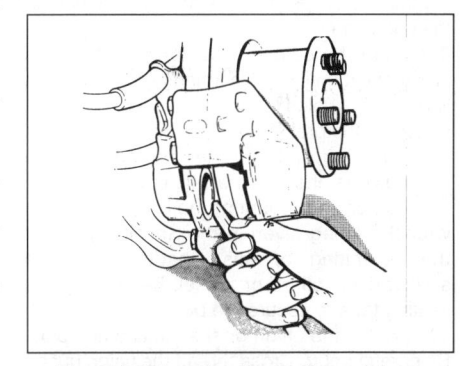

Fig. 9.4 Retracting the caliper pistons (Sec 4)

4.4 Front brake pad retaining pin spring clip

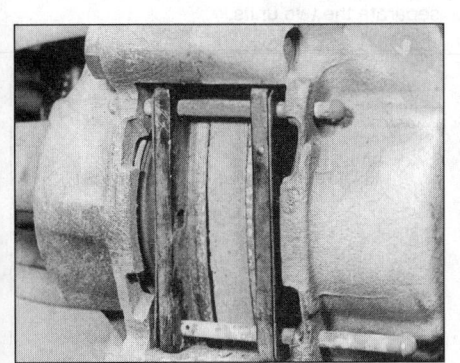

4.5 Removing the brake pad retaining pins

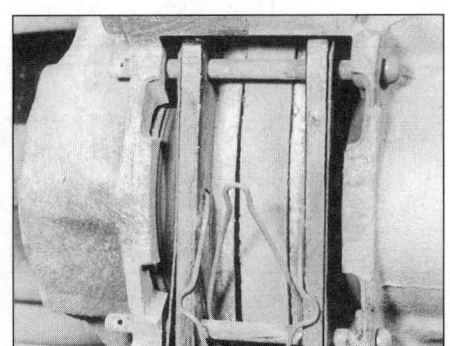

4.6 The wire anti-rattle spring retained by the lower pin

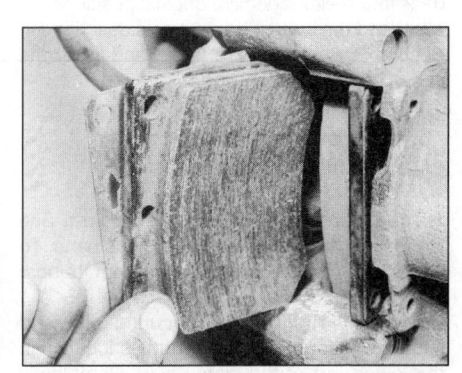

4.7 Lift away the friction pads and shims

9•4 Braking system

11 Pump the brake pedal several times to restore pressure and centralise the pads.
12 Refit the road wheel. Lower the car to the ground and tighten the wheel nuts. Renew the pads on the opposite wheel.
13 Check and if necessary top-up the hydraulic fluid reservoir in the master cylinder.

5 Front brake caliper - removal and refitting

1 Jack-up the front of the car and support on firmly based stands. Remove the front road wheel.
2 Remove the friction pads as described in Section 4.
3 If it is intended to fit new caliper pistons and/or seals, gently depress the brake pedal to move the pistons out of their cylinders and so assist subsequent removal.
4 Using a proprietary brake hose clamp, clamp the flexible hose. If a clamp is not available the end of the metal pipe will have to be securely plugged to prevent loss of fluid upon removal.
5 Unscrew the metal brake pipe union at the joint where it enters the caliper. It will be possible to lift the pipe from the caliper when the mounting bolts are removed.
6 Using a screwdriver or chisel, bend back the tabs on the locking plate and remove the two caliper retaining bolts. Lift away the locking plate and at the same time ease the brake pipe from the caliper.
7 Slide the caliper from its mounting flange on the hub carrier.
8 To refit the caliper, position over the disc and move it until the mounting bolt holes are in line with the holes in the hub carrier.
9 Locate the brake pipe into the caliper and fit the mounting bolts through the holes in the locking plate and into the caliper body.
10 Tighten the mounting bolts to the torque wrench setting given in the Specifications at the beginning of this Chapter. Using a screwdriver, pliers or chisel, bend back the locking tabs to secure the bolts.
11 Tighten the union on the metal brake pipe. Be careful not to cross thread the union nut on the initial turns.
12 Refer to Section 4 and refit the friction pads into their respective original positions.
13 Bleed the hydraulic system as described in Section 2. Refit the road wheel and lower the car.

6 Front brake caliper - dismantling and reassembly

1 The pistons should be removed first. See Fig. 9.6 for the location of the various parts. To do this, half withdraw the piston from its bore in the caliper body.
2 Carefully remove the securing clip and extract the sealing bellows from its location in the lower part of the piston skirt. Completely remove the piston.
3 If difficulty is experienced in withdrawing the

Fig. 9.5 Front brake caliper retaining bolts - arrowed (Sec 5)

pistons use a jet of compressed air or a foot pump to move it out of its bore.
4 Remove the sealing bellows from its location in the annular ring machined in the cylinder bore.
5 Remove the piston rubber sealing ring from the cylinder bore using a small screwdriver but do take care not to scratch the fine finish of the bore.
6 To remove the second piston repeat paragraphs 1 to 5 inclusive.
7 It is important that the two halves of the caliper are not separated under any circumstances. If hydraulic fluid leaks are evident from the joint, the caliper must be renewed.
8 Thoroughly wash all parts in methylated spirit or the correct hydraulic fluid. During reassembly new rubber seals must be fitted and these should be well lubricated with clean hydraulic fluid.
9 Inspect the pistons and bores for signs of wear, score marks or damage, and if evident, new parts, or a new caliper, should be obtained ready for fitting.
10 To reassemble, fit one of the piston seals into the annular groove in the cylinder bore.
11 Fit the rubber bellows to the cylinder bore groove so that the lip is turned outwards.
12 Lubricate the seal and rubber bellows with the correct hydraulic fluid. Push the piston, crown first, through the rubber sealing bellows

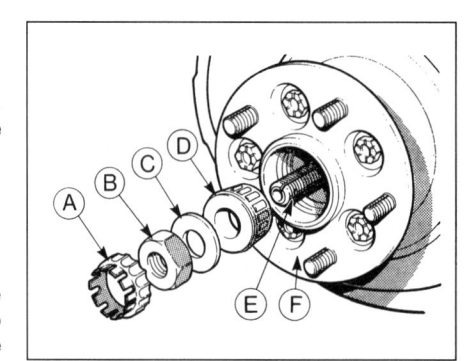

Fig. 9.7 Front hub bearing components (Sec 7)

A Nut retainer
B Locknut
C Thrust washer
D Outer bearing
E Stub axle
F Hub and disc assembly

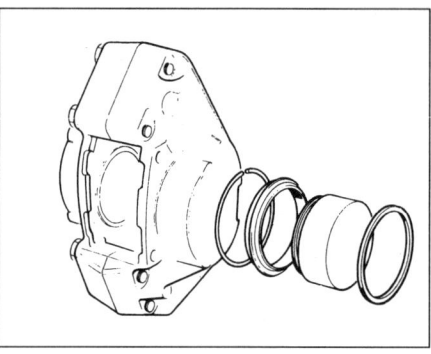

Fig. 9.6 Front brake caliper and piston assembly (Sec 6)

and then into the cylinder bore. Take care as it is easy for the piston to damage the rubber bellows.
13 With the piston half inserted into the cylinder bore fit the inner edge of the bellows into the annular groove in the piston skirt.
14 Push the piston down the bore as far as it will go. Secure the rubber bellows to the caliper with the circlip.
15 Repeat paragraphs 10 to 14 inclusive for the second piston.
16 The caliper is now ready for refitting. It is recommended that the hydraulic pipe end is temporarily plugged to stop any dirt entering whilst being refitted, before the pipe connection is made.

7 Front brake disc - removal, inspection and refitting

1 Jack-up the car and support on firmly based stands. Remove the road wheel.
2 Remove the brake caliper as described in Section 5.
3 By judicious tapping and levering, remove the dust cap from the centre of the hub.
4 Remove the split pin from the nut retainer and lift away the adjusting nut retainer.
5 Unscrew the adjusting nut and lift off the thrust washer and outer tapered bearing.
6 Pull off the complete hub and disc assembly from the stub axle.
7 To separate the disc from the hub mark the position of the hub and disc to assist correct refitment. Undo the five securing bolts and separate the two units.

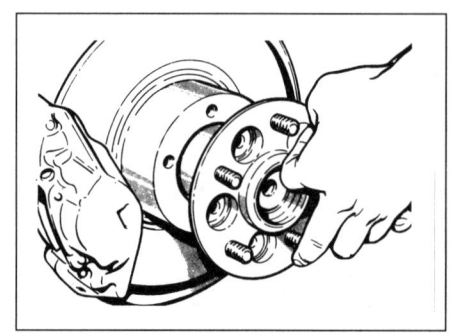

Fig. 9.8 Removing the hub from the disc (Sec 7)

Braking system 9•5

8 Thoroughly clean the disc and inspect for signs of deep scoring or excessive corrosion. If these are evident the disc must be renewed.
9 To reassemble make absolutely sure that the mating faces of the disc and hub are quite clean.
10 Place the disc on the hub and align any previously made marks. Refit the securing bolts and progressively tighten in a diagonal sequence to the torque figure given in the Specifications.
11 Refit the hub and disc assembly onto the stub axle and refit the outer bearing and thrust washer.
12 Refit the adjusting nut and tighten to a torque wrench setting of 27 lbf ft (36 Nm) whilst rotating the hub and disc to ensure free movement and centralise the bearings. Slacken the nut back by 90° or until 0.001 - 0.005 in (0.03 - 0.13 mm) endfloat of the hub assembly is obtained. Fit the nut retainer and a new split pin but at this stage do not lock the split pin.
13 If a dial indicator gauge is available, it is advisable to check the disc for run out. The measurements should be taken as near to the edge of the worn yet smooth part of the disc as possible, and must not exceed 0.002 inch (0.05 mm). If the figure obtained is found to be excessive, check the mating surfaces of the disc and hub for dirt or damage and also check the bearings and cups for excessive wear or damage.

14 If a dial indicator gauge is not available the run out can be checked by means of a feeler gauge placed between the casting of the caliper and the disc. Establish a reasonably tight fit between the top of the casting and the disc and rotate the disc and hub. Any high or low spot will immediately become obvious by extra tightness or looseness of the fit of the feeler gauge. The amount of run out can be checked by adding or subtracting feeler gauges as necessary. It is only fair to point out that this method is not as accurate as when using a dial indicator gauge owing to the rough nature of the caliper casting.
15 Once the disc run out has been checked and found to be correct bend the ends of the split pin back and refit the dust cap.
16 Reconnect the brake hydraulic pipe and bleed the brakes as described in Section 2 of this Chapter.

8 Drum brake shoes - inspection, removal and refitting

After high mileages it will be necessary to fit replacement shoes with new linings. A viewing aperture is provided on the rear of the backplate to allow lining thickness to be checked, but it is desirable to remove the brake drum and also inspect the operating mechanism at the same time. If the linings have worn to a thickness of less than 0.060 in (1.5 mm), replacement shoes must be fitted. On no account should only one, or a pair of shoes be renewed otherwise braking imbalance or poor braking performance may result.

⚠ **Always renew brake shoes in sets of four**

1 Apply the handbrake and chock the front wheels. Jack-up the rear of the car and place on firmly based axle-stands. Remove the road wheel.
2 Release the handbrake and unscrew the speed nut from the wheel stud (photo).

8.2 Rear brake drum retaining speed nut

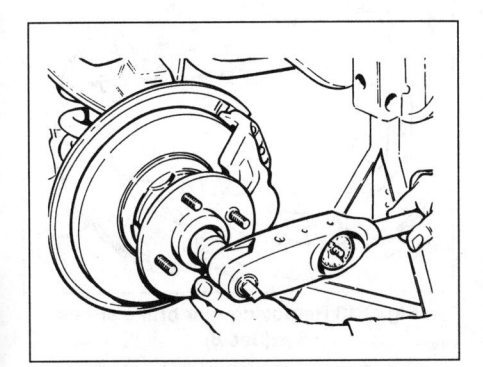

Fig. 9.9 Adjusting the hub bearing torque (Sec 7)

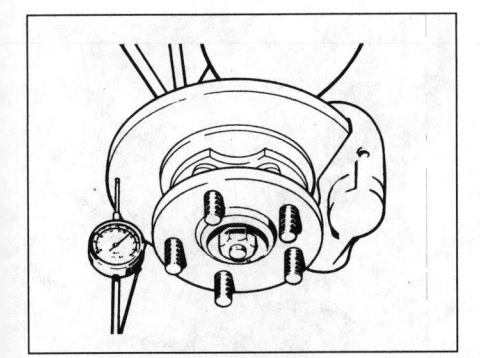

Fig. 9.10 Checking disc run-out using a dial gauge (Sec 7)

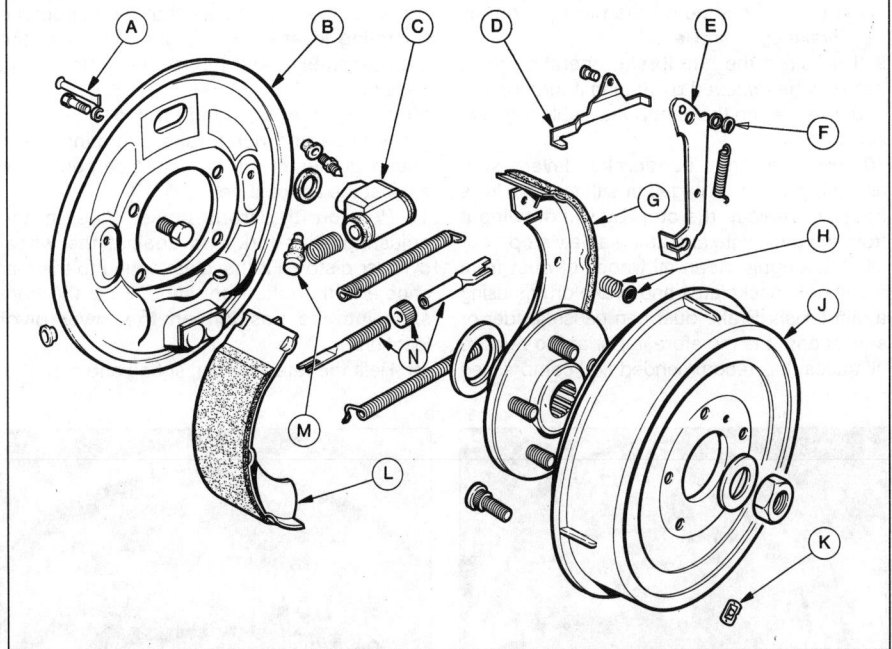

Fig. 9.11 Rear drum assembly - left-hand wheel (Sec 8)

A Brake shoe securing pin
B Brake backplate
C Wheel cylinder
D Self-adjusting lever
E Relay lever
F Spring clip
G Rear brake shoe
H Securing spring and cup
J Brake drum
K Speed nut
L Front brake shoe
M Adjustment plunger
N Automatic adjuster

9•6 Braking system

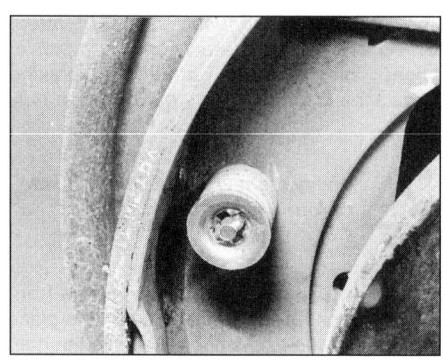

8.4a brake shoe securing pin, spring and cup

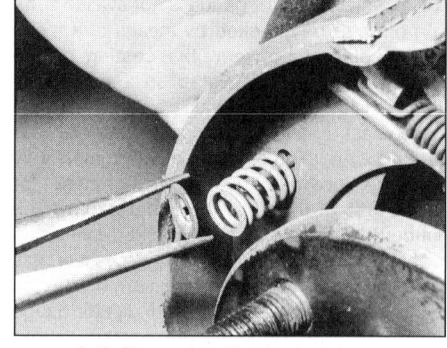

8.4b Removing the brake shoe securing cup

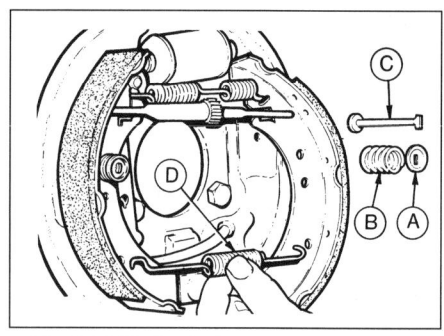

Fig. 9.12 Method of securing rear brake shoes (Sec 8)

A Cup   B Spring   C Pin   D Spring

3 Pull the brake drum off the wheel studs. If it is tight, gently tap its circumference with a soft faced hammer.

4 Remove the front brake shoe securing pin spring and cup (Fig. 9.12) by depressing the cup with a pair of pliers and turning it through 90°. Lift off the spring and remove the pin through the rear of the backplate (photos).

5 Pull the front brake shoe off its lower pivot location and detach the spring, from the front shoe (Fig. 9.12). By twisting slightly detach the spring from the rear shoe also.

6 Unhook the top spring from the rear shoe retaining bracket (Fig. 9.13). Lift away the front shoe, adjusting lever and top spring.

7 Remove the rear shoe retaining pin, spring and cup. Unclip the brake shoe retaining bracket by pushing it sideways in its slot and lifting off (Fig. 9.14).

8 Lift the brake shoe off the pivot pin on the handbrake operating lever.

9 If required the handbrake operating lever can now be removed by turning it through 90° and disengaging the handbrake cable from the slot in the lever.

10 Examine the handbrake lever stop assembly. If it is damaged or stiff to move in its housing, replace the stop by withdrawing it from the backplate and fitting a new stop.

11 Thoroughly clean all traces of dust from the shoes, backplates and brake drums using a stiff brush. Brake dust can cause judder or squeal and it is therefore important to remove all traces. It is recommended that compressed air is not used for this operation as this increases the possibility of the dust being inhaled.

12 Check that the pistons are free in the cylinder, that the rubber dust covers are undamaged and in position, and that there are no hydraulic fluid leaks.

13 Prior to reassembly smear a trace of brake grease on the shoe support pads, brake shoe pivots and on the ratchet wheel face and threads. If new shoes are being fitted, turn the automatic ratchet adjuster to its fully retracted position.

14 To reassemble, first refit the handbrake operating lever onto the cable end. Fit the rear brake shoe over the handbrake lever pivot pin and secure with the retaining bracket.

15 Position the rear shoe against its lower pivot location and against the wheel cylinder piston at the top. Ensure that the handbrake operating lever is correctly seated over the stop assembly. Refit the retaining pin, spring and cup.

16 Locate the adjusting strut over the rear brake shoe and handbrake operating lever. The longer leg of the fork must be positioned against the brake shoe.

17 Position the front brake shoe in the adjusting strut fork and against the wheel cylinder piston. Fit the lower spring to the rear shoe and then the front shoe. Ease the front shoe into its position on the lower pivot location.

18 Refit the retaining pin, spring and cup.

19 Fit the top spring to the oval hole in the front shoe and then to the retaining bracket on the rear shoe. Make sure the tag on the retaining bracket is located in the cut-out in the brake shoe.

20 Turn the ratchet wheel on the adjusting strut to remove all slack movement and to bring the adjusting arm into its correct position relative to the wheel.

21 Check that the brake shoes are central, the springs are fitted correctly and that the adjusting strut and ratchet wheel are correctly positioned (photos). Refit the brake drum. The shoes may have to be moved up or down slightly to allow the drum to be fitted. Secure the brake drum with the speed nut.

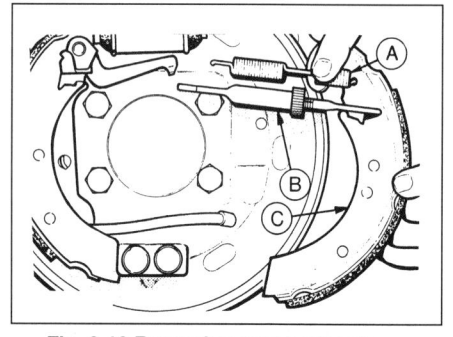

Fig. 9.13 Removing rear brake shoes (Sec 8)

A Spring   B Automatic adjuster   C Front brake shoe

8.21a Rear right-hand brake assembly correctly fitted

8.21b Correct fitment of adjusting strut, ratchet wheel and top spring

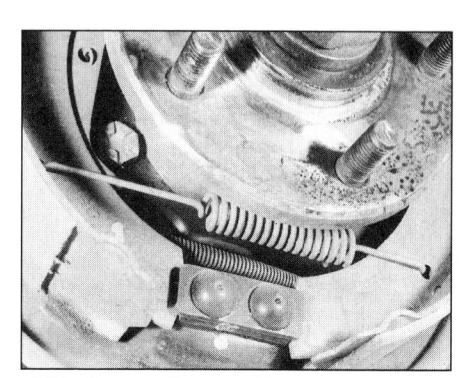

8.21c Correct fitment of lower springs and shoes

Braking system 9•7

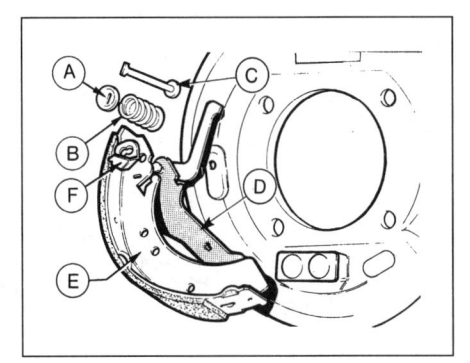

Fig. 9.14 Rear shoe components (Sec 8)

A Cup
B Spring
C Pin
D Relay lever
E Rear shoe
F Retaining bracket

22 Depress the brake pedal several times to centralise the shoes and expand the adjuster. The correct adjustment has been achieved when the pedal movement is constant for several consecutive applications.
23 Refit the roadwheel and lower the car. Road test to ensure correct operation of the rear brakes.

9 Drum brake wheel cylinder - removal, overhaul and refitting

If hydraulic fluid is leaking from the brake wheel cylinder, it will be necessary to dismantle and renew the seals or renew the complete cylinder. Should brake fluid be found running down the side of the wheel, or if it is noticed that a pool of liquid forms alongside one wheel or the level in the master cylinder drops, it is indicative that seal failure has reached an advanced stage. If any of these conditions are apparent the cause must be located and rectified immediately otherwise a complete failure of the rear braking system may occur.

1 Refer to Section 8 and remove the brake drum and shoes. Clean down the rear of the backplate using a stiff brush. Place a quantity of rag under the backplate to catch any hydraulic fluid which may issue from the open pipe or wheel cylinder.
2 Using a proprietary brake hose clamp, clamp the flexible pipe located above the rear

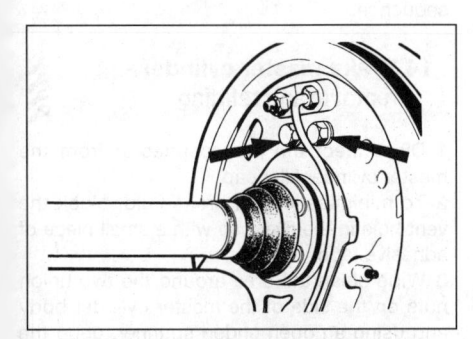

Fig. 9.16 Wheel cylinder mounting bolts - arrowed (Sec 9)

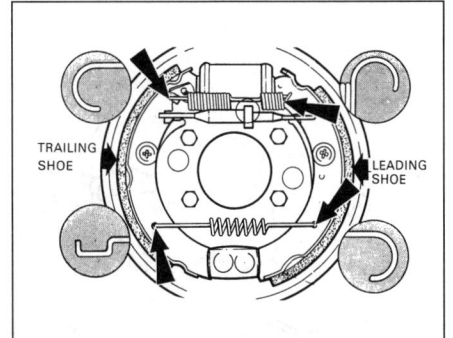

Fig. 9.15 Correct fitting of brake shoe springs - right-hand brake shown (Sec 8)

suspension arm. If a clamp is not available the end of the metal pipe will have to be plugged upon removal to prevent loss of fluid and ingress of dirt.
3 Using an open ended spanner, carefully undo the hydraulic pipe union to the rear of the wheel cylinder.
4 Undo and remove the two bolts and washers that secure the wheel cylinder to the brake backplate.
5 Withdraw the wheel cylinder from the front of the brake backplate.
6 To dismantle the wheel cylinder first ease off each rubber dust cover retaining ring and lift away each dust cover.
7 Carefully lift out each piston together with seal from the wheel cylinder bore. Recover the return spring.
8 Using the fingers only, remove the piston seal from each piston, noting which way round it is fitted. Do not use a metal screwdriver as this could scratch the piston.
9 Inspect the inside of the cylinder for score marks caused by impurities in the hydraulic fluid. If any are found, the cylinder and pistons will require renewal. **Note:** if the wheel cylinder requires renewal always ensure that the replacement is exactly the same as the one removed.
10 If the cylinder is sound, thoroughly clean it out with fresh hydraulic fluid.

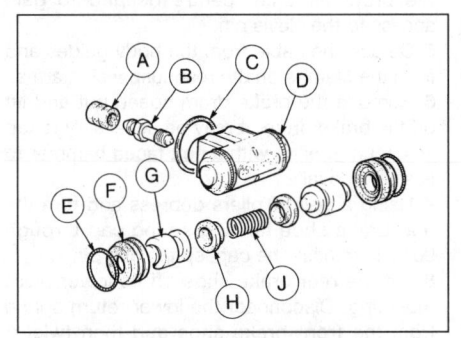

Fig. 9.17 Wheel cylinder components (Sec 9)

A Dust cap
B Bleed screw
C Sealing ring
D Wheel cylinder
E Boot retaining spring
F Rubber boot
G Piston
H Rubber seal
J Separating spring

11 Always fit new rubber seals. The old seals may appear satisfactory but once disturbed they may fail under pressure. Smear the rubber seals with hydraulic fluid and fit to the pistons. Make sure they are fitted with the lip of the seal towards the spring.
12 Liberally lubricate the cylinder bore with hydraulic fluid and carefully insert the piston, (seal end first) into the cylinder. Make sure that the seals do not roll over as they are initially fitted into the bore. Insert the spring and refit the second piston in the same manner.
13 Position the rubber boots on each end of the wheel cylinder and secure in position with the retaining rings.
14 Locate the ring seal on the back of the wheel cylinder and refit the cylinder to the backplate. Secure in position with the two bolts and spring washers.
15 Reconnect the brake pipe to the rear of the wheel cylinder, taking care not to cross thread the union nut.
16 Refit the brake shoes and drum as described in Section 8.
17 Remove the brake hose clamp and bleed the rear brake hydraulic system as described in Section 2.

10 Drum brake backplate - removal and refitting

1 Refer to Chapter 8 and remove the hub flange.
2 Refer to Section 8 and Section 9 of this Chapter and remove the brake shoes and wheel cylinder respectively.
3 Remove the handbrake cable from the backplate by pulling the cable from behind, through its location in the backplate.
4 Undo the four bolts and lift away the backplate.
5 Refitting is a reverse of the removal sequence.

11 Handbrake - adjustment

1 It is important to check that lack of adjustment is not caused by the cable becoming detached from the body mounted clips, that the equaliser bracket and pivot

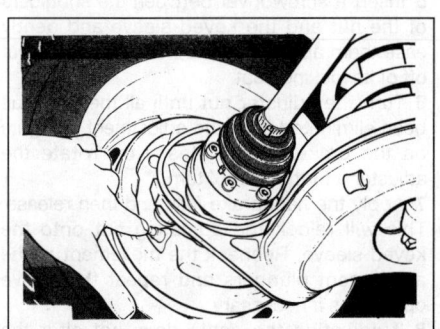

Fig. 9.18 Checking movement of handbrake adjustment plungers (Sec 11)

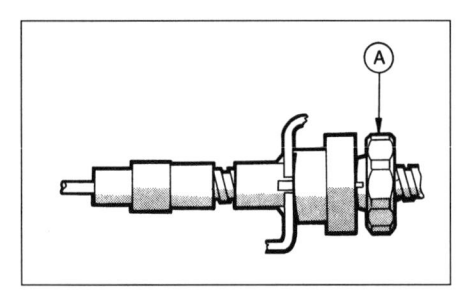

Fig. 9.19 Handbrake adjusting nut - A (Sec 11)

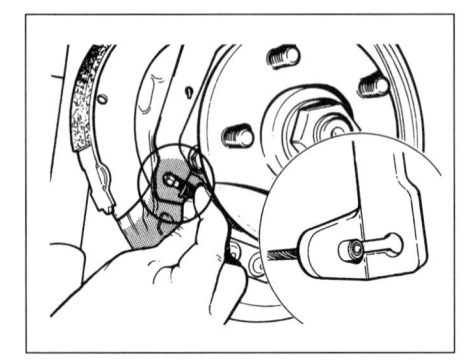

Fig. 9.20 Handbrake cable attachment in relay lever (Sec 12)

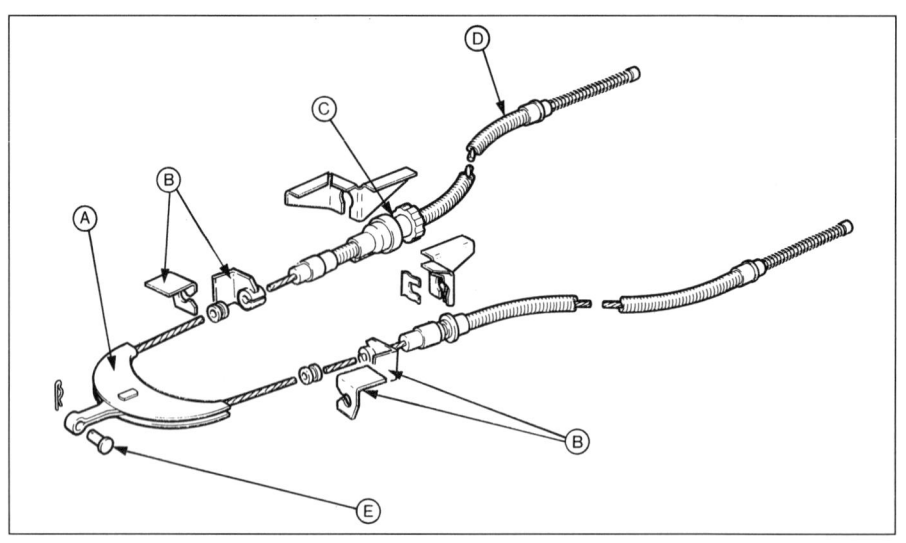

Fig. 9.21 The handbrake components and layout (Sec 12)

A Equaliser yoke   B Cable guides   C Adjuster   D Cable   E Clevis pin

points are adequately lubricated and that the rear shoe linings have not worn excessively.

2 Chock the front wheels, jack-up the rear of the car and support on firmly based axle-stands. Release the handbrake.

3 Check the adjustment of the handbrake by measuring the movement of the adjustment plunger located in each brake backplate. Grip each plunger between finger and thumb and move it in and out. The total movement of both plungers added together should not exceed 0.04 to 0.1 in (1.0 to 2.5 mm). If no movement is apparent or if the movement exceeds the dimension quoted above, the handbrake system requires adjustment. If the movement of the plungers differs from side to side, the cable may be centralised by moving it in its equaliser bracket.

4 To adjust the handbrake, first ensure that the keyed sleeve located on the right-hand cable in front of the adjusting nut, is fully engaged in its slot on the support bracket.

5 Insert a screwdriver between the shoulders of the nut and the keyed sleeve and gently ease them apart. This will lift the adjusting nut off of its locking seat.

6 Turn the adjuster nut until all the play has been eliminated from the adjustment plungers on the brake back plates. Now rotate the adjuster a further ¼ to ½ turn.

7 Apply the handbrake fully and then release. This will relock the adjusting nut onto the keyed sleeve. Recheck the movement of the adjustment plungers and repeat the above operations if necessary.

8 If adjusting the cable does not alter the adjustment plunger movement, this indicates that the cable is binding, there is a malfunction in the brake mechanism or the plungers are seized.

9 When adjustment is correct, remove the axle-stands, and lower the car.

12 Handbrake cable - removal and refitting

1 Chock the front wheels, jack-up the rear of the car and support on firmly based axle-stands. Remove the two rear road wheels. Release the handbrake.

2 Unlock the adjusting nut from the sleeve on the right-hand cable by inserting a screwdriver between their shoulders and gently prising apart.

3 Slacken the handbrake cable by rotating the adjusting nut until all tension has been released.

4 Remove the spring clip from the clevis pin and pull the pin from the handbrake lever and equaliser bracket. On some models it will be necessary to remove the two bolts and lower the propeller shaft centre bearing to gain access to the clevis pin.

5 Detach the cable from the body guides and from the clamps on the rear suspension arms.

6 Remove the brake drum speed nut and lift off the brake drum. It may be necessary to tap its circumference with a soft faced hammer to assist removal.

7 Using a pair of pliers depress and turn the rear brake shoe holding spring cap through 90° and release the cap, spring and pin.

8 Lift the rear brake shoe off its lower pivot mounting. Disconnect the lower return spring from the front brake shoe and then twist it clear of the rear shoe.

9 The handbrake cable can now be unclipped from the handbrake relay lever.

10 Repeat operations 6 to 10 for the other side.

11 Withdraw the cable from its locations in the brake backplates.

12 Refitting is a reverse of the removal sequence. Adjust the cable as described in Section 11.

13 Refit the rear road wheels, remove the stands and lower the car.

13 Handbrake lever - removal and refitting

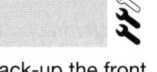

1 Chock the rear wheels and jack-up the front of the car. Release the handbrake.

2 From underneath the car remove the spring clip from the clevis pin securing the cable yoke to the handbrake lever. Lift out the clevis pin. On some models it may be necessary to remove the two bolts and lower the propeller shaft centre bearing to gain access to the clevis pin.

3 From inside the car, remove the centre console, if fitted, as described in Chapter 12.

4 Remove the six screws securing the top plate to the handbrake lever gaiter and lift off the plate.

5 Lift up the gaiter and unplug the connection to the handbrake warning light.

6 Undo the two mounting bolts and remove the handbrake lever.

7 Refitting is a reverse of the removal sequence.

14 Brake master cylinder - removal and refitting

1 Disconnect the plug connector from the master cylinder filler cap.

2 To minimise loss of brake fluid, block the vent hole in the filler cap with a small piece of adhesive tape.

3 Wipe clean the area around the two union nuts on the side of the master cylinder body and using an open ended spanner, undo the two nuts. Tape over the ends of the pipes to stop dirt entering.

Braking system 9•9

4 Undo and remove the two nuts and spring washers that secure the master cylinder to the servo unit. Lift away the master cylinder and seal.

⚠ *Take care that no hydraulic fluid is allowed to spill out onto the paintwork.*

5 Refitting is a reverse of the removal sequence. Always start the union nuts before finally tightening the master cylinder nuts.
6 Bleed the hydraulic system as described in Section 2.

15 Brake master cylinder - dismantling, overhaul and reassembly

If a new master cylinder is to be fitted, it will be necessary to lubricate the seals before fitting to the car as they are coated with a protective lubricant when originally assembled. Remove the blanking plugs from the hydraulic pipe union seatings. Inject clean hydraulic primary piston several times so that the fluid spreads over all the internal working surfaces. Note that on certain master cylinders the arrangement of seals and cups may be slightly different from that shown in Fig. 9.22.

1 Remove the master cylinder as described in Section 14. Invert the cylinder and drain the hydraulic fluid from the reservoir.
2 Remove the reservoir by pulling upwards and remove the rubber seals.
3 Push the primary piston (protruding from the end of the cylinder) in slightly to relieve pressure on the circlip. Using circlip pliers remove the circlip and withdraw the primary piston assembly.
4 Unscrew the piston stop screw from the top of the master cylinder body.
5 Gently tap the cylinder on a block of wood and remove the secondary piston assembly. Alternatively use a compressed air jet carefully applied to the rear outlet connection.
6 Lay out all parts on a clean work surface in their correct removal sequence. Examine the bore of the cylinder carefully for any signs of scores or ridges. If this is found to be smooth all over new seals can be fitted. If however there is any doubt about the condition of the bore, then a new cylinder must be fitted.
7 To dismantle the primary piston, unscrew and remove the screw and sleeve. Remove the spring, retainer and seals. Gently lever off the other seals, taking care not to scratch the piston.
8 To dismantle the secondary piston, remove the spring, retainer and seal. Gently lever off the other seals, taking care not to damage the piston.
9 Thoroughly clean all parts in clean hydraulic fluid or methylated spirit. Ensure that the bypass ports are clear.
10 All components should be assembled wet by dipping in clean brake fluid.

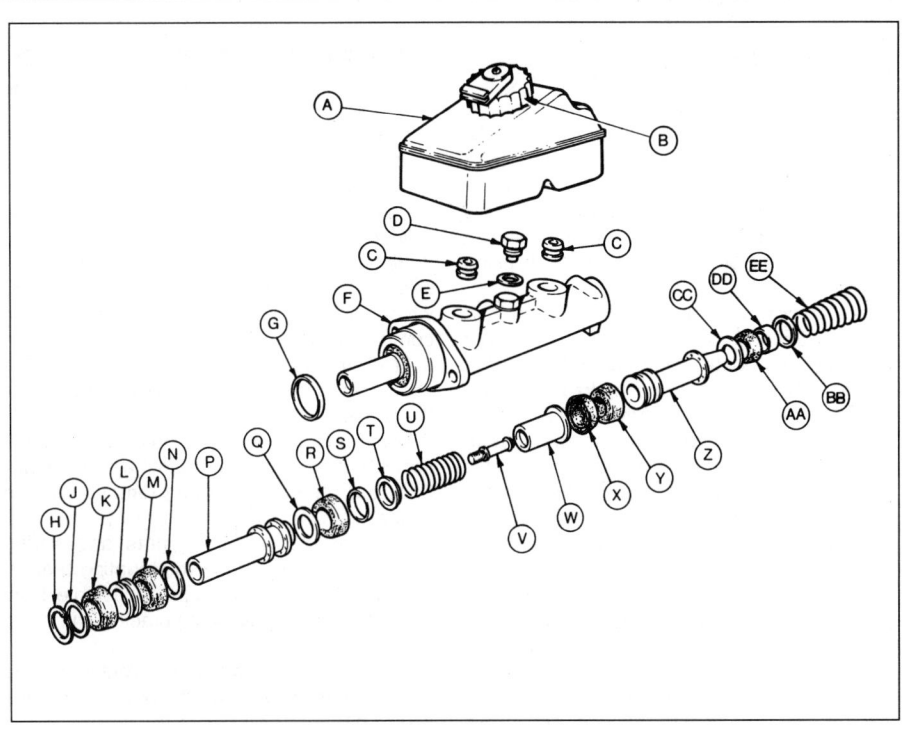

Fig. 9.22 Brake master cylinder components (Sec 15)

A Reservoir
B Filler cap
C Rubber plugs
D Secondary piston stop screw
E Stop screw washer
F Master cylinder (assembled)
G Fluid seal
H Circlip
J Washer
K Secondary cup
L Spacer
M Secondary cup
N Stop washer
P Primary piston
Q Shim
R Primary cup
S Thrust washer
T Spring seat
U Spring
V Special screw
W Spring retainer
X Primary cup
Y Secondary cup
Z Secondary piston
AA Primary cup
BB Spring seat
CC Shim
DD Thrust washer
EE Spring

11 Using the fingers only, fit the seals to the secondary piston, ensuring that they are the correct way round. Fit the retainer and spring. Check that the master cylinder bore is clean, and smear with clean brake fluid.
12 Wet the secondary piston assembly with clean fluid and insert into the master cylinder, spring first. Ease the lips of the seals into the cylinder bore taking care they do not roll over.
13 Using a long drift carefully press the secondary piston assembly inwards against its spring and refit the piston stop screw to the top of the cylinder. Release the piston and allow it to contact the screw.
14 Fit the seals to the primary piston, ensuring that they are the correct way round. Fit the retainer, spring, sleeve and screw.
15 Wet the primary piston assembly with clean fluid, and insert into the master cylinder, spring first. Ease the lips of the seals into the cylinder bore taking care they do not roll over.
16 Fit the circlip to retain the primary piston in the cylinder bore.
17 Check the condition of the front and rear reservoir seals. If there is any doubt as to their condition they must be renewed.
18 Refit the hydraulic fluid reservoir.
19 The master cylinder is now ready for refitting to the servo unit as described in Section 14.
20 Bleed the complete hydraulic system as described in Section 2 and road test the car.

16 Brake pedal - removal and refitting

1 Remove the screws securing the lower dash trim panel and lift out the panel.
2 Withdraw the spring clip from the clevis pin securing the servo pushrod to the pedal. Ease out the clevis pin and remove the two bushes.
3 Detach the brake pedal return spring from the pedal bracket.
4 Remove the spring clip and washer and carefully push the shaft through the pedal and bracket.
5 Lift away the brake pedal. Remove the half bushes from each side of the brake pedal.
6 Inspect the bushes for signs of wear and if evident, they must be renewed. Ensure that the key on each bush engages with the cut-out in the pedal.
7 Refitting the pedal is the reverse sequence to removal. Lubricate the bushes with molybdenum disulphide grease. Ensure that the spacer separating brake and clutch pedals is correctly positioned on the shaft.
8 Adjust the stop light switch if required as described in Section 20.

17 Vacuum servo unit - description

A vacuum servo unit is fitted into the brake hydraulic circuit in series with the master cylinder, to provide assistance to the driver when the brake pedal is depressed. This reduces the effort required by the driver to operate the brakes under all braking conditions.

The unit operates by vacuum obtained from the induction manifold and comprises basically a booster diaphragm and check valve. The servo unit and hydraulic master cylinder are connected together so that the servo unit piston rod acts as the master cylinder pushrod. The driver's braking effort is transmitted through another pushrod to the servo unit piston and its built in control system. The servo unit piston does not fit tightly into the cylinder, but has a strong diaphragm to keep its edges in constant contact with the cylinder wall, so assuring an air tight seal between the two parts. The forward chamber is held under vacuum conditions created in the inlet manifold of the engine and, during periods when the brake pedal is not in use, the controls open a passage to the rear chamber so placing it under vacuum conditions as well. When the brake pedal is depressed, the vacuum passage to the rear chamber is cut off and the chamber opened to atmospheric pressure. The consequent rush of air pushes the servo piston forward in the vacuum chamber and operates the main pushrod to the master cylinder.

The controls are designed so that assistance is given under all conditions and, when the brakes are not required, vacuum in the rear chamber is established when the brake pedal is released. All air from the atmosphere entering the rear chamber is passed through a small air filter.

Under normal operating conditions the vacuum servo unit is very reliable and does not require overhaul except at very high mileages. In this case it is far better to obtain a service exchange unit, rather than repair the original unit.

18 Vacuum servo unit - removal and refitting

1 Slacken the clips securing the vacuum hose to the servo unit and carefully draw the hose from its union.
2 Refer to Section 14 and remove the master cylinder.
3 Remove the nuts and washers securing the servo mounting bracket to the bulkhead, which are accessible from the engine compartment.
4 From inside the car, remove the screws securing the lower dash trim panel and lift out the panel.
5 Withdraw the spring clip from the clevis pin securing the servo pushrod to the brake pedal. Ease out the clevis pin and remove the two bushes.
6 Remove the remaining nuts and spring washers securing the servo mounting bracket to the bulkhead which are accessible from inside the car. Lift away the bracket and servo unit.
7 Undo and remove the four retaining nuts and washers and remove the bracket from the servo unit.
8 Refitting is a reverse of the removal sequence.
9 Bleed the braking system as described in Section 2.

19 Brake pressure control valve - removal and refitting

The brake pressure control valve is mounted on a bracket in the engine compartment and controls the flow of hydraulic fluid to the rear brakes. The valve is only serviced as a unit and therefore should not be dismantled.

1 Place a small piece of adhesive tape over the vent hole in the master cylinder to minimise the loss of hydraulic fluid.
2 Wipe the area around the two union nuts on the top of the valve body and using an open ended spanner undo the two nuts.
3 Remove the two bolts securing the valve to the bracket and remove the valve.

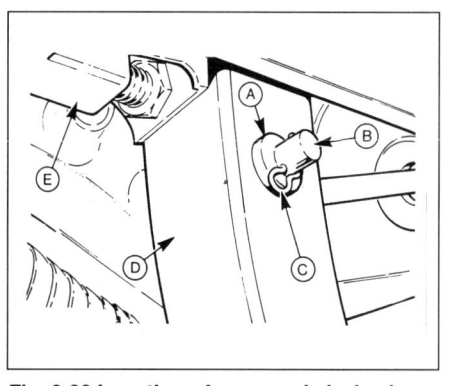

Fig. 9.23 Location of servo rod clevis pin on brake pedal (Sec 18)

A Bush
B Clevis pin
C Spring clip
D Brake pedal
E Stop light switch

HAYNES HiNT *Note the position of the pipes relative to one another and tape over their ends to prevent dirt entering.*

4 Refitting is a reverse of the removal sequence.
5 Bleed the rear brakes as described in Section 2.

20 Stop light switch - removal and refitting

1 Remove the screws securing the lower dash trim panel, fitted above the pedals, and remove the panel.
2 Disconnect the wiring from the switch, remove the locking nut and detach the switch.
3 Refitting the switch is a reverse of the removal sequence. Adjust the switch position so that when the brake pedal is at the rest position, the plunger in the switch is depressed half its total length. Tighten the locknut and reconnect the wiring.
4 Depress the brake pedal and check the operation of the switch and stop lamps.

Fault diagnosis - braking system

Excessive brake pedal travel

Brake shoes set too far from the drums (automatic adjusters seized)

Stopping ability poor, even though pedal pressure is firm

Linings, discs, or drums badly worn or scored
One or more hydraulic wheel cylinders seized
Brake linings contaminated with oil or hydraulic fluid
Wrong type of linings fitted (too hard)
Brake shoes wrongly assembled
Servo unit not functioning

Pedal feels spongy when the brakes are applied

Air present in the hydraulic system
Flexible hoses bulging under hydraulic pressure

Brakes judder when applied

Excessive run-out or distortion of brake discs or drums
Brake backplate mounting bolts loose
Suspension or steering joints loose or worn
Brake linings not bedded into the drums (after fitting new ones)

Pedal travel excessive with little or no resistance and brakes are virtually non-operative

Leak in hydraulic system resulting in lack of pressure for operating wheel cylinders
Master cylinder internal seals are failing to maintain pressure

Binding, juddering, overheating

One or a combination of reasons given in the foregoing Sections

Chapter 10 Electrical system

For modifications, and information applicable to later models, see Supplement at end of manual

Contents

Alternator - fault finding and repair	10
Alternator - general description	6
Alternator - removal and refitting	9
Alternator - routine maintenance	7
Alternator - special procedures	8
Alternator brushes (Bosch) - inspection, removal and refitting	12
Alternator brushes (Lucas) - inspection, removal and refitting	11
Alternator brushes (Motorola) - inspection, removal and refitting	13
Battery - charging	5
Battery - electrolyte replenishment	4
Battery - maintenance and inspection	3
Battery - removal and refitting	2
Central door locking system - general	54
Electrically operated windows - general	55
Fault diagnosis - electrical system	See end of Chapter
Flasher unit - fault tracing and rectification	21
Front direction indicator light assembly and bulb - removal and refitting	36
Fuel tank indicator unit - removal and refitting	45
Fuses	52
General description	1
Headlight and auxiliary light alignment	35
Headlight and sidelight bulbs - removal and refitting	34
Headlight assembly - removal and refitting	33
Headlight washer pump - removal and refitting	31
Horn - fault tracing and rectification	29
Ignition switch and lock - removal and refitting	47
Instruments, warning lights and illumination lights - removal and refitting	43
Instrument cluster and printed circuit - removal and refitting	41
Instrument panel and centre console switches - removal and refitting	48
Instrument voltage regulator - removal and refitting	42
Interior light - removal and refitting	40
Loudspeakers - removal and refitting	50
Radio - removal and refitting	49
Radio aerials - removal and refitting	51
Rear direction indicator, stop and tail light assembly - removal and refitting	38
Rear direction indicator, stop and tail light bulbs - removal and refitting	37
Rear number plate light assembly and bulb - removal and refitting	39
Rear window washer pump - removal and refitting	32
Relays	53
Speedometer inner and outer cable - removal and refitting	44
Starter motor (Bosch EF-0.85KW and GF-1.1 KW) - dismantling and reassembly	19
Starter motor (Bosch JF2KW) - dismantling and reassembly	20
Starter motor - general description	14
Starter motor (Lucas 5M90) - dismantling and reassembly	17
Starter motor (Lucas 2M100) - dismantling and reassembly	18
Starter motor - removal and refitting	16
Starter motor - testing on engine	15
Steering column switches - removal and refitting	46
Windscreen washer pump and reservoir - removal and fitting	30
Wiper arms - removal and refitting	24
Wiper blades - removal and refitting	23
Wiper mechanism - fault diagnosis and rectification	25
Wiper mechanism - maintenance	22
Wiper motor - dismantling, inspection and reassembly	28
Wiper motor and linkage (front) - removal and refitting	26
Wiper motor and linkage (rear) - removal and refitting	27
Wiring diagrams	57

Degrees of difficulty

Easy, suitable for novice with little experience	**Fairly easy,** suitable for beginner with some experience	**Fairly difficult,** suitable for competent DIY mechanic	**Difficult,** suitable for experienced DIY mechanic	**Very difficult,** suitable for expert DIY or professional

Specifications

Battery

Type .... Lead acid, 12 volt, negative earth
Capacity
  Manual transmission
    Standard .... 45 amp hour
    Optional .... 55 or 66 amp hour
  Automatic transmission
    Standard .... 55 amp hour
    Optional .... 66 amp hour

Wiper blades

Front .... Champion C45-01
Rear (Estate models only) .... Champion C38-01

Starter motor (Lucas manufacture)

Type	5M90 pre-engaged	2M100 pre-engaged
Number of brushes	4	4
Brush material	Carbon	Carbon
Minimum brush length	0.32 in (8.0 mm)	0.37 in (9.5 mm)
Brush spring pressure	28 oz (800 gm)	36 oz (1020 gm)
Minimum commutator thickness	0.08 in (2.05 mm)	0.08 in (2.05 mm)
Armature endfloat	0.010 in (0.25 mm)	0.012 in (0.30 mm)
Type of drive	Solenoid	Solenoid
Direction of rotation	Clockwise	Clockwise

Starter motor (Bosch manufacture)

Type	EF-0.85KW and GF-1.1 KW pre-engaged	JF 2KW pre-engaged
Number of brushes	4	4
Brush material	Carbon	Carbon
Minimum brush length	0.39 in (10 mm)	0.39 in (10 mm)
Brush spring pressure	32 to 46 oz (900 to 1300 gm)	32 to 46 oz (900 to 1300 gm)
Minimum commutator diameter	1.29 in (32.8 mm)	1.66 in (42.2 mm)
Armature endfloat	0.012 in (0.3 mm)	0.012 in (0.3 mm)
Type of drive	Solenoid	Solenoid
Direction of rotation	Clockwise	Clockwise

Alternator

Output .... 44 to 70 amp dependent on unit fitted and vehicle specification
Maximum continuous speed .... 15 000 rpm
Regulating voltage at 4000 rpm with 3 to 7 amp load ...... 13.7 to 14.6 volt

Bulb application

Bulb	Wattage	Fitting
Headlight	40/45	Retainer
Headlight (Halogen)	55/60	Retainer
Driving lights (Halogen)	55	Retainer
Sidelights	4	Bayonet
Direction indicators (front)	21	Bayonet
Direction indicators (rear)	21	Bayonet
Stop lights	21/5	Bayonet
Tail lights	5	Bayonet
Number plate light	4	Bayonet
Reversing lights	21	Bayonet
Rear foglight	21	Bayonet

1 General description

The electrical system is of the 12 volt negative earth type and the major components comprise a 12 volt battery of which the negative terminal is earthed, an alternator which is driven from the crankshaft pulley, and a starter motor.

The battery supplies a steady amount of current for the ignition, lighting and other electrical circuits and provides a reserve of electricity when the current consumed by the electrical equipment exceeds that being produced by the alternator.

The alternator has its own regulator which ensures a high output if the battery is in a low state of charge or the demand from the electrical equipment is high, and a low output if the battery is fully charged and there is little demand for the electrical equipment.

When fitting electrical accessories to cars with a negative earth system it is important, if they contain silicone diodes or transistors, that they are connected correctly, otherwise serious damage may result to the components concerned. Items such as radios, tape players, electronic ignition systems, electronic tachometer, automatic dipping etc, should all be checked for correct polarity.

It is important that the battery is disconnected before removing the alternator output lead as this is live at all times. Also if body repairs are to be carried out using electric arc welding equipment - the alternator must be disconnected otherwise serious damage can be caused to the more delicate instruments. Whenever the battery has to be disconnected it must always be reconnected with the negative terminal connected last and the cable well earthed. Do not disconnect the battery with the engine running. If "jumper cables" are used to start the car, they must be connected correctly - positive to positive and negative to negative.

On vehicles with a central door lock system, the relay used in the system causes a spark whenever the battery terminals are reconnected to the battery posts. For this reason, always cover the top of the battery with a thick cloth and connect the positive lead, followed by the negative lead - in that order.

2 Battery - removal and refitting

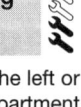

1 The battery is on a carrier fitted to the left or right wing valance of the engine compartment depending on model. It should be removed once every three months for cleaning and testing. Disconnect the negative and then the positive leads from the battery terminals by undoing and removing the nuts and bolts. Note that two cables are connected to the positive terminal.

2 Unscrew and remove the bolt, and plain washer that secures the battery clamp plate to the carrier. Lift away the clamp plate. Carefully lift the battery from its carrier and hold it vertically to ensure that none of the electrolyte is spilled.

3 Replacement is a direct reversal of this procedure.

HAYNES HINT *Refit the positive lead before the negative lead and smear the terminals with petroleum jelly to prevent corrosion. Never use an ordinary grease.*

3 Battery maintenance and inspection

1 Normal weekly battery maintenance consists of checking the electrolyte level of each cell to ensure that the separators are covered by ¼ inch (6 mm) of electrolyte. If the level has fallen, top-up the battery using distilled water only. Do not overfill. If a battery is overfilled, or any electrolyte spilled, immediately wipe away the excess as electrolyte attacks and corrodes any metal it comes into contact with very rapidly.

2 If the battery has an Auto-fill device, a special topping-up sequence is required. The white balls in the Auto-fill battery are part of the automatic topping-up device which ensures correct electrolyte level. The vent chamber should remain in position at all times except when topping-up or taking specific gravity readings. If the electrolyte level in any of the cells is below the bottom of the filling tube top-up as follows:

a) Lift off the vent chamber cover
b) With the battery level, pour distilled water into the trough until all the filling tubes and trough are full
c) Immediately replace the cover to allow the water in the trough and tubes to flow into the cells. Each cell will automatically receive the correct amount of water

3 As well as keeping the terminals clean and covered with petroleum jelly, the top of the battery, and especially the top of the cells, should be kept clean and dry. This helps prevent corrosion and ensures that the battery does not become partially discharged by leakage through dampness and dirt.

4 Once every three months remove the battery and inspect the battery securing bolts, the battery clamp plate, tray and battery leads for corrosion (white fluffy deposits on the metal which are brittle to touch). If any corrosion is found, clean off the deposit with ammonia and paint over the clean metal with an anti-rust/anti-acid paint.

5 At the same time inspect the battery case for cracks. Cracks are frequently caused to the top of the battery by pouring in distilled water in the middle of winter *after* instead of *before* a run. This gives the water no chance to mix with the electrolyte and so the former freezes and splits the battery case.

6 If topping-up the battery becomes excessive and the case has been inspected for cracks that could cause leakage, but none are found, the battery is being overcharged.

7 With the battery on the bench at the three monthly interval check, measure the specific gravity with a hydrometer to determine the state of charge and condition of the electrolyte. There should be very little variation between the different cells and, if a variation in excess of 0.025 is present it will be due to either:

a) Loss of electrolyte from the battery at some time caused by spillage or a leak, resulting in a drop in the specific gravity of the electrolyte when the deficiency was replaced with distilled water instead of fresh electrolyte
b) An internal short circuit caused by buckling of the plates or similar fault pointing to the likelihood of total battery failure in the near future

8 The specific gravity of the electrolyte for fully charged conditions, at the electrolyte temperature indicated, is listed in Table A. The specific gravity of a fully discharged battery at different temperatures of the electrolyte is given in Table B.

Table A - Specific gravity - battery fully charged

1.268 at 100°F or 38°C electrolyte temperature
1.272 at 90°F or 32°C electrolyte temperature
1.276 at 80°F or 27°C electrolyte temperature
1.280 at 70°F or 21°C electrolyte temperature
1.284 at 60°F or 16°C electrolyte temperature
1.288 at 50°F or 10°C electrolyte temperature
1.292 at 40°F or 4°C electrolyte temperature
1.296 at 30°F or - 1.5°C electrolyte temperature

Table B - Specific gravity - battery fully discharged

1.098 at 100°F or 38°C electrolyte temperature
1.102 at 90°F or 32°C electrolyte temperature
1.106 at 80°F or 27°C electrolyte temperature
1.110 at 70°F or 21°C electrolyte temperature
1.114 at 60°F or 16°C electrolyte temperature
1.118 at 50°F or 10°C electrolyte temperature
1.122 at 40°F or 4°C electrolyte temperature
1.126 at 30°F or -1.5°C electrolyte temperature

4 Battery - electrolyte replenishment

1 If the battery is in a fully charged state and one of the cells maintains a specific gravity reading which is 0.025 or more lower than the others, it is likely that electrolyte has been lost from the cell at some time.

2 Top-up the cell with a solution of 1 part sulphuric acid to 2.5 parts of water obtainable ready mixed from your garage. If the cell is already fully topped-up syphon some electrolyte out.

5 Battery - charging

1 In winter time when heavy demand is placed upon the battery, such as when starting from cold, and much electrical equipment is continually in use, it is a good idea to occasionally have the battery fully charged from an external source at the rate of 3.5 to 4 amps.

2 Continue to charge the battery at this rate, until no further rise in specific gravity is noted over a four hour period.

3 Alternatively, a trickle-charger charging at the rate of 1.5 amps can be safely used overnight.

4 Specially rapid "boost" charges, which are claimed to restore the power of the battery in 1 to 2 hours, are not recommended as they can cause serious damage to the battery plates through overheating.

5 While charging the battery note that the temperature of the electrolyte should not exceed 100° F (37.8°C).

6 Alternator - general description

The main advantages of an alternator are its high current output, its capacity for providing a high charge at low revolutions and its ability to withstand high rpm without damage.

The three makes of alternator fitted to the Granada variants generate alternating current (ac) which is changed to a direct current (dc) by an internal diode system. They all incorporate a regulator which limits the output to 14 volts maximum at all times. A warning lamp illuminates if the alternator fails to operate.

The alternator assembly basically consists of a fixed coil winding (stator) in an aluminium housing, which incorporates the mounting lugs. Inside the stator rotates a shaft wound coil (rotor). The shaft is supported at each end by ball bearings, which are lubricated for life.

Slip rings are used to conduct current to and from the rotor field coils via two carbon brushes which bear against them. By keeping the mean diameter of the slip rings to a minimum, relative speed between brushes and hence wear, are also minimal.

The rotor is belt-driven from the engine through a pulley keyed to the rotor shaft. A pressed steel fan adjacent to the pulley draws cooling air through the machine. This fan forms an integral part of the alternator specification. It has been designed to provide adequate air flow with a minimum of noise, and to withstand the high stresses associated with the maximum speed. Rotation is clockwise viewed on the drive end.

The brush gear is housed in a moulding, screwed to the outside of the slip ring and bracket. This moulding thus encloses the slip ring and brush gear assembly, and together with the shielded bearing, protects the assembly against the entry of dust and moisture.

The regulator is set during manufacture and requires no further attention.

Electrical connections to external circuits are brought out to Lucar connector blades, these being grouped to accept a moulded connector socket which ensures correct connection. Detail design differences are shown in Fig. 10.1 10.2 and 10.3

10•4 Electrical system

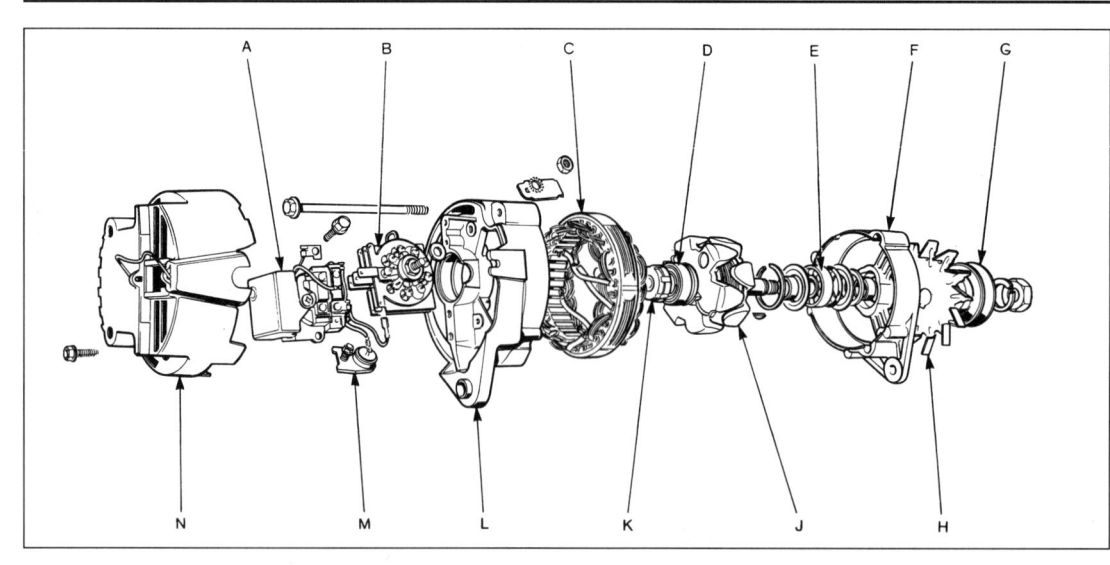

Fig. 10.1 Lucas 18 ACR alternator (Sec 6)

A Regulator
B Rectifier
C Stator
D Bearing
E Bearing
F End housing
G Pulley
H Fan
J Rotor
K Slip ring
L End housing
M Surge protective diode
N End cover

Fig. 10.2 Bosch alternator (Sec 6)

A Fan
B Spacer
C End housing
D Bearing retainer
E Bearing
F End housing
G Brush box and regulator
H Rectifier
J Stator
K Slip ring
L Rotor
M End bearing
N Spacer
O Pulley

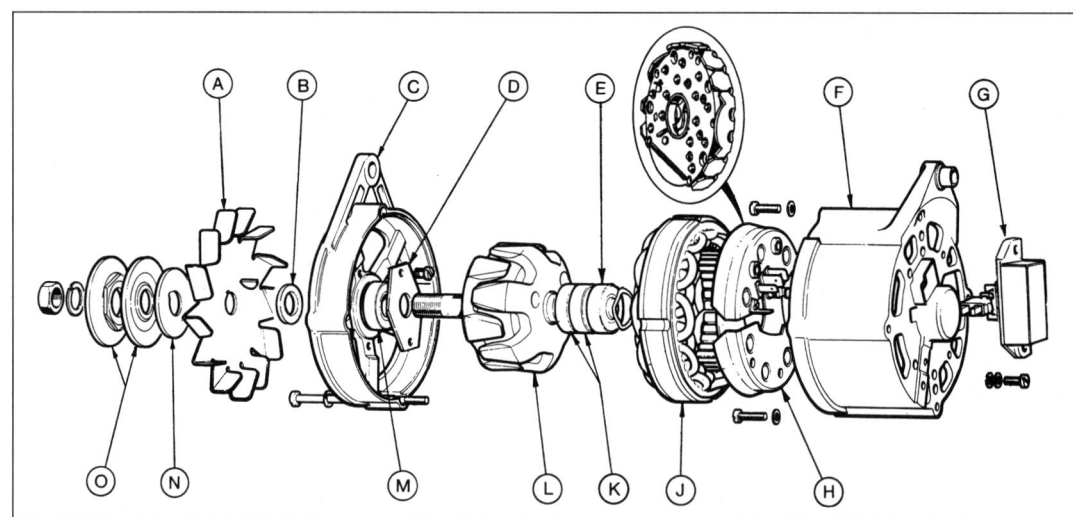

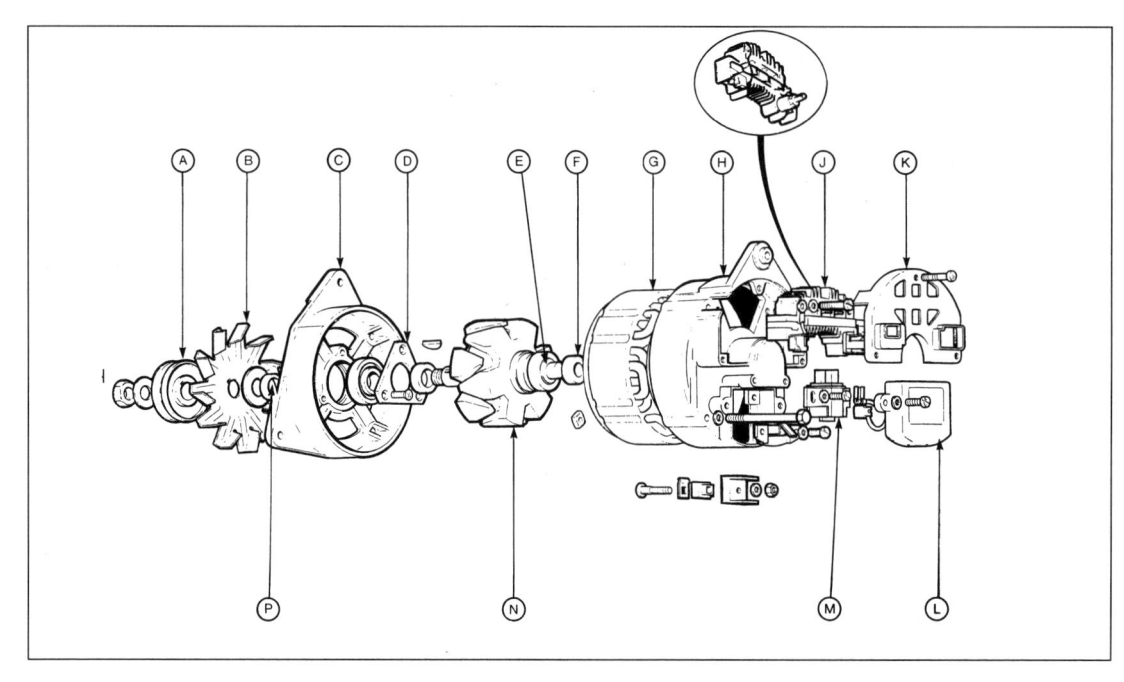

Fig. 10.3 Motorola alternator (Sec 6)

A Pulley
B Fan
C End cover
D Bearing retainer
E Slip ring
F Bearing
G Stator
H End housing
J Diode bridge
K End cover
L Regulator
M Brush box
N Rotor
P Spacer

Electrical system 10•5

7 Alternator - routine maintenance

1 The equipment has been designed for the minimum amount of maintenance in service, the only items subject to wear being the brushes and bearings.
2 Brushes should be examined after about 75 000 miles (120 000 km) and renewed if necessary. The bearings are pre-packed with grease for life, and should not require further attention.
3 Check the fan belt every 3000 miles (5000 km) for correct adjustment which should be 0.5 inch (13 mm) total movement at the centre of the run between the alternator and water pump pulleys.

8 Alternator - special procedures

Whenever the electrical system of the car is being attended to, or external means of starting the engine are used, there are certain precautions that must be taken otherwise serious and expensive damage can result.
1 Always make sure that the negative terminal of the battery is earthed. If the terminal connections are accidentally reversed or if the battery has been reverse charged, the alternator diodes will burn out.
2 The output terminal on the alternator marked "BAT" or B+ must never be earthed but should always be connected directly to the positive terminal of the battery.
3 Whenever the alternator is to be removed, or when disconnecting the terminals of the alternator circuit, always disconnect the battery earth terminal first.
4 The alternator must never be operated without the battery to alternator cable connected.
5 Should it be necessary to use a booster charger or booster battery to start the engine, always double check that the negative cable is connected to negative terminal and the positive cable to positive terminal.

9 Alternator - removal and refitting

1 Disconnect the battery leads.
2 Release the spring clip and disconnect the multi-pin connect from the rear of the alternator (photo).
3 Loosen the alternator mounting bolts and push the alternator inwards towards the engine and remove the fan-belt from the pulley.
4 Remove the alternator mounting bolts and lift the alternator out of the engine compartment.
5 Take care not to knock or drop the alternator as this can cause irreparable damage.
6 Refitting the alternator is the reverse of the removal procedure. Adjust the tension of the fan-belt so that it has 0.5 in (13 mm) deflection at the centre of the run between the alternator and water pump pulley.

10 Alternator - fault finding and repair

Owing to the specialist knowledge and equipment required to test or service an alternator it is recommended that if the performance is suspect the car be taken to an automobile electrician who will have the facilities for such work. Because of this recommendation, information is limited to the inspection and renewal of the brushes. Should the alternator not charge or the system be suspect, the following points may be checked before seeking further assistance:
1 Check the fan-belt tension, as described in Section 9.
2 Check the battery, as described in Section 3.
3 Check all electrical cable connections for cleanliness and security.

11 Alternator brushes (Lucas) - inspection, removal and refitting

1 Refer to Fig.10.1 and undo and remove the two screws that hold on the end cover. Lift away the end cover.
2 Remove the brush retaining screws (Fig. 10.4) and withdraw the brushes from the brush box.

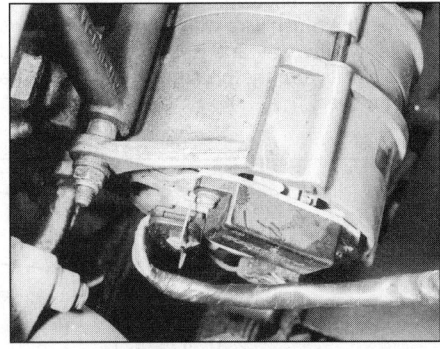

9.2 Alternator multi-plug connector

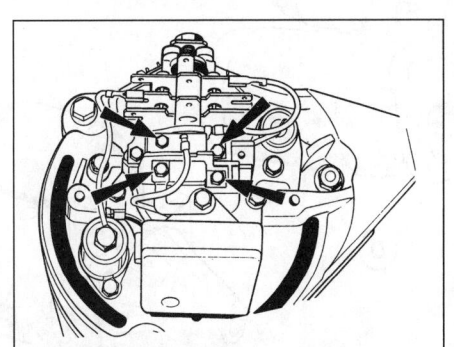

Fig. 10.4 Brush box attachment points - Lucas alternator (Sec 11)

3 Measure the length of the brushes and if they have worn down to 0.2 in (5 mm) or less, they must be renewed.
4 Insert the new brushes and check to make sure that they are free to move in their guides. If they bind, lightly polish with a very fine file.
5 Reassemble in the reverse order of dismantling. Make sure that leads which may have been connected to any of the screws are reconnected correctly.

12 Alternator brushes (Bosch) - inspection, removal and refitting

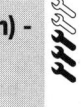

1 Undo and remove the two screws, spring washers and plain washers that secure the brush box to the rear of the brush end housing. Lift away the brush box.
2 Check that the carbon brushes are able to slide smoothly in their guides without any sign of binding.
3 Measure the length of the brushes and if they have worn down to 0.2 in (5 mm) or less, they must be renewed.
4 Hold the brush wire with a pair of engineer's pliers which will act as a heat sink and unsolder it from the brush box. Lift away the two brushes.
5 Insert the new brushes and check to make sure that they are free to move in their guides. If they bind, lightly polish with a very fine file.

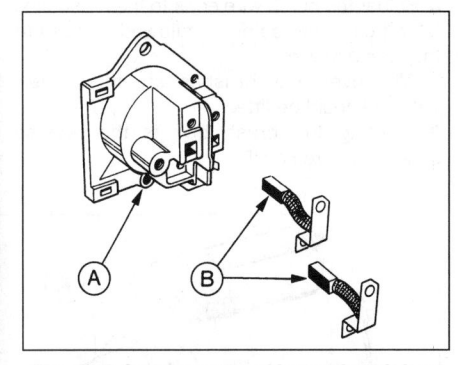

Fig. 10.5 Components of Lucas brush box (Sec 11)

A Brush box B Brushes

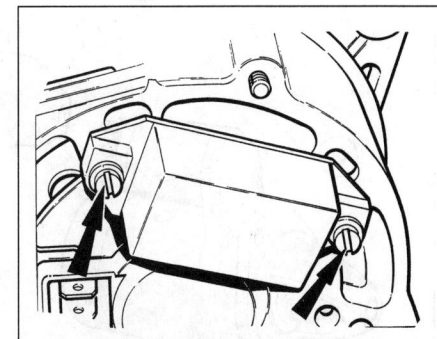

Fig. 10.6 Regulator securing screws - Bosch alternator (Sec 12)

10•6 Electrical system

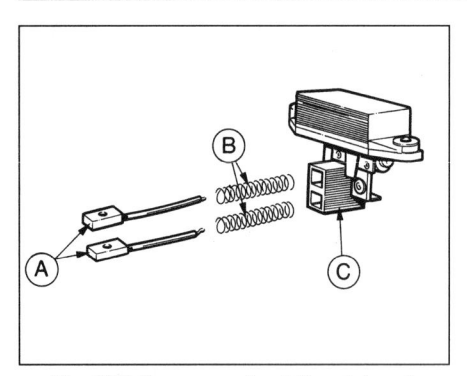

Fig. 10.7 Components of Bosch brush box/regulator assembly (Sec 12)

A Brushes B Springs C Brush box

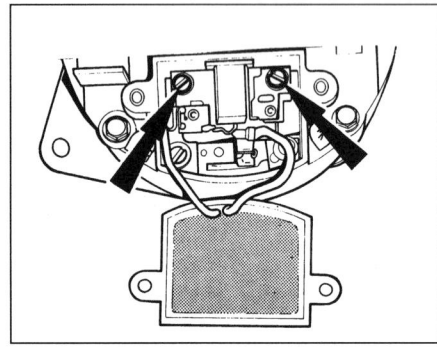

Fig. 10.8 Brush box securing screws - Motorola alternator (Sec 13)

6 Solder the brush wire ends to the brush box, taking care that solder is allowed to pass to the stranded wire.
7 Whenever new brushes are fitted, new springs should be fitted.
8 Refitting the brush box is the reverse sequence to removal.

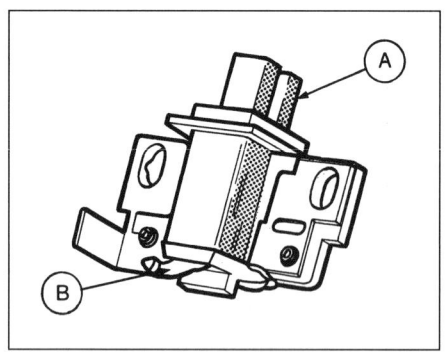

Fig. 10.9 Motorola alternator brush box (Sec 13)

A Brushes B Brush box

13 Alternator brushes (Motorola) - inspection, removal and refitting

1 Undo and remove the two screws and washers securing the regulator to the rear of the alternator. Detach the two regulator wires and lift off the regulator.
2 Undo and remove the two screws and washers that secure the brush box to the alternator and carefully turn the brush box outwards.
3 Check that the carbon brushes are able to slide smoothly in their guides without any sign of binding.
4 Measure the length of the brushes and if they have worn down to 0.2 in (5 mm) or less, they must be renewed.
5 Hold the brush wire with a pair of engineer's pliers which will act as a heat sink and unsolder it from the brush box. Lift away the two brushes.

6 Insert the new brushes and check to make sure that they are free to move in their guides. If they bind, lightly polish with a very fine file.
7 Solder the brush wire ends to the brush box, taking care that solder is not allowed to pass to the stranded wire.
8 Whenever new brushes are fitted, new springs should be fitted.
9 Refitting the brush box is the reverse sequence to removal.

14 Starter motor - general description

The Granada range of vehicles covered by this manual is fitted with pre-engaged starter motors of either Lucas or Bosch manufacture (Fig. 10.10). These motors are all of series wound, four pole, four brush design and are fully serviceable.

The operation of the starter motor is as follows: With the ignition switched on, current flows from the solenoid which is mounted on the top of the starter motor body. Contained within the solenoid is a plunger which moves inward so causing a centrally pivoted engagement lever to move in such a manner that the forked end pushes the drive pinion into mesh with the starter ring gear. When the solenoid plunger reaches the end of its travel, it closes an internal contact and full starting current flows to the starter field coils. The armature is then able to rotate the crankshaft so starting the engine.

A one way clutch is fitted to the starter drive pinion so that when the engine fires and starts to operate on its own, it does not drive the starter motor.

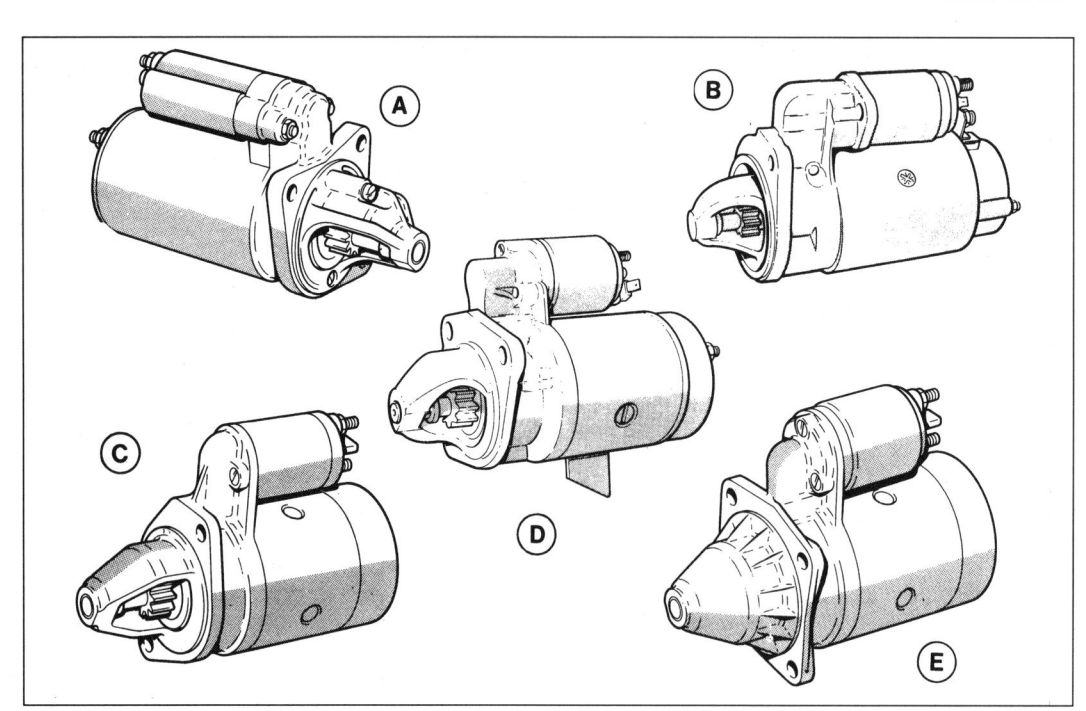

Fig. 10.10 Starter motor types (Sec 14)

A Lucas 5M90
B Lucas 2M100
C Bosch EF-0.85JW
D Bosch JF 2KW
E Bosch GR-1.1KW

Electrical system 10•7

15 Starter motor – testing on engine

1 If the starter motor fails to operate, then check the condition of the battery by turning on the headlamps. If they glow brightly for several seconds and then gradually dim, the battery is in a discharged condition.
2 If the headlights continue to glow brightly and it is obvious that the battery is in good condition, then check the tightness of the battery wiring connections, in particular the earth lead from the battery terminal to its connection on the body frame. If the positive terminal on the battery becomes hot when an attempt is made to work the starter, this is a sure sign of a poor connection on the battery terminal. To rectify, remove the terminal, clean the mating faces thoroughly and reconnect. Check the connections on the rear of the starter solenoid. Check the wiring with a voltmeter or test lamp for breaks or shorts.
3 To test the continuity of the solenoid windings, connect a test lamp circuit comprising a 12 volt battery and low wattage bulb between the STA terminal and the solenoid body. If the two windings are in order, the lamp will light. Next connect the test lamp (fitted with a high wattage bulb) between the solenoid main terminals. Energise the solenoid by applying a 12 volt supply between the unmarked "Lucar" terminal and the solenoid body. The solenoid should be heard to operate and the test bulb light. This indicates full closure of the solenoid contacts.
4 If the battery is fully charged, the wiring in order, and the starter/ignition switch working and the starter motor still fails to operate, then it will have to be removed from the car for examination.

16 Starter motor – removal and refitting

1 Disconnect the battery negative terminal.
2 Make a note of the cable connection to the rear of the solenoid and detach the cable terminals from the solenoid.
3 Undo and remove the starter motor securing nuts, bolts and spring washers and lift away the starter motor.
4 Refitting is the reverse of the removal sequence.

17 Starter motor (Lucas 5M90) – dismantling and reassembly

1 Clamp the starter motor in a vice with soft jaws, and remove the plastic cap from the commutator end plate.
2 Remove the retaining clip from the end of the armature shaft and discard the clip. Remove the thrust washer.
3 Disconnect and remove the connecting cable from the end of the solenoid. Remove the two securing nuts and washers and guide the solenoid away from the drive end housing.

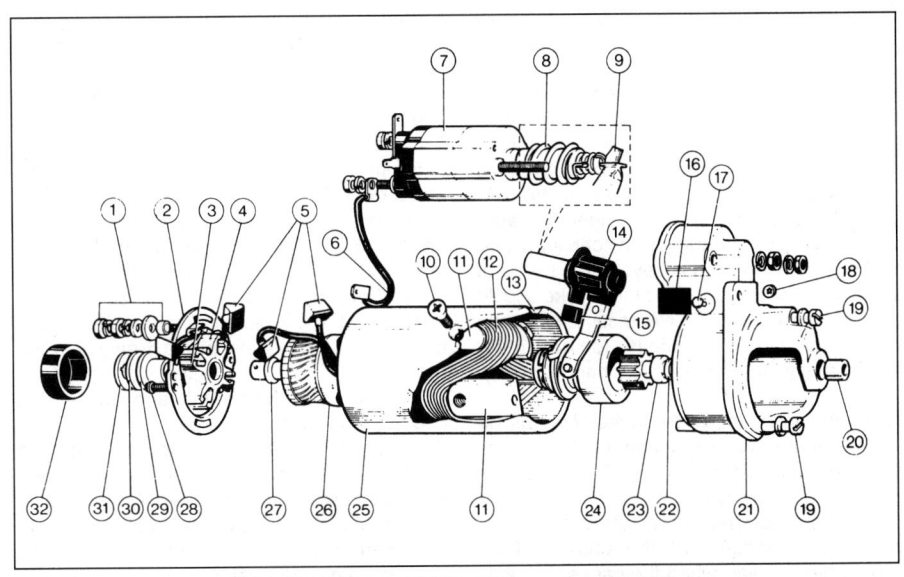

Fig. 10.11 Lucas 5M90 starter motor (Sec 17)

1 Terminal assembly	12 Field coils	23 Thrust collar
2 End plate	13 Earth connection	24 Drive assembly
3 Brush housing	14 Rubber seal	25 Main body (yoke)
4 Brush springs	15 Rubber dust pad	26 Armature
5 Brushes	16 Rubber dust cover	27 Thrust washer
6 Solenoid link wire	17 Pivot pin	28 Securing screws
7 Solenoid	18 Retaining clip	29 Bearing bush
8 Return spring	19 Securing screws	30 Thrust plate
9 Engagement lever	20 Bearing bush	31 Star clip
10 Pole screw	21 Drive end housing	32 Dust cover
11 Pole shoe	22 C-clip	

Unhook the solenoid armature from the actuating lever by moving it upwards and away from the lever.
4 Remove the two drive end housing screws and guide the housing and armature assembly away from the body.
5 Remove the armature from the end housing and unhook the actuating arm from the pinion assembly. Remove the rubber block and sleeve from the housing.
6 Drive the pivot pin from the end housing, and remove the actuating lever. Discard the pivot pin clip, which will be distorted.
7 If it is necessary to dismantle the starter pinion drive, place the armature between soft faces in a bench vice and using a universal puller draw the jump ring from the armature.
8 Tap down the circlip retaining cover and remove the washer, circlip and cover. The pinion assembly may now be removed.
9 Draw the actuating bush towards the pinion so as to expose the circlip. Remove the circlip, bush, spring and large washer. It is very important that the one way clutch is not gripped in the vice at the point adjacent to the pinion whilst this is being carried out otherwise the clutch will be damaged.
10 The drive pinion and one way clutch are serviced as a complete assembly, so if one part is worn or damaged a new assembly must be obtained.
11 Remove the four commutator end plate screws and carefully tap the plate free from the body. Withdraw the plate slightly, remove the two field winding brushes and remove the plate.
12 To renew the field winding brushes, their flexible connectors must be cut leaving 0.25 in (7 mm) attached to the field coils. Discard the old brushes. Solder new brushes to the flexible connector studs. Check that the new brushes move freely in their holders.
13 The main terminal stud and its two brushes are available as a unit. To remove, take off the nut, washer and insulator and push the stud and second insulator through the end plate.
14 If cleaning the commutator with petrol fails to remove all the burnt areas and spots, then wrap a piece of glass paper round the commutator and rotate the armature.
15 If the commutator is very badly worn, remove the drive gear, (if still in place on the armature), and mount the armature in a lathe. With the lathe turning at high speed take a very fine cut out of the commutator and finish the surface by polishing with glass paper. *Do not undercut the mica insulators between the commutator segments. The minimum commutator thickness must never be less than 0.08 in (2 mm).*
16 With the starter motor dismantled, test the four field coils for an open circuit. Connect a 12 volt battery, with a 12 volt bulb in one of the leads between the field terminal post and the tapping point of the field coils to which the brushes are connected. An open circuit is proved by the bulb not lighting.

10

10•8 Electrical system

17 If the bulb lights, it does not necessarily mean that the field coils are in order, as there is a possibility that one of the coils will be earthed to the starter yoke or pole shoes. To check this, remove the lead from the brush connector and place it against a clean portion of the starter yoke. If the bulb lights, then the field coils are earthing. Renewal of the field coils calls for the use of a wheel operated screwdriver, a soldering iron, caulking and riveting operations and is beyond the scope of the majority of owners. The starter yoke should be taken to a reputable electrical engineering works for new field coils to be fitted. Alternatively purchase an exchange starter motor.

18 If the armature is damaged this will be evident after visual inspection. Look for signs of burning, discolouration, and for conductors that have lifted away from the commutator.

19 With the starter motor stripped down, check the condition of the bushes. They should be renewed when they are sufficiently worn to allow visible side movement of the armature shaft.

20 The old bushes are simply driven out with a suitable drift and the new bushes inserted by the same method. As the bushes are of the phosphor bronze type it is essential that they are allowed to stand in engine oil for at least 24 hours before fitment. If time does not allow, place the bushes in oil at 100°C (212°F) for 2 hours.

21 Reassembly is the reverse sequence to dismantling, but the following points should be noted:

a) When refitting the drive end housing, the peg on the housing should align with the notch in the casing
b) **New** retaining clips should be fitted to the actuating arm pivot pin and the armature shaft
c) When fitting the clip to the armature shaft end it should be pressed home firmly to eliminate any endfloat in the shaft

18 Starter motor (Lucas 2M100) - dismantling and reassembly

1 Clamp the starter motor in a vice with protective soft jaws.

2 Remove the nut and washers securing the starter motor terminal to the solenoid lower terminal marked STA.

3 Undo and remove the nuts and washers that secure the solenoid to the starter motor body. Carefully withdraw the solenoid and seal leaving behind the operating plunger. Unhook the plunger from the actuating lever by pulling it inwards and lifting up.

4 To remove the commutator end housing pull off the rubber plug to give access to the spire retaining clip fitted over the end of the armature shaft. Ensure that a new clip is available as the only satisfactory way of removing the clip is to break its claws with a sharp chisel.

5 After removing the clip unscrew and remove the two through bolts. The end housing can now be withdrawn sufficiently to allow removal of the two field coil brushes. Now lift off the end housing.

6 If the brushes are worn down to 0.37 in (9.5 mm) or less they must be renewed. The two terminal brushes and their connector are renewed as a unit. To renew the field winding brushes their flexible connectors must be cut leaving 0.25 in (7 mm) attached to the field coils. Discard the old brushes. Solder new brushes to the flexible connector studs. Check that the new brushes move freely in their holders.

7 Withdraw the armature and drive end housing from the motor body.

8 The actuating lever may now be removed from the end housing. One of two methods will be used to retain the pivot pin. Either a solid pin secured with a star clip or a hollow pin with a peened over end. To remove the solid pin, drive it from the housing using a small drift. The retaining clip will distort and allow removal of the pin. To remove the hollow pin, bend the peened part straight and drive it from the housing in the same manner.

9 Now remove the drive pinion and clutch assembly from the armature shaft. Using a suitable sized tube drive the C-clip from the thrust collar and remove the thrust collar from its groove. Slide the complete assembly from the armature shaft.

10 The drive pinion and clutch assembly are serviced as a complete unit, so if one part is worn or damaged a new assembly must be obtained.

11 The armature should now be cleaned and inspected prior to testing. Clean the commutator with petrol and inspect the segments for burning or scoring.

12 If cleaning the commutator with petrol fails to remove all the burnt areas and spots, then wrap a piece of glass paper round the commutator and rotate the armature.

13 If the commutator is very badly worn, remove the drive gear, (if still in place on the armature), and have the commutator refinished on a lathe - a job for your auto-electrical company. *Do not undercut the mica insulators between the commutator segments. The minimum commutator thickness must never be less than 0.08 in (2 mm).*

14 With the starter motor dismantled, test the four field coils for an open circuit. Connect a 12 volt battery with a 12 volt bulb in one of the leads between the field terminal post and the tapping point of the field coils to which the brushes are connected. An open circuit is proved by the bulb not lighting.

15 If the bulb lights, it does not necessarily mean that the field coils are in order, as there is a possibility that one of the coils will be earthed to the starter yoke or pole shoes. To check this, remove the lead from the brush connector and place it against a clean portion of the starter yoke. If the bulb lights, then the field coils are earthing. Replacement of the field coils calls for the use of a wheel operated screwdriver, a soldering iron, caulking and riveting operations and is beyond the scope of the majority of owners. The starter yoke should be taken to a reputable electrical engineering works for new field coils to be fitted. Alternatively purchase an exchange starter motor.

16 If the armature is damaged this will be evident after visual inspection. Look for signs

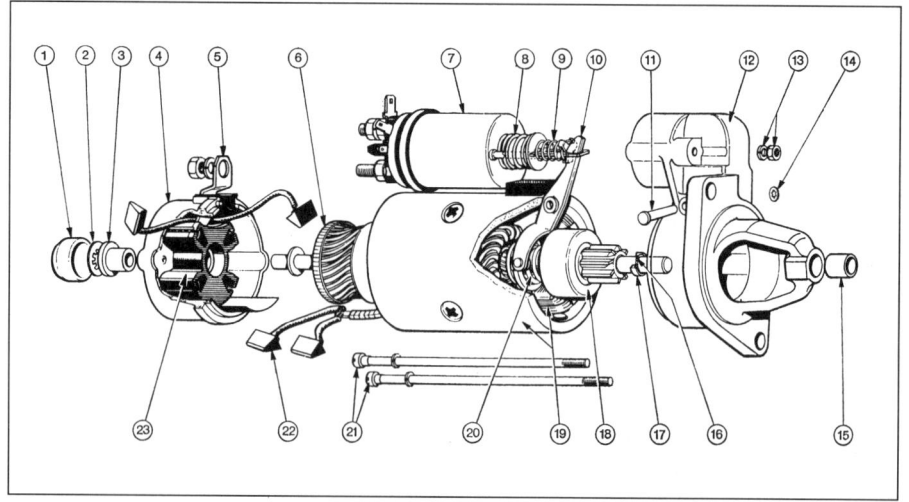

Fig. 10.12 Lucas 2M100 starter motor (Sec 18)

1 Rubber cover
2 Retaining clip
3 Bush
4 End housing
5 Terminal link
6 Armature
7 Solenoid
8 Return spring
9 Lose motion spring
10 Actuating lever
11 Pivot pin
12 Drive end housing
13 Retaining nut and washer
14 Retaining clip
15 Bush
16 C-clips
17 Thrust collar
18 Pinion gear
19 Pole shoe
20 Spring
21 Retaining screws
22 Brush
23 Brush box

Electrical system 10•9

of burning, discolouration, and for conductors that have lifted away from the commutator.

17 With the starter motor stripped down, check the condition of the bushes. They should be renewed when they are sufficiently worn to allow visible side movement of the armature shaft.

18 The old bushes are simply driven out with a suitable drift and the new bushes inserted by the same method. As the bushes are of the phosphor bronze type it is essential that they are allowed to stand in engine oil for at least 24 hours before fitment. If time does not allow, place the bushes in oil at 100°C (212°F) for 2 hours.

19 Reassembly is the reverse sequence to dismantling, but the following points should be noted:
a) When refitting the drive end housing, the peg on the housing should align with the notch in the casing
b) **New** retaining clips should be fitted to the actuating arm pivot pin and the armature shaft
c) When fitting the clip to the armature shaft end it should be pressed home firmly to eliminate any endfloat in the shaft

19 Starter motor (Bosch EF-0.85KW and GF-1.1 KW) - dismantling and reassembly

In addition to the standard equipment Lucas starter motors, either of the two Bosch units described in this Section or the heavy duty Bosch JF 2KW starter motor (Section 20) may be fitted to Granada vehicles depending on specification and engine type. The EF-0.85KW unit may be fitted to V6 engines and the GF-1.1 KW unit to in-line ohc engines. The dismantling, reassembly and test procedures are identical except where stated below.

1 Clamp the starter motor in a vice with protective soft jaws. Undo and remove the nut and washer securing the field winding cable to the solenoid and detach the cable from the stud.

2 **In-line ohc engines.** Undo and remove the three screws securing the solenoid to the starter motor body. Lift off the solenoid leaving the armature and spring behind. Unhook the armature from the actuating lever and lift away.

3 **V6 engines.** Undo and remove the two screws securing the solenoid to the starter motor body. Unhook the solenoid from the actuating arm and lift clear. Note the solenoid and armature are a complete unit and cannot be dismantled.

4 Take out the two screws that secure the commutator end housing cover and remove the cover and rubber seal.

5 Wipe clean the exposed end of the armature shaft and prise off the circlip and shims.

6 Undo and remove the two nuts and washers securing the end housing to the starter motor body and lift off the end housing. Note on some motors, fixing screws are used as an alternative to nuts.

7 The brushes and brush plate may now be removed from the motor. Carefully lift up the brush tensioning springs and pull the brushes clear of their guides. Now carefully slide the brush plate from its location on the armature.

8 Separate the armature and drive end housing from the starter motor body by gently tapping the end housing.

9 **In-line ohc engines.** Remove the rubber insert plug from the drive end housing. Undo and remove the nut from the end of the actuating arm pivot screw and withdraw the screw.

10 On starter motors that use studs to secure the end housing to the body, remove the studs.

11 **In-line ohc engines.** Withdraw the armature assembly complete with actuating arm from the drive end housing. Unhook the actuating arm from the drive pinion flange and lift off the arm.

12 **V6 engines.** Withdraw the armature assembly from the drive end housing. As it is pulled clear uncouple the actuating arm from the drive pinion flange. If it is required to remove the actuating arm from the drive end housing, proceed as described in paragraph 9.

13 The drive pinion assembly may be removed from the armature by tapping the thrust collar down the shaft with a suitable tube. This will expose the securing C-clip which may be withdrawn allowing the thrust collar and drive pinion to be slid off the end of the armature shaft. **Note.** Do not grip the one-way clutch in the vice during this operation as it is easily damaged.

14 Carefully inspect the dismantled components of the starter motor for wear or damage. Check the brushes for sticking in their holders. If necessary clean the brushes and brush plate with a petrol soaked cloth.

15 If the brushes are worn down to 0.39 in (10 mm) or less they should be renewed as a set.

16 To renew the brushes, cut the leads of the old brushes at a point midway between the brush and its base. Solder the leads of the new brushes at this point.

17 Clean the commutator with a petrol soaked rag and inspect for burning and scoring.

18 If cleaning the commutator with petrol fails to remove all the burnt areas and spots, then wrap a piece of glass paper round the commutator and rotate the armature.

19 If the commutator is very badly worn,

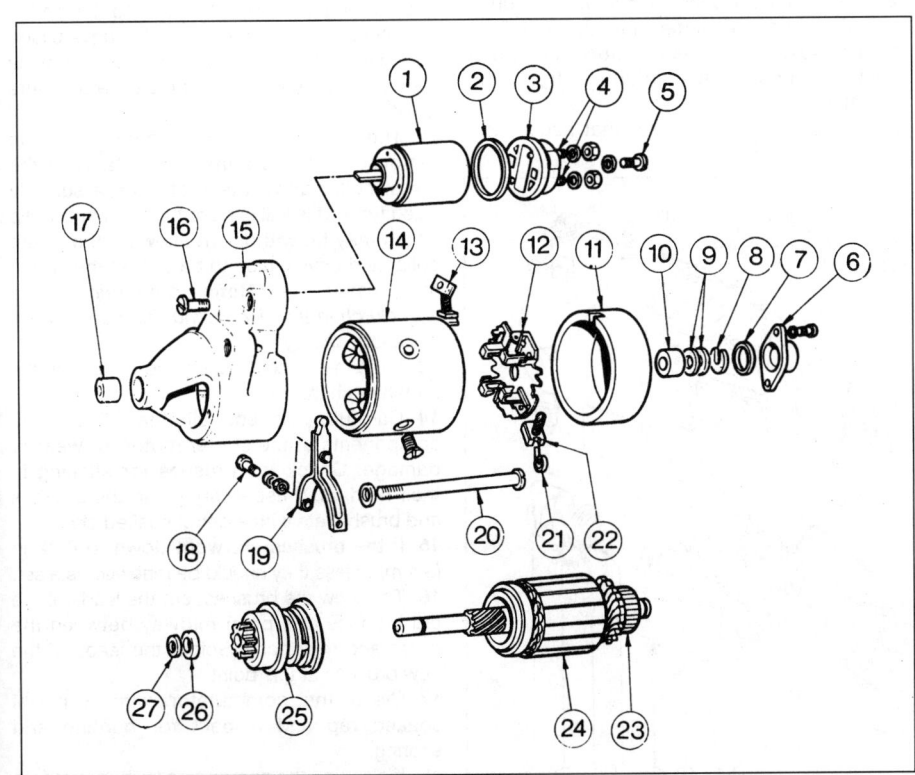

Fig. 10.13 Bosch EF-0.85 KW and GF-1.1 KW starter motor (Sec 19)

1 Solenoid body	10 Bush	19 Actuating lever
2 Gasket	11 End cover	20 Securing screw
3 Cover	12 Brushbox	21 Brush spring
4 Terminals	13 Braiding	22 Brush
5 Securing screw	14 Main body	23 Commutator
6 End cap	15 Drive end housing	24 Armature
7 Seal	16 Securing screw	25 Pinion assembly
8 C-clip	17 Bush	26 Bush
9 Shims	18 Pivot screw	27 Thrust washer

remove the drive gear, (if still in place on the armature), and have the commutator refinished on a lathe - a job for an auto-electrical company. With the lathe turning at high speed take a very fine cut-out of the commutator and finish the surface by polishing with glass paper. The minimum permissible diameter to which the commutator may be skimmed is 1.32 in (33.5 mm).

20 With the starter motor dismantled, test the four field coils for an open circuit. Connect a 12 volt battery with a 12 volt bulb in one of the leads between the field terminal post and the tapping point of the field coils to which the brushes are connected. An open circuit is proved by the bulb not lighting.

21 If the bulb lights, it does not necessarily mean that the field coils are in order, as there is a possibility that one of the coils will be earthed to the starter yoke or pole shoes. To check this, remove the lead from the brush connector and place it against a clean portion of the starter yoke. If the bulb lights, then the field coils are earthing. The field coils, poles and yoke are not serviced as separate items. If a fault is suspected on these units the yoke assembly will have to be renewed.

22 If the armature is damaged this will be evident after visual inspection. Look for signs of burning, discolouration, and for conductors that have lifted away from the commutator.

23 With the starter motor stripped down, check the condition of the bushes. They should be renewed when they are sufficiently worn to allow visible side movement of the armature shaft.

24 The old bushes are simply driven out with a suitable drift and the new bushes inserted by the same method. As the bushes are of the phosphor bronze type it is essential that they are allowed to stand in engine oil for 24 hours or if time is limited, oil at 100°C (212°F) for 2 hours.

25 Reassembly is now the reverse sequence to dismantling, but the following points should be noted:
   a) Smear all pivot points and bearing locations with grease before reassembly
   b) Fit sufficient shims to the armature shaft to eliminate excessive endfloat

20 Starter motor (Bosch JF 2KW) - dismantling and reassembly

1 Clamp the starter motor in a vice with protective soft jaws. Undo and remove the nut and washer securing the field winding cable to the solenoid and remove the cable from the stud.

2 Undo and remove the three screws securing the solenoid to the starter motor body. Lift off the solenoid leaving the armature and spring behind. Unhook the armature from the actuating arm and lift away.

3 Take out the two screws that secure the commutator end housing cover and remove the cover and rubber seal.

4 Wipe clean the exposed end of the armature shaft and prise off the circlip and shims.

5 Undo and remove the two nuts and washers securing the end housing to the starter motor body and lift off the end housing.

6 The brushes and brush plate may now be removed from the motor. Remove one brush from the brush plate assembly by bending out the retaining tabs with a pair of pliers so that the brush and spring can be withdrawn. Use a small screwdriver to retain the brush spring during this operation otherwise it will fly off and be lost. Repeat the above procedure for the remaining three brushes.

7 With the brushes removed the brush plate may be withdrawn from its location on the end of the armature shaft.

8 Separate the armature and drive end housing from the starter motor body by gently tapping the end housing.

9 Undo the nut from the end of the actuating arm pivot screw and withdraw the screw.

10 Unscrew and remove the two long studs that secure the end housing to the starter motor body.

11 Withdraw the armature assembly from the end housing. It will be necessary to detach the actuating arm forked end from the drive pinion and manipulate it in such a manner as to provide clearance for the drive pinion as the armature is withdrawn.

12 The drive pinion assembly may be removed from the armature by tapping the thrust collar down the shaft with a suitably sized tube. This will expose the securing C-clip which may be withdrawn allowing the thrust collar and drive pinion to be slid off the end of the armature shaft. **Note:** *Do not grip the one way clutch in the vice during this operation as it is easily damaged.*

13 Slide the centre bearing and plate from the armature shaft.

14 Carefully inspect all the dismantled components of the starter motor for wear or damage. Check the brushes for sticking in their holders. If necessary clean the brushes and brush plate with a petrol soaked cloth.

15 If the brushes are worn down to 0.32 in (8 mm) or less they should be renewed as a set.

16 To renew the brushes, cut the leads of the old brushes at a point midway between the brush and their base. Solder the leads of the new brushes at this point.

17 Clean the commutator with a petrol soaked rag and inspect for burning and scoring.

18 If cleaning the commutator with petrol fails to remove all the burnt areas and spots, then wrap a piece of glass paper round the commutator and rotate the armature.

19 If the commutator is very badly worn, remove the drive gear, (if still in place on the armature), and have the commutator refinished on a lathe - a job for your auto-electrical company. The minimum permissible diameter to which the commutator may be skimmed is 1.32 in (33.5 mm).

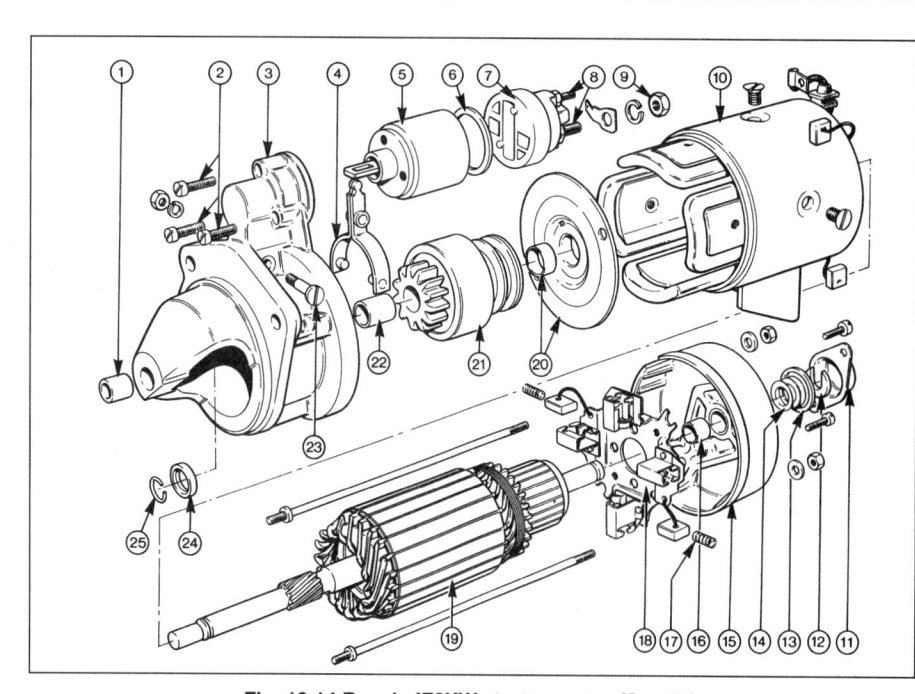

Fig. 10.14 Bosch JF2KW starter motor (Sec 20)

1 Bush
2 Securing screws
3 Drive end housing
4 Actuating fork
5 Solenoid body
6 Gasket
7 Cover
8 Terminals
9 Nut
10 Main body
11 End cap
12 C-clip
13 Seal
14 Shims
15 End cover
16 Bush
17 Brush spring
18 Brush box
19 Armature
20 Bearing and plate
21 Pinion assembly
22 Bush
23 Pivot screw
24 Bush
25 C-clip

Electrical system 10•11

20 With the starter motor dismantled, test the four field coils for an open circuit. Connect a 12 volt battery with a 12 volt bulb in one of the leads between the field terminal post and the tapping point of the field coils to which the brushes are connected. An open circuit is proved by the bulb not lighting.
21 If the bulb lights, it does not necessarily mean that the field coils are in order, as there is a possibility that one of the coils will be earthed to the starter yoke or pole shoes. To check this, remove the lead from the brush connector and place it against a clean portion of the starter yoke. If the bulb lights, then the field coils are earthing. The field coils, poles and yoke are not serviced as separate items. If a fault is suspected on these components the yoke assembly will have to be renewed.
22 If the armature is damaged this will be evident after visual inspection. Look for signs of burning, discolouration, and for conductors that have lifted away from the commutator.
23 With the starter motor stripped down, check the condition of the bushes. They should be renewed when they are sufficiently worn to allow visible side movement of the armature shaft.
24 The old bushes are simply driven out with a suitable drift and the new bushes inserted by the same method. As the bushes are of the phosphor bronze type it is essential that they are allowed to stand in engine oil for 24 hours or if time is limited, for 2 hours in oil at 100°C (212°F).
25 Reassembly is now the reverse sequence to dismantling, but the following points should be noted:
  a) Position the fluctuating arm fork on the drive pinion to facilitate assembly into the end housing
  b) Smear all pivot points and bearing locations with grease before reassembly
  c) Fit sufficient shims to the armature shaft to eliminate any excessive endfloat

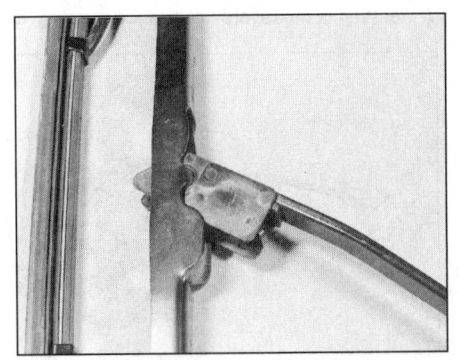

21 Flasher unit - fault tracing and rectification

1 The actual flasher unit consists of a relay located in the central electric box situated in the upper right-hand corner of the engine compartment (see Section 53 and photo 52.1).
2 If the flasher unit works twice as fast as usual, appears to flash once and then not at all or works on one side only this is an indication that there is a broken filament, in the front or rear indicator bulbs.
3 If the external flashers are working but the internal flasher warning light has ceased to function, check the filament of the warning bulb and renew if necessary.
4 If a flasher bulb is sound but does not work, check all the flasher circuit connections with the aid of a wiring diagram.
5 With the ignition switched on and the flasher operating on the suspect side, check that current is reaching the bulb holder and that there is not a build up of rust and corrosion.
6 With the flasher relay removed from the central electric panel, check that current is reaching the flasher unit socket. If the flasher unit is found to be faulty it must be renewed as a unit as it is not possible to dismantle and repair it.

22 Wiper mechanism - maintenance

1 Renew the screen wiper blades at intervals of 12 000 miles (20 000 km) or 12 months, or more frequently if found necessary.
2 The washer round the wheelbox spindle can be lubricated with several drops of glycerine every 6000 mlles (10 000 km). The screen wiper linkage pivots may be lubricated with a little engine oil.

23 Wiper blades - removal and refitting

1 Lift the wiper arm off the windscreen and move it through 90°.
2 Turn the wiper blade through 90° so it forms a "T" in relation to the arm (photo).
3 Compress the spring clip and slide the blade down the arm until it is clear of the hook.
4 Slide the blade back up the arm to clear the hook and lift off (photo).
5 Refitting is the reverse of the above procedure.

24 Wiper arms - removal and refitting

1 Before removing a wiper arm, turn the wiper switch on and off to ensure that the arms are in their normal parked position.
2 Lift up the pivoted cap to expose the wiper-arm retaining nut (photo).
3 Remove the retaining nut and lift off the washer and wiper arm (photo).
4 With the linkage in the parked position, fit the wiper arm on the pivot shaft. Fit the washer and retaining nut and press down the pivoted cap.

25 Wiper mechanism - fault diagnosis and rectification

1 Should the wipers fail, or work very slowly, then check the terminals on the motor for loose connections and make sure the insulation of all wiring has not been damaged, thus causing a short circuit. If this is in order, then check the current taken by the motor by connecting an ammeter in the circuit and turning on the wiper switch. Consumption should be between 2.3 and 3.1 amps.
2 If no current is passing through the motor, check that the switch is operating correctly.
3 If the wiper motor takes a very high current, check the wiper blades for freedom of movement. If this is satisfactory, check the gearbox cover and gear assembly for damage.
4 If the motor takes a very low current ensure that the battery is fully charged. Check the brush gear and ensure the brushes are bearing

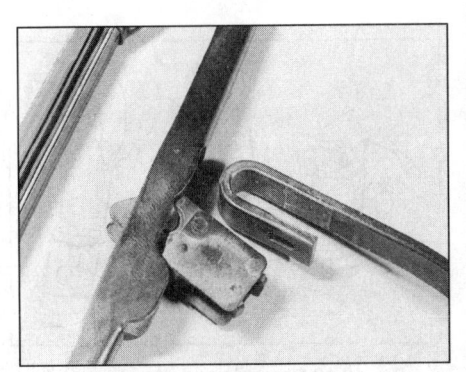

23.2 Wiper arm and blade

23.4 Removing wiper blade from arm

24.2 Retaining cap lifted to expose wiper arm securing nut

24.3 Removing wiper arm from shaft

10•12 Electrical system

26.5 Wiper motor mounting bracket securing bolt (arrowed)

on the commutator. If not, check the brushes for freedom of movement and, if necessary, renew the tension springs. If the brushes are very worn, they should be replaced with new ones. Check the armature by substitution, if this part is suspect.

26 Wiper motor and linkage (front) - removal and refitting

1 For safety reasons disconnect the battery.
2 Remove the wiper arms as described in Section 24.
3 Detach the multi-pin plug from the wiper motor.
4 Undo and remove the two nuts securing the wiper arm spindle to the body. Push the spindles through their locating holes in the bodywork.
5 Unscrew and remove the bolt securing the wiper motor mounting bracket to the bodywork (photo).
6 Disengage the wiper motor mounting bracket from its upper location grommet and withdraw the wiper motor, bracket and operating linkage from the car.
7 Undo and remove the nut securing the operating arm to the motor and the three bolts securing the motor to the bracket. Lift away the wiper motor.
8 Refitting is the reverse of the removal sequence.

27 Wiper motor and linkage (rear) - removal and refitting

1 For safety reasons disconnect the battery. Refer to Section 24 and remove the wiper arm and blade.
2 Lift off the spindle nut cover and remove the spindle nut, washer and seal.
3 Carefully remove the tailgate inner trim panel, and remove the three bolts securing the mounting bracket.
4 Remove the earth lead securing screw, disconnect the multi-plug and remove the motor and linkage assembly.
5 Prise the linkage from the motor link, remove the three bolts and detach the wiper motor.
6 Refitting is the reverse sequence of removal.

28 Wiper motor - dismantling, inspection and reassembly

1 Undo and remove the two crosshead screws that secure the gearbox cover plate to the gearbox.
2 Undo and remove the nut that secures the operating arm to the gear shaft. Lift away the operating arm, wave washers and plain washers. Note the position of the arm relative to the park cut-out segment of the gear contact plate.
3 Release the spring clips that secure the case and armature to the gearbox. Lift away the case and armature.
4 Wipe away all the grease from inside the gearbox and using a pair of circlip pliers remove the circlip that secures the gear to the shaft. Separate the gear from the shaft.
5 Undo and remove the screw that secures the brush mounting plate, detach the wiring loom plug and remove the brushes.
6 Clean all parts and then inspect the gears and brushes for wear or damage. Refit the spindle and check for wear in its bush in the gearbox body. Obtain new parts as necessary.
7 Reassembly is the reverse sequence to dismantling. Pack the gearbox with grease.

29 Horn - fault tracing and rectification

1 If the horn works badly or fails completely, first check the wiring leading to the horn for short circuits and loose connections. Also check that the horn is firmly secured and that there is nothing lying on the horn body.
2 Using a test lamp check the wiring to the number 13 fuse on the fuse box located in the engine compartment. Check that the fuse has not blown.
3 If the fault is an internal one it will be necessary to obtain a new horn.
4 To remove the horn, disconnect the battery and remove the radiator grille.
5 Detach the lead at the rear of the horn and then undo and remove the retaining bolt, spring, horn bracket and star-washer.
6 Refitting the horn is the reverse sequence of removal.

30 Windscreen washer pump and reservoir - removal and refitting

Note: *This reservoir also supplies the headlamp washers when fitted.*

1 For safety reasons disconnect the battery.
2 Drain or syphon the water from the reservoir.
3 Disconnect the electrical connections from the windscreen washer pump and headlamp washer pump if fitted.
4 Disconnect the water hoses from the pump(s).
5 Undo and remove the screw securing the retaining clamp around the reservoir and lift away the reservoir.
6 The pump(s) may now be detached from the reservoir if required.
7 Refitting is the reverse sequence to removal.

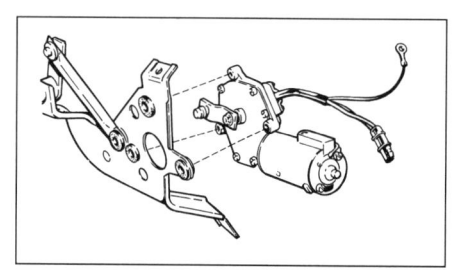

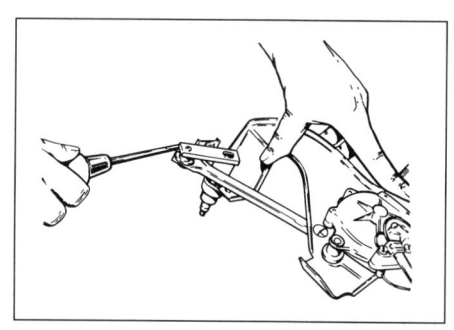

Fig. 10.15 Rear window wiper motor and mounting bracket (Sec 27)

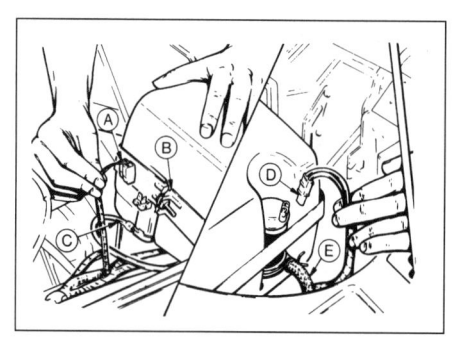

Fig. 10.16 Dismantling rear wiper linkage (Sec 27)

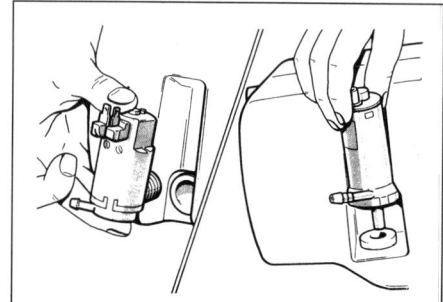

Fig. 10.17 Windscreen washer and pumps (Sec 30)

A Multi-plug
B Clamping bracket
C Washer supply
D Multi-plug
E Headlight washer supply

Fig. 10.18 Removing windscreen and headlight washer pumps (Sec 30)

Electrical system 10•13

31 Headlight washer pump - removal and refitting

The procedure for the removal of the washer pump is incorporated in Section 30.

32 Rear window washer pump - removal and refitting

1 For safety reasons, disconnect the battery.
2 Open the tailgate and take out the rear floor covering and the spare wheel cover.
3 Make a note of the position of the hoses and unplug them from the pump.
4 Disconnect the multi-plug from the pump. Undo and remove the securing screws and lift out the pump.
5 Refitting is the reverse sequence to removal.

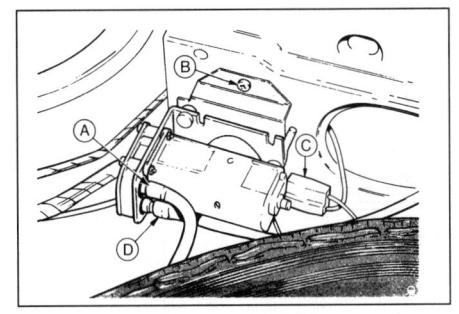

33 Headlight assembly - removal and refitting

1 If the headlight bulb only is to be replaced, refer to Section 34.
2 Open the bonnet and for safety reasons, disconnect the battery.
3 Undo and remove the screws securing the radiator grille to the body and detach the grille.
4 Rotate the headlight bulb cover anti-clockwise to expose the headlight and sidelight bulbs and lift off the cover.
5 Rotate the headlight bulb holder anti-

Fig. 10.19 Rear window washer pump (Sec 32)

A To window B Mounting screw
C Multi-plug D To reservoir

clockwise and withdraw the holder and bulbs from the headlight assembly (photo).
6 Unhook the spring, retaining the direction indicator assembly, from the rear of the headlight unit and allow it to hang.
7 Undo and remove the three headlight assembly securing screws and remove the headlight unit (photos).
8 Refitting is the reverse of the removal sequence.

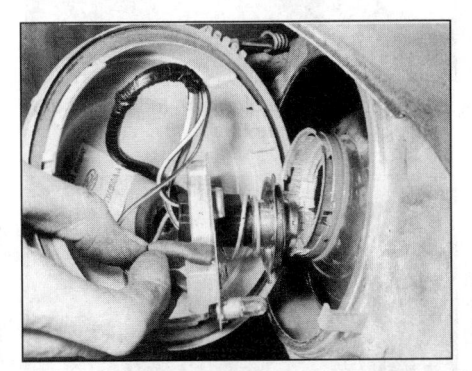

34 Headlight and sidelight bulbs - removal and refitting

1 For safety reasons, disconnect the battery.
2 Rotate the headlight cover anti-clockwise to expose the headlight and sidelight bulbs and lift off the cover.
3 Rotate the headlight bulb holder anti-clockwise and withdraw the holder and bulbs from the headlight assembly.
4 To remove the headlight bulb, pull off the multi-plug from the rear of the bulb and lift out the bulb (photo).

HAYNES HiNT *The glass of the headlight bulb should not be touched with the fingers, If it is touched with the fingers it should be washed with methylated spirit.*

5 To remove the side light bulb, push down slightly, turn anti-clockwise and lift out the bulb.
6 Refitting the bulbs is a reverse of the above procedure.

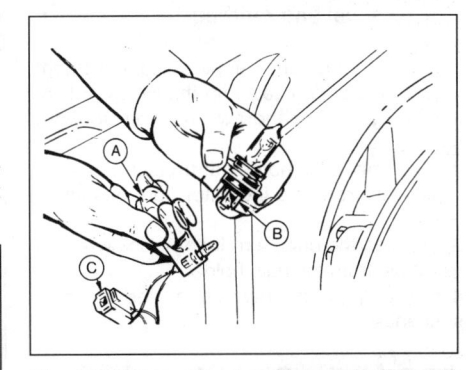

35 Headlight and auxiliary light alignment

1 It is always advisable to have the lights aligned using special optical beam setting equipment but if this is not available the following procedure may be used.
2 Position the car on level ground, 10 ft (3 m) in front of a dark wall or board. The wall or board must be at right angles to the centre-line of the car.
3 Draw a vertical line on the board in line with the centre-line of the car.
4 Bounce the car on its suspension to ensure correct settlement and then measure the height between the ground and the centre of the lights.
5 Measure the distance between the centres of the lights to be adjusted, and mark the board as show in Figs. 10.21 and 10.22.
6 *Headlights*. With the headlights on main beam, cover the other light(s) to prevent glare. By careful adjustments to the two adjusters at the rear of the headlight assembly, set the horizontal position until point "C" is at the cross on the aiming board. Adjust the vertical light position so that the top of the beam

Fig. 10.20 Removing headlight bulb (Sec 34)

A Retainer B Bulb C Multi-plug

33.5 Removing headlight and sidelight bulb holders

33.7a Headlight assembly securing bolts (arrowed)

33.7b Headlight unit removed

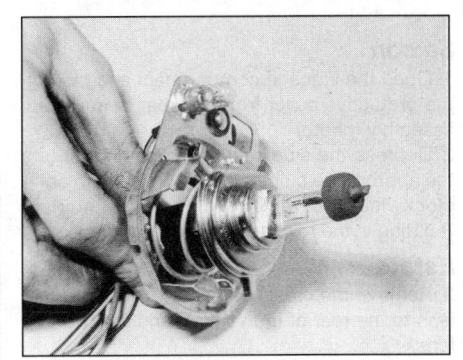

34.4 Headlight and sidelight bulb assemblies

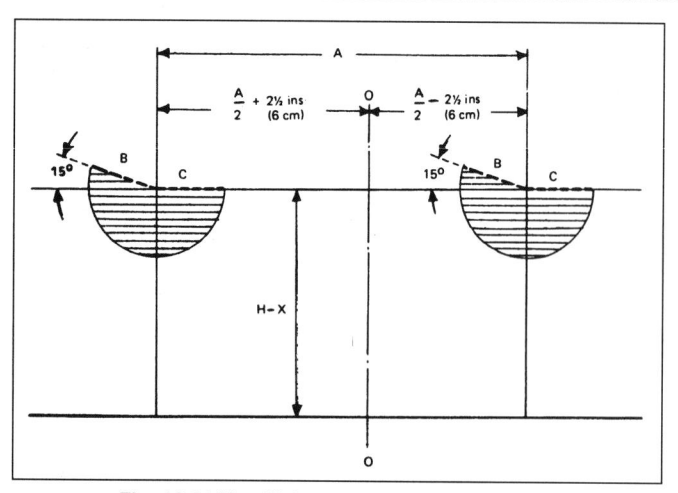

10•14 Electrical system

Fig. 10.21 Headlight alignment chart (Sec 35)

A Distance between light centres
B Dipped beam pattern
H Height from ground to headlight centre
X 3.0 in (76 mm)
OO Vehicle centre line

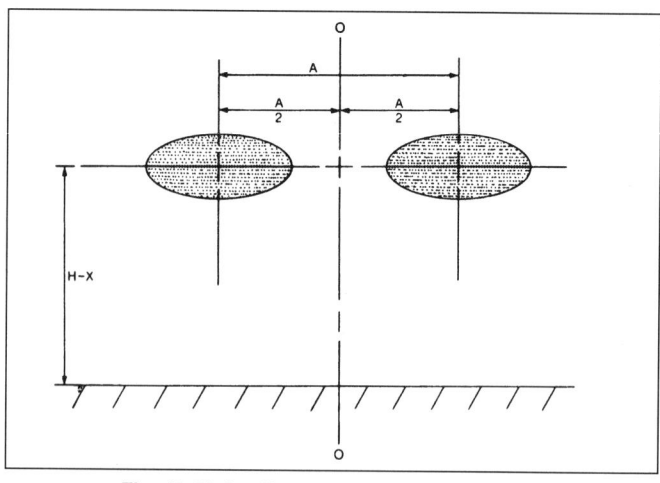

Fig. 10.22 Auxiliary light alignment (Sec 35)

A Distance between lamp centres
H Height from ground to lamp centre
X 2.0 in (50 mm)
OO Vehicle centre line

pattern just touches the dotted line (Fig. 10.21).

7 *Auxiliary lights*. Turn the auxiliary lights on and cover the other lights to prevent glare. Slacken the lamp retaining nut and adjust until the centre of illumination lies at the area shown in Fig. 10.22.

36 Front direction indicator light assembly and bulb - removal and refitting

1 From inside the engine compartment unhook the spring securing the indicator light assembly and withdraw the unit forward (photo).
2 Turn the bulb holder anti-clockwise and remove it from the indicator light assembly (photo).
3 Depress the bulb, turn it anti-clockwise and withdraw from the bulb holder.
4 Refitting is the reverse of the removal sequence.

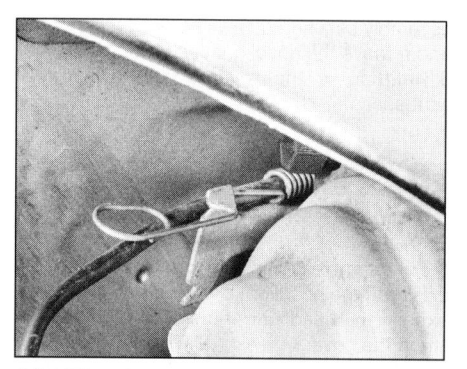

37 Rear direction indicator, stop and tail light bulbs - removal and refitting

Saloon

1 Open the luggage compartment and pull off the protective cover from the rear of the lamp assembly (photo).
2 Depress the side securing clips and lift out the appropriate bulb holder. Turn the bulb anti-clockwise and remove it from the holder. Refitting is the reverse of removal (photo).

Estate

3 Remove the crosshead screws securing the lens to the rear of the vehicle, and remove the lens.
4 Turn the bulb anti-clockwise to remove it from the holder. Refitting is the reverse of removal.

36.1 Direction indicator assembly securing spring

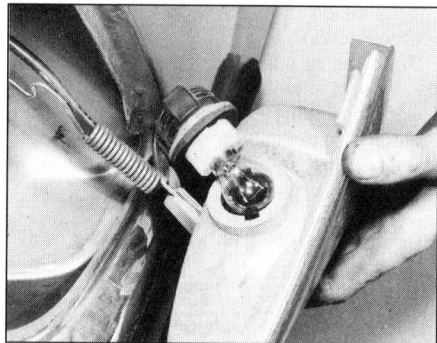

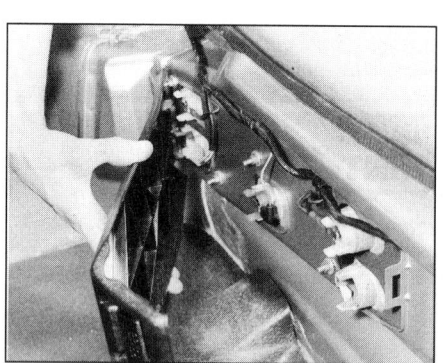

36.2 Removing the direction indicator bulb holder

37.1 Removing the cover from the rear lamp assembly

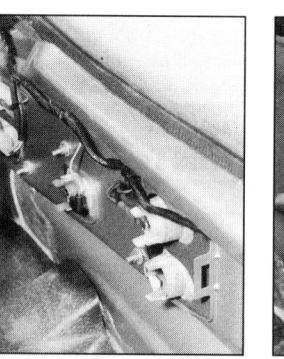

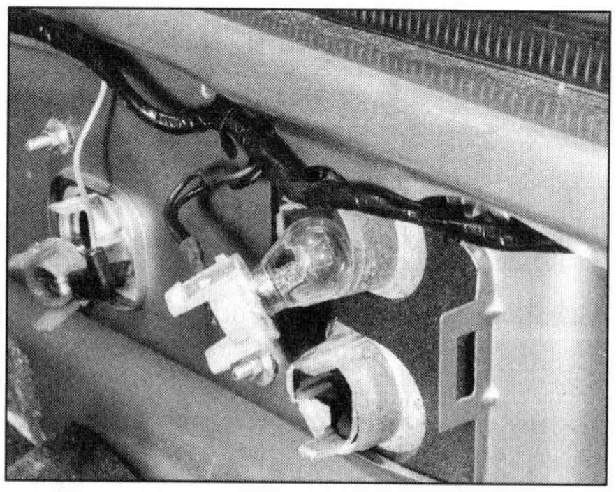

37.2 Rear light bulb and holder

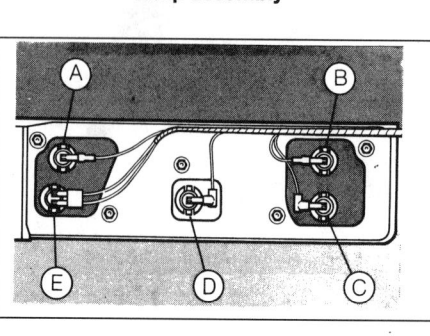

Fig. 10.23 Layout of rear lamp and bulbs - saloon (Sec 37)

A Direction indicator
B Reversing
C Rear fog
D Tail
E Stop/tail

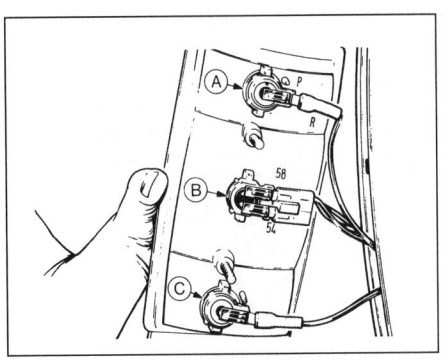

Electrical system 10•15

Fig. 10.24 Rear lamp assembly - estate (Sec 38)

A Direction indicator B Stop/tail
C Reversing

38 Rear direction indicator, stop and tail light assembly - removal and refitting

Saloon

1 Open the luggage compartment and pull off the protective cover from the rear of the lamp assembly.
2 Depress the side securing clips and lift out the bulb holders from the rear of the light assembly.
3 Undo and remove the five nuts and washers securing the light assembly to the body panel and lift off the light assembly (photo).

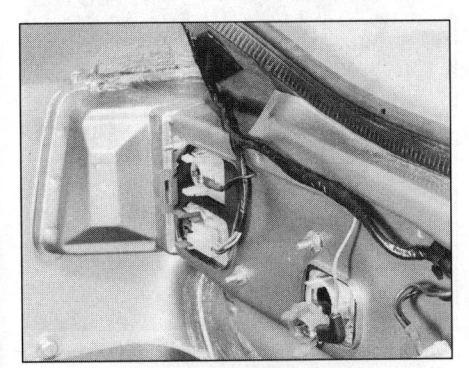

38.3 Rear light assembly bulb holders and securing screws

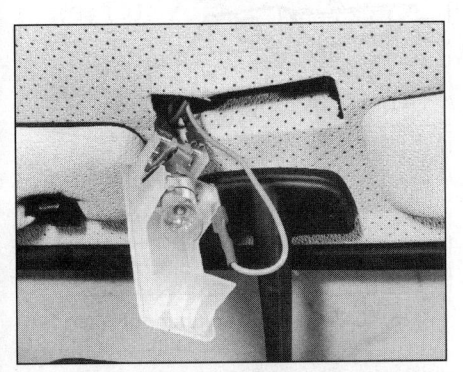

40.1 Interior light removed from roof panel

4 Refitting is the reverse of the removal procedure.

Estate

5 Unclip and remove the rear side trim panel.
6 Make a note of the wiring positions and pull the connectors off the bulb holders.
7 Undo and remove the two nuts and washers and lift off the light assembly.
8 Refitting is the reverse of the removal sequence.

39 Rear number plate light assembly and bulb - removal and refitting

1 From underneath the bumper gently squeeze the two plastic clips inwards and lift out the lamp (photo).
2 Pull off the wiring plug connector.
3 Gently prise off the lens cover which is retained by two lugs, lift away the lens and remove the bayonet bulb.
4 Refitting the bulb, lens and light assembly is the reverse sequence to removal.

HAYNES HiNT *Take care to ensure that the lens sealing washer is correctly fitted to prevent the entry of dirt and water.*

40 Interior light - removal and refitting

1 Very carefully pull the interior light assembly from the location in the roof panel (photo).
2 Disconnect the wires from the terminal connectors and lift away the light unit.
3 If the bulb requires renewal either lift the bulb from its clips (festoon type) or press and turn in an anti-clockwise direction (bayonet cap type).
4 Refitting the interior light is the reverse sequence to removal.

41 Instrument cluster and printed circuit - removal and refitting

1 Open the bonnet and disconnect the battery leads.
2 Undo and remove the screw securing the upper shroud to the steering column and lift off the shroud.
3 Withdraw the control knobs from the instrument panel. All are of the pull off type.
4 Very carefully pull off the facia trim panel. The panel is secured with six clip fixings and is easily broken if mistreated.
5 Undo and remove the screws securing the instrument cluster to the dash panel. From inside the engine compartment locate the

39.1 Rear number plate lamp and lens assembly

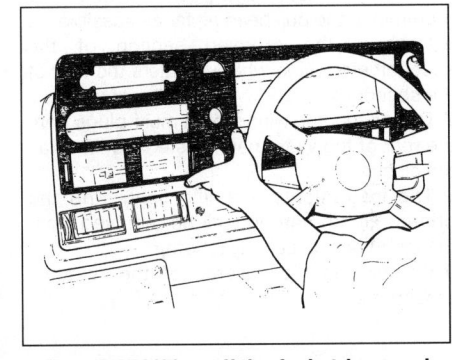

Fig. 10.25 Lifting off the facia trim panel (Sec 41)

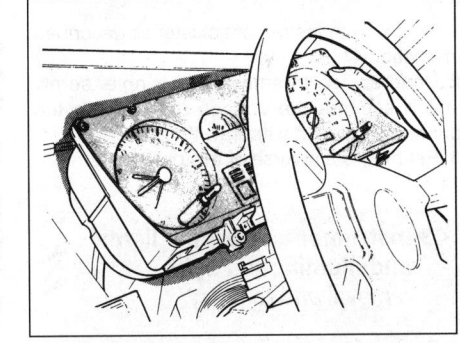

Fig. 10.26 Instrument cluster removal (Sec 41)

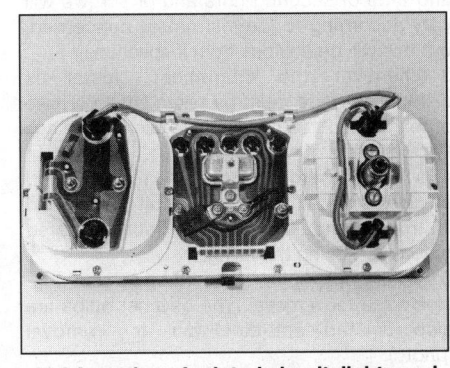

41.8 Location of printed circuit, lights and wiring on rear of instrument cluster

10•16 Electrical system

43.5 Removing instrument cluster warning light

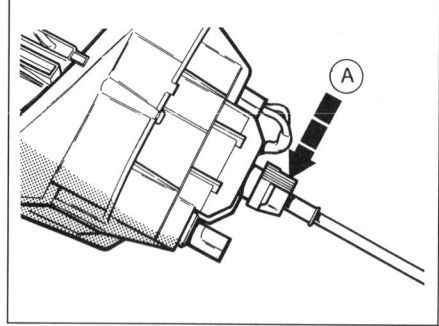

Fig. 10.27 Speedometer cable knurled pad - A (Sec 44)

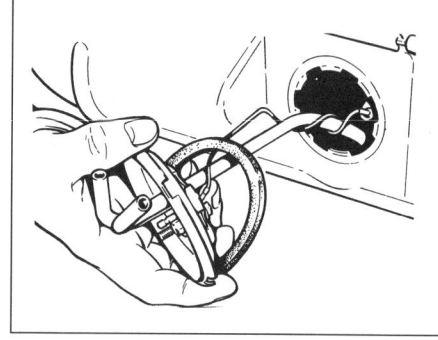

Fig. 10.28 Removing the fuel tank indicator unit (Sec 45)

speedometer cable and push it through the grommet in the bulkhead as far as possible.
6 Depress the grooved section of the speedometer cable where it enters the rear of the speedometer and withdraw it rearwards.
7 Disconnect the wiring loom multi-plugs from the rear of the instrument cluster and lift away the cluster.
8 The instruments, warning lights and the printed circuit can now all be removed from the instrument assembly (photo).
9 Refitting is the reverse of the removal procedure.

42 Instrument voltage regulator - removal and refitting

1 Remove the instrument cluster as described in Section 41.
2 Unscrew and remove the single screw which secures the voltage regulator to the printed circuit and withdraw it.
3 Refitting is a reverse of this procedure.

43 Instruments, warning lights and illumination light - removal and refitting

1 Remove the instrument cluster (refer to Section 41).
2 The exact details concerning the number and type of securing nuts and/or screws will vary according to the instrument concerned, but this will be obvious from inspection.
3 Whenever the instrument cluster is dismantled, care must be taken not to crease or tear the printed circuit films, used to replace large clusters of wires.
4 When reassembling the instrument cluster, ensure that all printed circuit films and instrument housings are correctly located before tightening the fixing.
5 All light bulb holders should be twisted anti-clockwise for removal. The bayonet bulbs are also twisted anti-clockwise for removal (photo).
6 Refitting the instrument cluster is the reverse sequence of removal.

44 Speedometer inner and outer cable - removal and refitting

1 *Manual gearbox.* Working under the car using a pair of circlip pliers, remove the circlip that secures the speedometer cable to the gearbox extension housing and withdraw the speedometer cable.
2 *Automatic transmission.* Undo and remove the bolt, and spring washer that secures the forked plate to the extension housing. Lift away the forked plate and withdraw the speedometer cable.
3 Remove the clip that secures the speedometer outer cable to the bulkhead.
4 Working under the facia depress the knurled pad on the speedometer cable ferrule and pull the cable from the speedometer head.
5 Remove the grommet that seals the cable to the bulkhead. Withdraw the speedometer cable.
6 Refitting the speedometer cable is the reverse sequence of removal.
7 It is possible to remove the inner cable, whilst the outer cable is still attached to the car. Follow the instructions in paragraphs 1 and 4 and pull the inner cable from the outer cable.

45 Fuel tank indicator unit - removal and refitting

1 With the level of petrol in the tank lower than the bottom of the indicator unit, wipe the area around the unit free of dirt.
2 Detach the indicator unit cables, fuel feed and return pipes, and note their correct location.
3 Using two crossed screwdrivers in the slots of the indicator unit body, or a soft metal drift carefully unlock the indicator unit. Lift away the indicator unit and sealing ring.
4 Refitting the indicator unit is the reverse sequence of removal. Always fit a new seal in the recess in the tank to ensure no leaks develop.

46 Steering column switches - removal and refitting

1 Open the bonnet and for safety reasons, disconnect the battery.
2 Undo and remove the screws securing the upper and lower shrouds to the steering column and lift off the shrouds.
3 Undo and remove the screws securing the multi switches to both sides of the steering column. Note the location of the brown earth wire (photo).
4 Disconnect the wiring loom multi-plugs and lift away the switches.
5 Refitting is the reverse sequence to removal.

46.3 Location of steering column switch earth wire (arrowed)

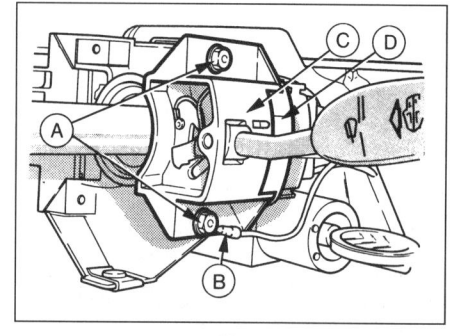

Fig. 10.29 Steering column right-hand switches (Sec 46)

A *Securing bolts*   C *Switch assembly*
B *Earth lead*   D *Loom multi-plug*

Electrical system 10•17

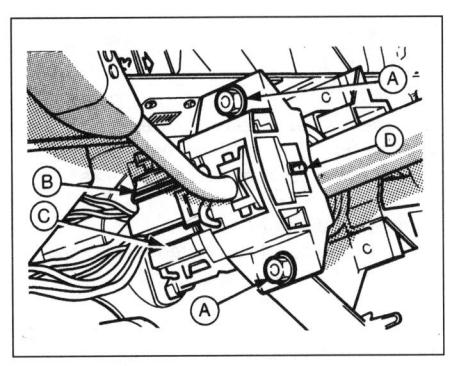

Fig. 10.30 Steering column left-hand switch (Sec 46)

A Securing bolts
B Loom multi-plug
C Loom multi-plug
D Switch assembly

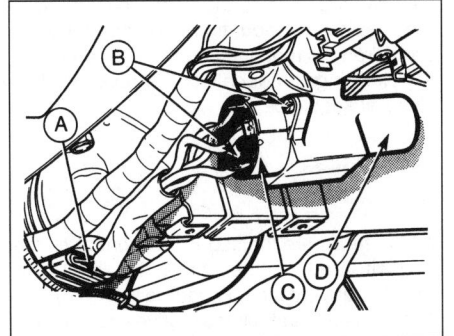

Fig. 10.31 Ignition switch and lock (Sec 47)

A Loom multi-plug
B Securing screws
C Ignition switch
D Ignition lock

47.2 The ignition switch and lock

47 Ignition switch and lock - removal and refitting

Ignition switch

1 The ignition switch can be removed by unscrewing the lower dash trim panel in the steering column area, and lowering the panel.
2 The switch is then removed by unscrewing the two screws, and disconnecting the multi-plug (photo).

Steering lock

3 To remove the steering lock refer to Chapter 11, Section 25 for the removal of the steering column.
4 Secure the steering column in a vice fitted with jaw protectors and drill out the two shear-bolts which retain the two halves of the steering column lock to the column.
5 Commence reassembly by fitting the new lock to the steering column so that the lock tongue engages in the cut-out in the column. Tighten the bolts slightly more than finger-tight and check the operation of the lock by inserting the ignition key. Move the lock assembly fractionally if necessary to ensure smooth and positive engagement when the key is turned.
6 Fully tighten the bolts until the heads shear.
7 Refit the steering column, refer to Chapter 11, Section 25.

48 Instrument panel and centre console switches - removal and refitting

Instrument panel switches

1 Remove the instrument cluster as described in Section 41.
2 The main round switches are removed by undoing the switch retaining ring, withdrawing the switch from the rear of the panel, removing the connections and lifting out. Refitting is the reverse of the removal procedure.
3 For square switches use a small screwdriver and prise the switch from its location. Remove the electrical connections and lift out the switch (Fig. 10.32). Refitting is the reverse of the removal procedure.

Centre Console

4 Carefully prise the required switch from the console using a small screwdriver.
5 Pull off the multi-plug from the rear of the switch and remove the switch. Take care that the multi-plug does not fall back into the centre console.
6 The switches are refitted in the reverse sequence to removal.

49 Radio - removal and refitting

1 Open the bonnet and disconnect the battery leads.
2 Undo and remove the screws securing the upper shroud to the steering column. Lift off the shroud.
3 Withdraw the control knobs from the instrument panel. All are of the pull off type.
4 Very carefully pull off the facia panel. The panel is secured with six clip fixings and is easily broken if mistreated.
5 Undo and remove the four screws securing the radio mounting plate to the instrument panel and withdraw the radio. On later models the radio is secured by two screws and two pop rivets, the rivets being used as an anti-theft measure. Carefully drill through the rivets with a 1/8 in (3.2 mm) drill.
6 Disconnect the wiring connections from the rear of the radio and remove the radio.
7 Refitting is the reverse of the above procedure. If rivets are used to secure the radio upper mounting, the size required is ⅛ in (3.2 mm) diameter by ¼ in (6.5 mm) long.

50 Loudspeakers - removal and refitting

1 Open the bonnet and disconnect the battery leads.

Facia mounted loudspeaker

2 Remove the instrument cluster as described in Section 41.

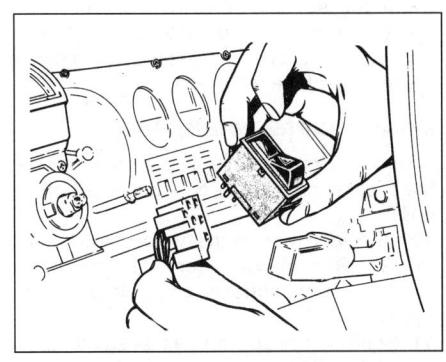

Fig. 10.32 Instrument panel switch removal (Sec 48)

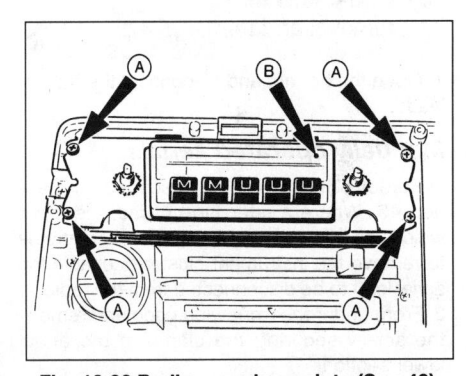

Fig. 10.33 Radio securing points (Sec 49)

A Radio securing screws
B Aerial trimming screw

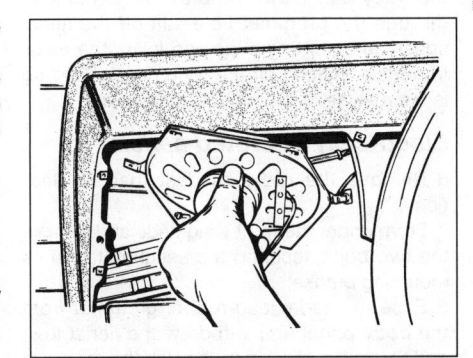

Fig. 10.34 Removing facia mounted speaker (Sec 50)

10•18 Electrical system

3 Disconnect the electrical connections to the speaker.
4 Using a small screwdriver or a screwdriver with an "L" shaped blade, unscrew and remove the two screws securing the speaker to the dash panel.
5 Withdraw the speaker from the dash panel through the instrument cluster aperture. Refitting is the reverse of the removal procedure.

Parcel shelf mounted loudspeaker

6 From inside the car undo and remove the four screws securing the speaker grille to the parcel shelf. Lift off the grille.
7 Undo and remove the four screws securing the speaker to the parcel shelf.
8 Lift the speaker from its location in the parcel shelf and detach the electrical connections. Lift away the speaker. Refitting is the reverse of the removal procedure

Door mounted loudspeaker

9 Undo and remove the two screws securing the speaker grille to the door panel. Lift off the grille.
10 Undo and remove the two screws securing the speaker to the door. Disconnect the electrical connections and withdraw the speaker.
11 Refitting the speakers is a reverse of the removal procedure.

51 Radio aerials - removal and refitting

1 Open the bonnet and disconnect the battery leads.

Manually operated aerials

2 Remove the radio as described in Section 49. Where a later type of radio is fitted which has DIN standard fittings, an alternative is to remove the instrument cluster to enable the aerial lead to be disconnected from the radio.
3 From under the front wing undo and remove the screw securing the clamp to the aerial lower section.
4 Undo and remove the upper retaining nut securing the aerial to the front wing.
5 Ease the aerial lead rubber grommet from the body panel and withdraw the aerial lead through the grommet hole. Lift off the aerial spherical mounting cups and lower the aerial from its mounting on the wing panel. Refitting is the reverse of the removal procedure.

Electrically operated aerials

6 Remove the radio as described in Section 49.
7 From under the front wing undo and remove the two bolts securing the aerial motor to its mounting bracket.
8 Ease the aerial lead rubber grommet from the body panel and withdraw the aerial lead and the two motor wires through the grommet hole.
9 Undo and remove the upper retaining nut securing the aerial to the front wing. Lift off the aerial spherical mounting cups and lower the aerial from its mounting on the wing panel.
10 Refitting the aerials is the reverse of the removal sequence.

52 Fuses

1 The fuses are located in the central electric box situated on the right-hand inner wing panel of the engine compartment (photo).
2 The fuses and their respective circuits are as follows.

Fuse	Size	Function
1	6 amp	Front and rear wiper motors, heater motor, rear washer pump
2	8 amp	Direction indicators, stop lights, reversing lights
3	8 amp	Number plate light (LHD only) and instrumentation illumination
4	8 amp	Number plate light (RHD only) and right-hand side and tail lights
5	8 amp	Left-hand side and tail lights, rear fog lights
6	8 amp	Auxiliary driving lamps
7	8 amp	Headlight main beam (left)
8	8 amp	Headlight main beam (right)
9	8 amp	Front and rear fog lamps
10	8 amp	Headlight dipped beam (left), rear fog light
11	8 amp	Headlight dipped beam (right)
12	8 amp	Heated rear screen
13	16 amp	Hazard flashers, horn, electric aerial, interior lights, cigarette lighter, clock
14	16 amp	Windscreen washer pump, headlamp washer pump

3 On certain models there is an additional fuse box located under the right-hand side of the facia panel, and contains the following fuses:

Fuse	Size	Function
1	16 amp	Electric windows (front)
2	16 amp	Electric windows (rear)
3	25 amp	Central locking system
4	16 amp	Air conditioning

4 On vehicles equipped with fuel injection there is a further "in-line" fuse clipped to the rear edge of the relay bracket.

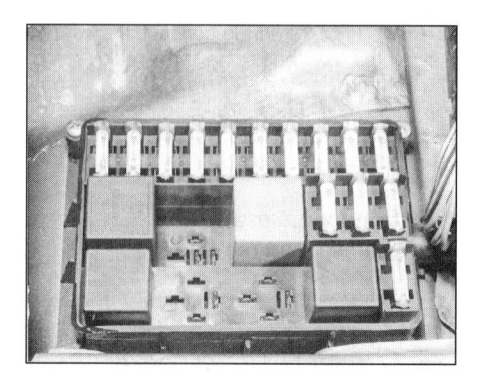

52.1 Location of fuses and relays in the central electric box

53 Relays

1 There are seven relays located in the central electric box. They may be removed by gently pulling upwards. Their functions are as follows:

Relay number Function
I   Power supply to heated rear screen, front and rear wipers, and heater motor
II Auxiliary driving lamps
III Front fog lamps
IV Starter inhibitor - automatic transmission
V   Direction indicators, hazard flashers
VI Headlamp washers
VII Intermittent wipe facility

2 On certain models there may be up to five additional relays located just above the auxiliary fuse box. These comprise:
A Air conditioning relay
B Electric window relay
C Fuel injection system control relay
D Fuel injection system supply relay
E Central locking system pulse relay
3 Radio and cassette players are protected by a "line" fuse in the supply line. Where an electric aerial is fitted there is an additional relay located behind the facia panel.

54 Central door locking system - general

Full information regarding removal and refitting of the central locking system solenoids, switches and locks will be found in Chapter 12 (Fig. 10.36).

55 Electrically operated windows - general

Full information regarding the removal and refitting of the electrically operated window mechanism and motors will be found in Chapter 13, Section 16. Refer also to Fig. 10.37 of this Chapter.

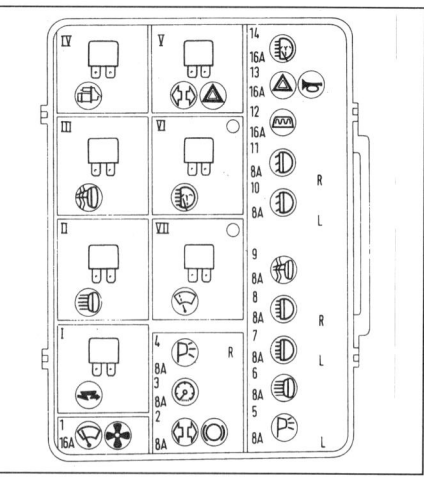

Fig. 10.35 Central electric box cover (Secs 52 and 53)

For key see Sections 52 and 53

Electrical system 10•19

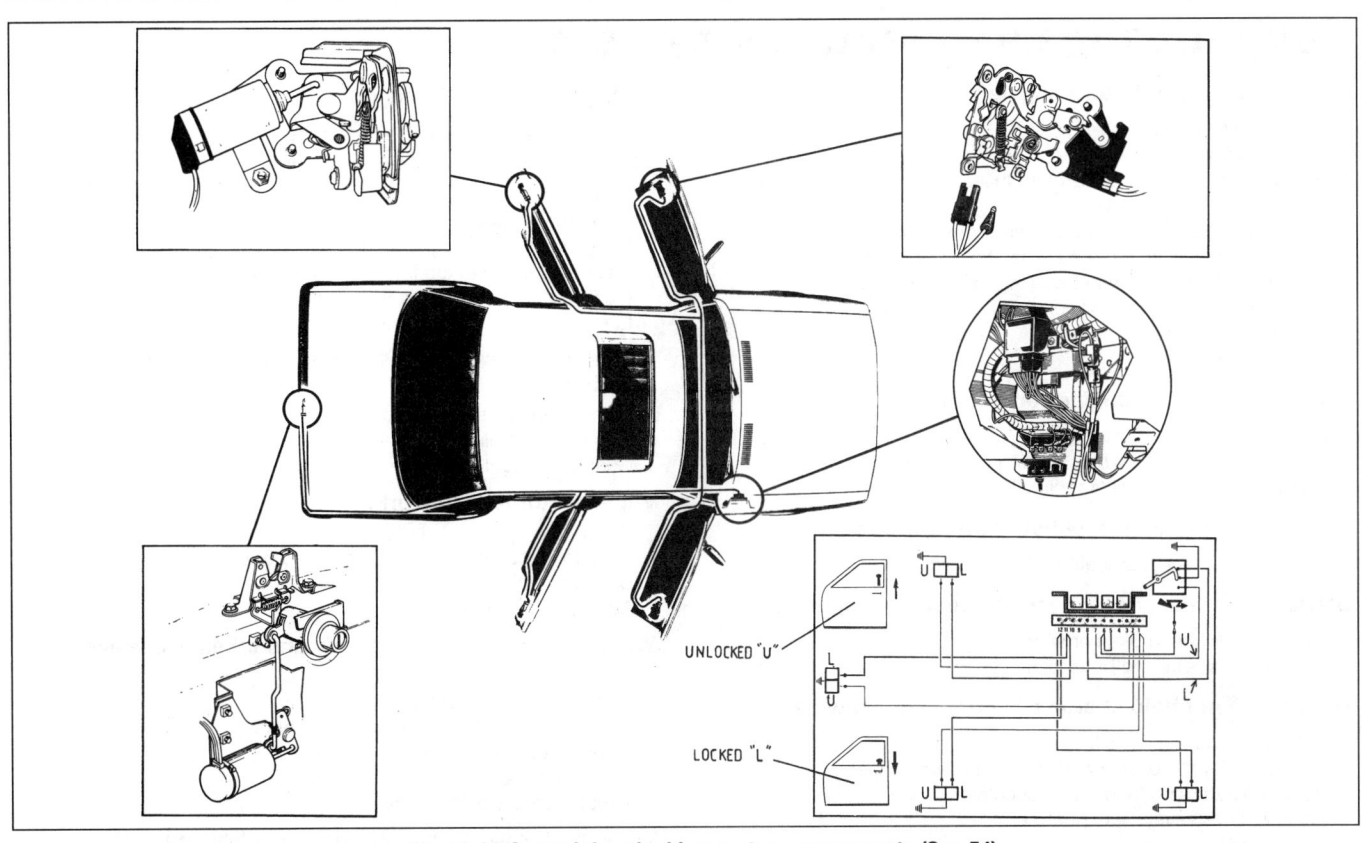

Fig. 10.36 Central door locking system components (Sec 54)

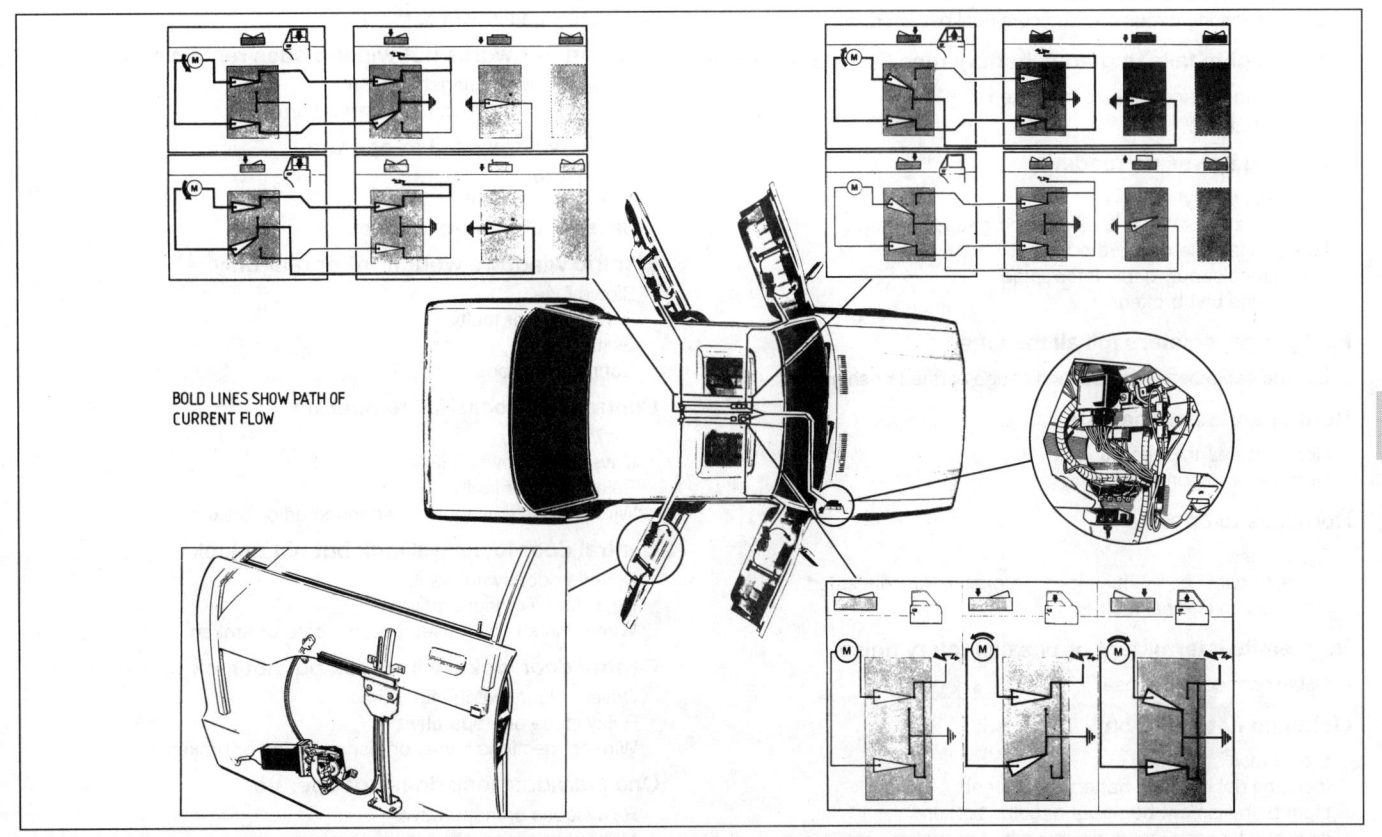

BOLD LINES SHOW PATH OF CURRENT FLOW

Fig. 10.37 Electrically operated windows (Sec 55)

Fault diagnosis - electrical system

Starter motor fails to turn engine
Battery discharged
Battery defective internally
Battery terminal leads loose or earth lead not securely attached to body
Loose or broken connection in starter motor circuit
Starter motor switch or solenoid faulty
Starter brushes badly worn, sticking, or brush wires loose
Commutator dirty, worn or burnt
Starter motor armature faulty
Field coils earthed

Starter motor turns engine very slowly
Battery in discharged condition
Starter brushes badly worn, sticking or brush wires loose
Loose wires in starter motor circuit

Starter motor operates without turning engine
Pinion or flywheel gear teeth broken or worn

Starter motor noisy or excessively rough engagement
Pinion or flywheel broken or worn
Starter motor retaining bolts loose

Battery will not hold charge for more than a few days
Battery defective internally
Electrolyte level too low or electrolyte too weak due to leakage
Plate separators no longer fully effective
Battery plates severely sulphated
Fan/alternator belt slipping
Battery terminal connections loose or corroded
Alternator not charging properly
Short in lighting circuit causing continual battery drain

Ignition lights fails to go out, battery runs flat in a few days
Fan belt loose and slipping or broken
Alternator faulty

Fuel gauge gives no reading
Fuel tank empty!
Electric cable between tank unit and gauge broken
Fuel gauge case not earthed
Fuel gauge supply cable interrupted
Fuel gauge unit broken

Fuel gauge registers full all the time
Electric cable between tank and gauge earthed or shorting

Horn operates all the time
Horn push earthed or stuck down
Horn cable to horn push earthed

Horn fails to operate
Blown fuse
Cable or cable connection loose, broken or disconnected
Horn has an internal fault

Horn emits intermittent or unsatisfactory noise
Cable connections loose

Lights do not come on
Blown fuse
If engine not running, battery discharged
Light bulb filament burnt out or bulbs broken
Wire connections loose, disconnected or broken
Light switch shorting or otherwise faulty

Lights come on but fade out
If engine not running battery discharged

Lights give very poor illumination
Poor earth connection
Reflector tarnished or dirty
Lamp badly out of adjustment
Incorrect bulb with too low wattage fitted
Existing bulbs old and badly discoloured
Electrical wiring too thin not allowing full current to pass

Lights work erratically - flashing on and off, especially over bumps
Battery terminals or earth connection loose
Lights not earthing properly
Contacts in light switch faulty

Wiper motor fails to work
Blown fuse
Wire connections loose, disconnected or broken
Brushes badly worn
Armature worn or faulty

Wiper motor works very slowly and takes excessive current
Commutator dirty, grease or burnt
Drive to wheelboxes bent or unlubricated
Wheelbox spindle binding or damaged
Armature bearings dry or unaligned
Armature badly worn or faulty

Wiper motor works slowly and takes little current
Brushes badly worn
Commutator dirty, greasy or burnt
Armature badly worn or faulty

Wiper motor works but wiper blades remain static
Wiper linkage damaged or worn
Wiper motor gearbox parts badly worn

Electric windows fail to operate
Blown fuse
Relay loose or faulty
Battery in discharged condition

Electric windows work front or rear only
Blown fuse
Relay loose or faulty
Switch faulty
Connections loose

Central door locks fail to operate
Blown fuse
Driver's door switch faulty
Relay loose or faulty
Wire connections loose, disconnected or broken

Central door locks will lock but not unlock
Driver's door switch faulty
Relay loose or inoperative
Wire connections loose, disconnected or broken

Central door locks will unlock but not lock
Driver's door switch faulty
Relay loose or inoperative
Wire connections loose, or disconnected or broken

One individual lock does not operate
Relay loose or inoperative
Multi-plug connections loose, broken or dirty
Door lock solenoid faulty

Chapter 11 Suspension and steering

For modifications, and information applicable to later models, see Supplement at end of manual

Contents

Fault diagnosis - suspension and steering	See end of Chapter
Front hub - removal and refitting	4
Front hub bearings - adjustment	3
Front hub bearings - removal and refitting	2
Front shock absorbers - removal and refitting	13
Front subframe mounting bushes - removal and refitting	7
Front suspension and subframe assembly - removal and refitting	5
Front suspension assembly - overhaul	6
General description	1
Lower suspension swinging arm - removal and refitting	9
Manual steering - lubrication	22
Power assisted steering - checking the fluid level	33
Power assisted steering control valve - checking and adjusting	36
Power assisted steering fluid cooler - removal and refitting	41
Power assisted steering pump - dismantling and reassembly	40
Power assisted steering pump - removal and refitting	39
Power assisted steering pump drivebelt - removal and refitting	38
Power assisted steering gear - dismantling and reassembly	35
Power assisted steering gear - removal and refitting	34
Power assisted steering gear - bellows renewal in situ	42
Power assisted steering system - bleeding	37
Rack and pinion steering gear - adjustments	29
Rack and pinion steering gear - dismantling and reassembly	30
Rack and pinion steering gear - removal and refitting	28
Rear axle and suspension assembly - removal and refitting	15
Rear axle casing rubber mounting - removal and refitting	20
Rear shock absorber - removal and refitting	18
Rear suspension bump rubber - removal and refitting	21
Rear suspension coil spring - removal and refitting	17
Rear suspension front rubber mounting - removal and refitting	19
Rear suspension arm - removal and refitting	16
Stabiliser bar - removal and refitting	10
Stabiliser bar connecting link bushes - removal and refitting	12
Stabiliser bar mounting bushes - removal and refitting	11
Steering column assembly - removal and refitting	25
Steering column flexible coupling and universal joint assembly - removal and refitting	27
Steering column shaft - removal and refitting	26
Steering rack rubber gaiter - removal and refitting (in situ)	31
Steering wheel - removal and refitting	24
Stub axle - removal and refitting	8
Track-rod end - removal and refitting	32
Tie-bar - removal and refitting	14
Wheel alignment and steering angles	23
Wheels and tyres	43

Degrees of difficulty

Easy, suitable for novice with little experience	Fairly easy, suitable for beginner with some experience	Fairly difficult, suitable for competent DIY mechanic	Difficult, suitable for experienced DIY mechanic	Very difficult, suitable for expert DIY or professional

Specifications

Front suspension

Type	Independent, coil spring, unequal length swinging arms. Double acting telescopic shock absorbers
Toe-in setting	0.00 to 0.23 in (0.00 to 6.0 mm)
Castor	1°30' +1°0' -0°45'
Camber	0°24' ± 0°45'
Maximum castor difference LH to RH	0°45'
Maximum camber difference LH to RH	1°0'
Coil spring identification	Colour coded. For replacement use springs of same colour
Hub bearing lubricant type/specification	Multi-purpose lithium based grease to NLGI 2

Rear suspension

Type	Semi-trailing arms. coil springs, double acting telescopic shock absorbers
Toe-in setting	0.00 to 0.09 in (0.00 to 2.40 mm)
Camber	-0°10' to +0°50'
Coil spring identification	Colour coded. For replacement use springs of same colour

Manual steering

Type	Rack and pinion
Turns lock-to-lock	4.39
Steering ratio	22:1
Steering gear adjustment	Shims
Lubricant capacity	0.26 pint (0.15 litre)
Lubricant type	Semi-fluid grease (SAM-1C-9106-A)

Suspension and steering

Power assisted steering

Type	Rack and pinion, power assisted
Turns lock-to-lock	3.25
Steering ratio	16.80:1
Pinion bearing adjustment	Adjustable bearing
Rack slipper adjustment	Shims
Rack lubricant capacity	0.33 pint (0.20 litre)
Rack lubricant type	SAE 40 engine oil
Power steering fluid capacity	0.9 pint (0.5 litre)
Power steering fluid type/specification	ATF to Ford spec SQM2C-9007-AA
Power assisted pump belt tension	0.5 inch (12 mm)

Roadwheels

Wheel size	5½ J x 14, 6J x 14 or 150TRX 390FH (S pack models)

Tyres

Tyre size

Tyre pressures (Saloon models) - lbf/in^2 (bar)

	Normally laden Front	Rear	Fully laden Front	Rear
175SR 14	24 (1.7)	24 (1.7)	27 (1.9)	33 (2.3)
175HR 14	26 (1.8)	26 (1.8)	27 (1.9)	36 (2.5)
185SR 14	24 (1.7)	24 (1.7)	27 (1.9)	33 (2.3)
185HR 14	24 (1.7)	24 (1.7)	27 (1.9)	33 (2.3)
190/65HR390	24 (1.7)	24 (1.7)	27 (1.9)	36 (2.5)

Tyre pressures (Estate models) - lbf/in^2 (bar)

	Front	Rear	Front	Rear
185SR 14	26 (1.8)	26 (1.8)	27 (1.9)	36 (2.5)
185HR 14	26 (1.8)	26 (1.8)	27 (1.9)	36 (2.5)
185SR 14 REIN	26 (1.8)	26 (1.8)	27 (1.9)	36 (2.5)
190/65HR390	24 (1.7)	24 (1.7)	27 (1.9)	36 (2.5)

Torque wrench settings

Front suspension

	lbf ft	Nm
Subframe to body bolts	44 to 52	60 to 70
Stabiliser bar clamp bolts	13 to 18	17 to 24
Engine mounting nuts	30 to 45	41 to 51
Stub axle balljoints		
First tightening	30 to 45	41 to 51
Final tightening	43 to 66	58 to 89
Tie-bar to lower arm	43 to 50	58 to 68
Tie-bar front bush nuts	46 to 72	62 to 97
Upper arm pivot bolts	52 to 70	70 to 95
Lower arm to subframe	52 to 59	70 to 80
Shock absorber upper mount	28 to 35	38 to 47
Shock absorber lower mount	6 to 9	8 to 12
Stabiliser bar to connecting link	7 to 9	10 to 12

Rear suspension

Axle casing cover bolts	33 to 37	45 to 50
Half shaft socket head bolts	28 to 31	38 to 43
Hub to suspension arm bolts	30 to 36	41 to 51
Hub centre nut	185 to 198	255 to 275
Axle mounting castellated nut	123 to 151	172 to 216
Crossmember to body centre bolt	50 to 63	68 to 88
Crossmember to body bracket bolts	29 to 36	40 to 50
Suspension arm to crossmember bolts	50 to 63	68 to 88
Shock absorber to suspension arm	29 to 36	40 to 50
Shock absorber to body	18 to 25	25 to 35

Manual steering

Steering gear to subframe	16 to 22	22 to 30
Track-rod end to steering arm	18 to 22	25 to 30
Steering coupling pinch bolt	12 to 15	16 to 20
Steering coupling clamp bolts	12 to 15	16 to 20
Steering wheel to shaft	29 to 36	40 to 50
Track-rod end locknut	42 to 50	57 to 68
Pinion preload cover	12 to 17	17 to 24
Rack slipper cover	12 to 17	17 to 24
Wheel nuts	50 to 65	70 to 90

Suspension and steering 11•3

Torque wrench settings (cont.)

	lbf ft	Nm
Power assisted-steering		
Rack end housing locating peg	9 to 12	12 to 16
Pinion bearing cover plate	12 to 15	16 to 20
Control valve housing to pinion housing	12 to 15	16 to 20
Rack slipper cover plate	12 to 15	16 to 20
Steering wheel to shaft	29 to 36	40 to 50
Feed hoses	19 to 23	26 to 31
Return hoses	12 to 15	16 to 20
Pulley hub securing bolt	12 to 15	16 to 20
Steering rack to subframe	16 to 22	22 to 30
Wheel nuts	50 to 65	70 to 90

1 General description

The front suspension is of the independent type comprising unequal length upper and lower swinging arms, coil springs and telescopic shock absorbers acting on the lower arms. The suspension and steering gear are mounted on a sub-frame which is bolted to the body side members. The suspension swinging arms are rubber bushed at their sub-frame mounting points and carry the stub axle balljoints at their outer ends. Rubber bump stops are mounted on the subframe to limit suspension upper travel.

Located on each stub axle are two taper roller bearings running in cups which are pressed into the wheel hubs. Located on the inner end of the hub is a spring loaded nitrile grease seal and on the outer hub flange are the five wheel studs which are a splined and pressed fit into the flange.

A tie-bar is bolted to each lower suspension swinging arm and mounted to the subframe through polyurethane bushes. The tie-bars provide adjustment for castor angle which along with the front wheel toe-in setting is the only adjustment required.

A stabilizer (anti-roll) bar links the two lower swinging arms. The stabilizer is mounted to the swinging arms by a through bolt, spacer and rubber bushes and to the subframe by split rubber bushes and clamping brackets.

Rear suspension is also of the independent type utilising two semi-trailing suspension arms, coil springs and telescopic shock absorbers. Progressive rubber bump stops are fitted to limit suspension upper travel. Drive is transmitted through two half shafts incorporating two constant velocity joints on each shaft to cater for suspension movement. Rear suspension toe-in setting is adjustable by means of a cam plate on each inner pivot bolt of the suspension arms.

The steering gear is of the conventional rack and pinion type and is mounted on the front subframe by two U-shaped clamps. The pinion is connected to the steering column by a splined collar and clamp bolt attached to a flexible coupling. Above the flexible coupling the column is split by a universal joint, designed to collapse on impact, thus minimizing driver injury in the event of an accident. Additionally a convoluted tube type outer column and shear retainers are used at the column mounting points.

Turning the steering wheel causes the rack to move in a lateral direction and the trackrods attached to either end of the rack pass this movement to the steering arms on the stub axle assemblies thereby moving the road wheels.

The two track-rods are adjustable in length to allow alteration of the toe-in setting and wheel lock angles. The rack assembly has internal lock stops and are not adjustable.

Where power assisted steering is fitted, steering effort is reduced by applying hydraulic pressure on a ram. The steering rack is enclosed by a tube acting as the power cylinder and the rack incorporates a double acting piston which is energised by fluid fed to either side by tubes from a spool valve. This valve, which is located in the pinion housing, is operated by helical grooves in the pinion assembly, which also includes a torsion bar to give graduated power assistance and good steering feel.

In the event of failure of the hydraulic system the steering can still be operated but more effort will be required.

2 Front hub bearings - removal and refitting

1 Refer to Chapter 9 and remove the disc brake caliper.
2 By judicious tapping and levering remove the dust cap from the centre of the hub.
3 Remove the split pin from the nut retainer and lift away the adjusting nut retainer.
4 Unscrew the adjusting nut and lift away the thrust washer and outer tapered bearing (photos).
5 Pull off the complete hub and disc assembly from the stub axle.
6 From the back of the hub assembly carefully prise out the grease seal noting which way round it is fitted. Lift away the inner tapered bearing.
7 Carefully clean out the hub and wash the bearings with petrol making sure that no grease or oil is allowed to get into the brake disc.
8 Using a soft metal drift carefully remove the inner and outer bearing cups.
9 To fit new cups make sure they are the right way round and using metal tubes of suitable size carefully drift them into position.
10 Pack the cone and roller assembly with grease working the grease well into the cage and rollers. **Note:** *Leave the hub and grease seal empty to allow for subsequent expansion of the grease.*
11 To reassemble the hub first fit the inner bearing and then gently tap the grease seal back into the hub. A new seal must always be fitted as during removal it was probably damaged. The lip must face inwards to the hub.
12 Refit the hub and disc assembly to the stub axle and slide on the outer bearing and thrust washer.
13 Refit the adjusting nut and tighten it as described in Section 3.
14 Refit the disc brake caliper as described in Chapter 9.

2.4a Lift off the thrust washer . . .

2.4b . . . and the outer tapered bearing

11•4 Suspension and steering

3 Front hub bearings - adjustment

1 To check the condition of the hub bearings, jack-up the front of the car and support on firmly based stands. Grasp the roadwheel at two opposite points to check for any rocking movement in the wheel hub. Watch carefully for any movement in the steering gear which can easily be mistaken for hub movement.
2 If the front wheel hub movement is excessive, this is adjusted by removing the hub cap and then tapping and levering the dust cap from the centre of the hub.
3 Extract the split pin from the nut retainer and lift away the adjusting nut retainer.
4 Tighten the adjusting nut to a torque wrench setting of 27 lbf ft (36 Nm) whilst rotating the wheel to ensure free movement and centralise the bearings. Slacken the nut back to 90° or until 0.001 to 0.005 in (0.03 to 0.13 mm) endfloat of the hub assembly is obtained. Refit the nut retainer and lock with a new split pin (photo).
5 Refit the dust cap to the centre of the hub.

4 Front hub - removal and refitting

1 Follow the instructions given in Section 2 up to and including paragraph 5.
2 Remove the five bolts securing the hub to the brake disc.
3 If a new hub assembly is being fitted it is supplied complete with the new cups and bearings. The bearing cups will already be fitted in the hub. It is essential to check that the cups and bearings are of the same manufacture; this can be done by reading the name on the bearings and by looking at the initial letter stamped on the hub. "T" stands for Timken and "S" for Skefco.
4 Clean with scrupulous care the mating surface of the hub and check for blemishes or damage. Any dirt or blemishes will almost certainly cause the disc to run out of true.
5 Refit the hub to the brake disc and progressively tighten the five bolts in a diagonal sequence to the torque figure given in the Specifications.
6 To grease and reassemble the hub assembly follow the instructions given in Section 2 paragraph 10 onwards.

3.4 Refitting nut retainer and split pin

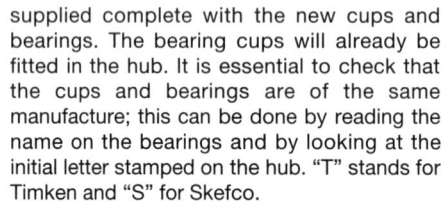

5.6 Left-hand engine mounting

5 Front suspension and subframe assembly - removal and refitting

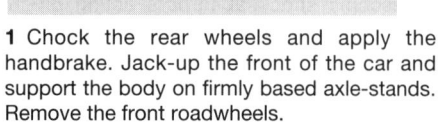

1 Chock the rear wheels and apply the handbrake. Jack-up the front of the car and support the body on firmly based axle-stands. Remove the front roadwheels.
2 Using a block of wood as a pad place a jack or stand under the clutch housing to support the engine and gearbox assembly.
3 Bend back the lock tabs, undo and remove the two bolts and separate the universal coupling from the steering column shaft.
4 Unclip the side splash panels and remove the five screws securing the engine lower splash shield. Lift off the splash shield.
5 Using a proprietary brake hose clamp, clamp the two brake flexible hoses, and disconnect them from the front brake calipers. If suitable clamps are not available plug the ends of the hoses upon removal to prevent loss of brake fluid.
6 Undo and remove the engine mounting securing nuts from the underside of the mountings (photo).
7 If power assisted steering is fitted, disconnect the control valve to reservoir and pump to control valve hoses. Plug the ends of the hoses to prevent loss of fluid and ingress of dirt.
8 Position a suitable jack beneath the engine crossmember and jackup sufficiently to just take the weight of the subframe assembly.

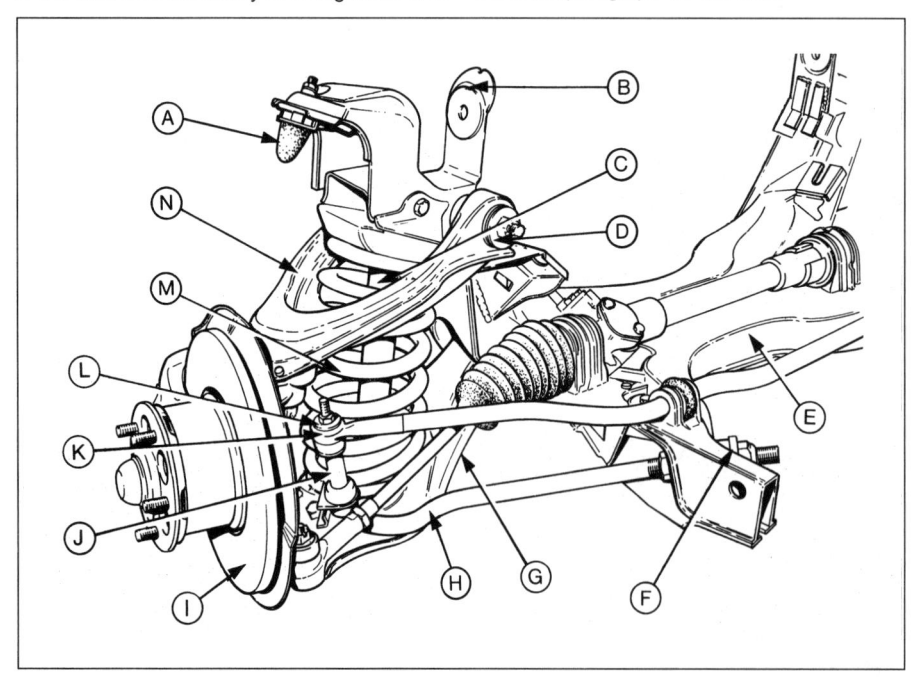

Fig. 11.1 Front suspension assembly (Sec 5)

A Bump rubber
B Body to subframe mount
C Shock absorber
D Upper pivot bolt
E Subframe
F Tie-bar front mounting
G Lower swinging arm
H Tie-bar
I Disc/hub assembly
J Connecting link
K Stabiliser bar
L Bush
M Coil spring
N Upper swinging arm

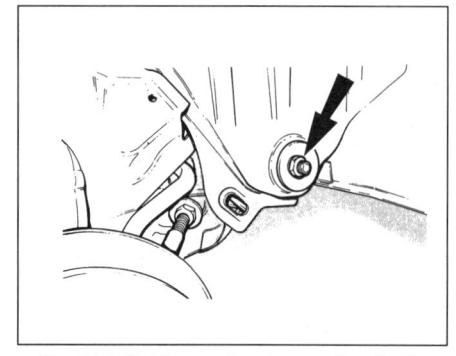

Fig. 11.2 Subframe front mounting bolt - arrowed (Sec 5)

Suspension and steering 11•5

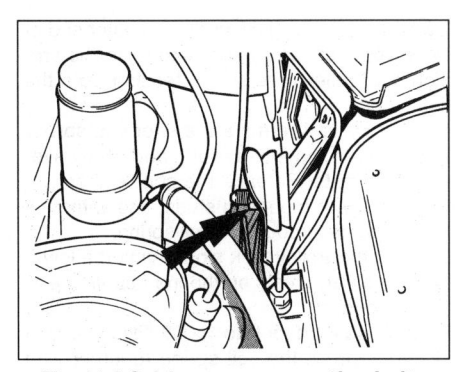

Fig. 11.3 Subframe rear mounting bolt - arrowed (Sec 5)

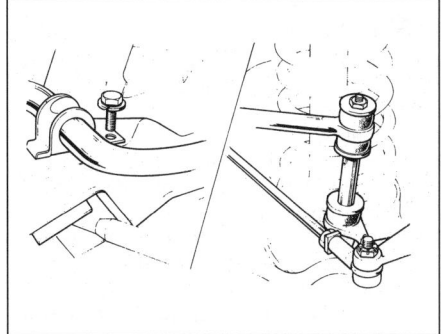

Fig. 11.4 Stabiliser bar mounting points (Sec 6)

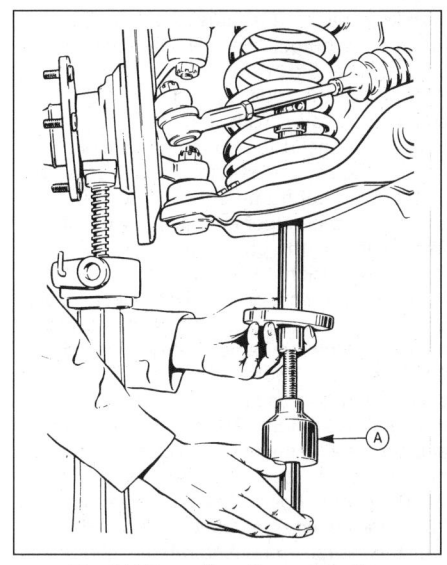

Fig. 11.5 Inserting the coil spring compressor tool "A" (Sec 6)

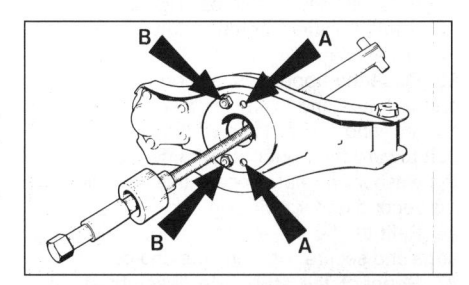

Fig. 11.6 Correct positioning of the coil spring compressor tool lower plate (Sec 6)

A Free holes   B Retaining nuts

9 Undo and remove the body side member to front frame retaining nuts and bolts from both sides.
10 Undo and remove the upper subframe to side member retaining nuts and bolts from both sides.
11 Carefully lower the complete assembly and draw forward from under the front of the vehicle.
12 Refitting is the reverse sequence to removal, bearing in mind the following points.
   a) Refer to Chapter 9 and bleed the front brake hydraulic system
   b) Where power assisted steering is fitted it will be necessary to bleed the system as described in Section 37
   c) Do not tighten the universal coupling clamp bolts until the weight of the car is on the wheels

6 Front suspension assembly - overhaul

After high mileages it may be considered necessary to overhaul the complete front suspension assembly. As certain suspension components can only be removed after partially or completely lowering the front suspension assembly, it is far better to remove the complete unit as described in Section 5 and dismantle it rather than to work on it still mounted on the car.

1 Refer to Chapter 9 and remove the disc brake caliper.
2 Prise off the hub dust cap and withdraw the split pin and nut retainer. Undo and remove the nut.
3 Carefully pull the hub and disc assembly from the stub axle.
4 Remove the split pins from top and bottom suspension arm balljoint securing nuts and remove the nuts. Using a universal balljoint separator, separate the balljoint pins from the stub axle. Alternatively, use a heavy duty bolt and nut fitted between the ends of the ballpins. Unscrew the nut to force the pins from the stub axle eyes.
5 Remove the split pin and nut from the track-rod end and separate the track-rod end from the steering arm using the balljoint separator. Lift off the stub axle.

6 Take out the split pin and undo and remove the nut from the upper suspension arm mounting bolt. Withdraw the bolt and lift off the arm.
7 Undo and remove the bolts that secure the tie bar to the lower arm.
8 Position a trolley jack under the outer end of the lower suspension arm, and raise the arm slightly to relieve the spring tension from the shock absorber.
9 Unscrew and remove the shock absorber upper mounting pivot bolt, and the lower mounting nuts. Release the shock absorber from the lower mounting studs, and withdraw it downwards through the coil spring.
10 Lower the trolley jack slowly until the lower suspension arm rests on the subframe at its inner end. It is now necessary to compress the coil spring together with the lower suspension arm. To do this obtain Ford tool No 14006 from a tool hire agent or Ford dealer. Insert the tool up through the coil spring and secure using the shock absorber upper mounting pivot bolt. Fit the tool lower plate with the holes marked "B" located on the shock absorber lower mounting studs. Fit and tighten the nuts. It is important that the free holes on the plate are towards the inner end of the lower suspension arm. Fit the tool spindle and tighten until the coil spring is compressed by approximately 2 in (50.8 mm).
11 Unscrew and remove the lower suspension arm inner pivot bolt and move the arm clear of the subframe. Unscrew the tool spindle and withdraw the lower suspension arm, coil spring and tool.
12 Bend back the lock tabs, then unscrew and remove the four bolts which secure the steering rack "U"-shaped brackets to the subframe. Lift away the steering rack assembly.
13 Undo and remove the two bolts and unhook the stabiliser mounting clamps from their slots in the frame. Remove the nuts, through bolts and spacers securing the stabiliser bar to the tie bars and lift off the stabiliser bar.
14 Undo and remove the nut, retainer and bush from the front of the tie-bar and withdraw the bar from the frame.
15 Repeat operations 1 to 14 and dismantle

the second front suspension assembly. It will be necessary to release the coil spring compressor.
16 Dismantling is now complete. Wash all parts and wipe dry ready for inspection. Inspect all bushes for signs of wear and all parts for damage or excessive corrosion, and if evident new parts will have to be obtained. Carefully inspect the coil springs and shock absorbers for signs of wear or leaks. If renewal is necessary they must be changed in pairs never individually. The bushes in the swinging arms may be removed using a suitable drift or pulled out with a metal tube, long bolt, packing washers and nut. Lubricate the new bushes in soapy water before refitting. If the suspension arm balljoints are worn, renew the arm as a complete assembly.
17 During reassembly it is important that none of the rubber mounted bolts are fully tightened until the weight of the vehicle is taken on the front wheels.
18 Fit new end bushes to the tie-bars and locate the bars in the subframe. Refit the outer bushes retainers and locknuts.
19 Fit new bushes to the stabiliser through bolt and refit the bolt, nut and spacers to the stabiliser and tie-bar.
20 Place two new split bushes over the

11•6 Suspension and steering

stabiliser bar and secure the bar with the retaining clamps. Ensure the clamps sit squarely over the bushes.
21 Locate the coil spring on the lower suspension arm with its feathered edge uppermost. Insert the spring compressor tool through the arm and spring, offer the assembly to the front subframe and secure using the shock absorber upper mounting pivot bolt.
22 Compress the coil spring, then align the lower suspension arm holes with the holes in the front subframe, oil the pivot bolt and insert it, making sure that the bolt head is towards the front of the subframe. Fit the nylon nut.
23 Raise the outer end of the lower suspension arm using a trolley jack. Remove the compressor tool. Refit the shock absorber and secure with the upper pivot bolt and lower mounting nuts. Lower the trolley jack.
24 Repeat the operations in paragraphs 19 to 23 on the other side. Check the condition of the steering rack mounting rubbers and obtain new if necessary. Place the steering rack on the subframe and secure to the mounting brackets with the U-clamps. Tighten the bolts to the torque setting given in the Specifications.
25 Place the upper suspension swinging arm on the frame and insert the pivot bolt through the arm and frame holes with the head of the bolt toward the front of the frame. Secure with the washer and nut. Repeat this operation for the second upper swinging arm.
26 Refit the tie-bar to the two lower swinging arms and secure with the nuts and bolts.
27 Connect the stub axle assembly to the suspension arm balljoints locate the track-rod ends to the stub axles and tighten all the nuts and fit new split pins.
28 Refer to Section 2 and refit the hub and disc assemblies.
29 Refer to Chapter 9 and refit the brake caliper.
30 The complete front suspension assembly may now be refitted to the car as described in Section 5. Tighten all rubber mounted bolts when the weight of the car has been taken by the front wheels.

7 Front sub-frame mounting bushes - removal and refitting

1 Refer to Section 5 and remove the front suspension assembly.
2 Using a piece of tube about 4 inches long of suitable diameter, a long bolt and nut and packing washers, draw the bushes from the side members.
3 New bushes may now be fitted using the reverse procedure for removal. It is important that the new bushes are fitted with the arrow on the flange pointing toward the dimple on the frame. Lubricate the bushes with soapy water to assist in fitting and draw the bushes into position from the engine compartment side of the subframe towards the wheel arch side.
4 Refit the front suspension assembly as described in Section 5.

8 Stub axle - removal and refitting

1 Refer to Section 2 and remove the front hub and disc assembly.
2 Undo and remove the three bolts and spring washers that secure the brake disc splash shield to the stub axle.
3 Extract the split pins then undo and remove the castellated nuts securing the three balljoint pins to the stub axle.
4 Using a universal balljoint separator, separate the balljoint pins from the stub axle. Lift away the stub axle.
5 Refitting the stub axle is the reverse sequence to removal. Tighten the balljoint securing nuts to the torque settings given in the Specifications and fit new split pins.
6 If a new stub axle has been fitted it is recommended that the steering geometry and front wheel toe-in setting be checked. Further information may be found in Section 23.

9 Lower suspension swinging arm - removal and refitting

1 Chock the rear wheels, jack up the front of the car and support on axle stands. Make sure there is sufficient ground clearance for fitting the coil spring compressor tool required later. Remove the roadwheel.
2 Unscrew and remove the two bolts securing the tie bar to the lower suspension arm.

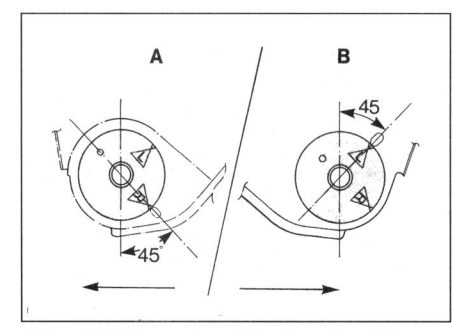

Fig. 11.7 Correct positioning of subframe front mounting bushes (Sec 7)

A Left-hand side   B Right-hand side

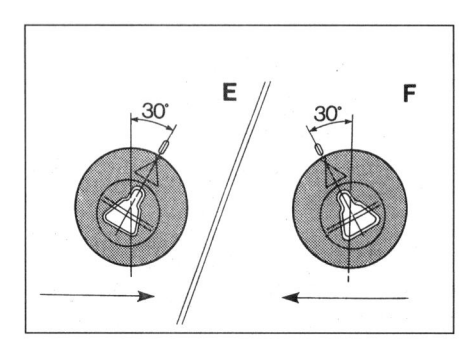

Fig. 11.8 Correct positioning of subframe rear mounting bushes (Sec 7)

E Right-hand side   F Left-hand side

3 Position a trolley jack under the outer end of the lower suspension arm, and raise the arm slightly to relieve the spring tension from the shock absorber.
4 Unscrew and remove the shock absorber upper mounting pivot bolt, and the lower mounting nuts. Release the shock absorber from the lower mounting studs, and withdraw it downwards through the coil spring.
5 Lower the trolley jack slowly until the lower suspension arm rests on the front subframe at its inner end.
6 Using the Ford tool No 14006 (Figs 11.5 and 11.6) compress the coil spring together with the lower suspension arm as described in Section 6.
7 Extract the split pin, then unscrew and remove the castellated nut which secures the lower suspension balljoint to the stub axle. Using a balljoint separator tool, separate the balljoint pin from the stub axle.
8 Unscrew and remove the lower suspension arm inner pivot bolt and move the arm clear of the front axle frame. Unscrew the tool spindle and withdraw the lower suspension arm, coil spring and tool.
9 To fit a new bush first remove the old bush by using a piece of tube about 4 inches (101.6 mm) long and suitable diameter, a long bolt and nut and packing washer, draw the bush from the lower suspension arm. Fitting a new bush is the reverse sequence to removal.
10 Refitting the lower suspension arm is the reverse sequence to removal. The lower arm retaining bolt must be tightened once the car has been lowered to the ground. Use new split pins where necessary.
11 If a new lower suspension arm has been fitted it is recommended that the steering geometry and front wheel toe-in be checked. Further information may be found in Section 23.

10 Stabiliser bar - removal and refitting

1 Chock the rear wheels, jack-up the front of the car and support the body on firmly based stands. Remove the front roadwheels.
2 Remove the side shields and the lower engine splash shield from under the car.
3 Undo and remove the bolt securing the

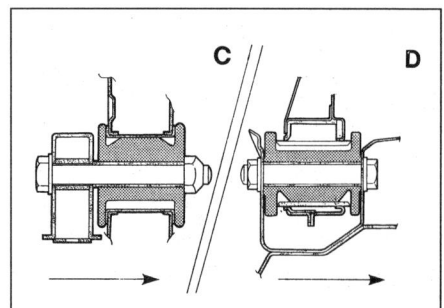

Fig. 11.9 Front (C) and (D) rubber mountings in position (Sec 7)

Suspension and steering 11•7

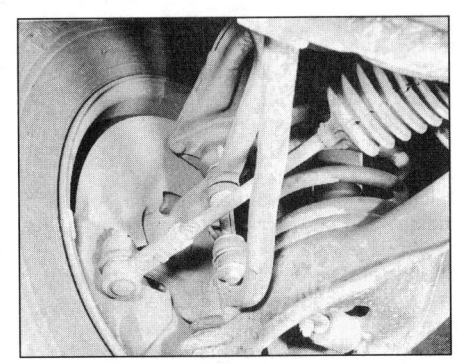

10.4 Stabiliser bar and connecting link bushes

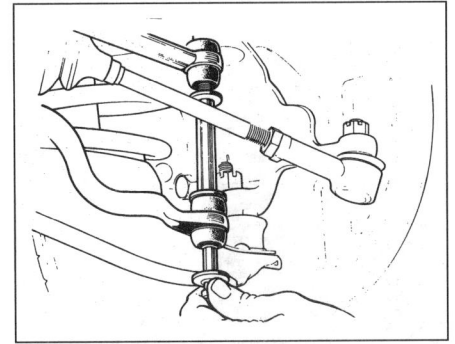

Fig. 11.10 Refitting the stabiliser bar connecting link (Sec 10)

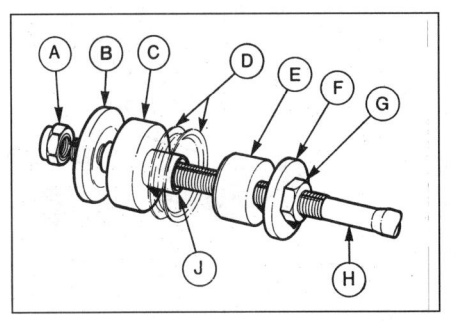

Fig. 11.11 Tie-bar front bush assembly (Sec 14)

A Locknut
B Retainer
C Front bush
D Subframe location
E Rear bush
F Retainer
G Locknut
H Tie-bar
J Spacer

stabiliser bar mounting clamps to the subframe.
4 Undo and remove the two nuts, dished washer and upper bushes and detach the connecting links from their locations in the ends of the stabiliser bar (photo). The stabiliser may now be lifted away from the underside of the car.
5 Refitting is the reverse sequence to removal (Fig. 11.10).

11 Stabiliser bar mounting bushes - removal and refitting

1 Undo and remove the bolts securing the stabiliser mounting clamps to the subframe.
2 Push off the old bushes from the stabiliser bar and place new bushes in their approximate locations.
3 Align the bushes with the mounting clamps and resecure the clamps to the subframe.

12 Stabiliser bar connecting link bushes - removal and refitting

1 Refer to Section 10 and remove the stabiliser bar.
2 Using a sharp knife or hacksaw blade cut the cone ends off the connecting link bushes and discard the bushes.
3 Using a bench vice, a piece of tube of suitable diameter and a socket, press in the new connecting link bushes.
4 Refit the stabiliser bar as described in Section 10.

13 Front shock absorbers - removal and refitting

1 Chock the rear wheels, jack-up the front of the car and place on firmly based stands. Remove the roadwheel.
2 Locate a small jack under the lower suspension arm and partially compress the coil spring.
3 Undo and remove the shock absorber top mounting bolt.
4 Undo and remove the two nuts that secure the shock absorber lower mounting. The shock absorber may now be lifted away

through the coil spring and lower arm aperture.
5 Examine the shock absorber for signs of damage to the body, distorted piston rod, loose mounting or hydraulic fluid leakage. If evident a new unit should be fitted.
6 To test for damping efficiency hold the unit in the vertical position and gradually extend and contract it between its maximum and minimum limits ten times. It should be apparent that there is equal resistance on both directions of movement. If this is not apparent a new unit should be fitted.

⚠ **Always renew shock absorbers in pairs.**

7 Refitting the shock absorbers is the reverse sequence to removal.

14 Tie-bar - removal and refitting

1 Chock the rear wheels, jack-up the front of the car and support the body on firmly based stands. Remove the front roadwheel.
2 Undo and remove the two nuts, bolts and spring washers securing the tie-bar to the lower suspension swinging arm.
3 Undo and remove the stabiliser connecting link nut and bolt.
4 Remove the nut, retainer and bush from the outer end of the tiebar and withdraw the tie-bar from the sub-frame.
5 Refitting is a reverse of the removal sequence. The steering geometry and front wheel toe-in setting must be checked after refitting. Further information will be found in Section 23.

15 Rear axle and suspension assembly - removal and refitting

1 Jack-up the rear of the car and support the body on firmly based stands.
2 Unscrew and remove the nuts securing the exhaust pipe mounting brackets to the rear crossmember and lift off the brackets.
3 Mark the propeller shaft and rear axle flanges and unscrew and remove the four

bolts securing the propeller shaft to the rear axle flange. Remove the two bolts, spring and plain washers holding the propeller shaft to the rear axle flange. Remove the two bolts, spring and plain washers holding the propeller shaft centre bearing to the floor pan and lower the rear section of the shaft to the ground.
4 Extract the spring clip and clevis pin from the handbrake cable yoke and separate the yoke from the handbrake lever. Detach the cable from the guides and support the brackets along the underside of the body and allow the cable to hang from the rear suspension crossmember.
5 Detach the flexible brake hoses from the brake pipe unions and mounting brackets on the rear suspension arms. Plug the ends of the hoses to prevent loss of fluid or preferably clamp the hoses with a proprietary brake hose clamp.
6 Using a hydraulic jack and a suitable block of wood, support the final drive housing and rear suspension arms and take the weight of the rear suspension assembly. Withdraw the split pin from the final drive mounting nut and unscrew and remove the nut and thrust plate. It will be necessary to hold the bolt from inside the luggage compartment (photo).
7 From inside the luggage compartment lift off the plastic covers and unscrew and remove the nut and washer from each shock absorber upper mounting. On Estate cars the shock

15.6 Final drive mounting nut secured with split pin

11•8 Suspension and steering

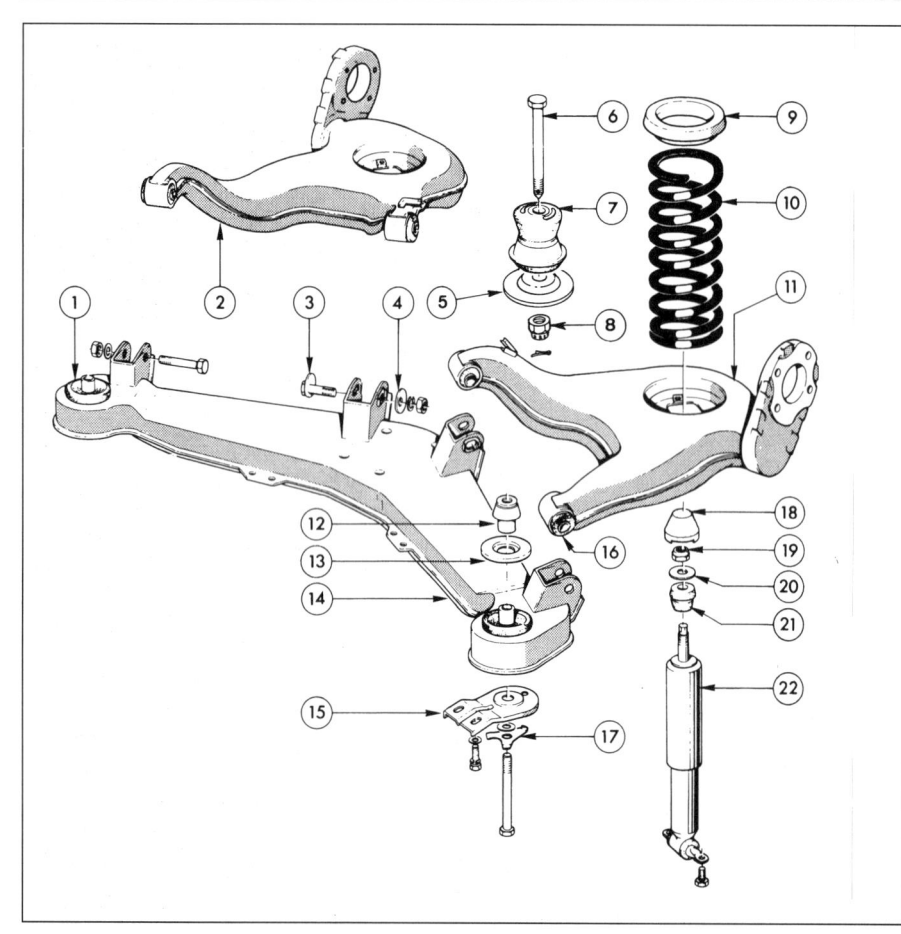

Fig. 11.12 Rear suspension components (Sec 15)

1 Front mounting
2 Suspension arm
3 Mounting and toe adjustment bolt (eccentric cam)
4 Cam disc
5 Thrust plate
6 Rear mounting bolt
7 Rear rubber mounting
8 Castellated nut
9 Spring seat
10 Coil spring
11 Suspension arm
12 Bush
13 Washer
14 Crossmember
15 Mounting plate
16 Suspension arm bush
17 Lock tab
18 Plastic cap
19 Locknut
20 Washer
21 Bush
22 Shock absorber

absorber upper mounting is located beneath a plate behind the rear seat secured with three self tapping screws.
8 Bend back the lock tabs and remove the guide plate centre bolts securing the crossmember front mountings to the floor pan. Undo and remove the two remaining bolts and

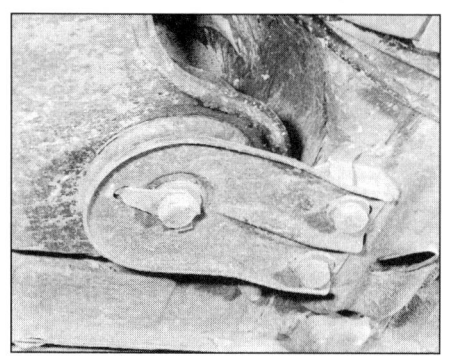

15.8 Rear suspension front mounting guide plate and bolts

guide plates from both front mountings (photo).
9 Carefully lower the complete rear suspension assembly onto its wheels and withdraw rearwards from under the car.
10 Refitting the rear suspension assembly is the reverse of the removal sequence. It will be necessary to bleed the rear hydraulic system as described in Chapter 9.

16 Rear suspension arm - removal and refitting

1 Refer to Section 15 and remove the rear axle and suspension assembly.
2 Undo and remove the hub nut and washer.
3 Release the brake drum retaining speed nut and withdraw the brake drum. If it is a tight fit gently tap its circumference with a soft faced mallet.
4 It is now necessary to draw off the hub flange. If the flange is an easy fit it may be drawn off with a universal three legged puller.

Otherwise the Ford special tool will have to be borrowed from your local Ford garage.
5 Wipe the area of the brake pipe union on the slave cylinder and unscrew the union. Tape the open end to stop dirt entering.
6 Detach the brake pipe from the lower suspension arm clips.
7 Undo and remove the four bolts retaining the hub and brake backplate and separate the two components.
8 Undo and remove the two bolts securing the rear shock absorber to the lower suspension arm.
9 Undo and remove the nuts and bolts securing the lower arm to the crossmember. Note that the inner bolt contains the cam plate for toe-in setting adjustment so mark its setting to facilitate reassembly. Lift the lower suspension arm off the crossmember and half shaft stub axle.
10 Inspect the suspension arm for signs of damage or excessive rust. Renew if these are evident.
11 Should it be necessary to renew the rubber bushes they may be withdrawn from the suspension arm using a length of tube of suitable diameter, a long bolt, nut and spacing washers.
12 Fitting new rubber bushes is the reverse sequence to removal. Lubricate the bushes in soapy water to assist fitting.
13 Refitting the lower suspension arm is the reverse sequence to removal. Tighten all bolts to the torque settings given in the Specifications and finally tighten the hub nut when the suspension assembly has been refitted to the car.
14 The rear suspension toe-in setting must be checked by your local Ford garage after refitting.

17 Rear suspension coil spring - removal and refitting

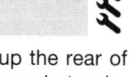

1 Chock the front wheels, jack-up the rear of the car and support on firmly based stands. Remove the rear wheel.
2 Undo and remove the six screws securing the half shaft to the stub axle. Using a piece of wire looped through the hole at the bottom of the wing panel support the weight of the half shaft.
3 Locate a jack under the suspension arm and raise the arm sufficiently to release the spring load from the shock absorber.
4 Working inside the luggage compartment remove the shock absorber mounting cap and then unscrew and remove the shock absorber securing nut. Lift away the retainer. On Estate models access is gained once the cover plate, retained by three self tapping screws, has been removed.
5 Undo and remove the two shock absorber lower mounting bolts and washers. Carefully lower and withdraw the shock absorber.
6 The jack should now be lowered until it is possible to lift away the coil spring. Do not lower too far otherwise the brake flexible hose will be strained.

Suspension and steering 11•9

7 Lift away the coil spring and recover the upper rubber ring.
8 Check the coil spring for signs of excessive corrosion, fracture or loss of tension. If any or all of these conditions exist, a pair of new springs should be fitted.
9 Inspect the upper rubber ring and if damaged or perished a new ring should be obtained.
10 Refitting the rear coil spring is the reverse sequence to removal.

18 Rear shock absorber - removal and refitting

1 Chock the front wheels, jack-up the rear of the car and support the body on firmly based stands. Place a jack under the rear suspension arm and slightly compress the spring.
2 From inside the luggage compartment remove the plastic cover and unscrew and remove the nut and washer from the shock absorber upper mounting. On Estate cars the shock absorber upper mounting is located beneath a plate behind the rear seat secured with three self tapping screws.
3 Undo and remove the two bolts and spring washers securing the shock absorber to the lower arm. Withdraw the shock absorber downwards through the spring aperture in the lower suspension arm (photo).
4 Examine the shock absorber for signs of damage to the body, distorted piston rod or hydraulic leakage. If evident, a new unit should be obtained.
5 To test for damping efficiency hold the unit in the vertical position and gradually extend and contract it between its maximum and minimum limits ten times. It should be apparent that there is equal resistance on both directions of movement. If this is not apparent a new unit should be fitted - always renew shock absorbers in pairs.
6 Refitting the shock absorber is the reverse sequence to removal.

19 Rear suspension front rubber mounting - removal and refitting

1 Chock the front wheels, jack-up the rear of the vehicle and support the body on firmly based stands.
2 Place a jack under the centre of the rear suspension crossmember and take the weight. Bend back the lock tab and remove the guide plate centre bolt securing the crossmember front mounting to the floor pan. Undo and remove the two remaining bolts and lift the guide plate off the front mounting.
3 Carefully lower the jack until a minimum clearance of 6 inches (152 mm) exists between the crossmember and the floor pan. It may be necessary to wedge a block of wood between the floor pan and crossmember to give the desired clearance. Ensure that the brake flexible hose is not strained and that the handbrake cable is not stretched.

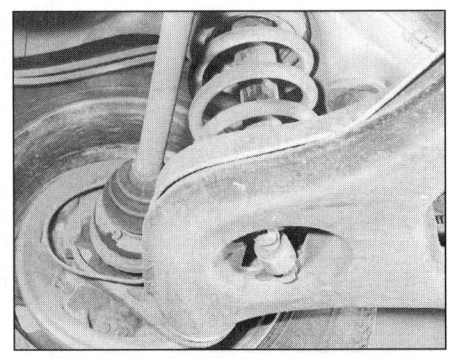

18.3 Rear shock absorber lower mounting

4 Using a large diameter tube, a long bolt, nut and spacing washers, draw the rubber mounting from its location in the crossmember. Reverse this procedure to fit the new mounting lubricating it with soapy water before fitting.
5 Refitting the crossmember is now the reverse sequence to removal. Tighten the mounting bolts to the torque setting given in the Specifications.

20 Rear axle casing rubber mounting - removal and refitting

1 Chock the rear wheels, jack-up the rear of the car and support the body on firmly based stands.
2 Place a jack under the rear axle casing and take the weight of the unit. Withdraw the split pin from the mounting nut and unscrew and remove the nut and thrust pad. It will be necessary to hold the bolt from inside the luggage compartment. Withdraw the mounting bolt.
3 Lower the jack until a clearance of approximately 6 inches (152 mm) exists between the axle casing and floor pan.
4 Place a container under the rear of the axle casing and wipe clean the area around the cover plate mounting bolts.
5 Undo and remove the ten bolts and spring washers securing the cover plate to the rear of the axle casing. Lift away the cover plate and recover the gasket.
6 The rubber mounting may now be drawn from the cover plate using a tube of suitable diameter, a long bolt, nut and packing washer.
7 To fit a new rubber mounting, first lubricate it with soapy water and place in position on the cover plate with the slots running parallel to the cover plate mating face. Using the tube long bolt, nut and washer used for removal, draw the new mounting into position in the cover plate.
8 Reassembly is now the reverse sequence to removal. Make sure the cover plate and axle casing mating surfaces are cleaned and use a new gasket if possible. Tighten the bolts to the torque figures given in the Specifications and refill the axle with the recommended grade of oil.

21 Rear suspension bump rubber - removal and refitting

Removal is simply a matter of pulling the bump rubber downwards from its location.

> ⚠ Lubricate the inner surface of the bump rubber with soapy water to facilitate fitting.

22 Manual steering - lubrication

1 The steering rack is filled with a semi-fluid grease (SAM-1C-9106-A) and is filled for life unless new bellows are fitted or the steering gear is overhauled.
2 If new bellows are fitted, the lubricant may be replenished by packing it into the ends of the rack and then moving the rack from lock-to-lock to distribute it. Apply some lubricant to the interior of the bellows before fitting them. Do not insert more lubricant than was removed with the old bellows.
3 If the steering rack has been overhauled, then the steering gear housing must be held vertically (pinion end uppermost) and lubricant packed into the open end of the housing.
4 The lubricant capacity is 0.26 pints (0.15 litres) for an overhauled assembly devoid of lubricant.

23 Wheel alignment and steering angles

Front wheels

1 Accurate front wheel alignment is essential to prevent excessive steering and tyre wear. Before considering the steering/suspension geometry, check that the tyres are correctly inflated, that the front wheels are not buckled, the hub bearings are not worn or incorrectly adjusted and that the steering linkage is in good order, without slackness or wear at the joints.
2 Wheel alignment consists of four factors:
Camber, which is the angle at which the front wheels are set from the vertical when viewed from the front of the car. Positive camber is the amount (in degrees) that the

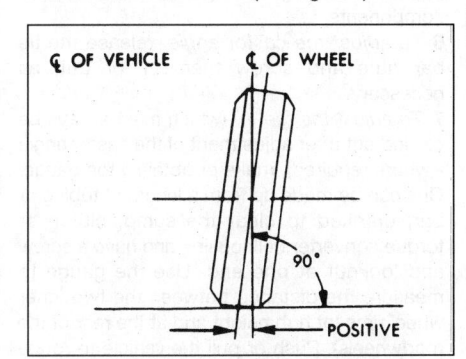

Fig. 11.13 Camber diagram (Sec 23)

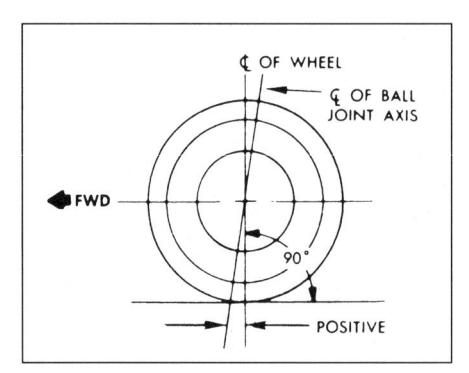

Fig. 11.14 Castor diagram (Sec 23)

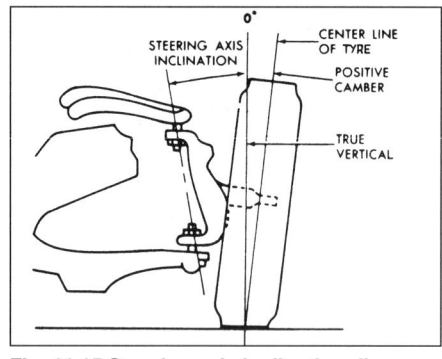

Fig. 11.15 Steering axis inclination diagram (Sec 23)

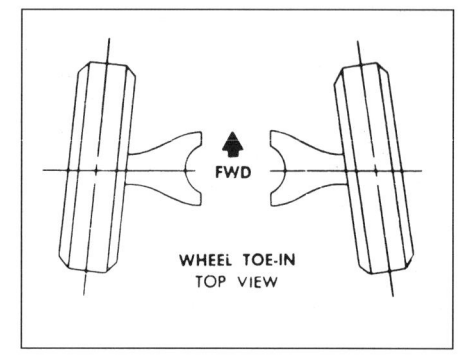

Fig. 11.16 Toe-in diagram (Sec 23)

wheels are tilted outwards at the top from the vertical.

Castor, is the angle between the steering axis and a vertical line when viewed from each side of the car. Positive castor is when the steering axis is inclined rearward.

Steering axis inclination, is the angle, when viewed from the front of the car between the vertical and an imaginary line drawn between the upper and lower suspension control arm balljoints (Fig. 11.15).

Toe-in setting, is the amount by which the distance between the front inside edges of the roadwheels (measured at hub height) differs from the diametrically opposite distance measured between the rear inside edges of the front roadwheels (Fig. 11.16).

3 Due to the need for special gauges it is not normally within the scope of the home mechanic to check and adjust any steering angle except toe-in. Where suitable equipment can be borrowed however, adjustment can be carried out in the following way, setting the tolerances to those given in the Specifications.

4 Before carrying out any adjustment, place the vehicle on level ground, tyres correctly inflated and the front roadwheels set in the "straight-ahead" position. Make sure that all suspension and steering components are securely attached and without wear in the moving parts.

5 *The camber and steering axis inclination angles* are set in production and they cannot be altered or adjusted. Any deviation from the angles specified must therefore be due to collision damage or gross wear in the components.

6 *To adjust the castor angle,* release the tie bar nuts and screw them in, or out, as necessary.

7 *To adjust the toe-in,* (which must always be carried out after adjustment of the castor angle - where required), make or obtain a toe gauge. One can be made up from a length of tubing or bar, cranked to clear the sump, clutch or torque converter bellhousing and have a screw and locknut at one end. Use the gauge to measure the distance between the two inner wheel rims (at hub height and at the rear of the roadwheels). Push or pull the vehicle to rotate the roadwheels through 180° (half a turn) and then measure the distance between the inner

wheel rims (at hub height and at the front of the roadwheels). This last measurement will differ from the first by the amount specified in the Specifications Section. This represent the correct toe-setting of the front wheels. Where the toe is found to be incorrect, loosen the locknuts on both the track-rod ends, also release the screws on the steering gaiter clips. Turn each track-rod in opposite directions by not more than one quarter turn at a time and then recheck the toe. When the adjustment is correct, tighten the locknuts without moving the track-rods and make sure that the track-rod ends are in their correct plane (centre position of arc of travel). Tighten the steering bellows clips making sure that the bellows have not twisted during the adjustment operations. It is important to always adjust each track-rod equally. Where new components have been fitted, adjust the height of each track-rod so that they are equal and the front roadwheels in approximately the "straight-ahead" position, before commencing final setting with the gauge.

Rear wheels

8 *The rear wheel camber,* is set in production and cannot be adjusted.

9 *The toe-in* characteristic of the rear wheels is adjustable by means of the eccentric cam plates located at the inboard pivots of the rear suspension arms (Fig. 11.12). Release the locknut before turning the camplate bolt.

24 Steering wheel - removal and refitting

1 With the front wheels in the straight-ahead position note the position of the spokes of the steering wheel and mark the hub of the steering wheel and inner shaft to ensure correct positioning upon refitting.

2 Carefully prise out the steering wheel insert and using a socket or box spanner of the correct size slacken the steering wheel nut but leave it two or three turns on the thread (photo).

3 Loosen the wheel by using a suitable puller. Do not strike the rear of the wheel rim or the safety type steering column may be damaged. Remove the nut and the wheel.

4 Refitting is the reverse procedure to removal. Correctly align the two marks previously made to ensure correct positioning of the spokes. Do not strike the steering wheel when refitting as it could cause the inner shaft to collapse. Refit the nut and tighten to the specified torque.

25 Steering column assembly - removal and refitting

1 Bend back the lockwasher tabs, undo and remove the two bolts that secure the clamp bar to the upper universal joint assembly. Lift away the lockwasher and clamp bar.

2 For safety reasons disconnect the battery.

3 Undo and remove the securing screws and take out the lower trim panel located under the dash.

4 Refer to Section 24 and remove the steering wheel.

5 Undo and remove the securing screws and lift off the steering column upper and lower shrouds.

6 Undo and remove the four bolts and spring washers securing the lights and windscreen wiper multi-function switches to the steering column. Allow the switches to hang from their wiring looms.

7 From under the dash pull off the section of air ducting that traverses the steering column.

8 Undo and remove the nuts, bolts and spring washers, securing the steering column support bar to its mounting brackets and lift away the bar.

24.2 Removing the steering wheel centre nut

Suspension and steering 11•11

9 Disconnect the ignition switch multi-plug and the wiring loom retaining strap from the column.
10 Remove the nuts and washers securing the steering column to the support bracket and lift the column out of the car.
11 Refitting is the reverse of the removal sequence, bearing in mind the following:
  a) Lubricate the bulkhead seal with soapy water to facilitate installation of the column
  b) Adjust the position of the column to give a clearance of 0.197 in (5 mm) between the column upper shroud and the instrument panel
  c) Do not tighten the steering column to universal coupling bolts until the full weight of car is on the roadwheels

26 Steering column shaft - removal and refitting

1 Refer to Section 25 and remove the steering column assembly.
2 Drill out the steering lock mounting bolt heads and remove the lock.
3 Using a pointed chisel carefully prise open the three dimples at the base of the column and remove the lower bearing.
4 Take off the steering shaft upper bearing sleeve. Withdraw the steering shaft complete with upper bearing from the top of the column.
5 Using a pair of circlip pliers withdraw the upper bearing retaining circlip from the top of the shaft.
6 Lift off the upper bearing, O-ring and lower circlip where fitted.
7 Refitting is the reverse of removal. To fit the steering lock, position the clamp ensuring that the lock tongue engages in the slot in the column tube. Tighten the new retaining bolts until the heads break off.

27 Steering column flexible coupling and universal joint assembly - removal and refitting

1 Undo and remove the nut, clamp bolt and spring washer that secures the flexible coupling bottom half to the pinion shaft (photo).
2 Bend back the locktabs and undo and remove the two bolts securing the universal joint lock bar to the lower steering shaft. Lift away the tab washer and lock bar (photo).
3 The lower steering shaft may now be lifted away.
4 To refit place the lower steering shaft in the approximate fitted position and align the master splines on the shaft and pinion. Connect the shaft to the pinion.
5 Position the triangular clamp on the bottom of the steering column and secure with the clamp bar bolts and tab washer. Tighten the bolts fully and lock by bending up the tabs.

27.1 Steering column flexible coupling

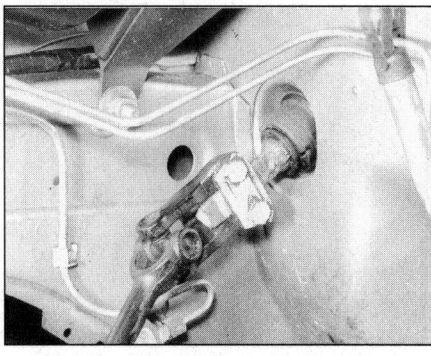

27.2 Steering column universal joint

6 Refit the flexible coupling bottom half clamp bolt, spring washer and nut. Tighten to the specified torque.

28 Rack and pinion steering gear - removal and refitting

1 Before starting this sequence set the steering wheel to the straight ahead position.
2 Jack-up the front of the car and place blocks under the wheels. Lower the car slightly so that the track-rods are in a near horizontal position.
3 Undo and remove the nut, bolt and spring washer that secures the flexible coupling bottom half clamp to the pinion shaft.

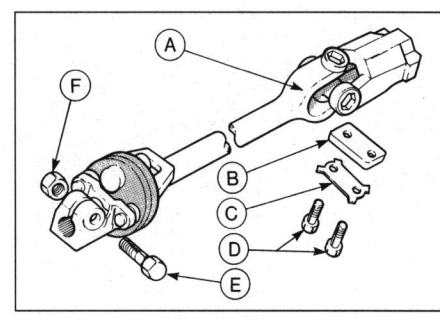

Fig. 11.17 Steering coupling and universal joint (Sec 27)

A Universal joint   D Clamp bolts
B Lock bar   E Clamp bolt
C Tab washer   F Locknut

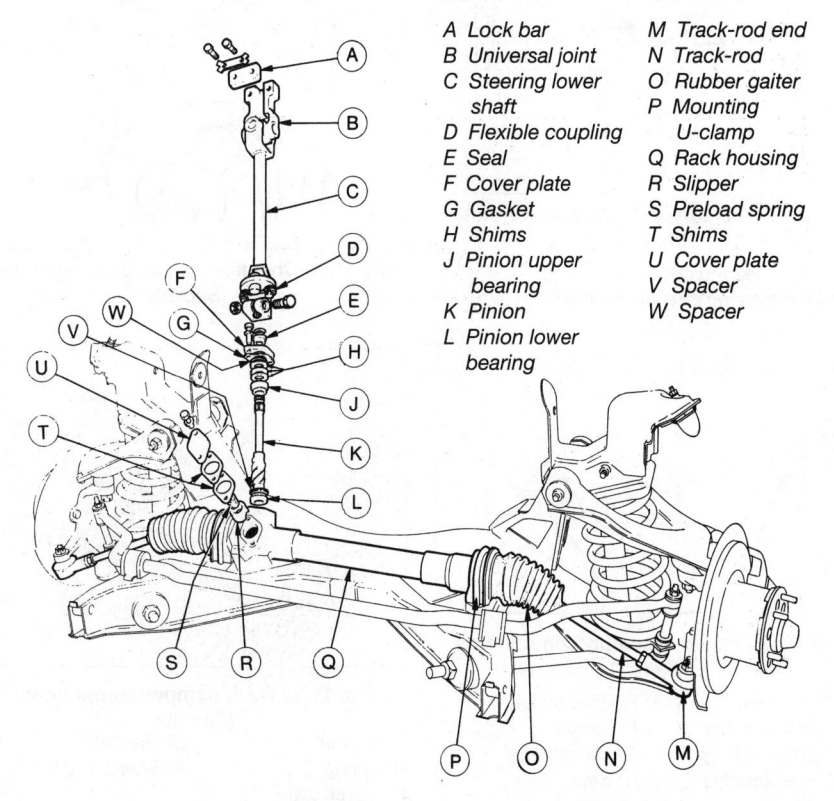

Fig. 11.18 Rack and pinion steering gear manual steering (Sec 28)

A Lock bar   M Track-rod end
B Universal joint   N Track-rod
C Steering lower   O Rubber gaiter
  shaft   P Mounting
D Flexible coupling   U-clamp
E Seal   Q Rack housing
F Cover plate   R Slipper
G Gasket   S Preload spring
H Shims   T Shims
J Pinion upper   U Cover plate
  bearing   V Spacer
K Pinion   W Spacer
L Pinion lower
  bearing

11•12 Suspension and steering

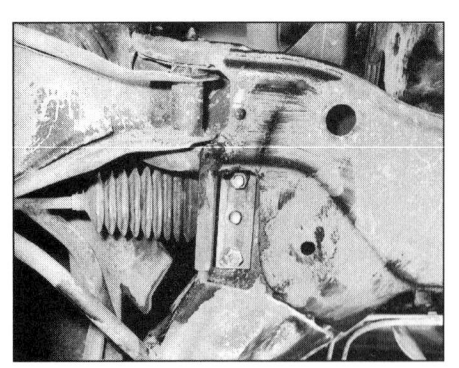

28.4 Steering gear left-hand mounting bolts

4 Bend back the lock tabs and then undo and remove the bolts securing the steering gear assembly to the mountings on the front axle frame. Lift away the bolts, lock washers and U-shaped clamps (photo).
5 Withdraw the split pins and undo and remove the castellated nuts from the ends of each track-rod where they join the steering arms. Using a universal balljoint separator, separate the track-rod ballpins from the steering arms and lower the steering gear assembly downwards out of the car.
6 Before refitting the steering gear assembly make sure the wheels have remained in the straight ahead position. Also check the condition of the mounting rubbers round the housing and if they appear worn or damaged they must be renewed.
7 Check that the steering rack is in the straight ahead position. This can be done by ensuring that the distances between the ends of both track-rods and the rack housing on both sides are the same.
8 Place the steering gear assembly in its location on the front axle frame and at the same time mate up the splines on the pinion shaft with the splines in the clamp on the steering column flexible coupling. There is a master spline so make sure these are in line.
9 The remainder of the refitting sequence may be carried out in the reverse order of removal. Tighten all nuts and bolts to the torque settings given in the Specifications. If the position of the track-rod ends has been altered the front wheel toe setting must be checked as described in Section 23.

29 Rack and pinion steering gear - adjustments

1 For the steering gear to function correctly two adjustments are necessary. These are pinion bearing preload and rack damper adjustment.
2 To carry out these adjustments remove the steering gear from the car as described in Section 28, then mount the steering gear assembly in a soft jawed vice so that the pinion is in a horizontal position and the rack damper cover plate is to the top.
3 Remove the rack damper cover plate by undoing and removing the two retaining bolts and spring washers. Lift away the cover plate gasket and shims. Also remove the small spring and the recessed slipper which bears onto the rack.
4 Now remove the pinion bearing cover plate from the base of the pinion, by undoing and removing the three bolts and spring washers. Lift away the cover plate, gasket and shims.
5 To set the pinion bearing preload, refit the shims, cover plate and bolts to the rack housing. The shim pack must contain at least three shims, one of them being the thick shim which must be placed immediately against the cover plate. Note that the plain gasket has been left out at this stage.
6 Tighten the retaining bolts evenly to compress the shim pack and then slacken off evenly until the cover plate is only just contacting the shim pack. Using feeler gauges measure the gap between the cover plate and the housing adjacent to each bolt. The gap should be equal to the thickness of the plain gasket plus 0.001 to 0.004 in (0.025 to 0.102 mm).
7 Build up a shim pack to the required thickness, ensuring that it contains at least two shims in addition to the thick shim. Shim thicknesses available are as listed.

Part No.	Thickness
72GB-3K544-BA	0.005 in (0.13 mm)
72GB-3K544-CA	0.007 in (0.20 mm)
72GB-3K544-DA	0.010 in (0.25 mm)

8 With the cover plate removed make sure that the pinion seal is packed with grease. Refit the shims plain gasket, cover plate and securing bolts and spring washers. Tighten the bolts to the torque settings given in the Specifications.
9 To reset the rack damper adjustment, refit the slipper in its location on the rack and make sure it is fully home. Using a straight edge and feeler gauges, record the maximum clearance obtained between the top of the slipper and the surface of the pinion housing, whilst the rack is moved from lock-to-lock by turning the pinion shaft.
10 Assemble a shim pack including the gasket to equal the thickness of the measurement obtained plus 0.002 to 0.005 in (0.050 to 0.125 mm). Shim thicknesses available are as listed.

Part No	Material	Thickness
72GB-3N597-BA	Steel	0.005 in (0.127 mm)
72GB-3N597-CA	Steel	0.010 in (0.25 mm)
72GB-3B597-AA	Brass	0.002 in (0.05 mm)
72GB-3N597-AA	Aluminium	0.0023 in (0.06 mm)
77GB-3N598-AA	BUNA coated flexoid	0.010 in (0.25mm)

11 Fit the spring into the recess in the slipper. Place the shim pack so that the gasket will be next to the cover plate and refit the cover plate. Refit the bolts and spring washers and tighten to the torque setting given in the Specifications.

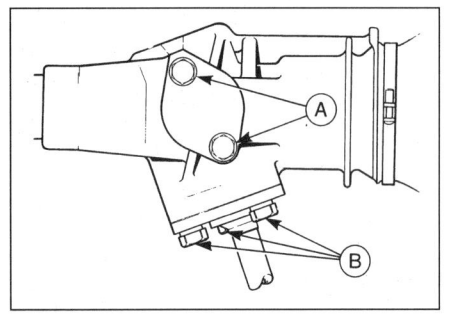

Fig. 11.19 Position of the rack cover plates (Sec 29)

A Slipper bearing cover plate bolts
B Pinion bearing cover plate bolts

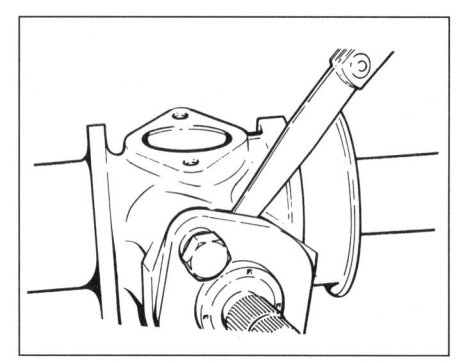

Fig. 11.20 Checking pinion bearing preload (Sec 29)

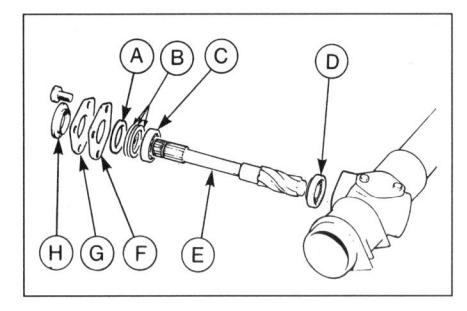

Fig. 11.21 Pinion assembly components (Sec 29)

A Thick shim
B Preload shims
C Upper bearing
D Lower bearing
E Pinion shaft
F Gasket
G Cover plate
H Seal

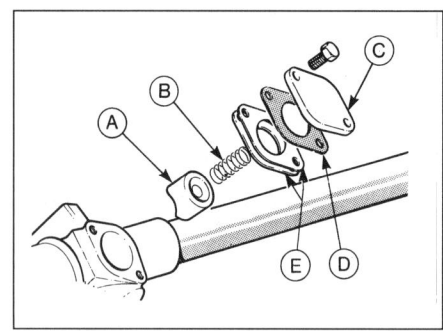

Fig. 11.22 Rack damper components (Sec 29)

A Slipper
B Spring
C Cover plate
D Gasket
E Shims

Suspension and steering 11•13

30 Rack and pinion steering gear - dismantling and reassembly

1 Remove the steering gear assembly from the car as described in Section 28.
2 Undo the track-rod balljoint locknuts and unscrew the balljoints. Lift away the plain washer and remove the locknut. To assist in obtaining an approximate correct setting for track-rod adjustment mark the threads or count the number of turns required to undo the balljoint.
3 Slacken off the clips securing the rubber gaiter to each track-rod and rack housing end. Carefully pull off the gaiters. Have a quantity of rag handy to catch the lubricant which will escape when the gaiters are removed. **Note:** *On some steering gear assemblies soft iron wire is used instead of clips. Always secure the gaiter with clips.*
4 To dismantle the steering gear assembly it is only necessary to remove the track-rod which is furthest away from the pinion.
5 To remove the track-rod place the steering gear assembly in a soft jawed vice. Working on the track-rod balljoint carefully drill out the pin that locks the ball housing to the locknut. Great care must be taken not to drill too deeply or the rack will be irreparably damaged. The hole should be about 0.375 in (9.525 mm) deep.
6 Hold the locknut with a spanner, then grip the ball housing with a mole wrench and undo it from the threads on the rack.
7 Take out the spring and ball seat from the recess in the end of the rack and then unscrew the locknut from the threads on the rack. The spring and ball seat must be renewed during reassembly.
8 Undo and remove the two bolts and spring washers that secure the pinion cover plate. Lift away the cover, shims and gaskets. Remove the pinion and upper bearing.
9 Undo and remove the two bolts and spring washers that secure the rack damper cover. Lift away the cover, gasket shims, springs and yokes.
10 With the pinion removed, withdraw the complete rack assembly with one track-rod still attached from the pinion end of the casing.
11 The remaining pinion bearing assembly may now be removed from the rack housing.
12 It is always advisable to withdraw the rack from the pinion end of the rack housing. This avoids passing the rack teeth through the bush at the other end of the casing and causing possible damage.
13 Carefully examine all parts for signs of wear or damage. Check the condition of the rack support bush at the opposite end of the casing from the pinion. If this is worn renew it. If the rack or pinion teeth are in any way damaged a new rack and pinion will have to be obtained.
14 Take the pinion seal off the top of the casing and replace it with a new seal.
15 To commence reassembly fit the lower pinion bearing and thrust washer into their recess in the casing.
16 Refit the rack in the housing from the pinion end and position it in the straight-ahead position by equalising the amount it protrudes at either end of the casing.
17 Refit the remaining pinion bearing and thrust washer onto the pinion and fit the pinion into its housing so that the larger masterspline on the pinion shaft is positioned as shown in Fig. 11.24.
18 Refit the rack damper yoke, springs, shims, gasket and cover plate.
19 To refit the track-rod that has been removed, start by fitting a new spring and ball seat to the recess in the end of the rack shaft and refit the locknut onto the threads of the rack.
20 Lubricate the ball, ball seat and bellhousing with a small amount of special lubricant (SAM-1C-9106-A). Then slide the ballhousing over the track-rod and screw the housing onto the rack threads keeping the track-rod in the horizontal position until the track-rod starts to become stiff to move.
21 Using a normal spring balance hook it round the track-rod 0.25 in (6 mm) from the end and check the effort required to move it from the horizontal position.
22 By adjusting the tightness of the ballhousing on the rack threads the effort required to move the track-rod must be set at 5 lb (2.8 kg).
23 Tighten the locknut up to the housing and then recheck that the effort required to move the track-rod is still correct at 5 lb (2.8 kg).
24 On the line where the locknut and ballhousing met, drill a 0.125 in (3.175 mm) diameter hole which must be 0.375 in (9.525 mm) deep. Even if the two halves of the old hole previously drilled out align, a new hole must be drilled.
25 Tap a new retaining pin into the hole and peen the end over to secure it.
26 Refit the rubber gaiters and track-rod ends ensuring that they are refitted in exactly the same position from which they were removed.
27 Fill the gear with special lubricant as described in Section 22 paragraphs 3 and 4. Then carry out both steering gear adjustments as described in Section 29.
28 After the steering gear has been refitted to the car the toe-in must be checked. Further information will be found in Section 23.

31 Steering rack rubber gaiter - removal and refitting (in situ)

1 Jack-up the front of the car and place blocks under the wheels. Lower the car slightly so that the track-rods are in a near horizontal position.
2 Withdraw the split pin and undo the castellated nut holding the balljoint taper pin to the steering arm. Using a universal balljoint separator part the taper pin from the steering arm.
3 Undo the track-rod balljoint locknut and unscrew the balljoint. To assist in obtaining an approximate correct setting for track-rod adjustment mark the threads or count the number of turns required to undo the balljoint.
4 Slacken off the clips securing the rubber gaiter to the track-rod and rack housing end. Carefully pull off the gaiter. Have a quantity of rag handy to wipe off the lubricant when the gaiters are removed. **Note:** *On some steering gear assemblies soft iron wire is used instead of clips. Always secure the gaiter with clips.*
5 Fitting a new rubber gaiter is now the

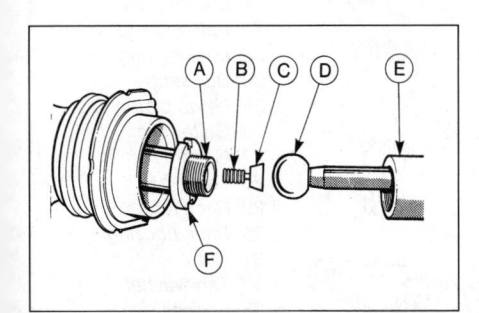

Fig. 11.23 Track-rod inboard balljoint components (Sec 30)

A Rack
B Spring
C Ball seating
D Track-rod arm and ball
E Ball housing
F Lock ring

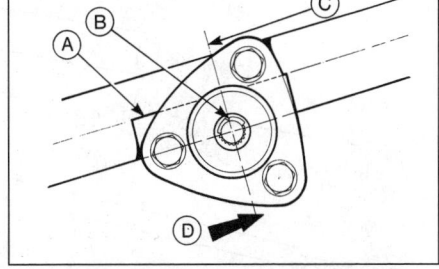

Fig. 11.24 Relation of pinion shaft master spline to rack housing (Sec 30)

A Rack centreline
B Master spline
C Pinion centre line
D Right-hand side of vehicle

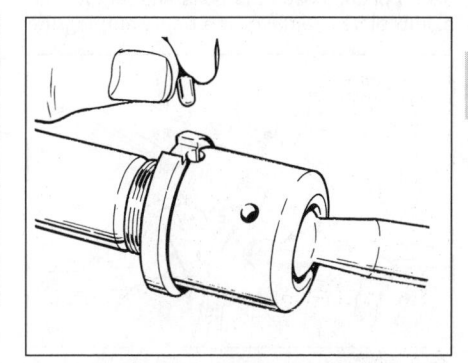

Fig. 11.25 Fitting a new retaining pin to the track-rod ball housing (Sec 30)

11•14 Suspension and steering

reverse sequence to removal. It will be necessary to refill the steering gear assembly with special lubricant. Full information will be found in Section 22.

6 It is recommended that the toe-setting be checked at the earliest opportunity. Further information will be found in Section 23.

32 Track-rod end - removal and refitting

Full information will be found in Section 31 omitting paragraphs 4 and 5.

33 Power assisted steering system - checking the fluid level

1 Remove the reservoir filler cap/dipstick. Wipe the dipstick with a non-fluffy cloth and refit the cap, then remove the cap and check the level of the fluid on the dipstick.

2 One side of the dipstick is marked *Full Cold* and the other side *Full Hot*. Top-up, if necessary, with the specified power steering fluid (automatic transmission fluid) to the appropriate mark on the dipstick. Always pour the fluid in slowly from a can which has not been shaken for the preceding 30 minutes.

3 Where topping-up is necessary, check all hoses and pipes in the system for leakage and rectify as required.

34 Power assisted steering gear - removal and refitting

1 Set the steering wheel to the straight ahead position.

2 Jack-up the front of the car and place blocks under the wheels then lower the car slightly so that the track-rods are in a near horizontal position. Remove the engine splash shield, if fitted.

3 Disconnect the fluid feed and return pipes from the control valve and drain the fluid into a container. Plug the ends of the pipes and the openings in the control valve to prevent the entry of dirt.

4 Bend back the lock tabs on the coupling clamp plate, loosen the bolts and remove the clamp plate. Remove the bolt clamping the lower end of the coupling assembly to the pinion shaft, disengage the assembly from the pinion shaft and lift it out.

5 Withdraw the split pins and remove the castellated nuts from the ends of the track-rods where they are attached to the steering arms. Using a universal balljoint separator, detach the track-rod ballpins from the steering arms.

6 Bend back the lock tabs, withdraw the split pins, then undo the nuts and remove the rack to crossmember securing bolts.

7 Remove the fan shroud securing screws and lift the shroud out of the engine compartment.

8 Remove the right-hand engine mounting securing nut and jack-up the engine approximately 5 in (125 mm). Move the rack to the left-hand side, and guide the right-hand track-rod over the top of the stabilizer bar then remove the steering gear assembly from the right-hand side.

9 Before refitting the steering gear assembly make sure the wheels are still in the straight-ahead position.

10 Check that the steering rack is in the centre of its travels. This can be done by ensuring that the distance between the ends of both trackrods and the rack housing on both sides are the same.

11 Slide the assembly in from the right-hand side of the car, and fit the steering coupling on the pinion shaft, with the master splines aligned, then loosely fit the clamp bolt and nut. Fit the steering gear assembly securing bolts using new locking tabs.

12 Fit the coupling clamp to the lower end of the steering shaft, fit new locking plate and refit the bolts loosely.

13 Tighten the steering gear to crossmember securing bolts to the specified torque and fit new split pins.

14 Reconnect the track-rod ends to the steering arms and tighten the castellated nuts to the specified torque. Fit new split pins.

15 Lower the engine and tighten the right-hand engine mounting nut to the specified torque.

16 Reconnect the power steering fluid feed and return pipes to the control valve.

17 Tighten the clamp bolts on the steering shaft flexible coupling and on the pinion shaft to the specified torque.

18 Refit the engine splash shield (if fitted) and the fan shroud.

19 Jack-up the car, remove the support blocks and lower the car to the ground.

20 Top-up the power steering fluid reservoir with the specified fluid and bleed the system as described In Section 37.

21 Check and adjust the toe-setting as described in Section 23.

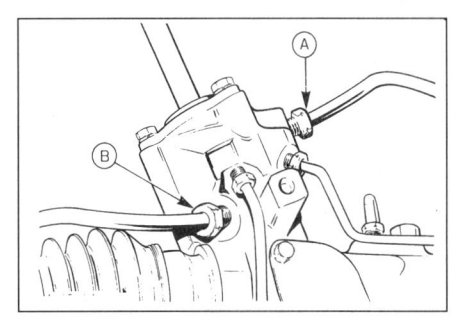

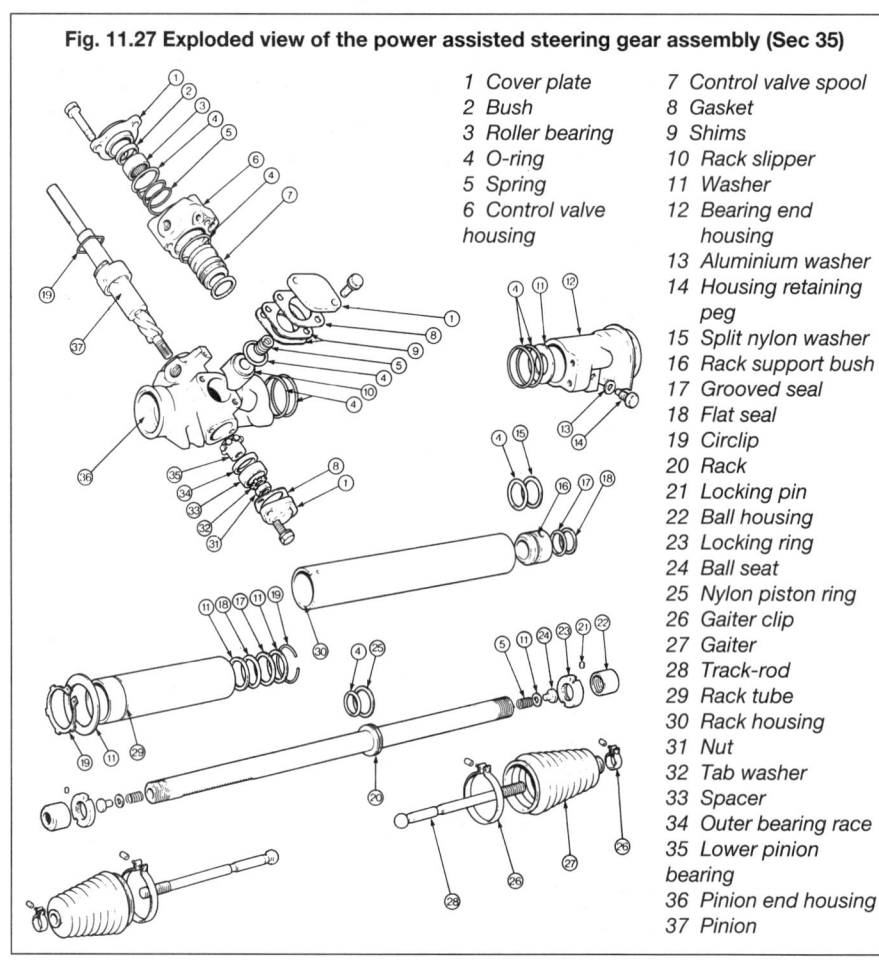

Fig. 11.26 Disconnecting the fluid feed and return pipes (power assisted steering) (Sec 34)

A Pressure feed pipe   B Return pipe

Fig. 11.27 Exploded view of the power assisted steering gear assembly (Sec 35)

1 Cover plate
2 Bush
3 Roller bearing
4 O-ring
5 Spring
6 Control valve housing
7 Control valve spool
8 Gasket
9 Shims
10 Rack slipper
11 Washer
12 Bearing end housing
13 Aluminium washer
14 Housing retaining peg
15 Split nylon washer
16 Rack support bush
17 Grooved seal
18 Flat seal
19 Circlip
20 Rack
21 Locking pin
22 Ball housing
23 Locking ring
24 Ball seat
25 Nylon piston ring
26 Gaiter clip
27 Gaiter
28 Track-rod
29 Rack tube
30 Rack housing
31 Nut
32 Tab washer
33 Spacer
34 Outer bearing race
35 Lower pinion bearing
36 Pinion end housing
37 Pinion

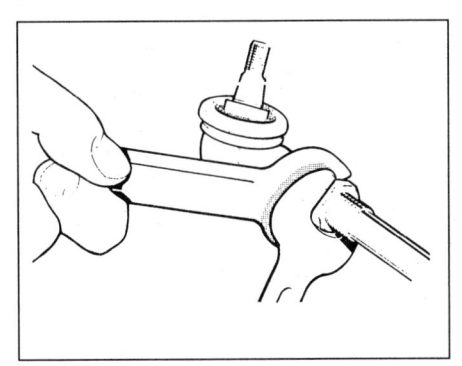

Suspension and steering 11•15

Fig. 11.28 Slackening the track-rod end locknut (Sec 35)

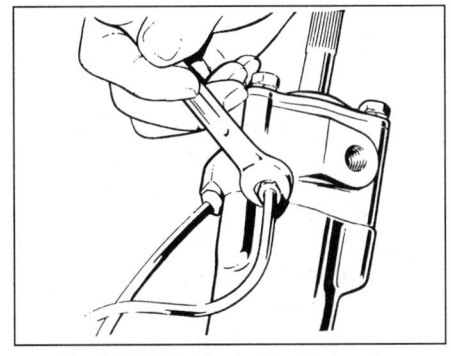

Fig. 11.29 Disconnecting the power steering transfer pipes (Sec 35)

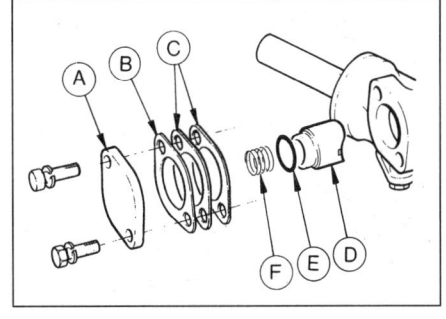

Fig. 11.30 Rack slipper and cover plate assembly (Sec 35)

A Cover   B Gasket   C Shims   D Slipper
E Sealing ring   F Spring

35 Power assisted steering gear - dismantling and reassembly

1 Loosen the track-rod balljoint locknuts and unscrew the balljoints. To assist in obtaining an approximate setting for track-rod adjustment, mark the position of the balljoint on the track-rod or count the number of turns required to unscrew the balljoint.

2 Slacken off the clips securing the rubber gaiters to each track-rod and rack housing end. Pull off the gaiters and collect the oil, which will drain out, in a container. **Note:** *Soft iron wire is used to secure the gaiters in production, instead of clips. At reassembly always use screw-type clips.*

3 Mount the steering gear assembly in a soft-jawed vice. Position the vice on an end casting, do not hold it by the tube as this is easily damaged.

4 Centre punch and drill out the pin securing each track-rod housing to locknut with a 0.16 in (4 mm) drill. Great care must be taken not to drill too deeply or the rack will be damaged. Do not drill deeper than 0.35 in (9 mm).

5 Hold the locknut with a spanner, then grip the ball housing with a mole wrench and unscrew it from the threads on the rack.

6 Collect the spring and ball seat from the recess in the end of the rack and then unscrew the locknut from the threads on the rack. The spring and ball seat must be renewed during reassembly.

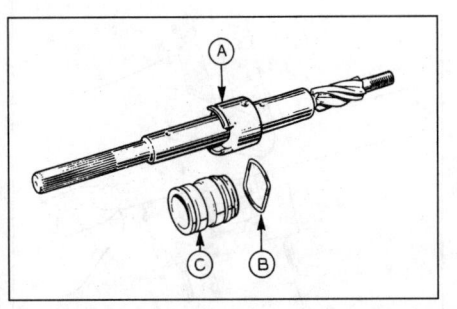

7 Remove the two control valve body to rack housing transfer pipes.

8 Undo and remove the two bolts securing the rack slipper cover plate, lift off the cover plate and remove the shim pack, sealing ring, spring and slipper.

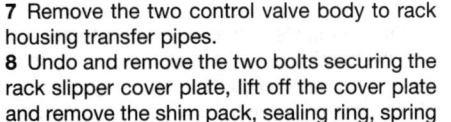

9 Undo the pinion lower bearing cover plate securing bolts and remove the cover plate and gasket. Bend back the lock tab and remove the lower bearing retaining nut. Protect the splines on the input shaft from damage when holding the shaft to remove the nut.

10 Remove the three securing bolts and lift off the control valve top cover, bearing and oil seal assembly. Take out the spool preload spring and withdraw the control valve body from the pinion shaft.

11 Unscrew the pinion lower bearing then withdraw the pinion, spacer and outer race from the housing (Fig. 11.32). Do not remove the middle bearing and oil seal unless the seal is suspected of being the cause of an oil leak or of being excessively worn. If either is suspect they both must be renewed.

12 Slide the spool off the pinion assembly after removing the retaining circlip (Fig. 11.33).

13 Remove the locating pegs from the end housings and using a rubber-faced hammer separate the end housings from the rack tube (Fig. 11.34). This will also free the transfer pipe.

14 Pull the rack in the direction of the bearing end housing and disengage the rack from the inner tube. Pull the support bearing off the rack (Fig. 11.35).

15 Remove the inner tube from the rack tube.

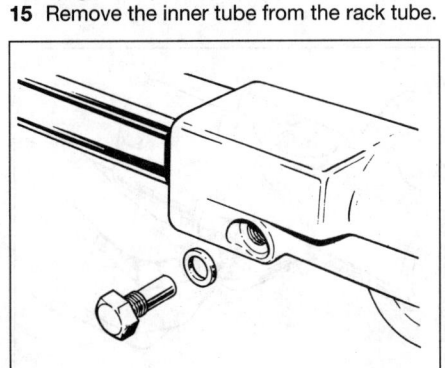

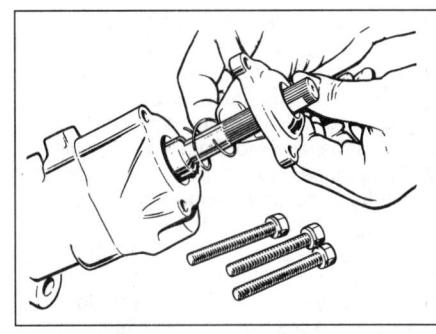

Fig. 11.31 Removing the spool preload spring (Sec 35)

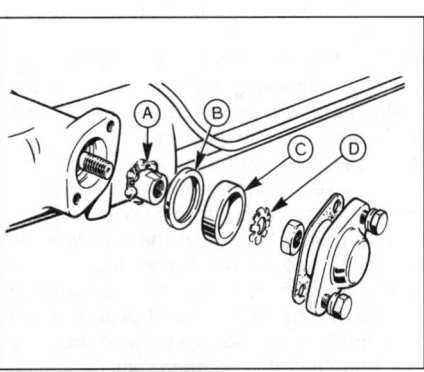

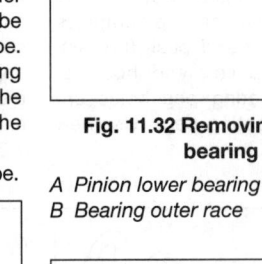

Fig. 11.32 Removing the pinion lower bearing (Sec 35)

A Pinion lower bearing   C Spacer
B Bearing outer race   D Tab washer

Fig. 11.33 Spool valve removal (Sec 35)

A Pinion assembly   B Retaining circlip
C Spool valve

Fig. 11.34 Removing the end housing locating pegs (Sec 35)

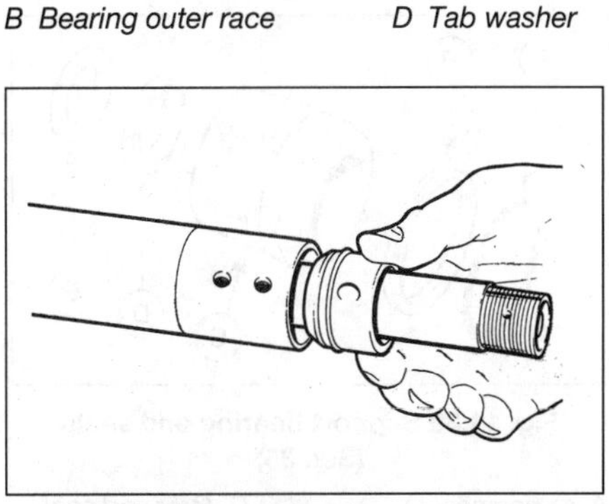

Fig. 11.35 Pull off the rack support bearing (Sec 35)

11

11•16 Suspension and steering

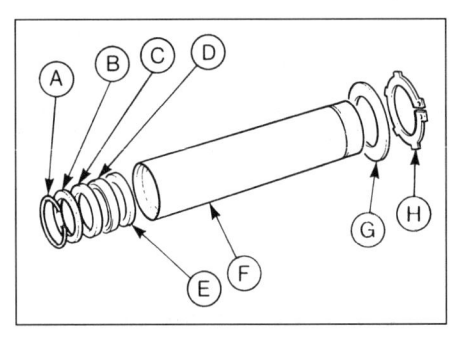

Fig. 11.36 Reassembly order of seals and clips (Sec 35)

- A Circlip
- B Washer
- C Grooved seal
- D Flat seal
- E Washer
- F Inner tab
- G Washer
- H Circlip

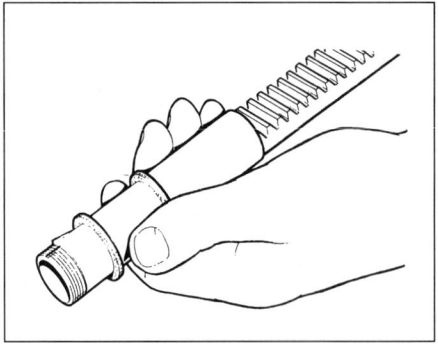

Fig. 11.37 Use a protective sleeve when fitting the seals (Sec 35)

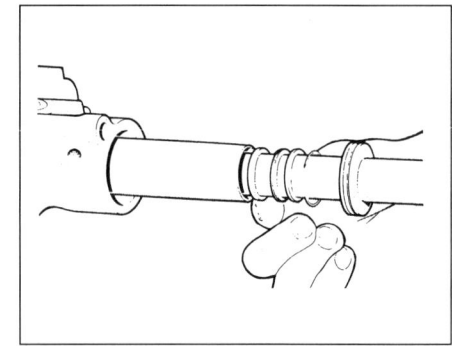

Fig. 11.38 Fit the seals and washer into the inner tube (Sec 35)

16 Clean and examine all parts for signs of wear and damage. If the rack or pinion teeth are in any way damaged a new rack and pinion will have to be obtained. If either the pinion valve spool or valve housing is defective, both parts must be renewed. Renew all oil seals.

17 Commence reassembly by sliding the inner tube seals and washers onto the rack from the toothed end in the following order: circlip, washer, grooved seal, flat seal and washer. A piece of thin foil or paper wrapped round the rack teeth will prevent possible damage to the seals as they are fitted.

18 Lubricate the two O-ring seals with power steering fluid and insert them in the pinion end housing. Fit the circlip and washer to the inner tube and then push the inner tube into the end housing.

19 Slide the rack through the inner tube and pinion end housing, then fit the seals and washers that are on the rack, into the inner tube and secure with the circlip.

20 Soak the piston O-ring and nylon piston ring in power steering fluid then fit the O-ring and piston ring in the piston groove.

21 Fit the rack tube over the rack with the end, with two round holes, towards the bearing support end. Take care to compress the piston ring carefully and push the rack tube firmly into the pinion end housing, checking that the locating peg holes are aligned, then fit the locating pegs. Use new aluminium washers.

22 Fit outer seals to the rack support bearing, slide the bearing over the rack then push it into the rack tube and align the locating peg holes.

23 Using a protective sleeve as described in paragraph 17, slide the inner seals over the rack, the grooved seal first, then fit the inner seals to the rack support bearing.

24 Fit the large washer and two O-ring seals in the end housing and the transfer pipe seals in their locations, then locate the transfer pipe in the pinion end housing.

25 Fit the bearing end housing to the rack, rack tube and transfer pipe, ensure the locating peg holes are aligned and then fit the locating pegs.

26 Slide the spool onto the pinion assembly and fit the retaining circlip in the lower groove of the valve spool.

27 If the middle bearing was removed during dismantling fit a new bearing and oil seal. Fit the lower bearing inner race in the housing.

28 Position the rack in the centre of its travel (equal amount of rack protruding from each end housing) and slide the control valve and pinion assembly into the housing to engage with the rack.

29 Screw the lower bearing onto the pinion and make a preliminary adjustment of the control valve to locate the spool in an approximately neutral position. To do this pull the pinion to remove endfloat from the bearing and adjust dimension A in Fig. 11.41 to 1.61 to 1.65 in (41 to 42 mm) by screwing the lower bearing in or out as necessary. Fit a new lockwasher and nut, but do not bend the lock tab at this stage.

30 To adjust the pinion bearing cover plate, fit the outer bearing race, spacer and end cover to the pinion housing without shims. Tighten the bolts evenly to 9 to 12 lbf ft (1.2 to 1.6 Nm), then using feeler gauges measure the gap between the cover plate and housing. Select shims to fit the measured gap. Remove the cover plate, fit the selected shims and then refit the cover plate and tighten the securing bolts to 12 to 15 lbf ft (1.7 to 2.1 Nm). During this adjustment it may be necessary to re-adjust dimension A in Fig.11.41.

31 Fit the control valve body, then pinion bearing preload spring (ensuring that the spring is not trapped in the roller bearing), the valve top cover assembly and the securing bolts. Tighten the bolts to the specified torque.

32 Fit the two transfer pipes connecting the control valve body to the rack housing.

33 For adjustment of the rack slipper a dial gauge and mounting block are required. Fit the dial gauge on the mounting block and with it positioned on a piece of glass adjust the dial gauge to zero. Fit the slipper in the housing and hold the mounting block on the slipper with the probe of the dial gauge contacting the rack housing. Move the rack from lock-to-lock by turning the pinion and note the maximum deflection of the dial gauge from zero. Make up a shim pack of thickness equal to the noted

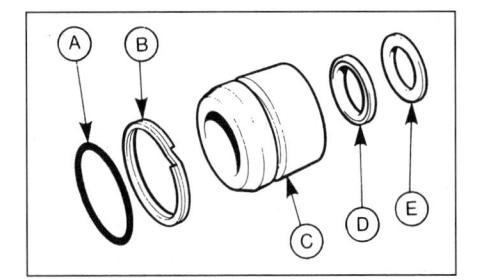

Fig. 11.39 Support bearing and seals (Sec 35)

- A O-ring seal
- B Nylon seal
- C Support bearing
- D Grooved seal
- E Flat seal

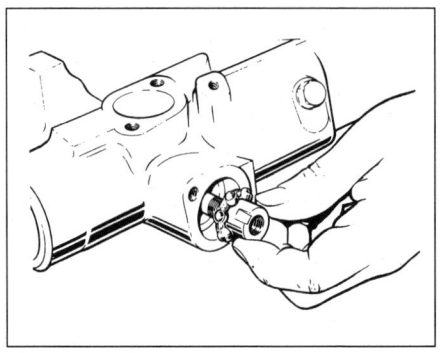

Fig. 11.40 Refitting pinion lower bearing (Sec 35)

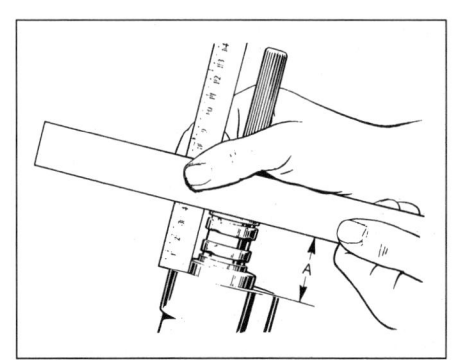

Fig. 11.41 Preliminary adjustment of pinion bearing (Sec 35)

Suspension and steering

deflection plus 0.001 to 0.005 in (0.025 to 0.125 mm).
34 Fit the spring, O-ring seal, the selected shims, gasket cover plate and securing bolts. Tighten the bolts to the specified torque.
35 Fit new springs and ball seats to the recess in the ends of the rack, and refit the locknuts onto the threads of the rack.
36 Lubricate the balls, ball seats and ball housings with a small amount of the specified oil, then slide the ball housings over the trackrod and screw them onto the rack threads.
37 Hook a normal type of spring balance round the track-rod (0.25 in (6 mm) from the end and check the effort required to move it from the horizontal position.
38 By adjusting the tightness of the ball housing on the rack thread, the effort required to move the track-rod must be set at 5 lb (2.1 kg).
39 On the line where the locknut and ball housing meet drill a 0.16 in (4 mm) diameter hole, 0.35 in (9 mm) deep. Even if the existing holes align a new hole must be drilled.
40 Fit a new retaining pin into the hole and peen the end over to secure it.
41 Refit the rubber gaiters and track-rod ends ensuring that they are refitted in the same position from which they were removed. When fitting the gaiters fill the rack with 0.33 pints (0.2 litres) of the specified engine oil. This is SAE40 engine oil **not** power steering fluid. Traverse the rack to distribute the oil. Do not over-fill. Tighten the gaiter securing clips.
42 After the steering gear assembly has been refitted to the car, the control valve adjustment must be finalised, and the toe-setting checked, as described in Sections 36 and 23.

36 Power assisted steering control valve - checking and adjusting

1 After overhaul the control valve must be finally checked and adjusted, if necessary, with the steering gear assembly fitted in the car. For this, a test gauge as shown in Fig. 11.43 will be required.
2 Connect the test gauge as shown in Fig. 11.43. Position the gauge in front of the windscreen so that it can be seen from the driver's seat. Fill the reservoir with fluid and bleed the system as described in Section 37.
3 Start the engine and run it until the fluid reaches normal operating temperature, while rotating the steering wheel slowly to each lock in turn a maximum of five times. Do not hold on full lock for more than 30 seconds or damage to the pump may occur.
4 Remove the steering wheel insert and connect a torque gauge to the steering wheel nut with a socket and extension.
5 With the engine running turn the steering wheel to each lock in turn and whilst on lock apply a torque of 34 lbf In (38 kgf cm) to the steering wheel nut and note the pressure registered on the gauge. Repeat two or three times on each lock. The pressure should be 160 lbf/in^2 (11.2 kgf/cm^2) and the difference between right and left lock should not exceed 12 lbf/in^2 (0.85 kgf/cm^2).
6 Adjustment is by means of the pinion lower bearing which is secured onto the pinion shaft. To adjust, proceed as follows:
  a) Jack-up the car and support on axle-stands
  b) Disconnect the track-rod ends from the steering arms, see Section 32
  c) Remove the rack to crossmember securing bolts and the right-hand engine mounting securing nut, then jack-up the engine approximately 5 in (125 mm), refer to Section 34
  d) Disconnect the steering coupling, see Section 27, and move the rack forward to gain access to the pinion lower cover plate. Remove the cover plate assembly. Collect the oil draining out in a container.
  e) The bearing lockwasher has an internal offset tab, by inverting the washer, adjustments in 18 steps can be made. Each step alters the pressure difference by approximately 25 lbf/in^2 (1.7 kgf/cm^2)
  f) Screwing the bearing onto the pinion shaft will increase the pressure on the left lock, screwing it out will increase the pressure on the right lock.
  g) When the correct adjustment has been obtained bend one tab of the lockwasher into the cut-out in the pinion bearing and one tab over the nut

  h) Refitting the steering gear assembly in position is the reverse of the removal procedure

7 Remove a rubber gaiter and top-up the steering gear with the same amount of specified SAE 40 engine oil as drained out when the lower cover plate was removed. Do not overfill.
8 Check the toe setting as described in Section 23.

37 Power assisted steering system - bleeding

1 Fill the reservoir to the maximum level, refer to Section 33 with the specified power steering (automatic transmission type) fluid, and jack-up the front of the car.
2 Wait for at least two minutes after filling the reservoir then start the engine and run it at approximately 1500 rpm.
3 Whilst an assistant slowly turns the steering wheel from lock-to-lock, top-up the reservoir until the level is stabilized and air bubbles can no longer be seen in the fluid. Fit the filler cap.
Note: *When the car is raised from the ground with the front wheels clear and suspended, do not use any force or rapid movement when moving the wheels from lock-to-lock as this will cause rack oil pressure to build up and burst or force off the rubber gaiters.*

38 Power assisted steering pump drivebelt - removal and refitting

1 *In-line ohc engines.* Slacken the bolts at the alternator mounting and adjusting arm locations, move the alternator towards the engine and lift the belts from the pulleys.
2 *V6 engines.* Slacken the two bolts securing the idler pulley bracket and slide the idler pulley towards the pump. Remove the belt from the pump, idler and crankshaft pulleys.
3 Refitting and adjustment is the same for both types of engines. Refit the belt(s) to the pulleys in the reverse sequence to removal. Adjust the position of the alternator or idler pulley until there is a deflection of 0.5 In (13 mm) at the midpoint of the longest run of the belt when moderate pressure is applied by hand.
4 Tighten the alternator or idler pulley securing bolts.

39 Power assisted steering pump assembly - removal and refitting

1 Remove the drivebelt, refer to Section 38.
2 Disconnect the fluid feed and return pipes from the pump and drain the fluid into a container. Plug the ends of the pipes and the openings in the pump.
3 To remove the pump from the mounting bracket requires a special wrench, but it is just as easy to undo the bracket to engine

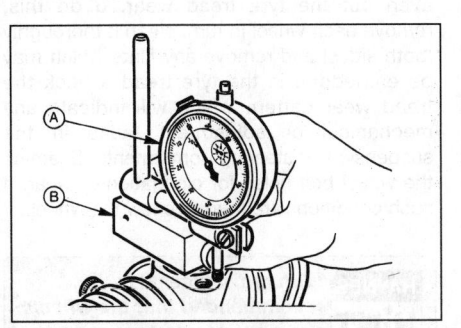

Fig. 11.42 Determining thickness of shim pack for rack slipper adjustment (Sec 35)

A Dial gauge   B Mounting block

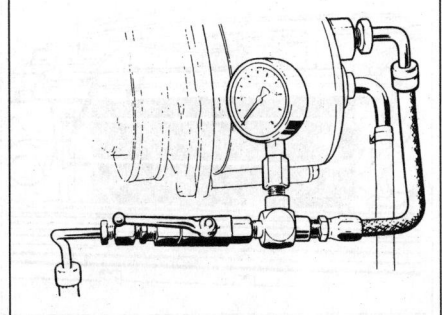

Fig. 11.43 Test gauge, control valve and fittings (Sec 36)

11•18 Suspension and steering

mounting bolts using a standard spanner and remove the pump and bracket as an assembly.
4 Refitting is the reverse of the removal procedure. Adjust the belt tension as described in Section 38. Top-up the reservoir with the specified power steering fluid and bleed the steering system as described in Section 37.

40 Power assisted steering pump - dismantling and reassembly

Before dismantling the pump and reservoir assembly clean the exterior to prevent dirt from contaminating the internal components of the pump.

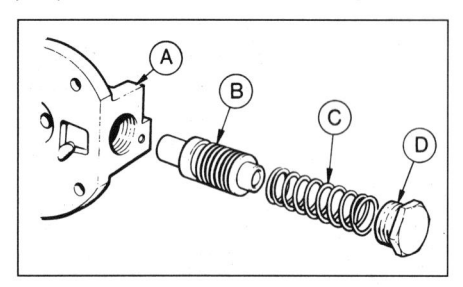

Fig. 11.44 Dismantling the pump valve (Sec 40)

A Housing B Valve C Spring D Valve cap

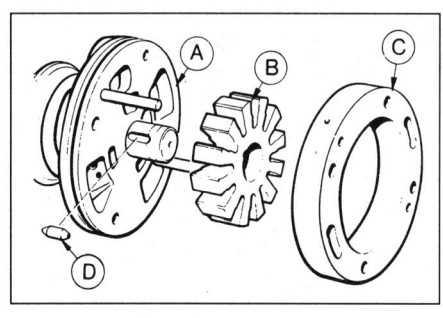

Fig. 11.45 Steering pump dismantled (Sec 40)

A Housing B Roller carrier C Body
D Drive pin

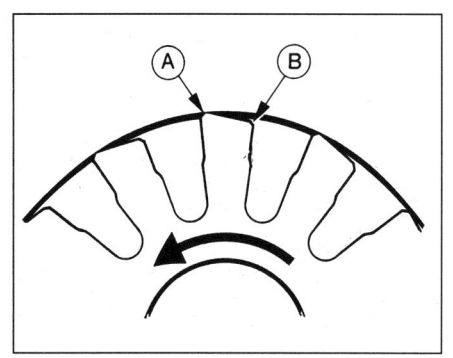

Fig. 11.46 Fitting the roller carrier (Sec 40)

A Leading edge   B Trailing edge

1 Undo the three securing bolts and remove the pulley.
2 Remove the outlet pipe adapter copper washer and O-ring.
3 Free the reservoir by tapping it at opposite points using a soft metal drift to avoid damaging the reservoir.
4 Remove the valve cap spring and valve then take out the four Allen screws and prise off the end housing.
5 Prise the cover plate from the locating pins and carefully tip out the rollers from the carrier.
6 Ease the main body off the front housing using two screwdrivers as levers then remove the roller carrier circlip and lift off the carrier and drive pin.
7 Pull the pump hub and shaft from the front housing and remove the locating pins.
8 Lever the oil seal out of the front housing.
9 Clean and examine all parts for signs of wear and damage. Renew as necessary. The front bush and housing also the hub and shaft are supplied as assemblies. Always fit a new seal at reassembly.
10 Fit the oil seal in the front housing with the inner lip facing to the rear.
11 Slide the pump shaft through the front housing.
12 Insert the locating pins in the front housing and fit the main body.
13 Fit the drive pin and roller carrier with the square indent towards the front. The leading edge of the carrier webs is longer than the trailing edge (Fig.11.46).
14 Fit the carrier retaining circlip and then locate the twelve rollers in the carrier recesses.
15 Fit the cover plate with the circlip groove facing the carrier then the end housing secured with the four Allen screws.
16 Refit the valve with the long spigot leading and then the spring and cap.
17 Fit a new O-ring in the front housing and ensuring that the hole in the reservoir is aligned with the pressure adapter hole tap the reservoir into its seating with a plastic hammer.
18 Fit a new O-ring copper washer and pressure adapter.
19 Fit the drivebelt pulley and securing bolts. Tighten the bolts to the specified torque.

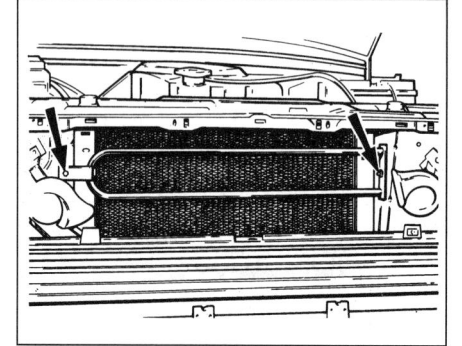

Fig. 11.47 Power assisted steering fluid cooler mountings - arrowed (Sec 41)

41 Power assisted steering fluid cooler - removal and refitting

1 Remove the five radiator grille securing screws and lift out the grille.
2 Loosen the power steering hoses securing clamps and pull the hoses off the cooler. Position the hoses with the ends pointing upwards to minimise fluid loss.
3 Remove the two bolts securing the cooler to the front panel and lift out the cooler.
4 Refitting is the reverse of the removal procedure. Top-up the reservoir and bleed the steering system as described in Section 37.

42 Power steering gear - bellows renewal in situ

1 It is possible to renew the bellows without removing the steering gear from the vehicle.
2 The procedure is virtually identical to that described for manual steering in Section 31 but it is important that after the original bellows have been removed the steering should be turned from lock-to-lock to expel all the old oil.
3 Refill the rack housing with 0.33 pints (0.20 litres) of SAE 40 engine oil by slipping the spout of an oil gun under each bellows neck in turn.

43 Wheels and tyres

1 The roadwheels are of the pressed steel or aluminium alloy type, and the tyres are of the radial ply type.
2 Check the tyre pressures weekly, including the spare.
3 The wheel nuts should be tightened to the appropriate torque as shown in the Specifications, and it is an advantage if a smear of grease is applied to the wheel stud threads.
4 Every 6000 miles (10 000 km) the roadwheels should be moved round the vehicle (this does not apply where the wheels have been balanced on the vehicle) in order to even out the tyre tread wear. To do this, remove each wheel in turn, clean it thoroughly (both sides) and remove any flints which may be embedded in the tyre tread. Check the tread wear pattern which will indicate any mechanical or adjustment faults in the suspension or steering components. Examine the wheel bolt holes for elongation or wear. If such conditions are found, renew the wheel.

HAYNES HINT *With radial types it is recommended that the wheels are moved front-to-rear and rear-to-front, not from side-to-side of the car.*

5 Renewal of the tyres should be carried out when the thickness of the tread pattern is worn to a minimum of 1/16 inch or the wear indicators (if incorporated), are visible.
6 Always adjust the front and rear tyre pressures after moving the wheels round as previously described.
7 All wheels are balanced initially, but have them done again halfway through the useful life of the tyres.
8 Cars fitted with the "S" handling pack have a special size wheel which has a metric designation. This is a 150TRX 390FH wheel and should only have Michelin TRX 190/65 HRX390 mm tyres fitted.

Fault diagnosis - suspension and steering

Unbalanced steering
   Worn rack and pinion
   Tyre pressures uneven
   Dampers worn
   Rack balljoints worn
   Suspension geometry incorrect
   Fluid pipe damaged (Power assisted steering)
   Control valve adjustment incorrect (Power assisted steering)

Lack of power assistance
   Pump belt needs adjustment
   Hoses or pipes restricted
   Fluid level low
   Hydraulic system may require bleeding of air
   Low fluid pressure perhaps caused by worn pump

Poor self centring
   Worn, damaged or badly adjusted suspension or steering
   Steering column badly misaligned
   Return fluid hose or pipe restricted (Power assisted steering)
   Control valve spool sticking (Power assisted steering)
   Stiff operation of rack caused by damaged rack piston or seals (Power assisted steering)

Noisy operation of steering pump
   Fluid level low
   Pump belt slack
   Flow control valve defective or worn pump components

Wheel wobble and vibration
   Worn balljoints or swivels
   Wheel nuts loose
   Hub bearings loose or worn
   Steering balljoints worn
   Front spring weak or broken

Notes

Chapter 12 Bodywork and fittings

For modifications, and information applicable to later models, see Supplement at end of manual

Contents

Air conditioning system	39
Bonnet lock - adjustment	20
Bonnet release cable - removal and refitting	21
Bonnet - removal and refitting	19
Boot lid lock - removal and refitting	23
Boot lid lock striker plate - removal and refitting	25
Boot lid - removal and refitting	22
Central locking system, boot lock solenoid - removal and refitting	24
Central locking system, door lock solenoid - removal and refitting	
Central locking system, tailgate lock solenoid - removal and refitting	28
Centre console - removal and refitting	38
Door glass and regulator - removal and refitting	18
Door lock assembly (front) - removal and refitting	14
Door lock assembly (rear) - removal and refitting	15
Door rattles - tracing and rectification	11
Door striker plate - removal, refitting and adjustment	17
Door trim - removal and refitting	13
Door window regulator (later models) - removal and refitting	40
Facia crash pad - removal and refitting	33
Front and rear doors - removal and refitting	12
Front bumper - removal and refitting	8
General description	1
Heater assembly - dismantling and reassembly	35
Heater assembly - removal and refitting	34
Heater controls - removal, refitting and adjustment	36
Heater motor - removal and refitting	37
Maintenance - bodywork and underframe	2
Maintenance - hinges and locks	7
Maintenance - PVC external roof covering	6
Maintenance - upholstery and carpets	3
Major body damage - repair	5
Minor body damage - repair	4
Radiator grille - removal and refitting	31
Rear bumper - removal and refitting	9
Rear window and tailgate glass - removal and refitting	30
Sliding roof - removal, refitting and adjustment	32
Tailgate assembly - removal and adjustment	32
Tailgate assembly - removal and refitting	26
Tailgate lock - removal and refitting	27
Tailgate lock striker plate - removal and refitting	29
Windscreen glass - removal and refitting	10

Degrees of difficulty

Easy, suitable for novice with little experience	Fairly easy, suitable for beginner with some experience	Fairly difficult, suitable for competent DIY mechanic	Difficult, suitable for experienced DIY mechanic	Very difficult, suitable for expert DIY or professional

Specifications

Overall length
Saloon .... 183.2 in (4650 mm)
Estate .... 187.4 in (4757 mm)

Overall width .... 70.6 in (1791 mm)

Overall height
Saloon .... 54.3 in (1378 mm)
Estate .... 54.4 in (1381 mm)

Wheelbase .... 109.1 in (2769 mm)

Track
Front .... 59.7 in (1515 mm)
Rear .... 60.5 in (1537 mm)

Luggage capacity
Saloon .... 14.3 cu ft (0.40 cu m)
Estate .... 41.7 cu ft (1.16 cu m)
Estate with seat folded .... 77 cu ft (2.15 cu m)

1 General description

The bodywork is of a monocoque, all steel, welded construction with impact absorbing front and rear sections.

The Granada is available in two or four door Saloon or five door Estate versions giving a seating capacity for five adults.

Toughened safety glass is fitted to all windows with the exception of the windscreen. A laminated windscreen is fitted as standard equipment on all models. This type of screen consists of two layers of glass sandwiching a layer of clear plastic. On impact this type of screen does not shatter but only cracks allowing virtually unimpeded vision.

A through-flow type of ventilation system is fitted. Air drawn in through a grille on the scuttle can either be heated or pass straight into the car. The air then passes out through a grille behind the rear side windows.

A central door locking system is fitted as standard equipment on the GL and Ghia models. This enables all the door locks and the boot/tailgate lock to be controlled internally or externally by the lock on the driver's door.

All models are fitted with a radio as standard equipment and electrically operated windows and a sun roof as options.

2 Maintenance - bodywork and underframe

The general condition of a vehicle's bodywork is the one thing that significantly affects its value. Maintenance is easy, but needs to be regular. Neglect, particularly after minor damage, can lead quickly to further deterioration and costly repair bills. It is important also to keep watch on those parts of the vehicle not immediately visible, for instance the underside, inside all the wheel arches, and the lower part of the engine compartment.

The basic maintenance routine for the bodywork is washing - preferably with a lot of water, from a hose. This will remove all the loose solids which may have stuck to the vehicle. It is important to flush these off in such a way as to prevent grit from scratching the finish. The wheel arches and underframe need washing in the same way, to remove any accumulated mud, which will retain moisture and tend to encourage rust. Paradoxically enough, the best time to clean the underframe and wheel arches is in wet weather, when the mud is thoroughly wet and soft. In very wet weather, the underframe is usually cleaned of large accumulations automatically, and this is a good time for inspection.

Periodically, except on vehicles with a wax-based underbody protective coating, it is a good idea to have the whole of the underframe of the vehicle steam-cleaned, engine compartment included, so that a thorough inspection can be carried out to see what minor repairs and renovations are necessary.

Steam-cleaning is available at many garages, and is necessary for the removal of the accumulation of oily grime, which sometimes is allowed to become thick in certain areas. If steam-cleaning facilities are not available, there are some excellent grease solvents available which can be brush-applied; the dirt can then be simply hosed off. Note that these methods should not be used on vehicles with wax-based underbody protective coating, or the coating will be removed. Such vehicles should be inspected annually, preferably just prior to Winter, when the underbody should be washed down, and any damage to the wax coating repaired. Ideally, a completely fresh coat should be applied. It would also be worth considering the use of such wax-based protection for injection into door panels, sills, box sections, etc, as an additional safeguard against rust damage, where such protection is not provided by the vehicle manufacturer.

After washing paintwork, wipe off with a chamois leather to give an unspotted clear finish. A coat of clear protective wax polish will give added protection against chemical pollutants in the air. If the paintwork sheen has dulled or oxidised, use a cleaner/polisher combination to restore the brilliance of the shine. This requires a little effort, but such dulling is usually caused because regular washing has been neglected. Care needs to be taken with metallic paintwork, as special non-abrasive cleaner/polisher is required to avoid damage to the finish. Always check that the door and ventilator opening drain holes and pipes are completely clear, so that water can be drained out. Brightwork should be treated in the same way as paintwork. Windscreens and windows can be kept clear of the smeary film which often appears, by the use of proprietary glass cleaner. Never use any form of wax or other body or chromium polish on glass.

3 Maintenance - upholstery and carpets

Mats and carpets should be brushed or vacuum-cleaned regularly, to keep them free of grit. If they are badly stained, remove them from the vehicle for scrubbing or sponging, and make quite sure they are dry before refitting. Seats and interior trim panels can be kept clean by wiping with a damp cloth. If they do become stained (which can be more apparent on light-coloured upholstery), use a little liquid detergent and a soft nail brush to scour the grime out of the grain of the material. Do not forget to keep the headlining clean in the same way as the upholstery. When using liquid cleaners inside the vehicle, do not over-wet the surfaces being cleaned. Excessive damp could get into the seams and padded interior, causing stains, offensive odours or even rot. If the inside of the vehicle gets wet accidentally, it is worthwhile taking some trouble to dry it out properly, particularly where carpets are involved. *Do not leave oil or electric heaters inside the vehicle for this purpose.*

4 Minor body damage - repair

Note: *For more detailed information about bodywork repair, Haynes Publishing produce a book by Lindsay Porter called "The Car Bodywork Repair Manual". This incorporates information on such aspects as rust treatment, painting and glass-fibre repairs, as well as details on more ambitious repairs involving welding and panel beating.*

Repairs of minor scratches in bodywork

If the scratch is very superficial, and does not penetrate to the metal of the bodywork, repair is very simple. Lightly rub the area of the scratch with a paintwork renovator, or a very fine cutting paste, to remove loose paint from the scratch, and to clear the surrounding bodywork of wax polish. Rinse the area with clean water.

Apply touch-up paint to the scratch using a fine paint brush; continue to apply fine layers of paint until the surface of the paint in the scratch is level with the surrounding paintwork. Allow the new paint at least two weeks to harden, then blend it into the surrounding paintwork by rubbing the scratch area with a paintwork renovator or a very fine cutting paste. Finally, apply wax polish.

Where the scratch has penetrated right through to the metal of the bodywork, causing the metal to rust, a different repair technique is required. Remove any loose rust from the bottom of the scratch with a penknife, then apply rust-inhibiting paint to prevent the formation of rust in the future. Using a rubber or nylon applicator, fill the scratch with bodystopper paste. If required, this paste can be mixed with cellulose thinners to provide a very thin paste which is ideal for filling narrow scratches. Before the stopper-paste in the scratch hardens, wrap a piece of smooth cotton rag around the top of a finger. Dip the finger in cellulose thinners, and quickly sweep it across the surface of the stopper-paste in the scratch; this will ensure that the surface of the stopper-paste is slightly hollowed. The scratch can now be painted over as described earlier in this Section.

Repairs of dents in bodywork

When deep denting of the vehicle's bodywork has taken place, the first task is to pull the dent out, until the affected bodywork almost attains its original shape. There is little point in trying to restore the original shape completely, as the metal in the damaged area will have stretched on impact, and cannot be reshaped fully to its original contour. It is better to bring the level of the dent up to a point which is about 3 mm below the level of the surrounding bodywork. In cases where the dent is very shallow anyway, it is not worth trying to pull it out at all. If the underside of the dent is accessible, it can be hammered out gently from behind, using a mallet with a

Bodywork and fittings 12•3

wooden or plastic head. Whilst doing this, hold a suitable block of wood firmly against the outside of the panel, to absorb the impact from the hammer blows and thus prevent a large area of the bodywork from being "belled-out".

Should the dent be in a section of the bodywork which has a double skin, or some other factor making it inaccessible from behind, a different technique is called for. Drill several small holes through the metal inside the area - particularly in the deeper section. Then screw long self-tapping screws into the holes, just sufficiently for them to gain a good purchase in the metal. Now the dent can be pulled out by pulling on the protruding heads of the screws with a pair of pliers.

The next stage of the repair is the removal of the paint from the damaged area, and from an inch or so of the surrounding "sound" bodywork. This is accomplished most easily by using a wire brush or abrasive pad on a power drill, although it can be done just as effectively by hand, using sheets of abrasive paper. To complete the preparation for filling, score the surface of the bare metal with a screwdriver or the tang of a file, or alternatively, drill small holes in the affected area. This will provide a really good "key" for the filler paste.

To complete the repair, see the Section on filling and respraying.

Repairs of rust holes or gashes in bodywork

Remove all paint from the affected area, and from an inch or so of the surrounding "sound" bodywork, using an abrasive pad or a wire brush on a power drill. If these are not available, a few sheets of abrasive paper will do the job most effectively. With the paint removed, you will be able to judge the severity of the corrosion, and therefore decide whether to renew the whole panel (if this is possible) or to repair the affected area. New body panels are not as expensive as most people think, and it is often quicker and more satisfactory to fit a new panel than to attempt to repair large areas of corrosion.

Remove all fittings from the affected area, except those which will act as a guide to the original shape of the damaged bodywork (eg headlight shells etc). Then, using tin snips or a hacksaw blade, remove all loose metal and any other metal badly affected by corrosion. Hammer the edges of the hole inwards, in order to create a slight depression for the filler paste.

Wire-brush the affected area to remove the powdery rust from the surface of the remaining metal. Paint the affected area with rust-inhibiting paint, if the back of the rusted area is accessible, treat this also.

Before filling can take place, it will be necessary to block the hole in some way. This can be achieved by the use of aluminium or plastic mesh, or aluminium tape.

Aluminium or plastic mesh, or glass-fibre matting, is probably the best material to use for a large hole. Cut a piece to the approximate size and shape of the hole to be filled, then position it in the hole so that its edges are below the level of the surrounding bodywork. It can be retained in position by several blobs of filler paste around its periphery.

Aluminium tape should be used for small or very narrow holes. Pull a piece off the roll, trim it to the approximate size and shape required, then pull off the backing paper (if used) and stick the tape over the hole; it can be overlapped if the thickness of one piece is insufficient. Burnish down the edges of the tape with the handle of a screwdriver or similar, to ensure that the tape is securely attached to the metal underneath.

Bodywork repairs - filling and respraying

Before using this Section, see the Sections on dent, deep scratch, rust holes and gash repairs.

Many types of bodyfiller are available, but generally speaking, those proprietary kits which contain a tin of filler paste and a tube of resin hardener are best for this type of repair. A wide, flexible plastic or nylon applicator will be found invaluable for imparting a smooth and well-contoured finish to the surface of the filler.

Mix up a little filler on a clean piece of card or board - measure the hardener carefully (follow the maker's instructions on the pack), otherwise the filler will set too rapidly or too slowly. Using the applicator, apply the filler paste to the prepared area; draw the applicator across the surface of the filler to achieve the correct contour and to level the surface. As soon as a contour that approximates to the correct one is achieved, stop working the paste - if you carry on too long, the paste will become sticky and begin to "pick-up" on the applicator. Continue to add thin layers of filler paste at 20-minute intervals, until the level of the filler is just proud of the surrounding bodywork.

Once the filler has hardened, the excess can be removed using a metal plane or file. From then on, progressively-finer grades of abrasive paper should be used, starting with a 40-grade production paper, and finishing with a 400-grade wet-and-dry paper. Always wrap the abrasive paper around a flat rubber, cork, or wooden block - otherwise the surface of the filler will not be completely flat. During the smoothing of the filler surface, the wet-and-dry paper should be periodically rinsed in water. This will ensure that a very smooth finish is imparted to the filler at the final stage.

At this stage, the "dent" should be surrounded by a ring of bare metal, which in turn should be encircled by the finely "feathered" edge of the good paintwork. Rinse the repair area with clean water, until all of the dust produced by the rubbing-down operation has gone.

Spray the whole area with a light coat of primer - this will show up any imperfections in the surface of the filler. Repair these imperfections with fresh filler paste or bodystopper, and once more smooth the surface with abrasive paper. If bodystopper is used, it can be mixed with cellulose thinners, to form a really thin paste which is ideal for filling small holes. Repeat this spray-and-repair procedure until you are satisfied that the surface of the filler, and the feathered edge of the paintwork, are perfect. Clean the repair area with clean water, and allow to dry fully.

The repair area is now ready for final spraying. Paint spraying must be carried out in a warm, dry, windless and dust-free atmosphere. This condition can be created artificially if you have access to a large indoor working area, but if you are forced to work in the open, you will have to pick your day very carefully. If you are working indoors, dousing the floor in the work area with water will help to settle the dust which would otherwise be in the atmosphere. If the repair area is confined to one body panel, mask off the surrounding panels; this will help to minimise the effects of a slight mis-match in paint colours. Bodywork fittings (eg chrome strips, door handles etc) will also need to be masked off. Use genuine masking tape, and several thicknesses of newspaper, for the masking operations.

Before commencing to spray, agitate the aerosol can thoroughly, then spray a test area (an old tin, or similar) until the technique is mastered. Cover the repair area with a thick coat of primer; the thickness should be built up using several thin layers of paint, rather than one thick one. Using 400-grade wet-and-dry paper, rub down the surface of the primer until it is really smooth. While doing this, the work area should be thoroughly doused with water, and the wet-and-dry paper periodically rinsed in water. Allow to dry before spraying on more paint.

Spray on the top coat, again building up the thickness by using several thin layers of paint. Start spraying at one edge of the repair area, and then, using a side-to-side motion, work until the whole repair area and about 2 inches of the surrounding original paintwork is covered. Remove all masking material 10 to 15 minutes after spraying on the final coat of paint.

Allow the new paint at least two weeks to harden, then, using a paintwork renovator, or a very fine cutting paste, blend the edges of the paint into the existing paintwork. Finally, apply wax polish.

5 Major body damage - repair

1 Because the car is built without a separate chassis frame and the body is therefore integral with the underframe, major damage must be repaired by competent mechanics with the necessary welding and hydraulic straightening equipment.

12•4 Bodywork and fittings

2 If the damage has been serious it is vital that the body is checked for correct alignment as otherwise the handling of the car will suffer and many other faults such as excessive tyre wear and wear in the transmission and steering may occur.
3 There is a special body jig which most large body repair shops have, and to ensure that all is correct it is important that the jig be used for all major repair work.

6 Maintenance - PVC external roof covering

Under no circumstances try to clean any external PVC roof covering with detergents, caustic soaps or spirit cleaners. Plain soap and water is all that is required, with a soft brush to clean dirt that may be ingrained. Wash the covering as frequently as the rest of the car.

7 Maintenance - hinges and locks

Once every 6000 miles (10 000 km) or 6 months, the door, bonnet and boot or tailgate hinges and locks should be given a few drops of oil from an oil can. The door striker plates can be given a thin smear of grease to reduce wear and to ensure free movement.

8 Front bumper - removal and refitting

1 Open the bonnet and remove the radiator grille as described in Section 31.
2 If a headlamp washer is fitted, detach the hoses from the overriders.
3 Undo and remove the bolt, washer and spacer assemblies that secure the wrap around ends of the bumper to the body. Unscrew and remove the nut and washer assemblies that secure the bumper and overriders to their mounting brackets.
4 The front bumper assembly may now be lifted away taking care not to scratch the paintwork on the front wings.
5 Refitting is a reverse of the removal sequence. Do not fully tighten the fixings until the bumper is perfectly straight and correctly located.

9 Rear bumper - removal and refitting

1 Press in the two plastic lugs and remove the number plate light. Disconnect the wiring plug and place the light to one side.
2 Remove the spare wheel from the luggage compartment to gain access to the bumper securing bolts.
3 Undo and remove the bolt, washer and spacer assemblies from the wrap around ends of the bumper. Unscrew and remove the nut and washer assemblies that secure the bumper and overrider to their brackets. On Estate cars, open the tailgate and remove the two socket headed screws securing the bumper brackets to the body.
4 The rear bumper assembly may now be lifted away taking care not to scratch the paintwork on the rear wings.
5 Refitting is a reverse of the removal sequence. Do not fully tighten the fixings until the bumper is perfectly straight and correctly located.

10 Windscreen glass - removal and refitting

1 If you are unfortunate enough to have your windscreen crack, or should you wish to renew your present windscreen, fitting a replacement is one of the jobs which the average owner is advised to leave to a professional. For the owner who wishes to attempt the job himself the following instructions are given.
2 Cover the bonnet with a blanket or cloth to prevent accidental damage and remove the windscreen wiper blades and arms as detailed in Chapter 10.
3 Put on a pair of lightweight shoes and get onto one of the front seats. An assistant should be ready to catch the glass as it is released from the body aperture.
4 Place a piece of soft cloth between the soles of your shoes and the windscreen glass and with both feet on one top corner of the windscreen push firmly.
5 When the weatherstrip has freed itself from the body aperture flange in that area repeat the process at frequent intervals along the top edge of the windscreen until from outside the car the glass and weatherstrip can be removed together.
6 If you have to renew your windscreen due to a cracked screen, remove all traces of sealing compound from the weatherstrip and body flange,
7 Carefully inspect the rubber moulding for signs of splitting or deterioration.
8 To refit the glass first fit the weatherstrip onto the glass with the joint at the lower edge.
9 Insert a piece of thick cord into the channel of the weatherstrip with the two ends protruding by at least 12 in (300 mm) at the bottom centre of the screen.
10 Mix a concentrated soap and water solution and apply to the flange of the windscreen aperture.
11 Offer up the screen to the aperture, and with an assistant press the rubber surround hard against one end of the cord, moving round the windscreen and so drawing the lip over the windscreen flange of the body. Keep the draw-cord parallel to the windscreen, Using the palms of the hands, thump on the glass from the outside to assist the lip in passing over the flange and to seat the screen correctly onto the aperture.
12 To ensure a good watertight joint apply some Sealastik SR51 between the weatherstrip and the body and press the weatherstrip against the body to give a good seal.
13 Any excess Sealastik may be removed with a petrol-moistened cloth,
14 Lubricate the finisher strip groove with the soap and water solution and insert the strip.
15 Refit the wiper arms and blades.

11 Door rattles - tracing and rectification

1 The most common cause of door rattle is a misaligned, loose or worn striker plate: however other causes may be:
  a) Loose door or window winder handles
  b) Loose or misaligned door lock components
  c) Loose or worn remote control mechanism

2 It is quite possible for door rattles to be the result of a combination of the above faults so a careful examination should be made to determine their exact cause.
3 If striker plate wear or misalignment is the cause, the plate should be renewed or adjusted as necessary. The procedure is detailed in Section 17.
4 Should the window winder handle rattle, this can be easily rectified by inserting a rubber washer between the escutcheon and door trim panel.
5 If the rattle is found to be emanating from the door lock it will in all probability mean that the lock is worn and therefore should be replaced with a new lock unit, as described in Sections 14 and 15.
6 Lastly, if it is worn hinge pins causing rattles they should be renewed.

12 Front and rear doors - removal and refitting

1 Where necessary remove the dash lower trim panel and disconnect the loudspeaker, electric window and central locking wiring connections.
2 Using a pencil accurately mark the outline of the hinge relative to the door. Remove the bolts that secure each door to the hinge and lift away the complete door.
3 For storage it is best to stand the door on an old blanket and allow it to lean against a wall also suitably padded at the top to stop scratching.
4 Refitting is a reverse sequence to removal. If, after refitting, adjustment is necessary, it should be done at the hinges to give correct alignment, or the striker plate if the door either moves up or down on final closure.

Bodywork and fittings 12•5

13.1 Removing the door window winder handle

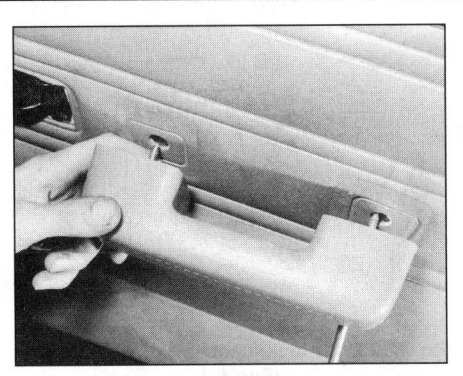

13.3 Lifting away the door arm rest

13.4 Removing the door interior handle surround

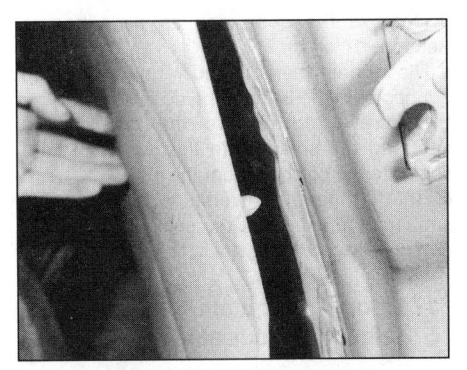

13.5a Easing away the trim panel

13.5b Door trim panel retaining buttons

13.7 Refitting the window winder handle

13 Door trim - removal and refitting

Note: *The following procedure is basically for vehicles with manually operated windows. Where electrically-operated windows are fitted, first disconnect the battery earth lead, then proceed to paragraph 3. Before the trim panel can be lifted away (para 5) it may be necessary to disconnect the switch, mirror and loudspeaker connectors.*

1 Remove the plastic trim on the window winder handle by sliding the trim away from the knob and lifting off. This will expose the handle retaining screw (photo).
2 Wind up the window and note the position of the handle. Undo and remove the crosshead screw and lift away the handle.
3 Undo and remove the crosshead screws that secure the door arm rest. (On some models a third attachment point is released by turning the arm rest through 90°.) Lift off the arm rest from the door (photo).
4 Depress the trim panel around the door handle and slide the trim panel forward to release the retaining lugs. Lift away the trim panel (photo).
5 Insert a thin strip of metal with all the sharp edges removed or a flat screwdriver between the door and trim panel. This will release one or two of the trim panel retaining buttons without damaging the trim. The panel can now be gently eased off by hand (photos).
6 Carefully remove the plastic weatherproof sheeting. Removal is now complete.
7 Refitting is generally a reverse of the removal sequence (photo). **Note:** When refitting the trim panel ensure that each of the retaining buttons is firmly located in its hole by sharply striking the panel with the palm of the hand in the approximate area of each button. This will make sure the trim is seated fully.

14 Door lock assembly (front) - removal and refitting

1 Refer to Section 13 and remove the door interior trim.
2 Remove the interior handle by sliding it forward to disengage it from the door (photo). Remove the interior knob by unscrewing.
3 Working inside the door shell carefully prise

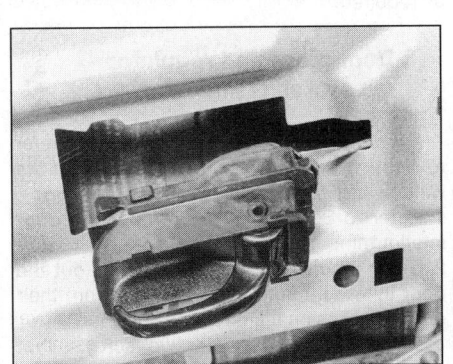

14.2 Remove the interior lock handle by sliding forward

14.4 Door lock securing screws

14.5 Exterior lock retaining clip

12

12•6 Bodywork and fittings

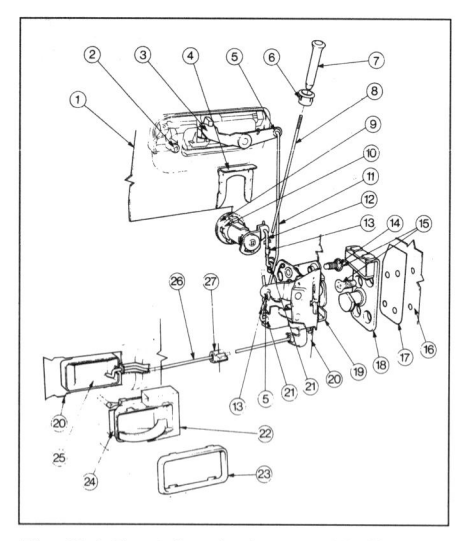

Fig. 12.1 Front door lock assembly (Sec 14)

1 Door outer panel
2 Screw
3 Exterior handle
4 Retaining clip
5 Bush
6 Grommet
7 Lock button
8 Control rod
9 Pad
10 Lock barrel
11 Control rod
12 Control rod
13 Spring clip
14 Washer
15 Screw
16 Door pillar
17 Pad
18 Striker plate
19 Lock assembly
20 Door panel
21 Bush
22 Foam pad
23 Trim panel
24 Interior handle
25 Housing
26 Control rod
27 Guide

the two exterior control rods from their locations in the lock assembly.
4 Remove the three lock securing screws and push the lock assembly into the door shell (photo). The assembly should be manoeuvred to disengage the interior lock rod, and the interior handle rod.
5 To remove the exterior handle, remove the two crosshead screws from inside the door shell. To remove the exterior lock, pull down and remove the retaining clip (photo).
6 Refitting the door lock assembly is the reverse sequence to removal. Lubricate all moving parts with a little grease.

17.2 Door striker plate

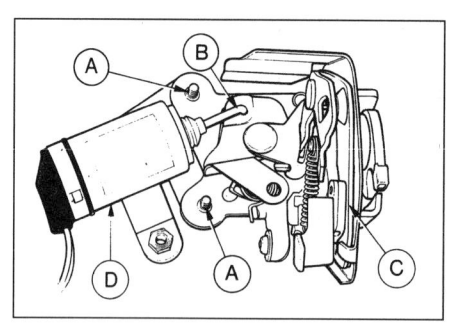

Fig. 12.2 Door lock, central locking system (Sec 16)

A Securing screws C Lock assembly
B Operating arm   D Solenoid

15 Door lock assembly (rear) - removal and refitting

1 Refer to Section 13 and remove the door interior trim.
2 Remove the interior lock rod crank from the door shell by drifting out the retaining pin with a suitable pin punch.
3 Prise the exterior handle rod from its location in the lock assembly. Remove the interior door handle by sliding it forward to disengage it from the door.
4 Remove the three lock securing screws and push the lock assembly into the door shell. Turn the interior handle rod to disengage it from the lock. Remove the lock.
5 The exterior handle may be removed after unscrewing the two crosshead screws from inside the door shell.
6 Refitting the door lock assembly is the reverse sequence. Lubricate all moving parts with a little grease.

16 Central locking system, door lock solenoid - removal and refitting

1 Remove the door trim panel as described in Section 13.
2 Unscrew and remove the crosshead screw securing the solenoid mounting bracket to the door panel (Fig. 12.2).

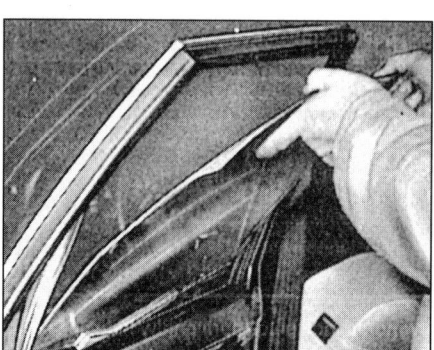

Fig. 12.4 Removing the door window glass to the inside (Sec 18)

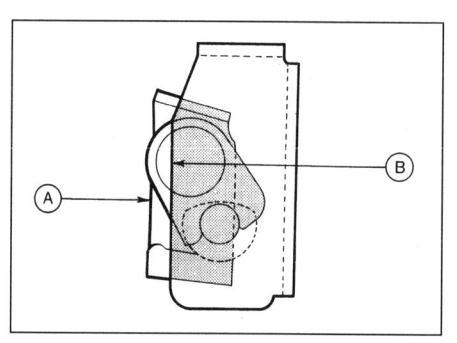

Fig. 12.3 Lock plate (A) and striker plate (B) in parallel condition (Sec 17)

3 Remove the door lock as described in Sections 14 or 15.
4 Disconnect the solenoid wiring multi-plug connector in the base of the door.
5 Unscrew and remove the two screws securing the solenoid to the door lock and separate the solenoid from the lock.
6 Refitting is the reverse of the removal sequence. Ensure that the solenoid operating arm is engaged in the slot in the door lock.

17 Door striker plate - removal, refitting and adjustment

1 If it is wished to renew a worn striker plate, mark its position on the door pillar so a new plate can be fitted in the same position.
2 To remove the plates simply undo and remove the four crosshead screws which hold the plate in position. Lift away the plate (photo).
3 Refitting of the door striker is the reverse sequence to removal.
4 To adjust the striker plate, close the door to the first of the two locking positions. Check that the edges of the lock plate and the striker plate are parallel, and that the rear edge of the door stands 025 in (6.0 mm) proud of the body. Move the striker plate as necessary.
5 With the lock in the open position, check the clearance at A in Fig. 12.3. This can be checked by placing a ball of plasticine on the striker post and carefully closing the door. The dimension A should be set to 0.08 in (2.0 mm) by carefully moving the striker plate vertically as required.

18 Door glass and regulator - removal and refitting

For models produced from June 1980, refer to Section 40 for removal and refitting of the window regulator mechanism.

1 Refer to Section 13 and remove the door interior trim.
2 Using a screwdriver carefully ease out the door inner and outer weatherstrips from their retaining clips on the door panel.
3 Undo and remove the two screws that secure the door glass to the window regulator. On rear doors remove the screws securing the rear channel bracket to the door and lift out

Bodywork and fittings 12•7

the channel and vent window assembly. Tilt the door glass and remove it upwards and out of the door.
4 Remove the screws securing the regulator to the door inner panel and lift out the regulator from aperture in the inner panel (photo).
5 Should it be necessary to remove the glass run channel, start at the front lower frame end and carefully ease the glass run channel from its location in the door frame.
6 Refitting the door glass and regulator is the reverse sequence to removal. Lubricate all moving parts with a little grease. Before refitting the trim panel check the operation and alignment of the glass and regulator and adjust as necessary. When all is correct fully tighten all securing bolts (photo).

19 Bonnet - removal and refitting

1 Open the bonnet and support it open using the bonnet stay.

HAYNES HiNT *To act as a datum for refitting, mark the position of the hinges relative to the bonnet inner panel.*

2 With the assistance of a second person hold the bonnet in the open position and release the stay.
3 Undo and remove the two bolts, spring washers and plain washers that secure each hinge to the bonnet taking care not to scratch the top of the wings (photo).
4 Lean the bonnet up against a wall, suitably padded to prevent scratching the paint.
5 Refitting the bonnet is the reverse sequence to removal. Any adjustments necessary can be made either at the hinges or the bonnet catch.

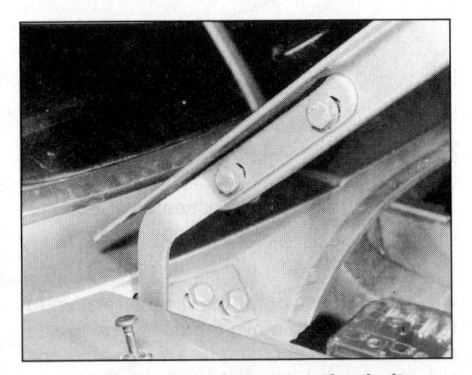

20 Bonnet lock - adjustment

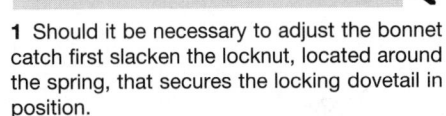

1 Should it be necessary to adjust the bonnet catch first slacken the locknut, located around the spring, that secures the locking dovetail in position.
2 Using a wide bladed screwdriver, screw the shaft in or out as necessary until the correct bonnet height is obtained. Tighten the locknut (photo).

21 Bonnet release cable - removal and refitting

1 Remove the radiator grille as described in Section 31.
2 Slacken off the cable adjustment clamp and remove the cable from the engine compartment clips along the cable run. Detach the cable from the bonnet lock spring.
3 Remove the dash lower trim panel. Unscrew and remove the screws securing the handle mounting bracket to the bulkhead (photo).
4 Pull the cable assembly through into the passenger compartment.

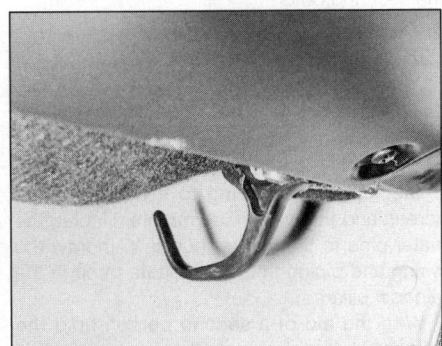

5 Tap out the roll pin securing the cable to the handle and lift off the handle.
6 Installation is the reverse of removal ensuring that the bulkhead grommet is correctly located.
7 The cable should be adjusted to give the dimension shown in Fig. 12.6 and the clamp bolt then tightened.

22 Boot lid - removal and refitting

1 Open the boot lid to its fullest extent. To act as a datum for refitting, mark the position of the hinge relative to the lid inner panel.
2 Remove the torsion bar anti-rattle clip then support the lid in the open position.
3 Release each torsion bar from its adjustment peg and hinge hook, and lift it away.

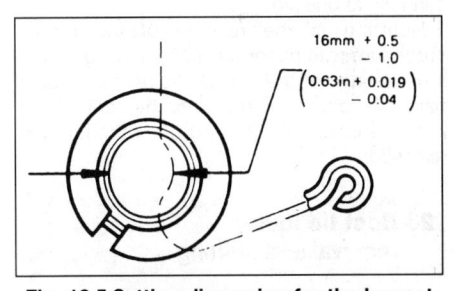

Fig. 12.5 Setting dimension for the bonnet release spring (Sec 21)

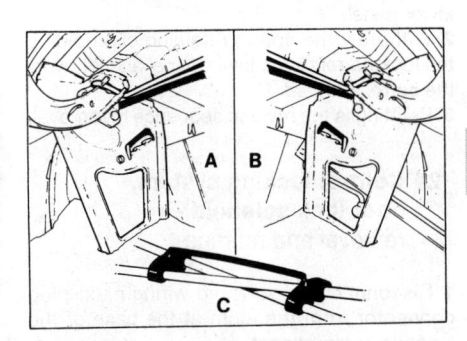

16mm + 0.5 − 1.0
(0.63in + 0.019 − 0.04)

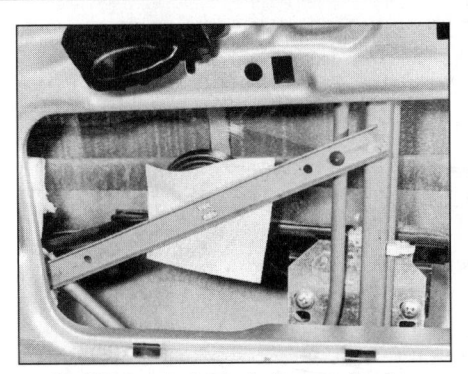

18.4 Door window winder mechanism (earlier models)

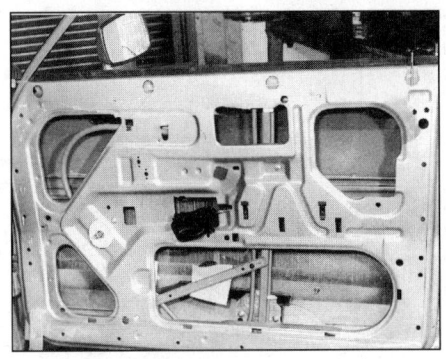

18.6 Correct fitting of door window regulator components (earlier type shown)

19.3 Bonnet to hinge mounting bolts

20.2 Bonnet lock dovetail and safety catch assemblies

21.3 The bonnet release handle located under the right-hand side of the dash panel

Fig. 12.6 Boot lid torsion bar arrangement (Sec 22)

A Left-hand hinge   B Right-hand hinge
C Anti-rattle clip

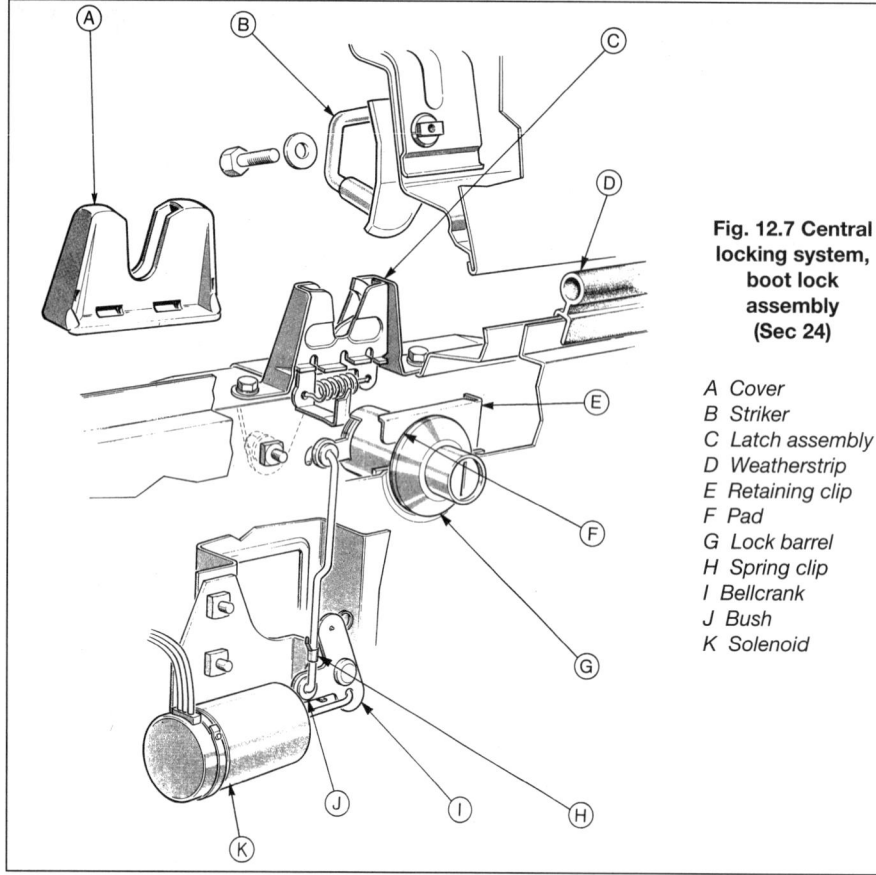

12•8 Bodywork and fittings

Fig. 12.7 Central locking system, boot lock assembly (Sec 24)

A Cover
B Striker
C Latch assembly
D Weatherstrip
E Retaining clip
F Pad
G Lock barrel
H Spring clip
I Bellcrank
J Bush
K Solenoid

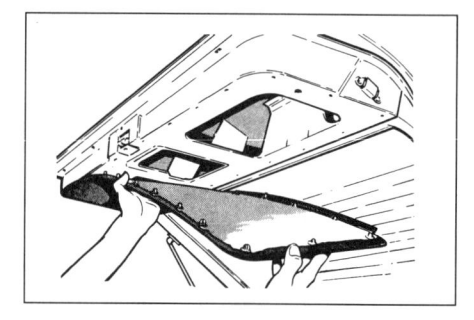

Fig. 12.8 Removing the trim panel from the tailgate (Sec 26)

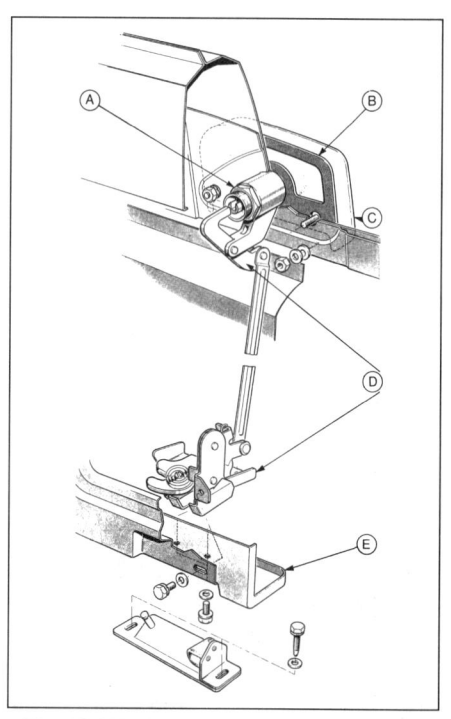

Fig. 12.9 Tailgate lock assembly (Sec 27)

A Lock barrel B Pad C Handle
D Linkage E Tailgate panel

4 With an assistant to help support the lid, remove the securing bolts and washers from the hinges and lift the lid away. Store it carefully to one side.
5 Refitting is the reverse of the above procedure with minor adjustment being made, if necessary, at the hinges. Adjust the torsion bars, one peg at a time, until the boot lid will just rise of its own accord once the latch is released.

23 Boot lid lock - removal and refitting

1 Open the boot lid and remove the two screws securing the latch assembly to the lower panel.
2 Withdraw the spring retaining clip holding the lock barrel to the lower panel and lift away the barrel and pad.
3 Refitting is the reverse sequence to removal.

24 Central locking system, boot lock solenoid - removal and refitting

1 Disconnect the solenoid wiring multi-plug connector from the loom at the base of the luggage compartment.
2 Undo and remove the two screws, spring and plain washers securing the solenoid to the lock support member.
3 Remove the spring clip securing the solenoid operating rod to the bellcrank and detach the operating lever. The solenoid can now be removed.
4 Refitting is a reverse of the removal sequence.

25 Boot lid lock striker plate - removal and refitting

1 Open the boot lid and with a pencil mark the outline of the striker plate relative to the boot lid to act as a datum for refitting.
2 Undo and remove the two bolts with spring and plain washers that secure the striker plate. Lift away the striker plate.
3 Refitting is a reverse sequence to removal. Line up the previously made marks and tighten the securing bolts.

26 Tailgate assembly - removal and refitting

1 Open the tailgate and remove the inner trim panel.
2 Disconnect the wiring to the heated rear screen and the wash/wipe motors. Unplug the water pipe to the washer pump. Withdraw the wiring and piping from the tailgate through the window pillars.
3 With the aid of a second person hold the tailgate in the open position and undo and remove the two crosshead screws securing each damper to the tailgate.
4 Mark the position of the hinges relative to the tailgate and remove the bolts, spring and plain washers securing the hinges to the tailgate. Carefully lift off the tailgate.
5 Refitting is the reverse sequence to removal. Any adjustment may be made at the hinges.

27 Tailgate lock - removal and refitting

1 Open the tailgate and carefully remove the inner trim panel.
2 Undo and remove the large hexagonal nut that retains the linkage to the lock barrel and lift off the linkage.
3 Undo and remove the two nuts, spring and plain washers from the handle and remove the handle and lock barrel assembly.
4 Remove the three bolts, spring and plain washers from the lower latch and remove the latch and linkage from the tailgate.
5 Refitting is the reverse sequence to removal.

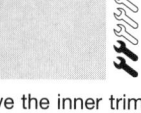

Bodywork and fittings 12•9

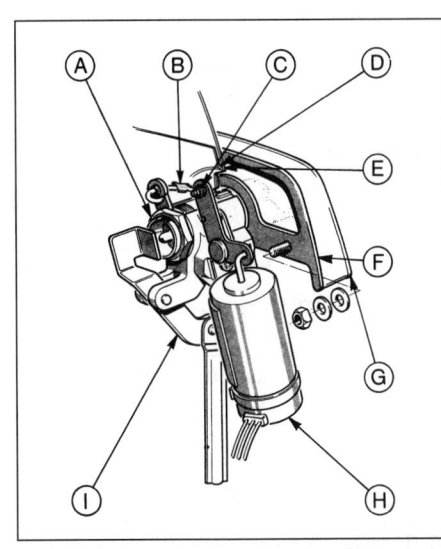

Fig. 12.10 Central locking system, tailgate lock (Sec 28)

A Hexagon nut
B Operating rod
C Bush
D Tailgate panel
E Reinforcement panel
F Pad
G Handle
H Solenoid
I Link

28 Central locking system, tailgate lock solenoid - removal and refitting

1 Remove the tailgate lock as described in Section 27.
2 Disconnect the solenoid wiring multi-plug connector at the tailgate.
3 Detach the spring clip securing the solenoid operating rod to the lock assembly.
4 Undo the two screws that attach the solenoid to the lock assembly and detach the solenoid from the lock.
5 Refitting the solenoid is the reverse of the removal sequence.

29 Tailgate lock striker plate - removal and refitting

1 Open the tailgate and with a pencil mark the outline of the striker plate relative to the luggage compartment floor.
2 Undo and remove the bolts, spring and plain washers that secure the striker plate and lift away the striker plate.
3 Refitting the striker plate is the reverse sequence to removal. Line up the striker plate with the previously made marks and tighten the bolts.

30 Rear window and tailgate glass - removal and refitting

Where a heated rear window is fitted, disconnect the electrical leads before commencing. Also take great care that the electrical element is not damaged during handling.

The procedure for removing and refitting the rear window and tailgate glass is similar to that for the windscreen glass. Refer to Section 10 paragraphs 3 to 14.

31 Radiator grille - removal and refitting

1 Open the bonnet and support it in the open position. Undo and remove the crosshead screws that secure the radiator grille to the front body panels.
2 Lift away the radiator grille.
3 Refitting is the reverse sequence to removal, but take care to locate the grille tabs in their respective slots in the front lower panel.

32 Sliding roof - removal, refitting and adjustment

It is strongly recommended that removal and refitting of the sliding roof is left to a specialist owing to the possibility of damage occurring to paintwork and trim. Adjustment procedures are covered in the Supplement to this Manual.

33 Facia crash pad - removal and refitting

1 Open the bonnet and disconnect the battery earth connection.
2 Refer to Chapter 11 Section 24 and remove the steering wheel.
3 Remove the two halves of the steering column shroud.
4 Pull the radio and switch knobs off their respective stalks, and using a screwdriver very carefully ease off the facia trim panel from the instrument cluster.

> **HAYNES HiNT**: *The facia trim panel is secured to the cluster with six push fit clips and is easily broken if mistreated.*

5 Remove the instrument cluster as described in Chapter 10 Section 41.
6 Unscrew and remove the screws securing the instrument cluster frame assembly complete with the radio and switches. Pull the frame assembly off the crash pad leaving the heater controls in position.
7 Remove the centre console as described in Section 38 and the four screws (B) securing the lower edge of the crash pad (Fig. 12.12).
8 Undo and remove the three screws (B) and two nuts (A) securing the package tray to the crash pad (Fig. 12.13).
9 Undo and remove the four nuts (A) securing the crash pad to the belt rail and lower the instrument panel (Fig. 12.13).
10 Remove the nuts securing the steering column to the lower bracket to give access to the remaining nut (C), (Fig. 12.12) securing the crash pad to the steering column bracket. Remove this nut and lift off the crash pad.
11 Refitting is a reverse of the removal procedure.

34 Heater assembly - removal and refitting

1 Open the bonnet and disconnect the battery earth connection.
2 Drain the cooling system as described in Chapter 2.
3 Slacken the two clamps on the heater feed and return hoses in the engine compartment and noting their positions, pull off the hoses. It is advisable to place a suitable container beneath the car during this operation as a small quantity of coolant will be lost when removing the hoses.

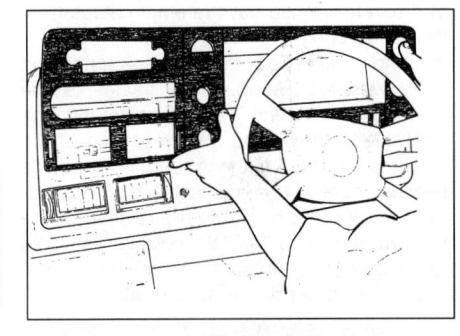

Fig. 12.11 Removing the facia trim panel from the instrument cluster (Sec 33)

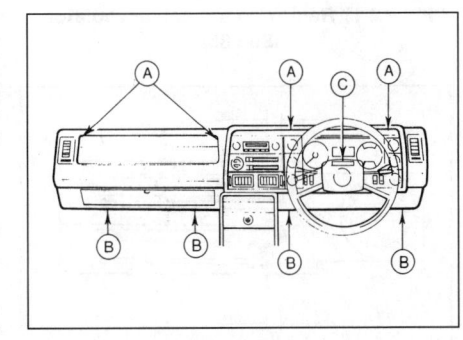

Fig. 12.12 Location of crash pad securing screws (Sec 33)

For key see text

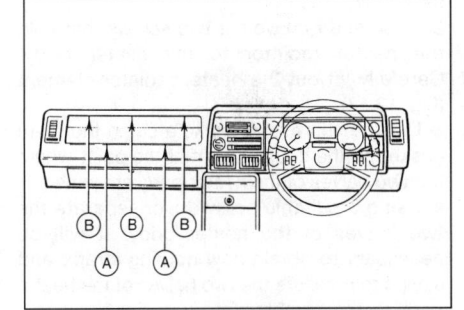

Fig. 12.13 Package tray and crash pad screw locations (Sec 33)

For key see text

12•10 Bodywork and fittings

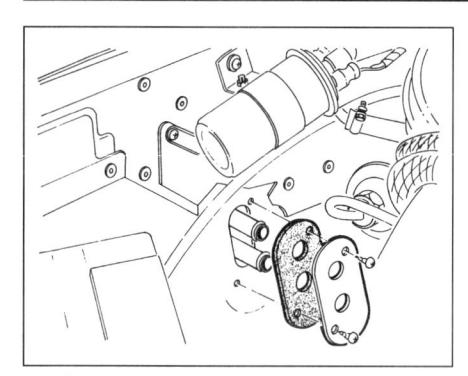

Fig. 12.14 Removing the heater hose connection cover and gasket (Sec 34)

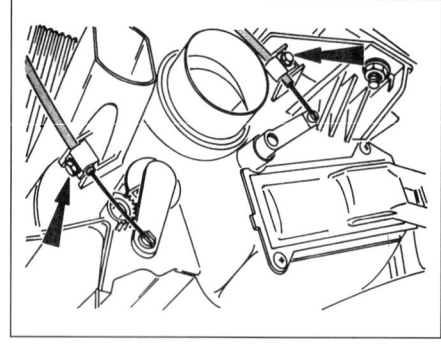

Fig. 12.15 Heater control cable clamp bolts - arrowed (Sec 34)

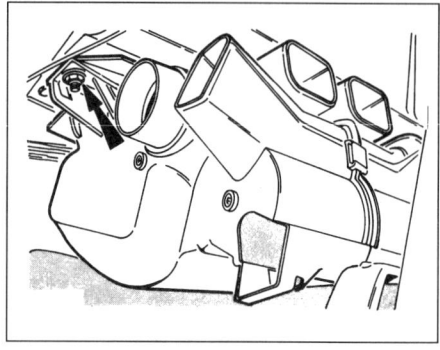

Fig. 12.16 Heater attachment points (Sec 34)

4 Remove the two screws and detach the cover plate with gasket from the water outlet connections.
5 Remove the centre console as described in Section 38.
6 Undo the six screws each side and take out the two interior trim panels located beneath the dash panel.
7 Undo the two screws and remove the glovebox.
8 Remove the two screws on the lower left side of the instrument panel cover and take off the cover.
9 Detach the four air hoses and the two centre vent nozzle air hoses from the heater assembly.
10 Undo the two screws securing the heater outer control cables to their mounting brackets and carefully prise out the inner cables from the heater operating levers.
11 Undo and remove the two nuts, spring and plain washers securing the heater to the bottom of the fairing. Pull the heater inwards until the water outlets are drawn out of the opening in the bulkhead and remove the heater assembly from the nearside of the car.
12 Refitting the heater is the reverse of the removal sequence. Refill the cooling system as described in Chapter 2. Adjust the heater controls as described in Section 36.

35 Heater assembly - dismantling and reassembly

1 Refer to Section 34 and remove the heater assembly.
2 Undo and remove the two screws that hold the heater radiator to the heater body. Carefully lift out the heater radiator element (Fig. 12.17).
3 Using a sharp knife make a cut in the foam gasket on the housing joint at the point where the two halves of the heater body are joined.
4 Using a suitable screwdriver separate the two halves of the heater body. It will be necessary to obtain new holding straps and clasps to resecure the two halves of the heater body on reassembly.
5 Take out the control valves and levers from the body. Disassembly is now complete.
6 Reassembly is the reverse of the dismantling sequence. Resecure the two halves of the heater with new holding straps and clasps.

36 Heater controls - removal, refitting and adjustment

1 Open the bonnet and disconnect the battery earth connection.
2 Remove the right-hand lower trim panel located beneath the dash panel.
3 Undo and remove the single screw securing the steering column upper shroud and lift the shroud off the column.
4 Pull the radio and switch knobs off their respective stalks and using a screwdriver very carefully ease off the trim panel from the instrument cluster. The trim panel is secured to the cluster with six push fit clips and is easily broken if mistreated.
5 Undo and remove the two bolts securing the heater control cables to the heater assembly.

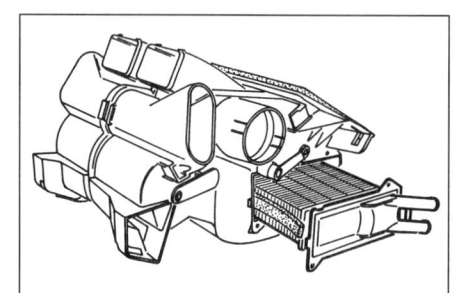

Fig. 12.17 Removing the heater radiator (Sec 35)

Ease the inner cables from the clips retaining them in the heater operating levers and allow the cables to hang clear of the levers.
6 Remove the three screws from the front of the heater control panel and withdraw the panel forward into the car. As the panel comes clear of the instrument cluster, disconnect the heater fan multi-plug and the panel illuminating bulb holder from the rear of the control assembly. Withdraw the unit complete with control cables into the car.
7 Refitting is the reverse of the removal sequence. When the unit is refitted into the car adjust the control cables as follows.
8 Set the two heater controls to a point approximately 0.078 in (2 mm) from their respective stops. Slacken the two bolts

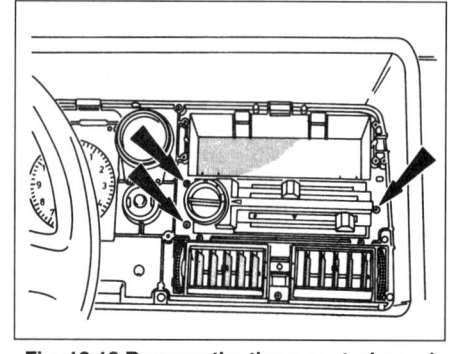

Fig. 12.18 Remove the three control panel retaining screws - arrowed (Sec 36)

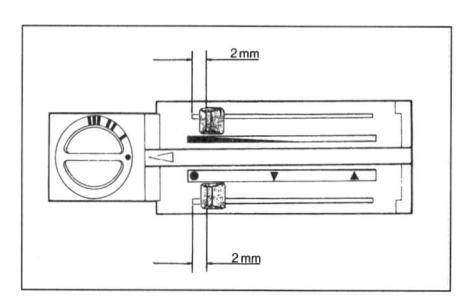

Fig. 12.19 Heater control adjustment (Sec 36)

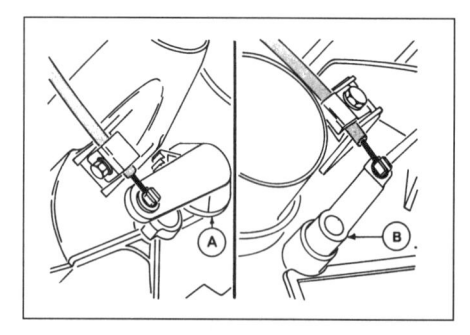

Fig. 12.20 Move operating levers to end position and tighten clamps (Sec 36)

A Air distributor valve
B Temperature control valve

Bodywork and fittings 12•11

37.4 Location of the heater motor in the engine compartment

securing the outer control cables to the heater body and move the two operating levers to their end positions. Tighten the two clamp bolts and test the heater controls. The controls must have the same tension at each end of their travel.

37 Heater motor - removal and refitting

1 Open the bonnet and disconnect the battery earth connection.
2 Remove the four screws and lift off the fan motor cover grille.
3 Disconnect the multiplug connector from the fan motor resistance plate, and undo and remove the screw securing the earth wire to the body.
4 Undo and remove the screws securing the fan housing and lift out the complete motor assembly (photo).
5 Remove both fan covers from their brackets by prising apart with a suitable screwdriver.
6 Slacken the motor retaining bracket and lift out the electric motor. **Note:** *The motor is balanced with the fan blades in position. The fan blades should therefore never be removed from the motor shaft.*
7 Refitting the motor is a reverse of the removal sequence.

38 Centre console - removal and refitting

1 Open the bonnet and disconnect the battery earth connection.

Short console

2 Unscrew the gear lever knob on the automatic transmission T-handle.
3 Pull out the ash tray and remove the two screws exposed by the removal of the ash tray.
4 Carefully prise the front panel from the centre console body and disconnect the leads to the switches, cigar lighter and radio as the panel comes clear.
5 Unscrew and remove the remaining four screws securing the console to the mounting bracket and lift the console from the transmission tunnel. Refitting is the reverse of the removal procedure.

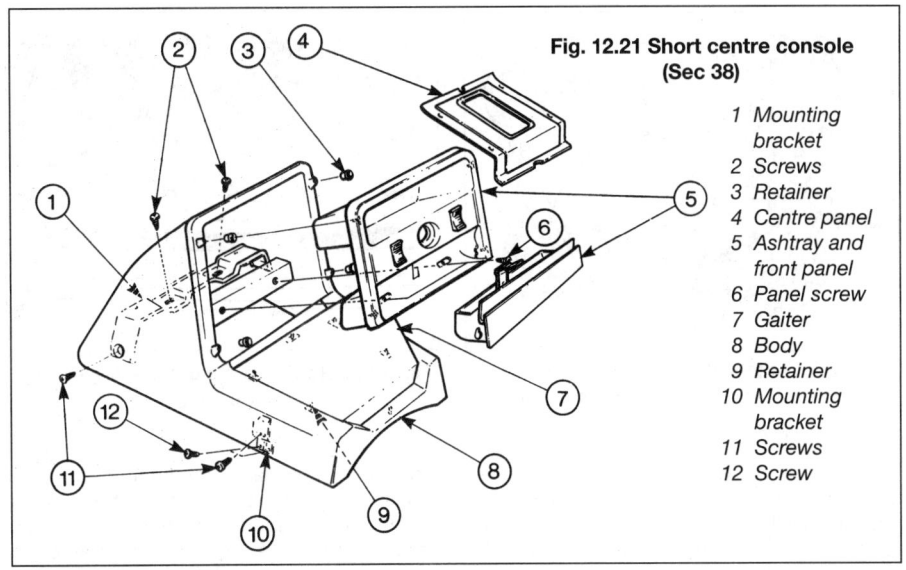

Full length console

6 Remove both front seats, gear lever knob or the automatic transmission T-handle.
7 Where fitted remove the ashtray from the centre console front panel and remove the two screws exposed by the removal of the ashtray.
8 Carefully prise the front panel from the console body and disconnect the leads to the switches, cigar lighter and radio as the panel comes clear.
9 Carefully prise up the centre panel off the console body and lift it up over the gear lever. Detach the leads to the electric window switches where fitted.
10 Take out the rear ashtray and unscrew and remove the screws securing the rear panel to the centre console. Lift off the panel and detach the leads to the electric window switches and cigar lighter where fitted.
11 Undo and remove the three screws located on each side of the console and lift the console off its brackets on the transmission tunnel.
Note: *On Ghia models the front and centre fixings are located beneath flaps in the carpet.*
12 Refitting the centre consoles is the reverse sequence to removal.

39 Air conditioning system

1 The air conditioner optionally fitted to certain models is of a conventional type incorporating a belt driven compressor, condenser, evaporator and allied components.
2 The refrigerant used is quite safe under normal operating conditions but once it escapes into atmosphere it evaporates so quickly that it will freeze the skin on contact and displace air so quickly that suffocation can occur. In addition, when released near a naked flame or hot metal components it will produce a poisonous gas - Phosgene.
3 In view of the foregoing, never disconnect any part of the air conditioning system unless the system has first been discharged

Fig. 12.21 Short centre console (Sec 38)

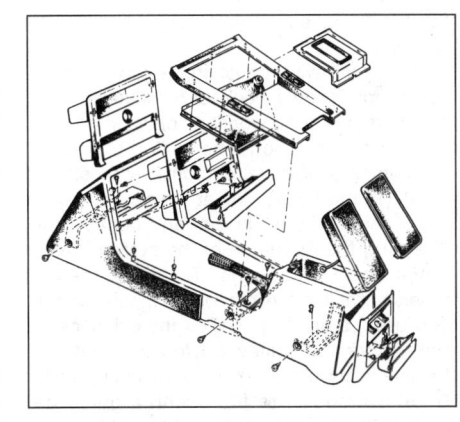

1 Mounting bracket
2 Screws
3 Retainer
4 Centre panel
5 Ashtray and front panel
6 Panel screw
7 Gaiter
8 Body
9 Retainer
10 Mounting bracket
11 Screws
12 Screw

Fig. 12.22 Full length centre console (Sec 38)

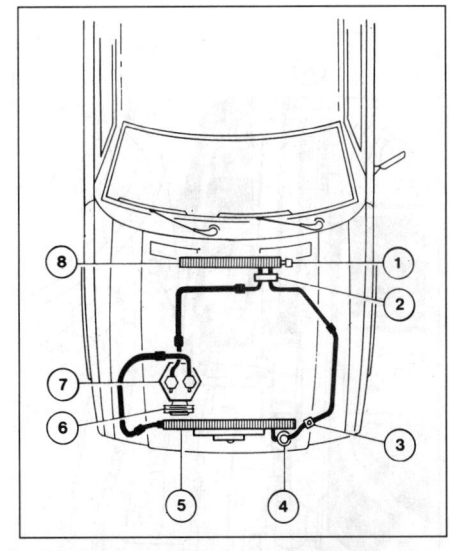

Fig. 12.23 Air conditioner layout (Sec 39)

1 De-icing thermostat
2 Expansion valve
3 Sight glass
4 Dehydrator
5 Condenser
6 Compressor clutch
7 Compressor
8 Evaporator

12•12 Bodywork and fittings

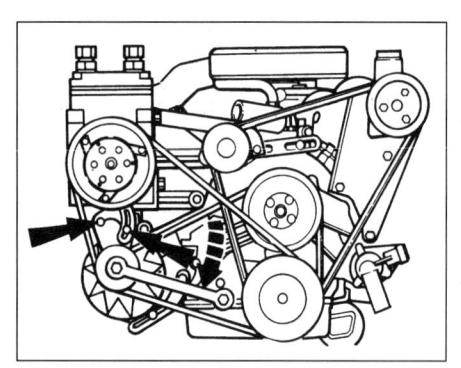

Fig. 12.24 Drivebelt arrangement with compressor pump fitted (Sec 39)

Adjusting bolts arrowed

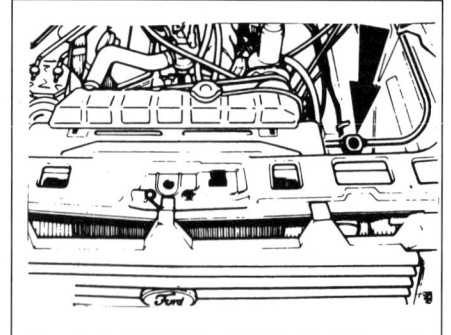

Fig. 12.25 Refrigerant sight glass - arrowed (Sec 39)

(evacuated) by your dealer or a qualified refrigeration engineer. It is permissible to disconnect the mountings of the compressor pump and the condenser and to move the units within the limits of movement of their flexible hoses as an aid to engine or radiator removal

4 During the winter season, switch on the air conditioner once a week for 5 minutes to keep the system in good working condition.

5 Maintenance should be limited to keeping the compressor drivebelt correctly tensioned, 0.4 in (10 mm) deflection at the centre point of the longest run of the belt and occasionally checking the condition of the system hoses, pipes and connections. To adjust the belt tension, slacken the idler pulley mounting bolts and using a ring spanner on the pulley centre bolt move the pulley downwards until the tension is correct. Tighten the mounting bolts.

6 Later models are fitted with a guide belt roller to prevent drivebelt rumble and flap. This assembly is available for fitment to earlier models if required. Its location is shown in Fig. 12.26. The drivebelt tension with this fitting remains the same, as does the method of adjustment. When fitted, the guide roller must be correctly aligned with the back of the drivebelt.

7 Keep the fins of the condenser free from flies and dirt by brushing or hosing with cold water.

8 It is advisable to check the refrigerant quantity before the summer begins. To do this, start the engine when it is cold and let it idle. Switch on the air conditioner and select blower control position III. Now observe the sight glass which should show no sign of bubbles after 10 seconds of engine idling from starting. If bubbles are visible after this period, have the system recharged by your dealer after he has traced the source of the leakage.

9 Occasionally check that the compressor bracket fixing bolts are secure.

40 Door window regulator (later models) - removal and refitting

From June 1980 a redesigned window regulator mechanism was fitted to both front and rear doors to provide easier manual operation. To remove and refit this later type mechanism proceed as follows. Note that the old and new type mechanisms are not interchangeable.

1 Remove the door trim panel as given in Chapter 13, Section 18.

2 With the window in the lowered position, remove the regulator lower fixing screw(s).

3 Now remove the regulator upper fixing screws.

4 Remove the glass lower channel bracket screws.

5 Unscrew and remove the regulator mechanism retaining screws and withdraw the regulator.

6 Refitting is a direct reversal of the removal procedure, but note the following additional points:
   a) On the front doors, once the regulator is loosely attached to the window bracket, wind up the window and apply a slight preload to the top rear corner of the window (applying downwards pressure). Then tighten the glass bracket screws (with window wound up on rear doors)
   b) On the front door it will be necessary to press the upper regulator mechanism bracket rearwards when tightening the retaining screw

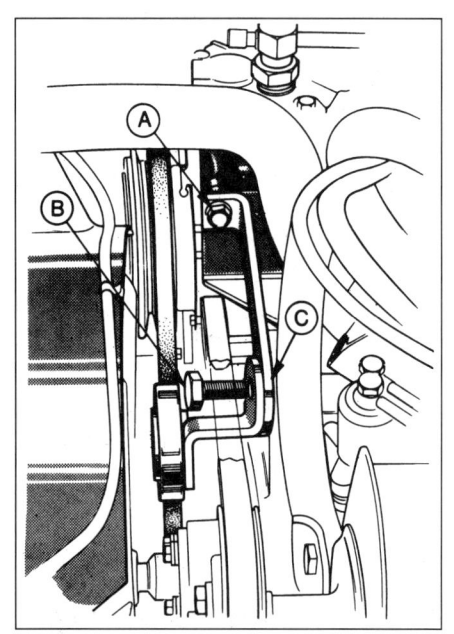

Fig. 12.26 Air conditioning drivebelt guide roller fitted to later models (Sec 39)

A Existing bolt   B New bolt
C Brace and roller assembly

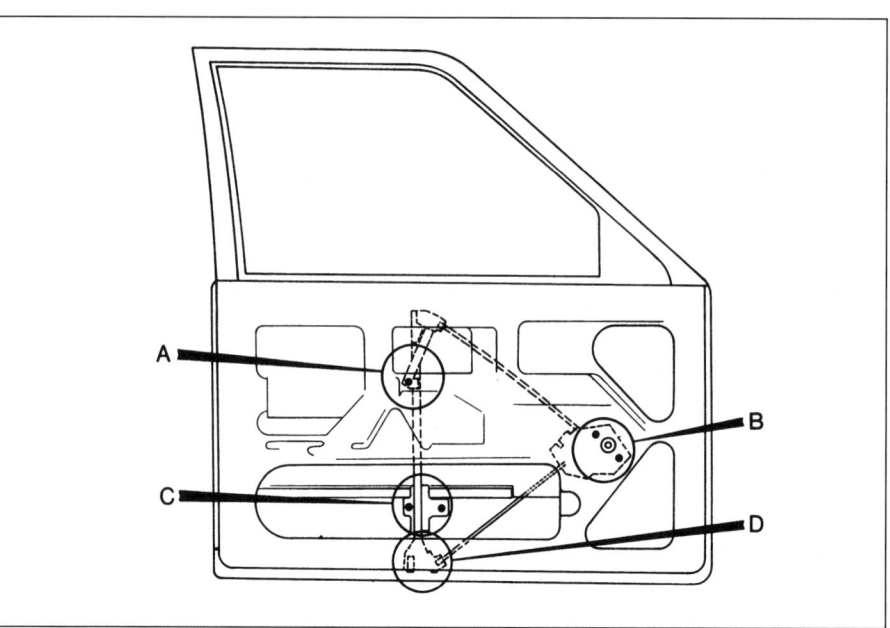

Fig. 12.27 Later type door window regulator (Sec 40)

A Upper retaining point   C Regulator-to-window bracket
B Handle location   D Lower fixing point

Chapter 13 Supplement: Revisions and information on later models

Contents

Introduction 1
Specifications 2
Maintenance intervals - later models 3
In-line ohc engine 4
   Camshaft drivebelt - refitting and timing
   Revised top cover gasket
   Cylinder head bolts
   Rocker ball stud locknuts
V6 engines 5
   Engine removal - later models
   Engine dismantling - later models
   Rocker shaft springs
   Oil cooler - removal and refitting
   Crankshaft reconditioning - all 2.8 litre engines
   Revised main bearing caps and bolts - later 2.8 litre engines
   Improved rocker covers and gaskets - later models
Cooling system 6
   Radiator repairs - later models
   Leaks from radiator blanking plug
   Coolant hoses chafing - later ohc models
   Fan removal - later ohc models
   Modified radiator (V6 engine)
Fuel system - carburettor models 7
   Accelerator cable adjustment - all models
   Weber carburettor top cover securing screws - all models
   Solex/Pierburg carburettor
   Solex/Pierburg carburettor electrically-heated choke - description of operation
   Solex/Pierburg carburettor throttle damper - description and adjustment
   Solex/Pierburg carburettor (1982 on) - idle speed and mixture adjustment
   Fuel tank
Fuel system - fuel injection models (1982 on) 8
   General description
   Control modules - removal and refitting
   Safety control module - bypassing
   Depressurising the fuel injection system
   Fuel pump - removal and refitting
   Fuel accumulator - removal and refitting
   Fuel filter - removal and refitting
   Fuel pump one-way valve - renewal
   Fuel tank
Ignition system 9
   General description (1982 on)
   Breakerless ignition (all types) - precautions
   Breakerless distributor (all types) - maintenance
   Breakerless distributor (1982 on) - removal and refitting
   Ignition timing (1982 on) - checking and adjusting
   Ignition coil (all types) - testing, removal and refitting
   Ignition amplifier (all types) - testing, removal and refitting
   Spark plugs (all models) - removal and refitting
Clutch 10
   Clutch cable noise (ohc models)
   Clutch cable removal and refitting (all models)
Manual gearbox 11
   Gearbox input shaft spigot diameter - all models
   5-speed gearbox (type N) - description
   5-speed gearbox (type N) - removal and refitting
   5-speed gearbox (type N) - alternative lubricant type
Automatic transmission 12
   Fluid level checking - all models
   Brake band adjustment - all models
Propeller shaft 13
   Propeller shaft noise - 1982 models (except 5-speed)
Rear axle 14
   Limited-slip differential
Braking system 15
   Brake pad renewal - 1982 on
   Brake shoe renewal - 1982 on
   Handbrake adjustment - later models
   Brake pressure control valve - all models
   Additional bleeding methods
   Bleeding - using one-way valve kit
   Bleeding - using a pressure bleeding kit
   All methods
Electrical system 16
   Maintenance-free battery - description and precautions
   Alternator brushes (Lucas type A133) - renewal
   Starter motor - alternative types
   Headlight washer jets - removal, refitting and adjusting
   Sidelight bulb renewal - 1982 on
   Auxiliary light bulb renewal (typical)
   Rear light bulb renewal - Saloon, 1982 on
   Rear light cluster unit removal and refitting - Saloon, 1982 on
   Rear foglamp removal and refitting - Estate models
   Number plate light bulb renewal - 1982 on
   Interior light bulb renewal - 1982 on
   Control illumination bulb renewal -1982 on
   Warning light bulb renewal -1982 on
   Oil pressure gauge and/or ammeter - removal and refitting (1982 on)
   Digital clock - removal and refitting
   Minor switches - removal and refitting (1982 on)
   Seat operating motor - removal and refitting
   Sliding roof motor - removal and refitting
   Auxiliary warning system - description
   Auxiliary warning system components - removal and refitting
   Trip computer - description
   Trip computer components - removal and refitting
   Radio removal - 1984 models
   Fuses and relays - 1982 on
   Roof console (auxiliary warning system) - removal and refitting
   Electrically-operated aerial mast - renewal
   Electrically-operated windows

Degrees of difficulty

| **Easy**, suitable for novice with little experience | **Fairly easy**, suitable for beginner with some experience | **Fairly difficult**, suitable for competent DIY mechanic | **Difficult**, suitable for experienced DIY mechanic | **Very difficult**, suitable for expert DIY or professional |

13•2 Supplement: Revisions and information on later models

Suspension and steering 17
   Front hub bearings adjustment (1982 on)
   Stabilizer bar link washers - all models
   Steering lock removal and refitting - 1982 on
   Steering wheel removal - 1982 on
   Manual steering gear lubrication - (1982 on)
   Power-assisted steering gear (1982 on) - summary of changes
   Power-assisted steering system bleeding - (1982 on)
   Power steering pump (1982 on) - removal and refitting
   Power steering pump (1982 on) - overhaul
   Wheels and tyres - general care and maintenance
Bodywork and fittings 18
   Door hinge lubrication (pre-1982 models)
   Bumper - removal and refitting (1982 on)
   Radiator grille - removal and refitting (1982 on)
   Glovebox - removal and refitting (all models)
   Front spoiler - removal and refitting
   Rear spoiler - removal and refitting
   Seat heating pads - removal and refitting

Air conditioning compressor drivebelt - 1982 on
Front seat - removal and refitting
Front seat slide - removal and refitting
Front seat head restraints - removal and refitting
Rear seat cushion - removal and refitting
Rear seat backrest (Saloon) - removal and refitting
Rear seat back locking mechanism (Estate) - dismantling
   and reassembly
Rear seat catch striker (Estate) - removal and refitting
Seat belts - maintenance
Seat belts (front) - removal and refitting
Seat belts (rear) Saloon - removal and refitting
Seat belts (rear) Estate - removal and refitting
Seat belts - general
Rear view mirrors - removal and refitting
Body mouldings - removal and refitting
Sliding roof - adjustment
Facia trim panel (1982 on) - removal and refitting
Door trim - removal and refitting

1 Introduction

This Supplement contains information which has become available since the manual was first published. Most of the information relates to the "new" Granada models introduced for the 1982 model year, but there are some items which apply to all models.

The Sections in the Supplement are arranged in the same order as the main Chapters to which they relate. The updated Specifications are all grouped at the front of the Supplement for ease of reference, but they too are arranged in Chapter order.

Before starting any procedure described in this manual, it is suggested that the reader refers to the appropriate Section(s) of the Supplement and then to the relevant main Chapter(s). If no information to the contrary is found in the Supplement, the procedure given in the main Chapter may be assumed to apply to all models.

2 Specifications

The Specifications below are revisions of, or supplementary to, the Specifications at the beginning of the preceding Chapters.

In-line ohc engine - 1982 on

General
Maximum continuous engine speed	5800 rpm
Maximum torque (DIN)	113 lbf ft (153 Nm) at 4000 rpm
Compression pressure (hot, at cranking speed)	157 to 185 lbf/in^2 (11 to 13 kgf/cm^2)

Auxiliary shaft
Endfloat	0.0016 to 0.0067 in (0.04 to 0.17 mm)

Cylinder head
Cast identification mark	6
Camshaft bearing bush diameters:	
Front	1.7745 to 1.7757 in (45.072 to 45.102 mm)
Centre	1.8776 to 1.8788 in (47.692 to 47.722 mm)
Rear	1.8926 to 1.8938 in (48.072 to 48.102 mm)

Valves and valve springs
Inlet valve length	4.356 to 4.396 in (110.65 to 111.65 mm)
Exhaust valve length	4.335 to 4.411 in (110.10 to 112.05 mm)
Inlet valve spring free length	1.85 in (47.0 mm)

Lubrication system
Minimum oil pressure at 80°C and 750 rpm	14 lbf/in^2 (1.0 kgf/cm^2)

Torque wrench settings

	lbf ft	Nm
Cylinder head bolts (cold) - ohc engines:		
Splined type:		
Stage 1	32	44
Stage 2	50	68
Stage 3 Wait 10 minutes	57	78
Stage 4 Warm engine (15 mins at 1000 rev/min)	75	102
Torx type:		
Stage 1	28	38
Stage 2	53	72
Stage 3 Wait 5 minutes	Turn each bolt through 90°	
Rocker ball stud locknuts (ohc engines)	40	55

Supplement: Revisions and information on later models

V6 engines - 1982 on

General
Maximum power output (DIN):
- 2.3 litre .... 84 kW at 5300 rpm
- 2.8 litre carburettor .... 99 kW at 5200 rpm
- 2.8 litre fuel injection .... 110 kW at 5700 rpm

Maximum torque (DIN):
- 2.3 litre .... 130 lbf ft (176 Nm) at 3000 rpm
- 2.8 lite carburettor .... 159 lbf ft (216 Nm) at 3000 rpm
- 2.8 litre fuel injection .... 159 lbf ft (216 Nm) at 4000 rpm

Crankshaft
Backlash .... 0.0067 to 0.0106 in (0.17 to 0.27 mm)
Main bearing clearance:
- 2.3 litre .... 0.0003 to 0.0024 in (0.008 to 0.062 mm)
- 2.8 litre .... 0.0007 to 0.0028 in (0.018 to 0.072 mm)

Camshaft
Backlash (2.3 litre) .... 0.0067 to 0.0106 in (0.17 to 0.27 mm)
Cam lift (exhaust, 2.8 litre fuel injection) .... 0.2638 in (6.700 mm)

Connecting rods
Big-end bearing internal diameter (fitted) .... 2.1262 to 2.1278 in (54.006 to 54.046 mm)

Valve springs
Free length:
- All carburettor models .... 2.087 in (53.0 mm)
- Fuel injection models .... 2.067 in (52.5 mm)

Torque wrench settings
	lbf ft	Nm
Camshaft thrust plate bolts	15	20
Timing cover to intermediate plate	11	15
Intermediate plate to block	15	20
Oil cooler threaded sleeve	22	30

Cooling system - 1982 on

Thermostat
Opening temperature:
- In-line engine .... 185° to 192°F (85° to 89°C)
- V6 engines .... 174° to 181°F (79° to 83°C)

System capacity (approx)
- In-line engine .... 13.4 pints (7.6 litres)
- V6 engine, carburettor .... 16.7 pints (9.5 litres)
- V6 engine, fuel injection .... 18.5 pints (10.5 litres)

Torque wrench settings
	lbf ft	Nm
Thermostat housing bolts (V6)	15	20
Water outlet connector bolts	15	20
Water pump pulley bolts (ohc)	18	25
Fan clutch-to-water pump hub bolts (V6)	7	10

Fuel and exhaust systems - 1982 on

Weber carburettor
Vacuum pull-down setting .... 0.25 to 0.27 in (6.25 to 6.75 mm)

Solex/Pierburg carburettor
Venturi diameter (2.3 litre) .... 1.02 in (26 mm)
Main jet:
- 2.3 litre 137.5
- 2.8 litre 150

Idle jet:
- 2.3 litre 47.5
- 2.8 litre 50

Idle speed .... 800 ± 20 rpm
Fast idle speed .... 3000 ± 100 rpm
Exhaust gas CO content at idle .... 1.5 ± 0.25%
Float level setting .... 0.43 in (11.0 mm)
Pump stroke direction setting .... 0.08 to 0.20 in (2.0 to 5.0 mm)

Solex/Pierburg carburettor (cont.)

Choke plate pull-down:
- 2.3 litre ..... 0.126 in (3.2 mm)
- 2.8 litre, manual transmission ..... 0.177 in (4.5 mm)
- 2.8 litre, automatic transmission ..... 0.165 in (4.2 mm)

Throttle damper setting (see text):
- 2.3 litre ..... 3.00 turns
- 2.8 litre ..... 2.75 turns

Torque wrench settings

	lbf ft	Nm
Fuel injection components		
Fuel distributor securing screws	26	35
Main pressure regulator	18	25
Banjo unions:		
Mixture control inlet and outlet	15	20
Warm-up regulator inlet	11	15
Warm-up regulator outlet	6	8
Fuel flow sensor (when fitted)	11	15
Fuel pump	15	20
Fuel filter	15	20
Fuel accumulator	15	20
Injection pipes and cold start valve	6	8
Fuel pump one-way valve	15	20
Exhaust system		
Manifold-to-downpipe clamp nuts	27	37
U-bolt clamp nuts	30	40

Ignition system - 1982 on

System type - all models
Breakerless

Distributor
Make ..... Motorcraft or Bosch

Shaft endfloat:
- Motorcraft ..... 0.024 to 0.039 in (0.60 to 1.00 mm)
- Bosch ..... 0.021 to 0.051 in (0.54 to 1.30 mm)

Multi-plug grease
To Ford spec ESB-MJC158-A or equivalent

HT leads
Resistance (maximum) ..... 30 000 ohms per lead

Spark plugs
Type (In-line engine, 1983 model year on) ..... Champion F7YCC or F7YC

Electrode gap:
- F7YCC spark plug ..... 0.032 in (0.8 mm)
- F7YC spark plug ..... 0.028 in (0.7 mm)

Manual gearbox (N Type) - 1982 on

Gearbox type
Five forward gears and one reverse. Synchromesh on all forward gears

Gear ratios (typical)

	2.0/2.3 litre	2.8 litre
First	3.650:1	3.360:1
Second	1.970:1	1.810:1
Third	1.370:1	1.260:1
Fourth	1.000:1	1.000:1
Fifth	0.816:1	0.825:1
Reverse	3.660:1	3.365:1

Lubrication
Lubricant type/specification ..... Gear oil, viscosity SAE 80EP, to Ford spec SOM-2C-9008-A, or semi-synthetic gear oil to Ford spec ESD-M2C-175-A (see text)

Lubricant capacity ..... 2.9 pints (1.66 litres) approx

Torque wrench settings

	lbf ft	Nm
Guide sleeve to transmission	7	10
Clutch housing to transmission	52	70
Extension housing to transmission	33	45
Transmission housing top cover	7	10
Selector locking mechanism	13	18
Oil filter/level plug	18	25
5th gear collar nut	100	135
5th gear locking plate	18	25
Gear lever to extension housing	18	25

Supplement: Revisions and information on later models 13•5

Rear axle - 1982 on

Lubrication
Lubricant capacity	2.5 pints (1.4 litres) approx
Correct level of lubricant	0.39 to 0.47 in (10 to 12 mm) below lower edge of filler hole

Braking system - 1982 on

Front brakes
Disc type	Solid or ventilated according to model
Maximum disc run-out	0.0035 in (0.09 mm) including hub

Rear brakes
Shoe width (Saloon and Estate)	2.25 in (57.2 mm)
Wheel cylinder diameter (Saloon and Estate)	0.874 in (22.2 mm)

Electrical system - 1982 on

Starter motor (Lucas manufacture)
Alternative types	9M90 and M35J
Overhaul data	As for type 5M90 (see Chapter 10)

Starter motor (Bosch manufacture)
Alternative types:
Long frame	0.8 kW
Short frame	0.8 kW, 0.85 kW and 1.1 kW
Overhaul data - long frame	As for type EF-0.85 kW (see Chapter 10)

Overhaul data - short frame:

	0.8/0.85 kW	**1.1 kW**
Minimum brush length	0.32 in (8.0 mm)	0.39 in (10.0 mm)
Brush spring pressure	56 oz (1600 g)	56 oz (1600 g)
Commutator minimum diameter	1.29 in (32.8 mm)	1.29 in (32.8 mm)
Armature endfloat	0.01 in (0.3 mm)	0.01 in (0.3 mm)

Starter motor (Cajavec manufacture)
Type	0.85 kW
Overhaul data	As for Bosch 1.1 kW short frame (see above)

Starter motor (Nippondenso manufacture)
Type	0.6 kW and 0.8 kW

Overhaul data:

	0.6 kW	**0.8 kW**
Number of brushes	2	4
Minimum brush length	0.39 in (10.0 mm)	0.39 in (10.0 mm)
Brush spring pressure	53 oz (1500 g)	42 oz (1200 g)
Commutator minimum thickness	0.02 in (0.6 mm)	0.02 in (0.6 mm)
Armature endfloat	0.02 in (0.6 mm)	0.02 in (0.6 mm)

Suspension and steering - 1982 on

Front suspension
Toe setting:
Checking value	0.08 in (2 mm) toe-out to 0.20 in (5 mm) toe-in
If outside checking limits, adjust to	0.04 to 0.12 in (1 to 3 mm) toe-in
Castor	2°00' nominal + 1°00' -0°45'

Manual steering
Pinion turning torque (unit on bench)	5 to 18 lbf in (0.6 to 2.0 Nm)

Power-assisted steering
Turns lock-to-lock	3.5
Rack lubricant capacity (bellows)	0.5 litres each
Rack lubricant type	Semi-fluid grease to Ford spec SQMIC-9106-A
Power steering fluid type/specification	ATF to Ford spec SQM-2C-9010-A

Tyre pressures - lbf/in² (bar)

	Normally laden		Fully laden	
Saloon models:	**Front**	**Rear**	**Front**	**Rear**
175SR 14	24 (1.7)	24 (1.7)	27 (1.9)	33 (2.3)
175HR 14	26 (1.8)	26 (1.8)	27 (1.9)	36 (2.5)
185SR 14	24 (1.7)	24 (1.7)	24 (1.7)	33 (2.3)
185HR 14	24 (1.7)	24 (1.7)	27 (1.9)	33 (2.3)
195/70HR 14	24 (1.7)	24 (1.7)	27 (1.9)	33 (2.3)
190/65HR 390	24 (1.7)	24 (1.7)	27 (1.9)	36 (2.5)

Tyre pressures - lbf/in² (bar) (cont.)

Estate models:	Normally laden Front	Normally laden Rear	Fully laden Front	Fully laden Rear
185SR 14	26 (1.8)	26 (1.8)	27 (1.9)	36 (2.5)
185HR 14	26 (1.8)	26 (1.8)	27 (1.9)	36 (2.5)
185HR 14 Rein	26 (1.8)	26 (1.8)	27 (1.9)	40 (2.8)
195/70HR 14	24 (1.7)	24 (1.7)	27 (1.9)	36 (2.5)
190/65HR 390	24 (1.7)	24 (1.7)	27 (1.9)	36 (2.5)

Torque wrench settings

	lbf ft	Nm
Front suspension		
Subframe-to-body bolts	66	90
Upper arm pivot bolts	77	105
Lower arm to subframe	66	90
Power-assisted steering		
Steering coupling clamp bolts	13	18
Steering coupling pinch-bolts	13	18
Rack slipper cover plate	8	11
Quick-fit connector coupling nuts	17	23
Pressure hose to pump	21	28
Return hose to pump	13	18
Pulley hub bolt	8	11
Steering rack-to-crossmember bolts	19	26
Track rod end locknuts	46	62
Track rod end to steering arm	20	27
Steering column clamp/brace nuts	15	20
Wheels		
Wheel nuts (steel or aluminium wheels)	63	85

Bodywork (all models)
Torque wrench settings

	lbf ft	Nm
Front seat belt stalk (Torx)	20	27
All other seat belt reel and anchor plate bolts	25	34

3 Maintenance intervals - later models

1 The manufacturer's recommended service intervals for later models (model year 1982 onwards) are given below. It will be noted that much of the work previously specified for the 6000-mile service has been transferred to the 12 000 mile service.

2 The home mechanic, who does not have labour costs to consider, may prefer to attend to some maintenance tasks more frequently than specified. Vehicles which are well worn through high mileage and/or hard use will also need more frequent attention.

3 Engine oil should be changed at least twice a year, regardless of mileage, since it deteriorates with time as well as with use.

Every 6000 miles (10 000 km)

Change engine oil and renew filter
Clean oil filler cap
Check idle speed
Check brake fluid level warning circuit
Check brake pads and shoes for wear; renew as necessary
Inspect brake hydraulic pipes and hoses
Check wheel nut tightness
Check tyre condition and pressures (including spare)
Inspect engine bay and under vehicle for oil, coolant or hydraulic leaks

Every 12 000 miles (20 000 km)

In addition to the previous items:

Renew spark plugs
Check drivebelts, adjust or renew as necessary
Check valve clearances, adjust as necessary
Check tightness of inlet manifold bolts (V6 only)
Check power steering fluid level (if applicable)
Check manual gearbox oil or automatic transmission fluid level
Check rear axle oil level
Lubricate handbrake linkage and (if applicable) automatic transmission control linkage
Inspect underside of vehicle for rust or other damage
Check steering and suspension components for security, freedom from leaks, condition of dust covers, balljoints and gaiters
Check that automatic choke releases fully with engine hot
Check idle mixture (CO level) adjustment
Check air conditioning refrigerant charge and clean debris from condenser (when applicable)
Check condition and security of exhaust system
Lubricate hinges, catches and locks, and check for correct operation
Check operation of all lights, warning systems, trip computer etc (as applicable)

Every 24 000 miles (40 000 km)

In addition to the previous items:

Check ignition timing and adjust if necessary
Lubricate distributor, clean and inspect distributor cap, coil tower and HT leads
Renew air cleaner element
Renew crankcase ventilation valve
Renew fuel filter (fuel injection models)
Adjust automatic transmission brake band (refer to this Supplement)

Every 36 000 miles (60 000 km) or two years, whichever comes first

Drain cooling system, flush and renew coolant

Every 36 000 miles (60 000 km) or three years, whichever comes first

Renew brake hydraulic fluid by bleeding

4 In-line ohc engine

Camshaft drivebelt - refitting and timing

1 The procedure in Chapter 1, Section 56 applies also to later models. However, the crankshaft timing mark has been changed - see Fig. 13.1.

2 When tightening the belt tensioner bolts, tighten the hexagon-headed bolt first, followed by the special bolt.

Supplement: Revisions and information on later models 13•7

Revised top cover gasket

3 From September 1983, a wider top cover gasket has been fitted. The gasket mating face on the cylinder head has been widened and the top cover and timing belt covers have been modified to accept the new gasket. The object of these modifications is to reduce the likelihood of distorting the top cover when tightening its securing bolts.

4 Should problems be experienced with oil leaks from an old type top cover, your Ford dealer can supply a kit of oil seals and reinforcing plates which should be fitted as shown in Fig. 13.2. A new gasket (of the old pattern) must be used and any distortion in the cover corrected.

5 The new type of cover is also secured with reinforcing plates under the bolt heads. The location of these plates is shown in Fig. 13.3.

Cylinder head bolts

6 As from April 1984, the cylinder head bolts are of T55 Torx socket-headed type instead of the earlier splined bolts.

7 The Torx type bolts must always be renewed at the time of overhaul. Never mix the earlier and later type bolts.

8 Refer to the Specifications Section of this Supplement for details of the revised tightening procedure for all bolts.

9 Further tightening of the cylinder head bolts at the first service interval after overhaul is no longer required.

Rocker ball stud locknuts

10 As from February 1984, the locknuts used on the rocker arm ball studs have been increased in thickness to 8.0 mm and in consequence, the torque tightening figure has also been increased (see Specifications at the beginning of this Supplement).

11 Where evidence is found of the earlier, thinner locknuts becoming loose, they should be replaced with the later type.

5 V6 engines

Engine removal - later models

1 The procedure in Chapter 1, Section 65 or 66, still applies. The following points should be noted:
   a) Disconnect the electrical leads from the automatic choke cover (carburettor models)
   b) Disconnect the coolant hoses from the oil cooler, when fitted

Engine dismantling - later models

2 From 1980 model year, the carburettor intermediate flange is retained by a socket-headed screw in addition to the carburettor nuts. This screw must be removed before the flange can be withdrawn.

Rocker shaft springs

3 Commencing January 1984, modified rocker shaft springs have been fitted.

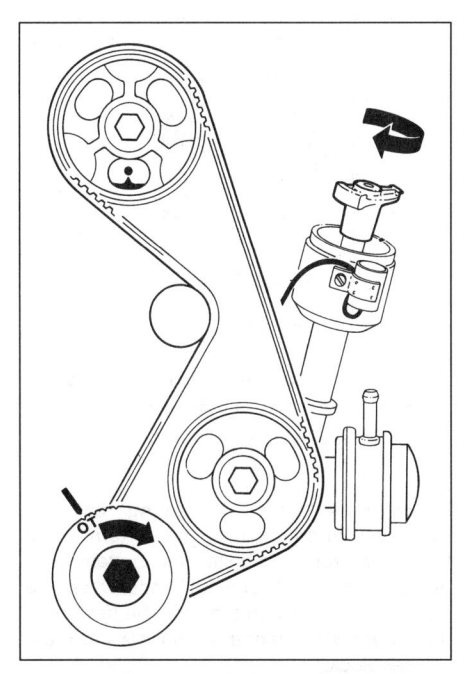

Fig. 13.1 Sprocket timing marks aligned - 1982 and later ohc models (Sec 4)

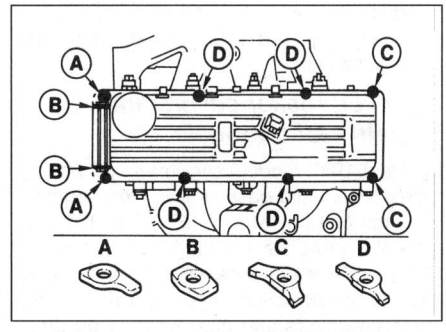

Fig. 13.3 Top cover reinforcing plate locations (ABCD) on later type cover (Sec 4)

4 Where tappet noise is evident on earlier models between engine speeds of 1500 and 2000 rev/min, the later type of springs may be fitted.

Oil cooler - removal and refitting

5 When an engine oil cooler is fitted, it is located between the oil filter cartridge and the cylinder block.

6 To remove the oil cooler, first remove the oil filter cartridge. Disconnect the coolant hoses from the oil cooler and plug them.

7 Use a spanner on the hexagonal section of the threaded sleeve and unscrew the sleeve. The oil cooler and gasket can now be removed. If the threaded bush is removed, or if it comes out with the threaded sleeve, it must be renewed.

8 Commence refitting by screwing the new bush into the cylinder block. Apply Omnifit Activator "Rapid" (to Ford specification SSM-998-9000-AA) to the exposed threads of the bush and to the inside of the threaded sleeve.

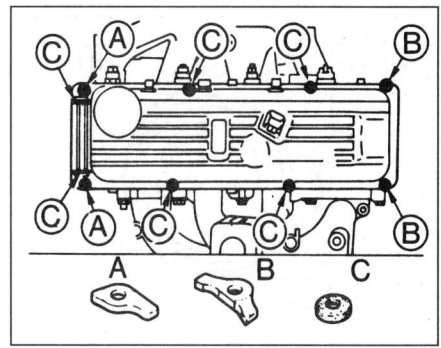

Fig. 13.2 Top cover oil seal (C) and reinforcing plate (A and B) locations - old type cover (Sec 4)

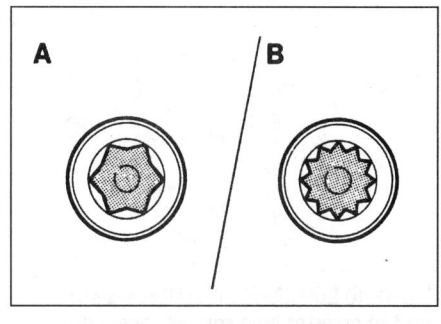

Fig. 13.4 Types of cylinder head bolt (Sec 4)

A T55 Torx   B Splined socket

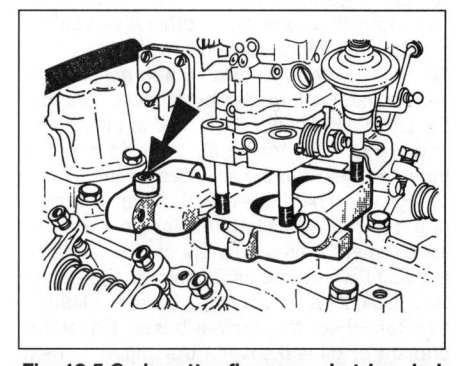

Fig. 13.5 Carburettor flange socket-headed screw, on later V6 engines (Sec 5)

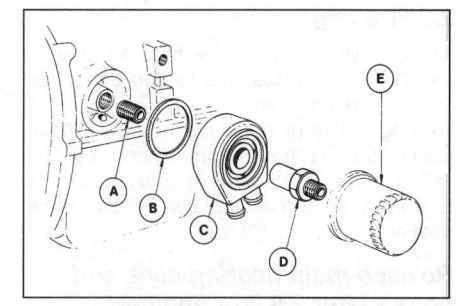

Fig. 13.6 Oil cooler components (Sec 5)

A Threaded bush B Seal C Cooler
D Sleeve E Oil filter

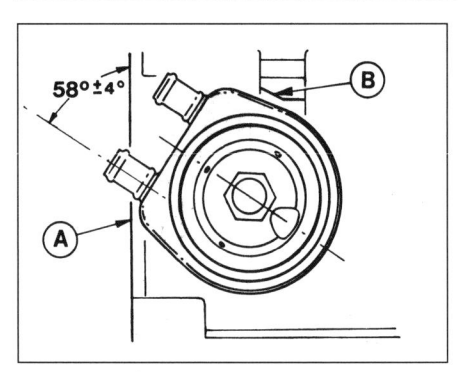

Fig. 13.7 Oil cooler installed position on 2.8 litre carburettor engine (Sec 5)

A Rear face of cylinder block
B Clearance (cooler to boss)

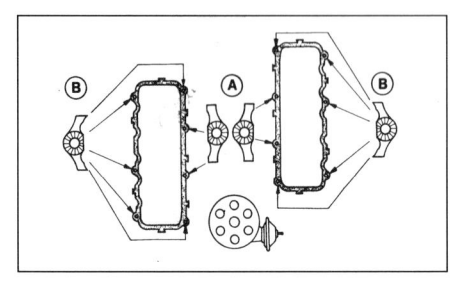

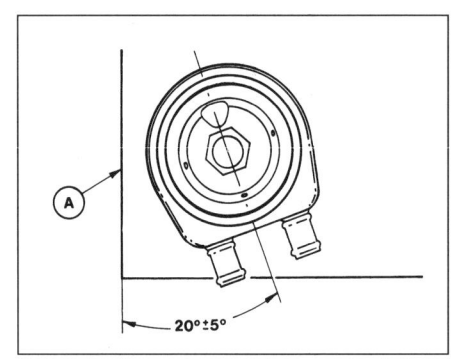

Fig. 13.8 Oil cooler installed position on 2.8 litre fuel injection engine (Sec 5)

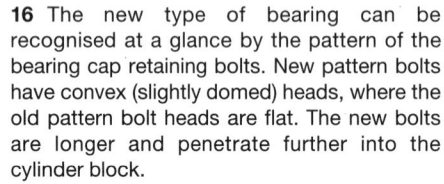

A Rear face of cylinder block

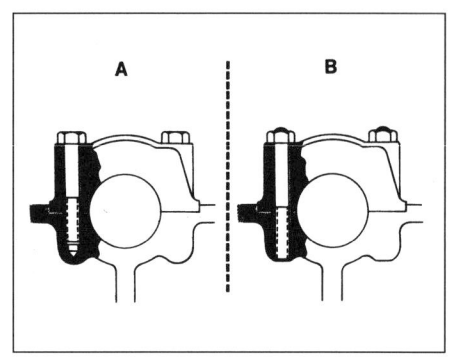

Fig. 13.9 Old (A) and new (B) type main bearing cap and bolts (Sec 5)

Fig. 13.10 Locations of reinforcing strips (A and B) on later type rocker covers (Sec 5)

9 Apply one drop of Omnifit Sealant "300 Rapid" (to Ford specification SSM-4G-9003-AA) to the leading threads of the bush. **Do not use more than one drop**, otherwise sealant may get into the lubrication circuit.
10 Fit the oil cooler, using a new gasket, and secure with the threaded bush. Make sure that the coolant pipes are positioned at the correct angle (Fig. 13.7 or 13.8), then tighten the threaded sleeve to the specified torque.
11 Fit a new oil filter element, oiling its sealing ring prior to installation. Tighten the filter approximately three-quarters of a turn beyond the point where the seal contacts the cooler face. Do not use any tool to tighten the filter.
12 Reconnect the coolant hoses. Check for coolant or oil leaks when the engine is next run, and check coolant and oil levels when it has been stopped.

Crankshaft reconditioning - all 2.8 litre engines

13 Specially selected crankshafts are fitted in the 2.8 litre engines; production undersize bearings are not used.
14 Regrinding or other machining of these crankshafts is not recommended by the makers. If the degree of wear warrants regrinding, an exchange crankshaft should be obtained.

Revised main bearing caps and bolts - later 2.8 litre engines

15 From 1982, all 2.8 litre engines have a modified design of main bearing at No 2 and No 3 journals.

16 The new type of bearing can be recognised at a glance by the pattern of the bearing cap retaining bolts. New pattern bolts have convex (slightly domed) heads, where the old pattern bolt heads are flat. The new bolts are longer and penetrate further into the cylinder block.
17 When rebuilding an engine fitted with the new pattern main bearing caps and bolts, take care to fit the convex headed bolts only to No 2 and No 3 caps. Use of these bolts in other locations may lead to thread damage and subsequent failure.

Improved rocker covers and gaskets - later models

18 From 1982 model year, improved pattern rocker covers and gaskets are fitted, together with reinforcing strips (Fig. 13.10). The securing screws are also longer.
19 If problems are experienced with oil leaks from the rocker covers on earlier engines, the new type covers and gaskets may be fitted with the appropriate screws and reinforcing strips. Use sealer if necessary on the areas where the inlet manifold joins the cylinder heads.
20 Do not overtighten the rocker cover securing screws. 6 lbf ft (8 Nm) is sufficiently tight.

6 Cooling system

Radiator repairs - later models

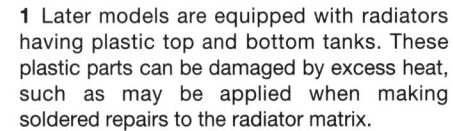

1 Later models are equipped with radiators having plastic top and bottom tanks. These plastic parts can be damaged by excess heat, such as may be applied when making soldered repairs to the radiator matrix.
2 You are strongly recommended to entrust radiator repairs to a competent specialist. DIY repairs should be restricted to "cold" techniques using plastic filler or similar material.

Leaks from radiator blanking plug

3 If leaks of coolant are observed from the radiator top tank blanking plug, these may be dealt with as follows.
4 On 1978/79 models, remove the radiator pressure cap (taking appropriate precautions against scalding). Drain or siphon coolant from the radiator until the level lies below the plug hole, then unscrew the blanking plug.
5 Clean and dry the threads of the plug and its hole. Wrap several turns of PTFE tape (obtainable from ironmongers and plumbers' suppliers) around the plug and refit it. Tighten the plug to 20 lbf ft (27 Nm).
6 On 1982 and later models a revised type of blanking plug, with gasket, is available. Follow the procedure given above for older models, but without using PTFE tape. The models to which this applies have a V6 engine, automatic transmission (or manual if fuel injected) and no air conditioning.

Coolant hoses chafing - later ohc models

7 Coolant hose chafing may occur on 1982 models equipped with an ohc engine and power-assisted steering. The points to check are as follows.
8 The heater return hose may chafe against the steering pump bracket. Shorten the hose by approximately 0.4 in (10 mm) to prevent this occurring; renew the hose if it is damaged.
9 It is also not unknown for the accessory drivebelt to touch the radiator bottom hose. This condition may be recognised by scuffing or even severing of the bottom hose in the area next to the drivebelt.
10 Further damage to the bottom hose may be prevented by installing a shield and a longer alternator mounting bolt, both available from your Ford dealer.

Fan removal - later ohc models

11 On later ohc models, the viscous fan clutch is secured to the water pump pulley spindle by a large nut between the pulley and the clutch. This nut has a left-hand thread.

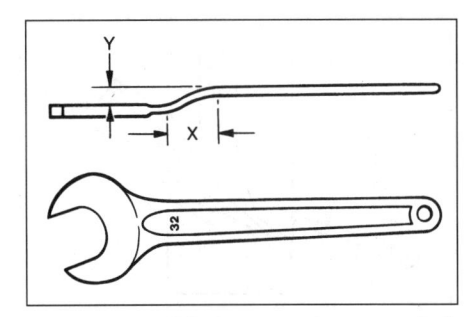

Supplement: Revisions and information on later models 13•9

Fig. 13.11 Modified spanner for removal of ohc fan (Sec 6)

X 1. in (25. mm)   Y 0.5 in (12.5 mm)

12 To gain access to this nut, it will be necessary to obtain a 32 mm (or 1¼ in AF) open-ended spanner and bend it slightly as shown in Fig. 13.11. The thickness of the spanner jaws must not exceed 0.2 in (5 mm). Heat the spanner before attempting to bend it. Your local blacksmith or welding shop may be able to do the job for you.

13 When using the modified spanner to undo the fan clutch securing nut, grip the water pump pulley to prevent it turning, using an old drivebelt and a self-gripping wrench (take the existing drivebelt off first). Strike the spanner with a soft-faced hammer if necessary to release the nut. Remember that the nut has a left-hand thread so it unscrews clockwise, ie in the direction of engine rotation.

14 With the fan and clutch removed, the two components may be separated by removing the four securing bolts.

Modified radiator (V6 engine)

15 Commencing February 1984, the radiator fitted to this model no longer incorporates a filler neck or pressure cap.

16 Fill or top up through the expansion tank and remember that the later type of radiator can only be fitted to earlier models if the modified expansion tank and hose are also fitted.

7 Fuel system - carburettor models

Accelerator cable adjustment - all models

1 The procedures in Chapter 3, Section 13 or 31, still apply, but note the following points.

2 Certain 1983 models are equipped with a "progressive" throttle linkage. The effect of this type of linkage is to give greater pedal movement at light and intermediate throttle openings than at wide throttle. Do not mistake this effect for a fault in the cable adjustment or a loss of performance.

3 Where problems are encountered with excessive pedal effort, a common sense approach should first be adopted. Make sure that the cable run is free of kinks and sharp bends. Disconnect the carburettor end of the cable to determine whether resistance is due

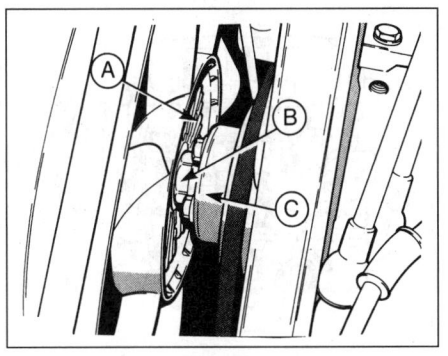

Fig. 13.12 Viscous cooling fan (ohc engine) (Sec 6)

A Fan and clutch   B Nut
C Water pump pulley

to the cable (lubricate or renew) or to the carburettor linkage (partially dismantle and lubricate). It is also possible to obtain a lighter return spring from your Ford dealer.

4 It is emphasised that when the accelerator linkage is correctly adjusted, wide-open throttle must not be obtained before the pedal contacts its stop. Great strain can otherwise be put on the pedal, cable, carburettor and associated brackets.

Weber carburettor top cover securing screws - all models

5 If the top cover securing screws are disturbed for any reason on the Weber carburettor, or if they are found to be loose, thread locking compound should be used on reassembly.

Solex/Pierburg carburettor

6 Carburettors fitted to later V6 engine models will almost certainly have the Solex/Pierburg or Pierburg identification. This is principally a change of name, with only minor changes to the technical specification.

Solex/Pierburg carburettor electrically-heated choke - description of operation

7 The electrically-heated choke fitted to later type Solex/Pierburg carburettors works in the same way as the coolant-heated type, except that electricity and not coolant is the source of heat.

8 Two heating elements are employed. One element is powered from the alternator and begins to heat up as soon as the engine is turning, The second element is switched on, by a thermoswitch located in the cooling system, only when a certain minimum coolant temperature has been reached.

9 This type of choke is quicker to respond to engine warm-up than the coolant-heated type, providing that both heating elements are connected and working correctly. If either element is not working, the choke will remain on even though the engine has warmed up.

10 Adjustments and checks are carried out as described for the coolant heated choke in

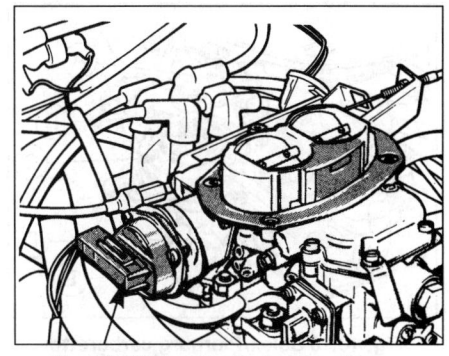

Fig. 13.13 Electrically-heater choke - Solex/Pierburg carburettor (Sec 7)

A Choke housing   B Thermoswitch

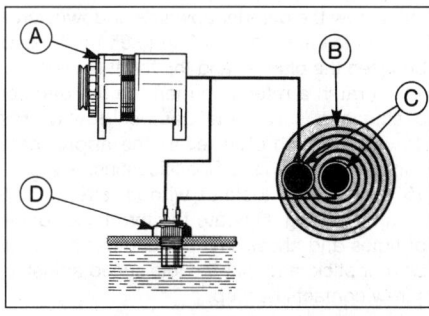

Fig. 13.14 Schematic wiring diagram for electrically-heated choke (Sec 7)

A Alternator   C Heating elements
B Bi-metallic coil   D Thermoswitch

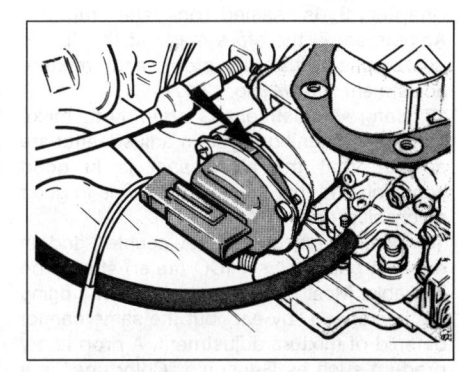

Fig. 13.15 Electrically-heated choke alignment marks (arrowed) (Sec 7)

Chapter 3. Refer to the Specifications at the beginning of this Chapter for revised adjustment values.

Solex/Pierburg carburettor throttle damper - description and adjustment

11 The function of the throttle damper fitted to later models in conjunction with the electrically-heated choke is to smooth the change from drive to overrun when the throttle is suddenly released.

12 Before the damper is adjusted, the engine must be at normal operating temperature and the idle speed and mixture must be correct.

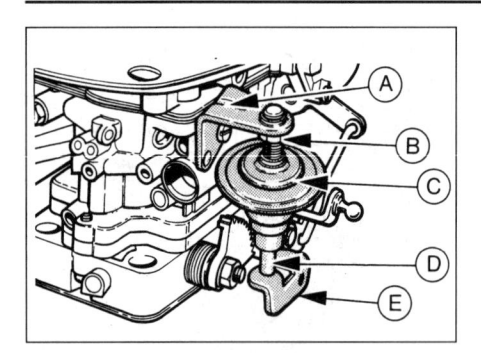

13•10 Supplement: Revisions and information on later models

Fig. 13.16 Solex/Pierburg carburettor throttle damper (Sec 7)

A Mounting bracket B Locknut C Damper
D Plunger E Throttle lever

13 Refer to Fig. 13.16. Slacken the locknut and screw the damper upwards or downwards until a clearance of 0.002 in (0.05 mm) exists between the plunger and the throttle lever.
14 Scratch a reference mark on the damper casing and screw the damper downwards (towards the throttle lever) the appropriate number of turns - see Specifications.
15 Tighten the locknut without altering the damper setting. Operate the throttle a couple of times and check that the damper does not bind or stick, and that the idle speed adjusting screw contacts its stop.

Solex/Pierburg carburettor (1982 on) - idle speed and mixture adjustment

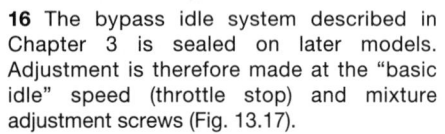

16 The bypass idle system described in Chapter 3 is sealed on later models. Adjustment is therefore made at the "basic idle" speed (throttle stop) and mixture adjustment screws (Fig. 13.17).
17 Before adjustment is attempted, make sure that all ignition system adjustments are correct, that the air cleaner is in good condition and that the engine itself is in good mechanical order.
18 Ideally a tachometer (rev counter) and an exhaust gas analyser (CO meter) should be available. Most people are capable of judging the idle speed "by ear" but the same cannot be said of mixture adjustment. A proprietary product such as Gunson's "Colortune" is a good substitute for an exhaust gas analyser. Use such a product in accordance with its maker's instructions.
19 Connect the measuring instruments to the engine. Start the engine and run it at 3000 rpm for half a minute or so, then allow it to idle. Turn the idle speed adjustment screw clockwise to increase or anti-clockwise to decrease the idle speed.
20 When the idle speed is correct, check the exhaust gas CO level. If adjustment is found to be necessary, proceed as follows.
21 Stop the engine and remove the tamperproof plugs which cover the two mixture adjustment screws (Fig. 13.18). Note that these plugs are intended to discourage and detect adjustment by unqualified operators, and that in some European countries (though not yet in the UK) it is illegal to operate a vehicle with the tamperproof plugs removed. The plugs are removed by prising them out with a sharp instrument and may then be discarded.
22 Screw both mixture adjustment screws fully home, without forcing them, then unscrew each one 5 complete turns. *All adjustments must be made to **both** screws in equal amounts.*
23 Start the engine again and run it at 3000 rpm for 30 seconds, then allow it to idle. Turn the idle mixture adjusting screws by equal amounts until the exhaust gas CO content is within the specified limits. Readjust the idle speed if necessary.
24 When adjustment is correct, stop the engine and disconnect the measuring instruments. Fit new tamperproof plugs where these are required by law.

Fuel tank

25 Fuel tank and sender unit removal procedures are similar to those given in Chapter 3, except that the sender unit is mounted on the top of the tank. The fuel feed and return pipes are either top-mounted or front-mounted, according to the model.

8 Fuel system - fuel injection models (1982 on)

General description

1 The description given in Part C of Chapter 3 is largely correct for 1982 and later models. Certain changes have been made in the areas detailed below.

Fuel start valve system

2 To improve hot starting, the fuel start valve is pulsed when the engine is cranking. The device responsible for pulsing the valve under these conditions is known as the impulse module.
3 The fuel start valve is still controlled by the thermo time switch when the engine is cold.

Safety control module

4 The safety switch described in Chapter 3 has been replaced by a safety control module.

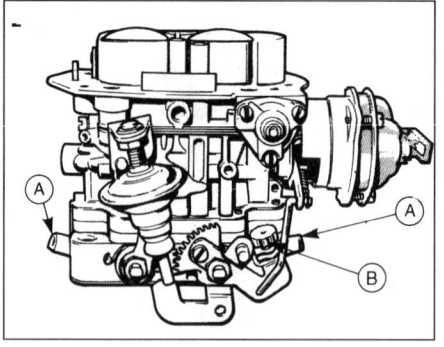

Fig. 13.17 Adjusting screws on later Solex/Pierburg carburettor (Sec 7)

A Idle mixture screws B Idle speed screw

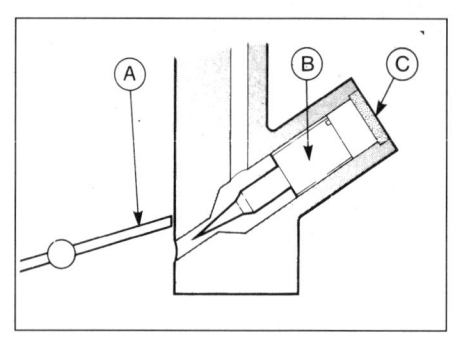

Fig. 13.18 Sectional view of idle mixture screw (Sec 7)

A Throttle valve plate B Idle mixture screw
C Tamperproof plug

This module supplies power to the fuel pump provided that it is receiving pulses from the ignition coil, these pulses being evidence that the engine is running. When the engine stops for whatever reason, the safety control module cuts off power to the fuel pump.

Control modules - removal and refitting

5 Remove the glovebox as described in Section 18.
6 Remove the left-hand air vent hose.
7 Identify the modules. The safety control module is coloured purple; the impulse module is coloured green.

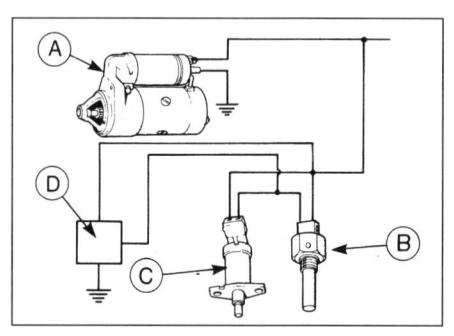

Fig. 13.19 Fuel start valve circuit diagram (Sec 8)

A Starter motor C Fuel start valve
B Thermo time switch D Impulse module

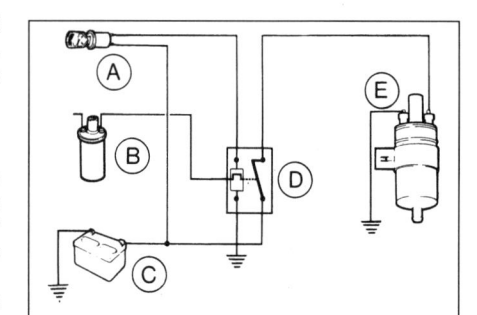

Fig. 13.20 Safety control module circuit diagram (Sec 8)

A Ignition switch B Ignition coil C Battery
D Safety module E Fuel pump

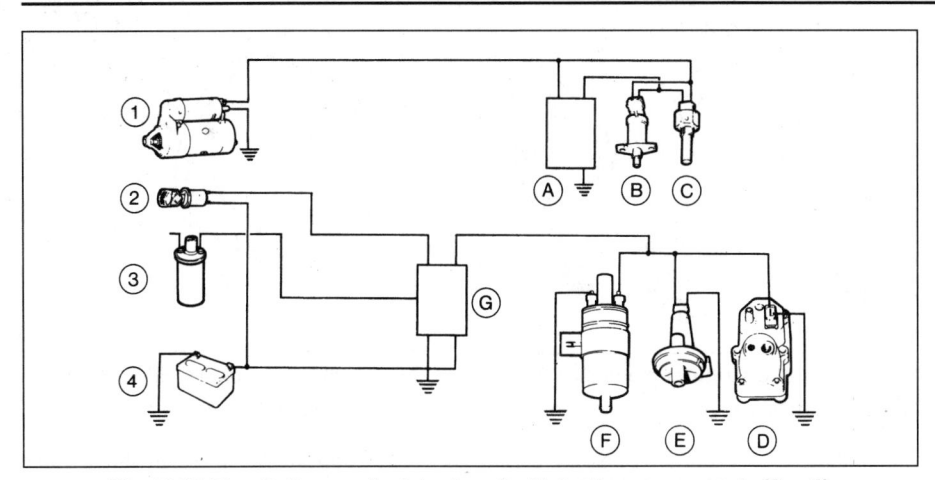

Supplement: Revisions and information on later models 13•11

Fig. 13.21 Circuit diagram for later type fuel injection components (Sec 8)

1 Starter motor
2 Ignition switch
3 Ignition coil
4 Battery
A Impulse module
B Fuel start valve
C Thermo time switch
D Warm up regulator
E Auxiliary air device
F Fuel pump
G Safety module

8 Remove the securing screw (if used) and unplug the appropriate module.
9 Refitting is a reversal of the removal procedure. No repair is possible: testing is by substitution of a known good unit.

Safety control module - bypassing

10 If it is wished to bypass the safety module - perhaps to check the fuel pump delivery with the engine stopped - this is easily done by substituting a standard headlamp relay for the safety control module.
11 With a headlamp relay fitted in place of the safety control module, the fuel pump should be heard running whenever the ignition is switched on.
12 Do not drive the car with the safety control module bypassed.

Depressurizing the fuel injection system

13 Before disconnecting any fuel line which is liable to contain fuel under pressure - eg when removing the fuel pump, fuel accumulator or fuel filter - the system should be depressurized.

14 The easiest way of releasing the pressure in the fuel lines is to slacken the warm-up regulator union at the fuel distributor (Fig. 13.23). Have some cloth ready to absorb the fuel which will be released, and take appropriate precautions against fire.
15 Tighten the union when the fuel pressure has been released.

Fuel pump - removal and refitting

16 Depressurize the fuel injection system as described above.
17 The procedure in Chapter 3, Section 44, may now be followed. Note however that the pump securing U-bolt has been superseded by a clamp (Fig. 13.24).
18 Refitting is a reversal of removal. Make sure that the rubber sleeve is fitted under the clamp and that the locating shoulder is up against the clamp (Fig. 13.25).

Fuel accumulator - removal and refitting

19 Depressurize the fuel injection system.
20 The procedure in Chapter 3, Section 44, may now be followed.

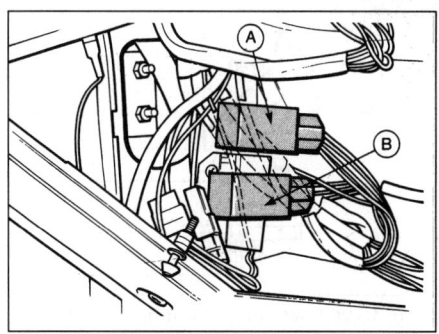

Fig. 13.22 Location of fuel injection control modules (Sec 8)

A Safety control module (purple)
B Impulse module (green)

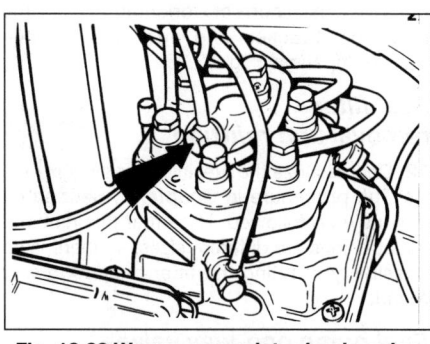

Fig. 13.23 Warm-up regulator banjo union bolt (arrowed) on fuel distributor (Sec 8)

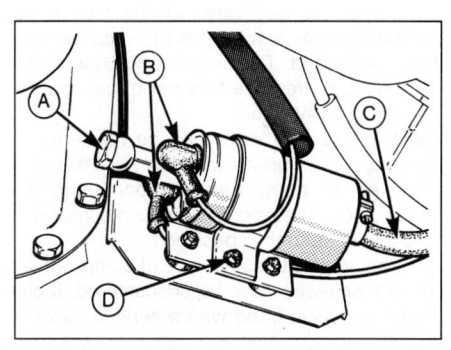

Fig. 13.24 Later type fuel pump mounting (Sec 8)

A Outlet union
B Electrical connections
C Inlet pipe
D Clamp nut

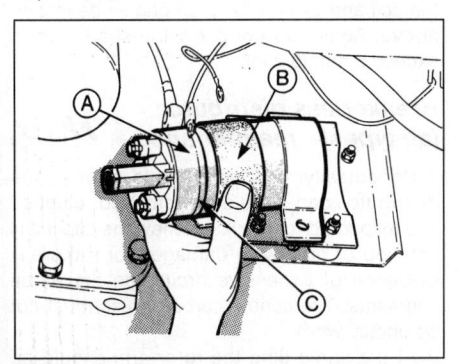

Fig. 13.25 Fitting later type fuel pump (Sec 8)

A Pump B Rubber sleeve
C Locating shoulder

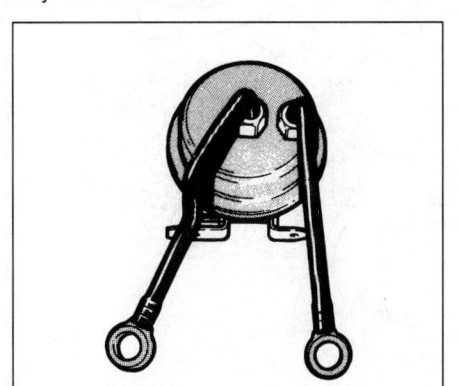

Fig. 13.26 Fuel accumulator with pipes and bracket correctly positioned (Sec 8)

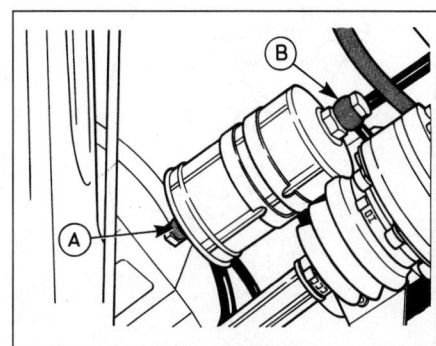

Fig. 13.27 Different size fuel filter unions (Sec 8)

A Outlet B Inlet

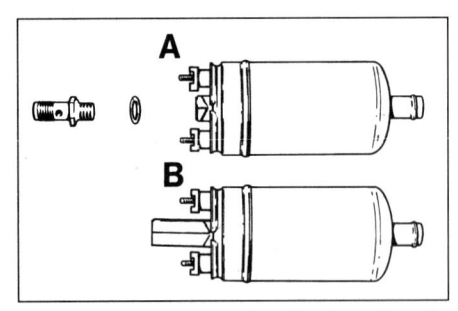

13•12 Supplement: Revisions and information on later models

Fig. 13.28 Fuel pump identification (Sec 8)

A Pump with detachable one-way valve
B Pump with integral one-way valve

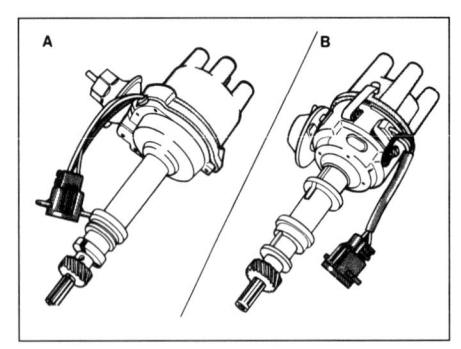

Fig. 13.29 Distributor types (Sec 9)

A Motorcraft (blue cap)   B Bosch (red cap)

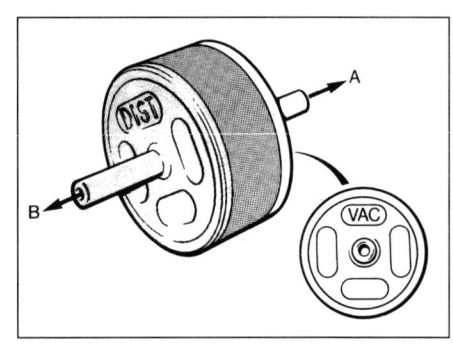

Fig. 13.30 Spark sustain valve (Sec 9)

A To carburettor   B To distributor

21 If the accumulator is to be renewed, observe the correct orientation of the mounting bracket and fuel pipes when transferring these (Fig. 13.26).

Fuel filter - removal and refitting

22 Depressurize the fuel injection system.
23 The procedure in Chapter 3, Section 44 may now be followed. Note that the inlet and outlet unions are different sizes, so there is no possibility of fitting the filter the wrong way round.

Fuel pump one-way valve - renewal

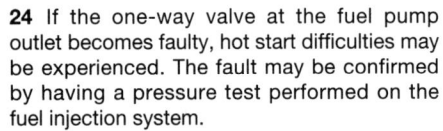

24 If the one-way valve at the fuel pump outlet becomes faulty, hot start difficulties may be experienced. The fault may be confirmed by having a pressure test performed on the fuel injection system.
25 The one-way valve can be renewed on most pumps by unscrewing the valve from the body of the pump after disconnecting the outlet union. The new valve is then screwed in and tightened to the specified torque.
26 Should an alternative type of pump with an integral one-way valve be encountered, it can if necessary be fitted with a service valve at the outlet. This valve is available from your Ford dealer and must be used in conjunction with a union adaptor.

Fuel tank

27 Refer to Section 7, paragraph 25.

9 Ignition system

General description (1982 on)

1 A breakerless (electronic) ignition system is now fitted to all Granada models. The Motorcraft system described in Chapter 4 is still fitted to some V6 models. Other V6 models, and all ohc models, are fitted with a Bosch ignition system. The two systems may be distinguished at a glance by the colour of the distributor cap: the Motorcraft cap is blue and the Bosch cap is red.
2 Certain ohc models are fitted with a spark sustain valve in the vacuum advance line. This valve has the effect of prolonging vacuum advance, thus improving driveability when the engine is cold. A device known as a ported vacuum switch (PVS) bypasses the spark sustain valve when normal operating temperature has been reached.

3 A fuel trap is also fitted in the vacuum advance line on some models. This prevents neat fuel entering the distributor vacuum unit. The fuel trap ends are marked to indicate which way round it should be fitted.

Breakerless ignition (all types) - precautions

4 The HT voltage generated by the breakerless ignition system is considerably higher than that in conventional systems. Great care should therefore be taken to avoid personal electric shocks. Particularly, do not grasp the coil, distributor or HT leads when the engine is running. Always use insulated tools to hold live HT leads.
5 If the distributor is knocked when the ignition is switched on, a single HT pulse may be produced at one spark plug lead. In some circumstances this could cause the engine to turn.
6 Do not energise the ignition system unless all the HT leads are properly connected to the coil, distributor cap and spark plugs. Checking for HT voltage by causing a spark to jump from the end of a plug lead to earth is permissible, but the spark gap must not exceed 5 mm (0.2 in) or module damage may result. Use the centre section of an old spark plug, or a small bolt, to extend the HT lead conductor from inside the plug cap.
7 When checking for HT voltage in the coil-to-distributor lead, connect a spark plug lead to the coil and extend the plug cap as described above. Again, do not allow the spark gap to exceed 5 mm (0.2 in).

Breakerless distributor (all types) - maintenance

8 Periodically, or if misfiring is evident, the distributor cap should be removed, cleaned and inspected for cracks. Renew the cap if it is cracked or otherwise damaged, or if there is evidence of excessive arcing on the metal segments. The central carbon brush must not be unduly worn.
9 At the same time the rotor arm should be removed and its metal track cleaned with fine abrasive paper. Renew the rotor arm if it is cracked or badly burnt
10 With the rotor arm removed, lubricate the

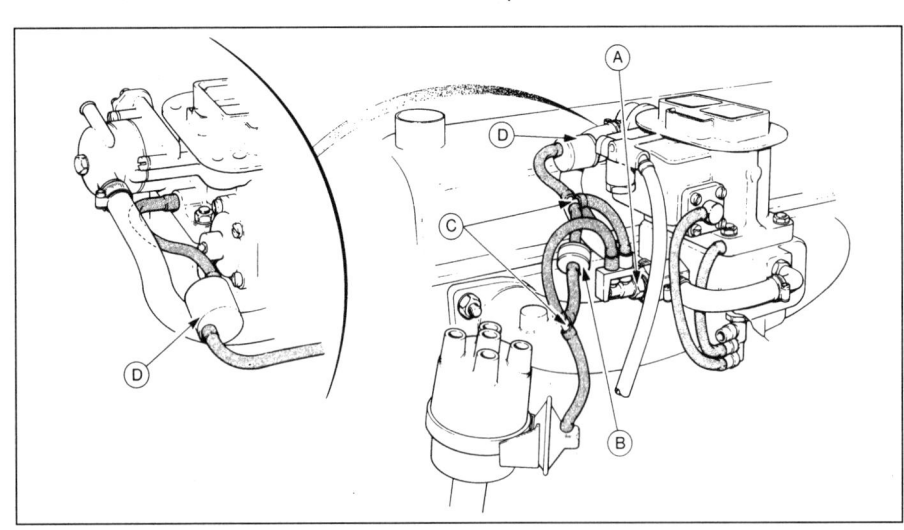

Fig. 13.31 Typical vacuum advance connections (Sec 9)

A Ported vacuum switch   B Spark sustain valve   C Tee piece connectors   D Fuel trap

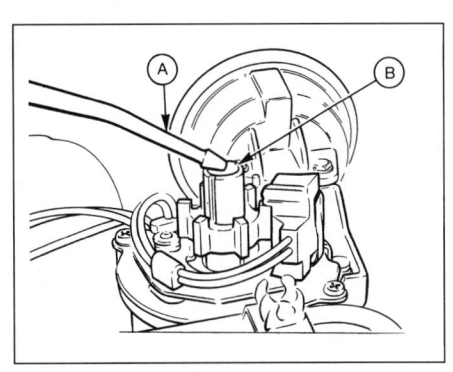

Supplement: Revisions and information on later models 13•13

Fig. 13.32 Lubricating distributor shaft wick (Sec 9)

A Oil can B Wick

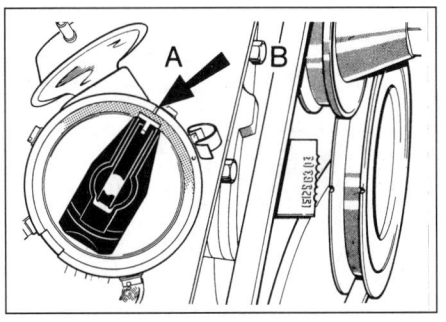

Fig. 13.33 Timing marks when No 1 piston at TDC on V6 engine (Sec 9)

A Rotor/body alignment mark
B Crankshaft pulley timing mark

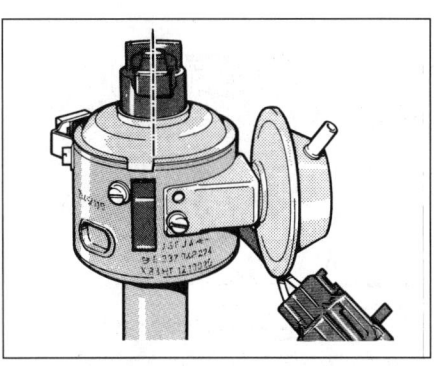

Fig. 13.34 Bosch distributor rotor aligned with edge of body rim cut-out (ohc engine) (Sec 9)

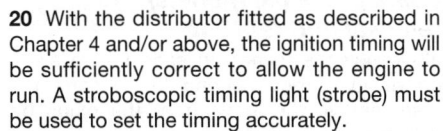

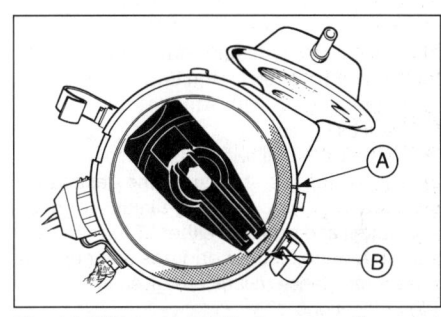

Fig. 13.35 Bosch distributor rotor alignment marks (V6 engine) (Sec 9)

A No 1 piston firing point (scribed line on body rim) - distributor installed
B Rotor position (dot punch on body rim) before distributor installed

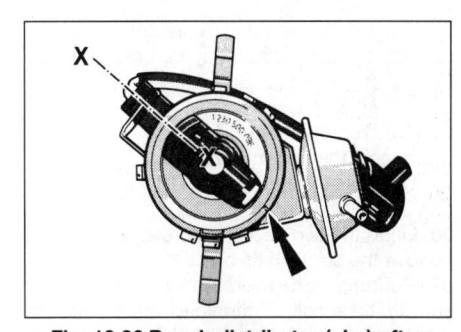

Fig. 13.36 Bosch distributor (ohc) after installation (Sec 9)

No 1 firing mark (arrowed)
Line X - X at 90° to crankshaft centre line

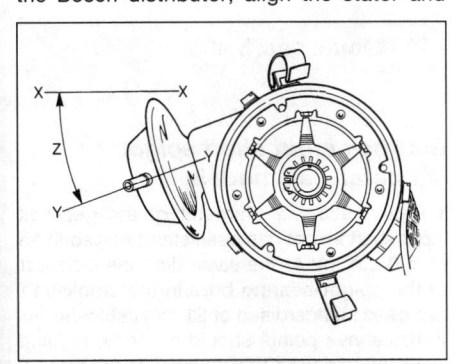

Fig. 13.37 Bosch distributor (V6) after installation (Sec 9)

Line X - X Rear face of cylinder block
Line Y - Y should be between 0 and 14°

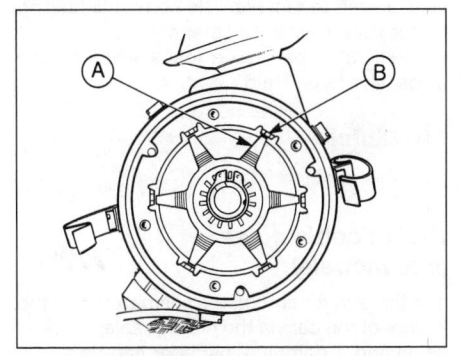

Fig. 13.38 Stator and trigger wheel alignment - Bosch distributor (V6 engine) (Sec 9)

A Trigger wheel spoke B Stator notch

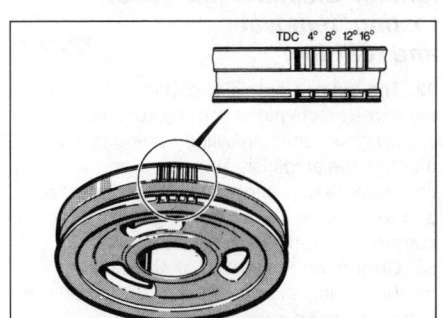

Fig. 13.39 Crankshaft pulley (cast) timing marks on ohc engine (Sec 9)

wick in the centre of the distributor shaft with two drops of engine oil, Fig. 13.32.

11 Refit the rotor arm and the distributor cap, then check the ignition timing as described in Chapter 4 for later in this Section.

12 No other maintenance or repair operations are possible on the breakerless distributor.

Breakerless distributor (1982 on) - removal and refitting

13 Removal of all types of breakerless distributor is basically as described in Chapter 4, Section 13. There is no need to remove the air cleaner on ohc models.

14 Bosch distributors have a line scribed on the edge of the body; when the tip of the rotor arm points to this line, the distributor is at the firing point for No 1 cylinder.

15 Refitting of Motorcraft distributors is as described in Chapter 4.

16 Prior to refitting a Bosch distributor, align the rotor arm as shown in Fig. 13.34 (ohc engine) or Fig 13.35 (V6 engine). On V6 engines the rotor will turn as the distributor is installed owing to the meshing of the drive gears. After refitting, the rotor arm tip should point to the scribed line, and the distributor itself should be aligned as shown (Fig. 13.36 or 13.37).

17 Before tightening the clamp bolt on the Bosch distributor, align the stator and trigger wheel as shown in Fig. 13.38

18 On all models, repack the distributor multi-plug with suitable grease if necessary.

19 Check the ignition timing and adjust if necessary.

Ignition timing (1982 on) - checking and adjusting

20 With the distributor fitted as described in Chapter 4 and/or above, the ignition timing will be sufficiently correct to allow the engine to run. A stroboscopic timing light (strobe) must be used to set the timing accurately.

21 Various types of strobe are available. All make some connection to No 1 spark plug lead. The brighter lights will also require a connection to be made to the car battery or to a mains electricity supply.

22 The procedure is as described in Chapter 4, Section 12 or 15. Refer to Figs 13.39 and 13.40 for details of the different types of timing marks which may be encountered on ohc models.

Ignition coil (all types) - testing, removal and refitting

23 The ignition coil is mounted on the rear bulkhead (V6 models) or on the left-hand inner wing (ohc models). It may be tested *in situ* if an ohmmeter is available; otherwise, the simplest test is by substitution of a known good unit.

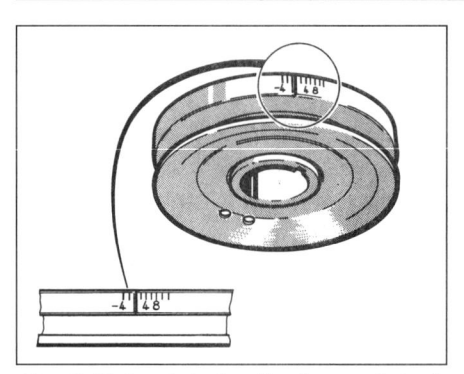

13•14 Supplement: Revisions and information on later models

Fig. 13.40 Crankshaft pulley (pressed steel) timing marks on ohc engine (Sec 9)

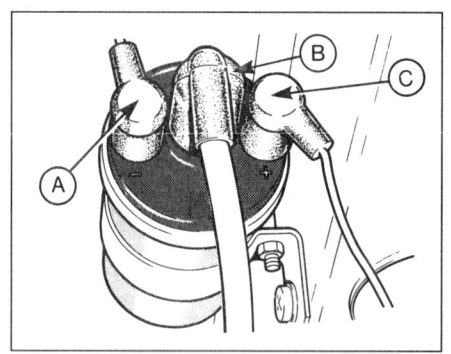

Fig. 13.41 Ignition coil connections (Sec 9)
A LT negative (to distributor or amplifier)
B HT (to distributor cap)
C LT positive (to ignition switch)

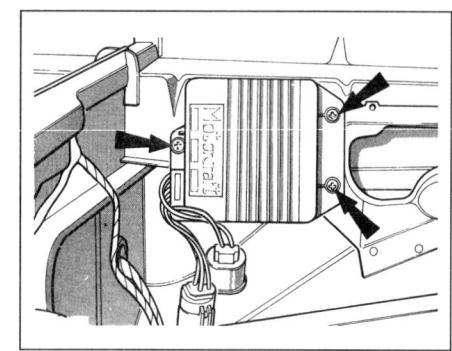

Fig. 13.42 Ignition amplifier module fixing screws (Sec 9)

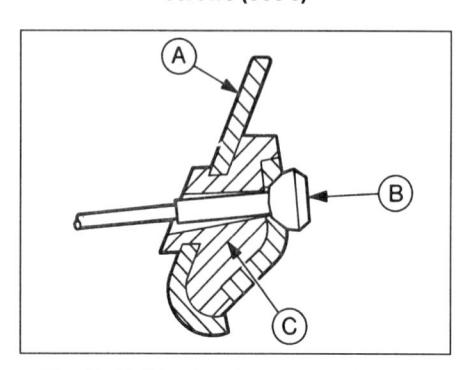

Fig. 13.43 Clutch cable damper (Sec 10)
A Release lever   B Cable
C Damper insulator

24 Ignition coils used with conventional (contact breaker) ignition are not interchangeable with those used with breakerless ignition.
25 To test the coil windings, disconnect the LT and HT leads. Measure the resistance between the two LT terminals (primary resistance) and between either LT terminal and the HT terminal (secondary resistance). The correct resistance values are given in Chapter 4 Specifications.
26 Unless very accurate measuring equipment is available, all that the DIY mechanic can hope to establish when testing the coil is that the primary winding is not open-circuit (high or infinite resistance) and that the secondary winding is not short-circuited (low or zero resistance).
27 Even if the winding resistances appear correct, the coil output may be unsatisfactory for other reasons. Specialised diagnostic equipment is required for further testing.
28 To remove the ignition coil, first disconnect the battery earth lead.
29 Disconnect the LT and HT leads from the coil.
30 Undo the two securing screws or nuts and remove the coil with its bracket.
31 Refitting is a reversal of removal. Make sure that the coil LT terminals are connected the right way round. Usually they are different sizes so that incorrect connection is impossible.

Ignition amplifier (all types) - testing, removal and refitting

32 The ignition amplifier cannot be tested by the home mechanic except by substitution of a known good unit. Unskilled testing is liable to damage the amplifier. The amplifier is normally very reliable and should only be suspected of a fault when all other ignition system components have been found good.
33 Commence removal of the amplifier by disconnecting the battery. On V6 models the battery must be removed.
34 Disconnect the two wiring multi-plugs by pulling on the wires, not on the plugs themselves.

35 Undo the securing screws and remove the amplifier.
36 Refitting is a reversal of removal. Pack the multi-plugs with the specified grease if necessary. Firm pressure will be needed to engage the plugs fully.

Spark plugs (all models) - removal and refitting

37 Identify the spark plug leads if there is any possibility of reconnecting them incorrectly, then remove the leads from the plugs by pulling on the plugs caps.
38 Clean the area around the seat of each plug so that dirt does not fall into the cylinders.
39 Unscrew the plugs using a proper plug spanner or socket spanner. Take care not to apply any sideways force to the plug, or the insulator may be cracked.
40 Inspect the plugs as described in Chapter 4, Section 16. Check the electrode gaps as a matter of course, even on new plugs.
41 Apply a light smear of anti-seize compound or grease to the threads of each plug and screw them into their holes. It should be possible to screw the plugs nearly all the way home by hand if not, the thread may be crossed or damaged.
42 Ideally the plugs should be tightened using a torque wrench. It is particularly important that the taper seat plugs used in ohc engines are not overtightened, as they can then be very difficult to remove. The correct tightening torques are given in Chapter 4.
43 Refit the spark plug leads when all the plugs have been fitted.

10 Clutch

Clutch cable noise (ohc models)

1 If there is much noise transmitted into the interior of the car via the clutch cable, a longer cable and a damping insulator can be fitted. (These items are fitted as original equipment on models built from February 1982.)
2 Obtain the longer cable and the insulator from your Ford dealer.

3 Remove and discard the old cable and fit the new cable (see Chapter 5, Section 5, and below).
4 Fit the insulator to the clutch release lever end of the cable. Pull the cable through the slot in the top of the insulator so that its end rests in the depression in the metal plate; make sure that the insulator locates properly in the clutch release lever (Fig. 13.43).

Clutch cable removal and refitting (all models)

5 Disconnecting the cable from the pedal end will be made easier if the pedal is wedged up with a block of wood. Remove the block on completion.

11 Manual gearbox

Gearbox input shaft spigot diameter - all models

1 From 1983 the diameter of the gearbox input shaft spigot has been standardised at 15 mm (0.591 in). At the same time the diameter of the spigot bearing bore in the crankshaft has been standardised at 21 mm (0.827 in).
2 The above points should be borne in mind when considering the renewal of a gearbox, engine or spigot bearing. A range of bearings is available to enable any combination of the two diameters previously used to mate successfully.

Supplement: Revisions and information on later models

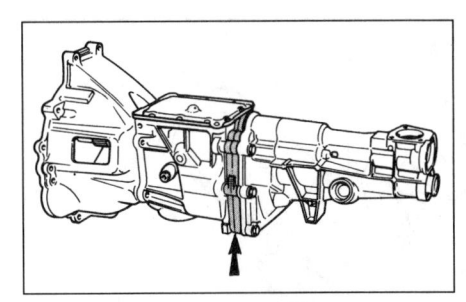

Fig. 13.44 Type N five-speed gearbox. Intermediate plate (arrowed) (Sec 11)

5-speed gearbox (type N) - description

3 The type N 5-speed gearbox available on later models is similar to the type B and E 4-speed gearboxes described in Chapter 6. The type N gearbox can be distinguished externally by the presence of an intermediate plate between the gearcase and the rear extension. Refer to Chapter 6, Section 1 for information on lubricant topping-up.

5-speed gearbox (type N) - removal and refitting

4 Removal of the type N gearbox is basically as described in Chapter 6, Section 2. Note, however that the gear lever is retained by 3 splined head ("Torx") screws. A splined key will be needed to undo these screws.
5 On fuel injection models, refer also to Section 8 of Chapter 6.
6 Refitting is a reversal of removal.

5-speed gearbox (type N) - alternative lubricant type

7 If persistent difficulty is experienced in engaging 1st or 2nd gear, the use of a semi-synthetic oil (see Specifications) may improve matters.

HAYNES HiNT *Measure the volume of gearbox oil removed to be certain that it has all been recovered.*

8 Since no drain plug is fitted, it will be necessary to syphon the old oil out via the filler/level plug.
9 Take care not to overfill the gearbox with oil, as this may itself cause gearshift problems.

12 Automatic transmission

Fluid level checking - all models

1 The procedure given in Chapter 6, Section 6, still applies to later models, but the dipstick markings are slightly different (Fig. 13.46).
2 On all models, the quantity of ATF required to raise the level from the low to the high mark on the dipstick is roughly half a pint (0.3 litre).
3 Frequent need for topping up suggests a leak which requires investigation.

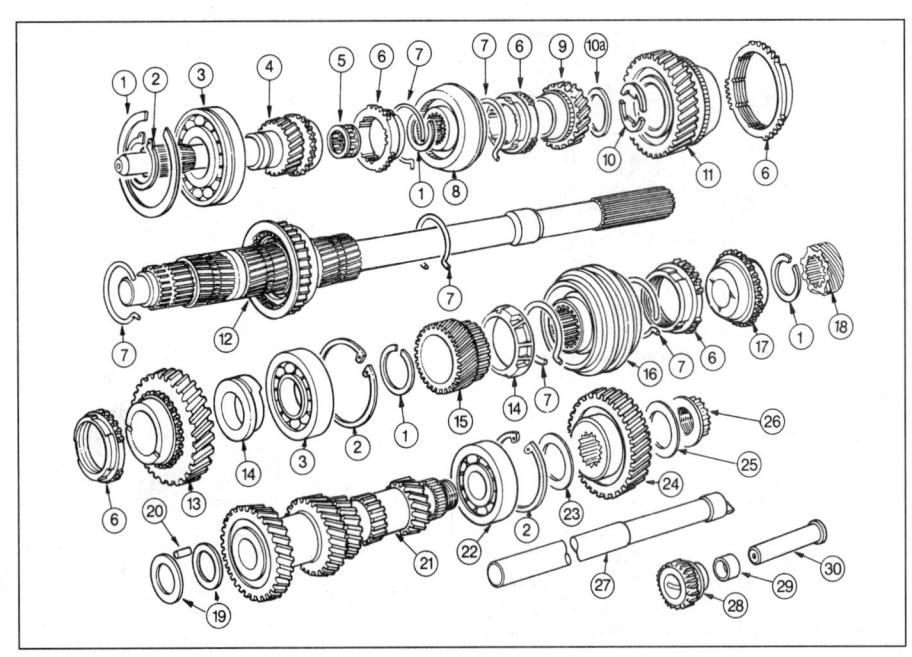

Fig. 13.45 Five-speed gearbox internal components (Sec 11)

1 Circlips	10 Thrust washers	15 5th gear (mainshaft)
2 Circlips	10a Thrust washer retaining ring	16 5th gear synchro unit
3 Ball bearing	11 2nd gear	17 5th gear synchro hub
4 Input shaft	12 Mainshaft and 1st/2nd synchro hub	18 Speedometer drivegear
5 Needle roller bearing	13 1st gear	19 Spacers
6 Synchro baulk rings	14 Oil scoop rings	20 Needle rollers
7 Synchro springs		21 Layshaft gear
8 3rd/4th synchro unit		22 Roller bearing
9 3rd gear		23 Spacer
		24 5th gear (layshaft)
		25 Washer
		26 Nut
		27 Layshaft
		28 Reverse idler gear
		29 Bush
		30 Idler gear shaft

Brake band adjustment - all models

4 Disconnect the downshift cable at the transmission.
5 Slacken the brake band adjuster screw locknut and unscrew the adjuster screw a few turns.
6 Tighten the adjuster screw to 10 lbf ft (14 Nm) using a suitable torque wrench. From this position slacken the adjuster screw by exactly 1½ turns (pre-1983 models) or 2 turns (1983 and later models), then tighten the locknut without disturbing the position of the adjuster screw.

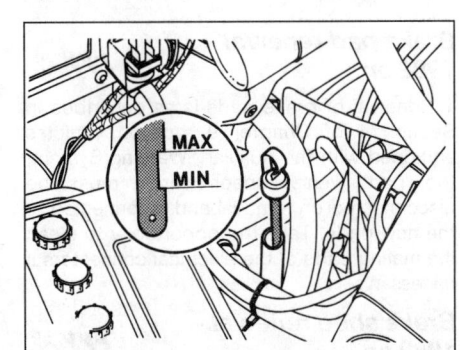

Fig. 13.46 Automatic transmission dipstick markings on later models (Sec 12)

7 Reconnect the downshift cable. Adjust if necessary as described in Chapter 6, Section 19.

13 Propeller shaft

Propeller shaft noise - 1982 models (except 5-speed)

1 If boom is experienced on manual transmission models at engine speeds of 3000 to 4000 rpm, this may be reduced by fitting a

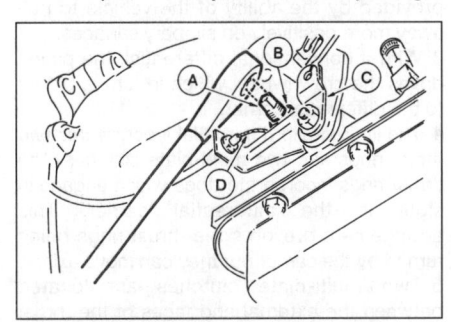

Fig. 13.47 Automatic transmission front brake band adjuster (Sec 12)

A Adjuster screw C Downshift lever
B Locknut D Downshift cable

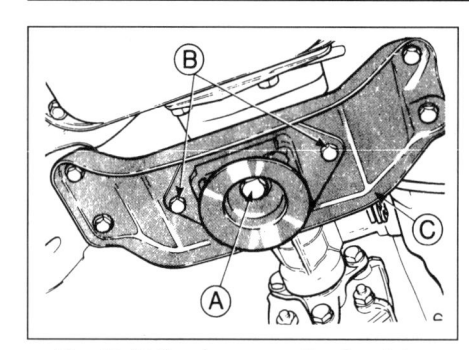

13•16 Supplement: Revisions and information on later models

Fig. 13.48 Gearbox mounting insulator (Sec 13)

A Centre bolt B Crossmember bolts
C Crossmember

propeller shaft having a rubber coupling (Guibo joint) at its forward end.
2 Vehicles equipped with Guibo jointed propeller shafts may produce a drumming noise from the driveline at speeds around 50 mph (80 km/h). This can be reduced by fitting a later pattern rear mounting insulator as follows.
3 Support the gearbox with a jack and a block of wood.
4 Remove and discard the insulator centre bolt. A new bolt must be used on reassembly.
5 Unbolt the insulator from the crossmember and discard it.
6 Fit the new insulator in the reverse sequence to removal. Tighten the bolts securing the insulator to the mounting before inserting and tightening the new centre bolt. Refer to Chapter 6 Specifications for the correct tightening torques.

14 Rear axle

Limited-slip differential

1 This is fitted as an option on some models or may be substituted at any time for the standard differential.
2 The purpose of the limited-slip differential is to transmit greater driving force to the inside roadwheel when driving round bends. This stabilises steering, reduces the risk of skidding and counteracts oversteer. A further benefit is provided by the ability of the vehicle to pull away more positively on slippery surfaces.
3 With a conventional differential, the pinion drives the crownwheel which in turn is bolted to the differential carrier.
4 The limited-slip differential incorporates two thrust rings and two multi-plate clutches. The thrust rings incorporate dogs which engage in slots in the differential carrier. This arrangement prevents the thrust rings being turned by the carrier but they can move axially.
5 Two multi-plate clutches are located between the external end faces of the thrust rings and the corresponding faces of the differential carrier. The clutches comprise inner and outer plates, the inner plate is located on the axle pinions while the dogs on the outer

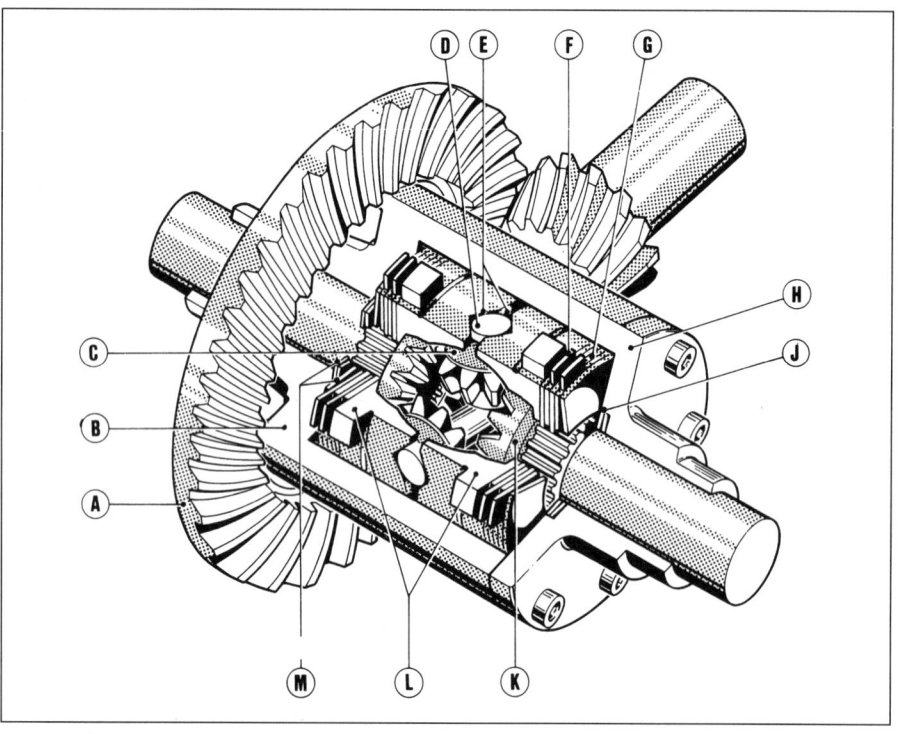

Fig. 13.49 Cutaway view of limited-slip differential (Sec 14)

A Crown wheel   E Thrust ring tapered   G Inner plate (on axle   K Axle pinion
B Differential carrier   faces   pinion)   L Thrust rings
C Pinion gear   F Outer plate (on   H End cover   M Spring washers
D Shaft   differential carrier) J Thrust washer

plates engage in the slots of the carrier.
6 Where a limited-slip differential is fitted as a replacement for a standard unit, note that the original half-shaft flange circlips must be changed for special ones. These are available in thicknesses between 1.21 and 2.05 mm.
7 The limited-slip type differential must never be dismantled, but only renewed as a unit.
8 Never balance wheels whilst they are on the car if a limited-slip differential is fitted.
9 For topping up purposes always use the specified oil for rear axles (see *Recommended lubricants and fluids* on page 11 of this Manual).

15 Braking system

Brake pad renewal - 1982 on

1 Renewal of brake pads is as described in Section 4 of Chapter 9, but on vehicles equipped with the Auxiliary Warning System, the pad wear sensor wires must be disconnected on removal and reconnected to the new pads. Take the opportunity to clean the mating parts of the electrical connectors if necessary.

Brake shoe renewal - 1982 on

2 Renewal of the rear brake shoes is essentially as described in Section 8 of Chapter 9, but the relationship of the handbrake relay lever and the self-adjusting lever is slightly different. Refer to Fig. 13.50 for details.

Handbrake adjustment - later models

3 An alternative type of handbrake cable adjuster to that described in Chapter 9 may be encountered. See Fig. 13.51. Adjustment is carried out as followed.
4 Chock the front wheels and raise and securely support the rear of the vehicle until the rear wheels are clear of the ground. Place the transmission in neutral and release the handbrake.
5 Release the locknut and adjuster sleeve nut by unscrewing one from the other. Do not attempt to prise them apart: damage will result as they are both threaded.
6 Apply the footbrake firmly several times in order to operate the self-adjusting mechanism.
7 Adjust the threaded sleeve until the total movement of both adjuster plungers added together is between 0.02 and 0.08 in (0.5 and 2.0 mm). Refer to Chapter 9, Fig. 9.18 for the location of the adjuster plungers. Grip the plungers between finger and thumb and move them in and out.
8 Tighten the locknut against the adjuster nut as tightly as possible by hand, then tighten a further two clicks with a suitable tool.
9 Apply the handbrake and make sure that the wheels lock, then release it and check that the

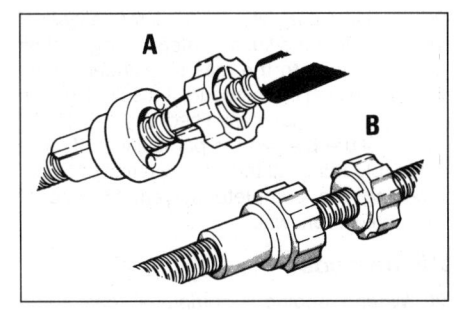

Supplement: Revisions and information on later models 13•17

Fig. 13.51 Earlier (A) and later (B) type handbrake cable adjuster (Sec 15)

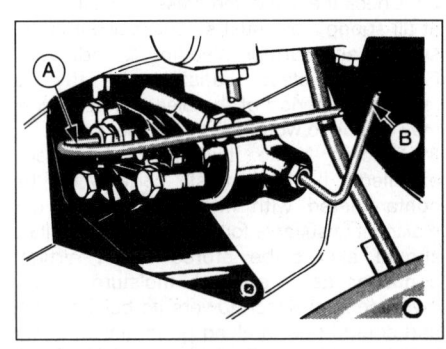

Fig. 13.52 Later type brake pressure control valve (Sec 15)
A Outlet B Inlet

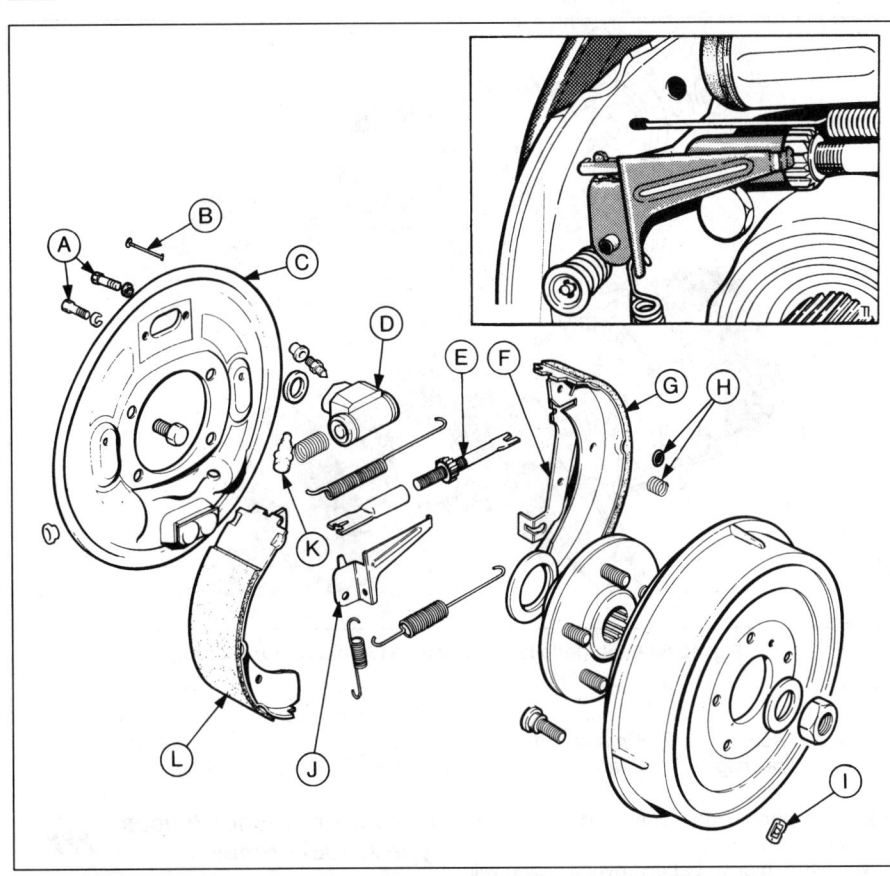

Fig. 13.50 Later type drum brake components with self-adjusting mechanism (inset) (Sec 15)

A Wheel cylinder bolts
B Shoe hold-down pin
C Backplate
D Wheel cylinder
E Self-adjusting strut
F Handbrake relay lever
G Trailing (rear shoe)
H Hold-down spring and clip
I Drum clip
J Self-adjusting lever
K Adjustment plunger
L Leading (front) shoe

wheels are free to rotate. Lower the vehicle to the ground when adjustment is complete.

Brake pressure control valve - all models

10 A revised pattern of brake pressure control valve has been introduced on later models. The old type of valve is no longer available, so if it is necessary to renew an old type valve, the new type will have to be used.
11 On RHD vehicles the outlet port on the new valve has been increased in diameter from 10 mm to 12 mm. On LHD vehicles the inlet port diameter too has been increased.
12 Make sure that the appropriate unions are obtained with a replacement valve, and observe the correct fitting of inlet and outlet connections (Fig. 13.52).

Additional bleeding methods

13 If the master cylinder or the pressure regulating valve has been disconnected and reconnected then the complete system (both circuits) must be bled.
14 If a component of one circuit has been disturbed then only that particular circuit need be bled.
15 Unless the pressure bleeding method is being used, do not forget to keep the fluid level in the master cylinder reservoir topped up to prevent air from being drawn into the system which would make any work done worthless.
Bleed the brakes in the following sequence:
Right-hand front
Left-hand front
Right-hand rear
Left-hand rear
16 Before commencing operations, check that all system hoses and pipes are in good condition with all unions tight and free from leaks.
17 Take great care not to allow hydraulic fluid to come into contact with the vehicle paintwork as it is an effective paint stripper. Wash off any spilled fluid immediately with cold water
18 As the system incorporates a vacuum servo, destroy the vacuum by giving several applications of the brake pedal in quick succession.

Bleeding - using one way valve kits

19 There are a number of one-man, one-way brake bleeding kits available from motor accessory shops. It is recommended that one of these kits is used wherever possible as it will greatly simplify the bleeding operation and also reduce the risk of air or fluid being drawn back into the system quite apart from being able to do the work without the help of an assistant.
20 To use the kit, connect the tube to the bleedscrew and open the screw one half a turn.
21 Depress the brake pedal fully and slowly release it. The one-way valve in the kit will prevent expelled air from returning at the end of each pedal downstroke. Repeat this operation several times to be sure of ejecting all air from the system. Some kits include a translucent container which can be positioned so that the air bubbles can actually be seen being ejected from the system.
22 Tighten the bleed screw, remove the tube and repeat the operations on the remaining brakes.
23 On completion, depress the brake pedal. If it still feels spongy repeat the bleeding operations as air must still be trapped in the system.

Bleeding - using a pressure bleeding kit

24 These kits too are available from motor accessory shops and are usually operated by air pressure from the spare tyre.
25 By connecting a pressurised container to the master cylinder fluid reservoir, bleeding is then carried out by simply opening each bleed screw in turn and allowing the fluid to run out, rather like turning on a tap, until no air is visible in the expelled fluid.

13•18 Supplement: Revisions and information on later models

26 By using this method, the large reserve of hydraulic fluid provides a safeguard against air being drawn into the master cylinder during bleeding which often occurs if the fluid level in the reservoir is not maintained.

27 Pressure bleeding is particularly effective when bleeding "difficult" systems or when bleeding the complete system at time of routine fluid renewal.

All methods

28 When bleeding is completed, check and top up the fluid level in the master cylinder reservoir.

29 Check the feel of the brake pedal. If it feels at all spongy, air must still be present in the system and further bleeding is indicated. Failure to bleed satisfactorily after a reasonable period of the bleeding operation, may be due to worn master cylinder seals.

30 Discard brake fluid which has been expelled. It is almost certain to be contaminated with moisture, air and dirt making it unsuitable for further use. Clean fluid should always be stored in an airtight container as it absorbs moisture readily (hygroscopic) which lowers its boiling point and could affect braking performance under severe conditions.

16 Electrical system

Maintenance-free battery - description and precautions

1 Maintenance-free or low-maintenance batteries are standard equipment on the latest Granada models and may also be encountered as replacements for older types.

2 Maintenance-free batteries will not need to be topped up with distilled water during normal service. On some such batteries there is no provision for topping up, ie the battery is completely sealed apart from a vent hole. On other designs the cell covers can still be removed if it is wished to check the electrolyte level or specific gravity.

3 Low-maintenance batteries are designed to require topping up no more often than once a year. This should be done as described in Chapter 10, Section 3.

4 With either type of battery, overcharging will cause excessive gassing and possible electrolyte loss. This is particularly to be avoided with the completely sealed type of battery.

5 Do not allow unskilled operators to boost charge a maintenance-free battery. Observe any further precautions (eg with regard to jump starting) detailed on the battery itself

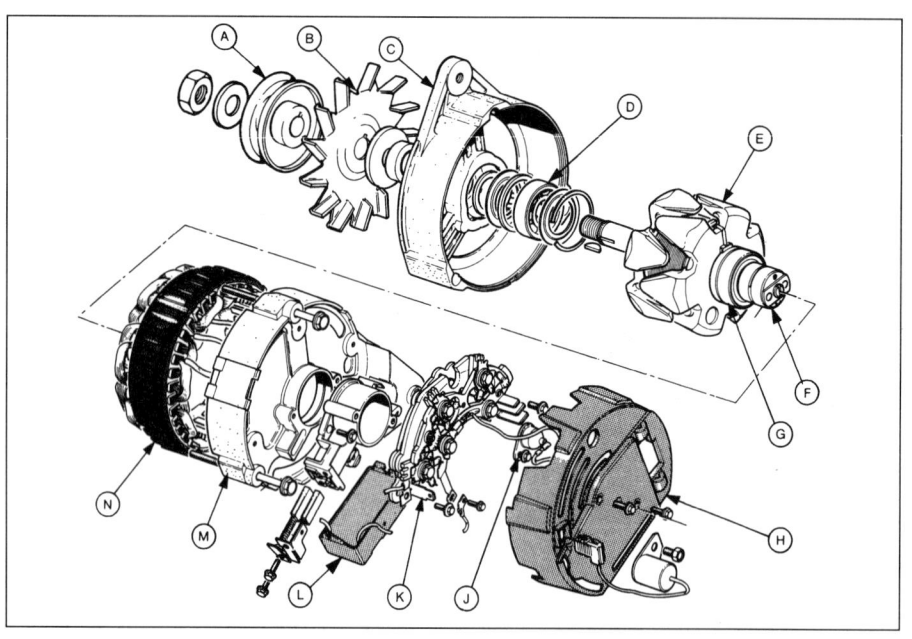

Fig. 13.53 Exploded view of Lucas A133 alternator (Sec 16)

A Pulley
B Fan
C Drive end housing
D Drive end bearing
E Rotor
F Slip rings
G Slip ring end bearing
H End cover
J Surge protection diode
K Diode pack
L Regulator
M Slip ring end housing
N Stator

Alternator brushes (Lucas type A 133) - renewal

6 If the alternator is still in the vehicle, disconnect the battery earth lead and remove the multi-plug from the rear of the alternator.

7 Undo the retaining screws and remove the rear cover from the alternator. Note the position of the radio interference suppressor, if fitted.

8 Undo the brush retaining screws and

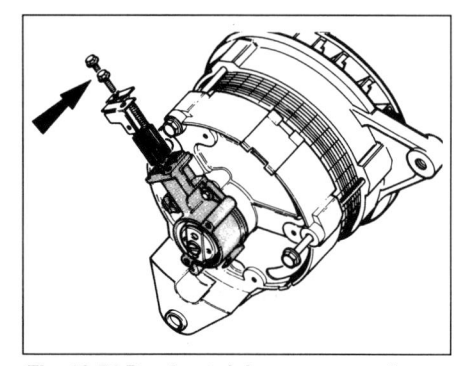

Fig. 13.54 Brush retaining screws on Lucas A133 alternator (Sec 16)

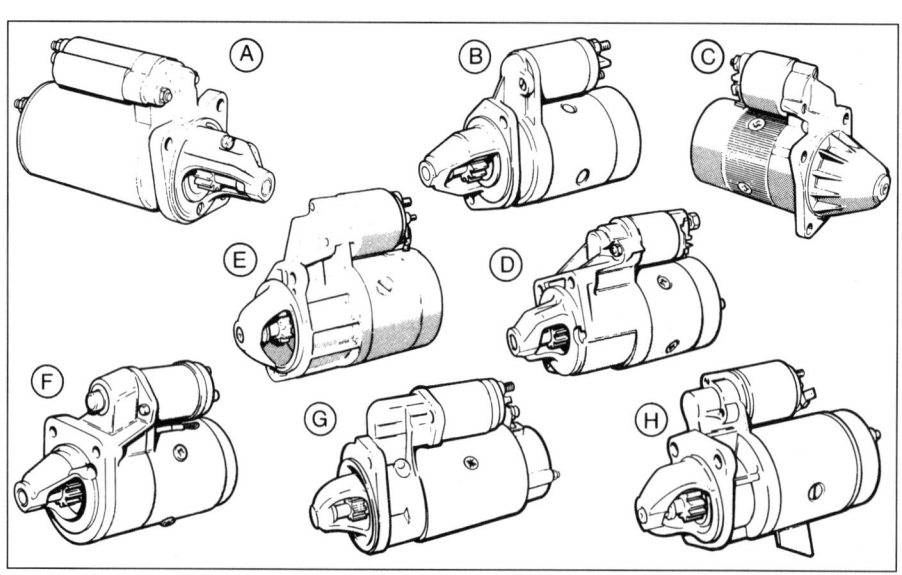

Fig. 13.55 Starter motor identification (Sec 16)

A Lucas M35J/5M90
B Bosch long frame
C Cajavec
D Nippondenso
E Bosch short frame
F Lucas 9M90
G Lucas 2M100
H Bosch JF 2kW

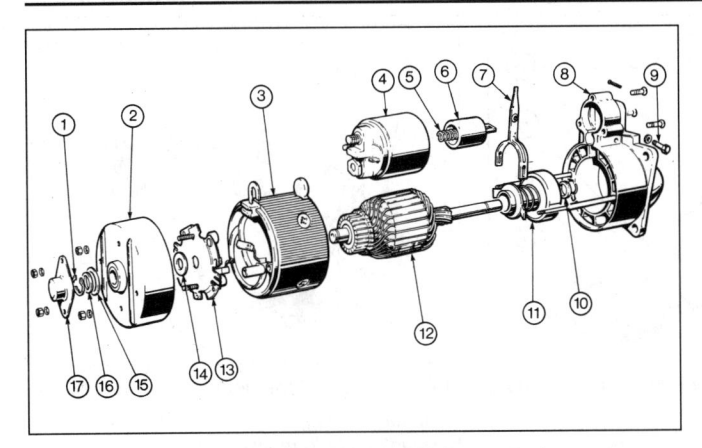

Supplement: Revisions and information on later models 13•19

Fig. 13.56 Exploded view of Cajavec starter motor (Sec 16)

1 E-clip	7 Actuating lever	12 Armature
2 Commutator end cover	8 Drive end housing	13 Brushplate
3 Main casing	9 Pivot pin	14 Brake ring
4 Solenoid body	10 Thrust collar	15 Sealing ring
5 Spring	11 Pinion and clutch assembly	16 Spacer
6 Armature		17 End cap

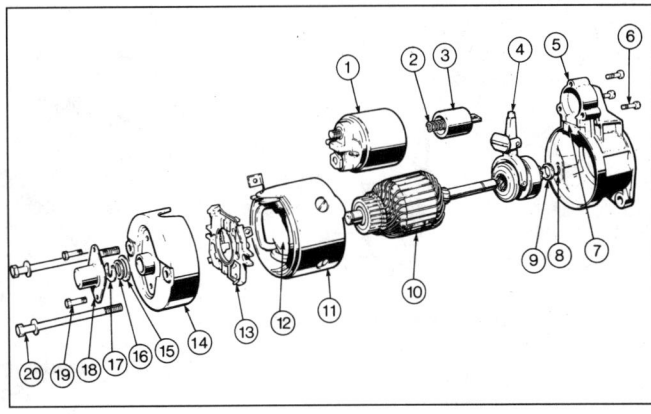

Fig. 13.57 Exploded view of Bosch short frame starter motor (Sec 16)

1 Solenoid	8 C-clip	15 Sealing ring
2 Spring	9 Thrust collar	16 Spacer
3 Armature	10 Armature	17 C-clip
4 Actuating lever	11 Main casing	18 End cover
5 Drive end housing	12 Pole shoe	19 End cover bolt
6 Solenoid bolts	13 Brushplate	20 Through-bolt
7 Rubber block	14 Commutator end housing	

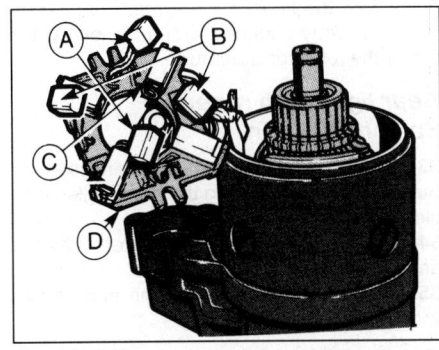

Fig. 13.58 Brush plate removed from Bosch short frame starter motor (Sec 16)

A Field brushes C Brush holders
B Terminal brushes D Brushplate

remove the retaining plates, brush springs and brushes from the brush box.

9 Make sure that the new brushes can slide freely in their holders. Remove the brush box if necessary to check this point.

10 Reassemble in the reverse order to dismantling.

Starter motor - alternative types

11 Various alternative types of starter motor have been added to those described in Chapter 10. Reference to Fig. 13.55 may help in identifying a particular motor.

12 Removal and refitting procedures are as described in Section 16 of Chapter 10.

13 Overhaul procedures are also as described in Chapter 10, with certain minor differences. The following points should be noted.

Lucas M35J and 9M90

14 Overhaul is as described in Chapter 10, Section 17, for the Lucas 5M90 starter motor.

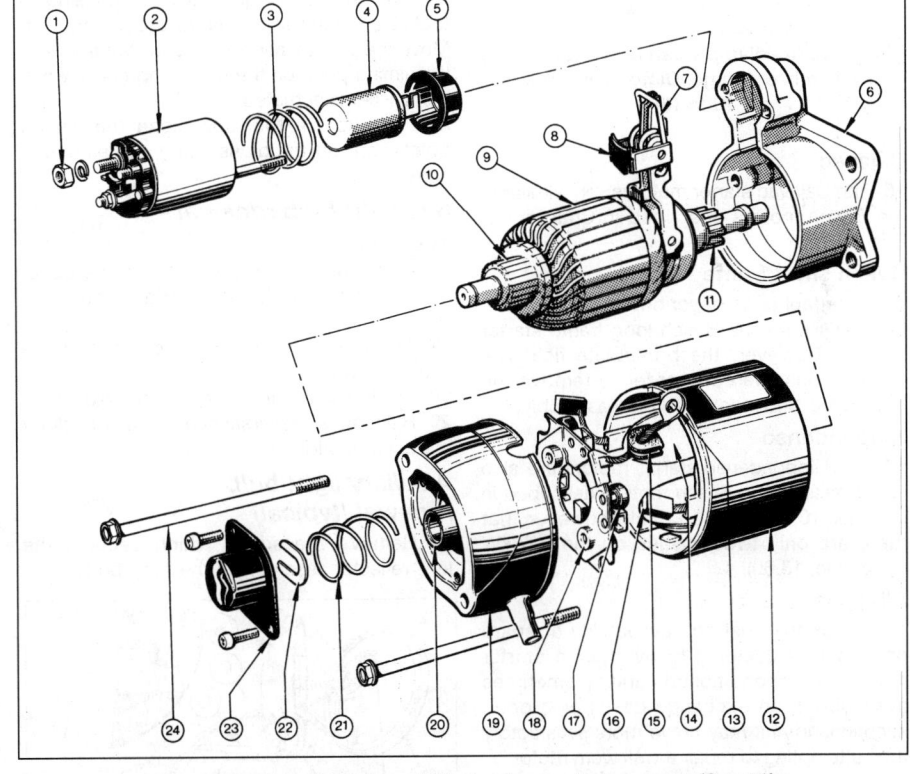

Fig. 13.59 Exploded view of Nippondenso starter motor (Sec 16)

1 Terminal nut	10 Commutator	18 Brushplate
2 Solenoid body	11 Pinion and clutch assembly	19 Commutator end housing
3 Spring	12 Main casing	20 Bush
4 Armature	13 Connecting link	21 Spring
5 Seal	14 Pole shoe	22 C-clip
6 Drive end housing	15 Seal	23 End cover
7 Actuating lever	16 Brush	24 Through-bolt
8 Pivot	17 Brush spring	
9 Armature		

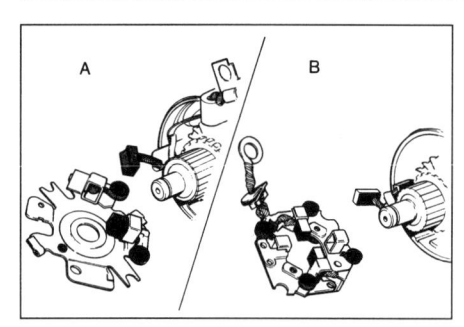

13•20 Supplement: Revisions and information on later models

Fig. 13.60 Nippondenso starter motor brushplate (Sec 16)

A 0.6 kW (two brushes)
B 0.8 kW (four brushes)

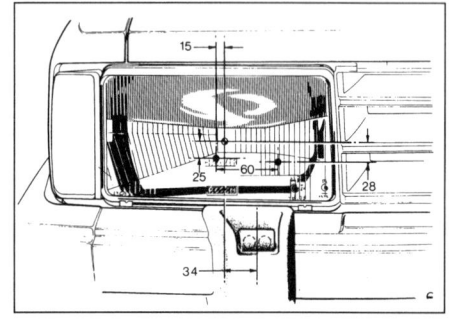

Fig. 13.62 Headlamp wash jet pattern (RHS) LHS is mirror image. Dimensions in mm (Sec 16)

Cajavec

15 The Cajavac starter motor is very similar to the Bosch motors described in Chapter 10, Section 19.

Bosch short frame

16 Overhaul is as described in Chapter 10, Section 19, for the Bosch long frame starter motors. However, the brushplate field link must be unsoldered in order to remove the brushplate, and resoldered on reassembly.

Nippondenso

17 The Nippondenso starter motors are also very similar to the Bosch motors described in Chapter 10, Section 19. Note however that there are only two brushes on the 0.6 kW motor (Fig. 13.60).

All types

18 Check the cost and availability of spare parts before deciding to overhaul a starter motor. A reconditioned unit (sometimes available on an exchange basis) or a good secondhand unit may prove more satisfactory than attempting to repair a well worn motor.

Headlight washer jets - removal, refitting and adjusting

19 Open the bonnet and remove the radiator grille.
20 Disconnect the hose from the jet union at the rear of the overrider by pulling it out.
21 Unscrew the two bolts from behind the bumper and remove the overrider complete

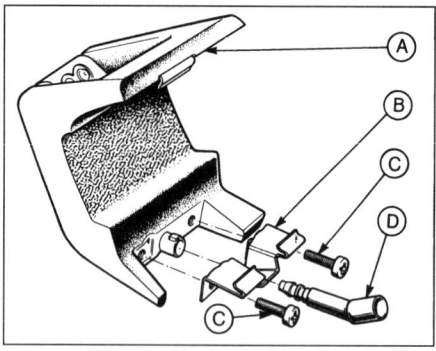

Fig. 13.61 Headlamp washer components (Sec 16)

A Override containing jet C Screws
B Clamp D Hose connector

with jet. The jet and overrider cannot be separated.
22 Refit the overrider and jet in the reverse order to removal, making sure that the hose is correctly routed through the bumper. Adjust the jet as follows.
23 Refer to Fig. 13.62 to determine the aiming point for the jet. Mark the aiming point with a soft crayon or a fragment of sticky tape.
24 Insert a suitable tool into the jet housing so that it engages in the slots around the jet. Move the tool as necessary until water strikes the aiming point on the headlamp glass, when the washer is actuated.
25 Remove the tool and clean the aiming point mark off the glass when adjustment is correct.

Sidelight bulb renewal - 1982 on

26 Open the bonnet and remove the cover from the back of the appropriate headlamp unit.
27 Pull the sidelight bulbholder out of the headlamp reflector.
28 Pull the sidelight bulb out of its holder.
29 Refitting is a reversal of removal. Check for correct operation on completion.

Auxiliary light bulb renewal (typical)

30 Remove the screw which secures the lens/reflector assembly to the lamp body.

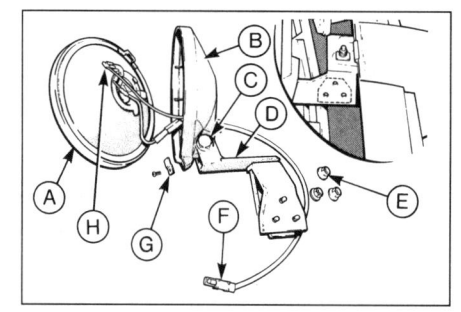

Fig. 13.64 Typical auxiliary lamp (Sec 16)

A Reflector E Nuts
B Body F Connector
C Nut G Lens clip
D Bracket H Earth connector

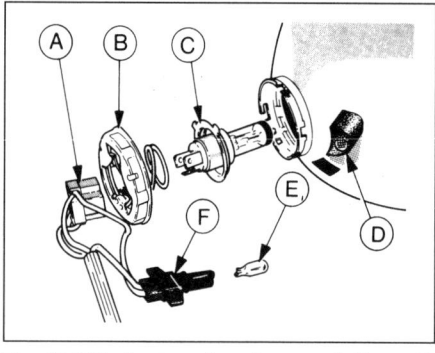

Fig. 13.63 Later type headlamp unit (Sec 16)

A Headlight plug
B Headlight bulb retainer
C Headlight bulb
D Sidelight bulb retainer
E Sidelight bulb
F Sidelight bulbholder

31 Pull the reflector away from the lamp body far enough to gain access to the bulbholder. Release the spring clip and withdraw the bulb. Disconnect the electrical lead.
32 Fit the new bulb in the reverse order to removal, taking care not to touch the bulb glass with the fingers. Note that the plug on the lamp body engages with a cut-out at the top of the reflector assembly.

Rear light bulb renewal - Saloon, 1982 on

33 Working from inside the boot, release the bulbholder unit by turning the two toggles anti-clockwise.
34 Remove the failed bulb from the holder and fit a new one.
35 Offer up the bulbholder unit and secure with the toggles.

Rear light cluster unit removal and refitting - Saloon, 1982 on

36 Release the bulbholder as described above and disconnect its multi-plug.
37 Remove the four nuts from inside the boot to free the outer part of the light cluster.
38 Take off the large rubber gasket if it is wished to separate the lens from the reflector. The two are held together by six screws.

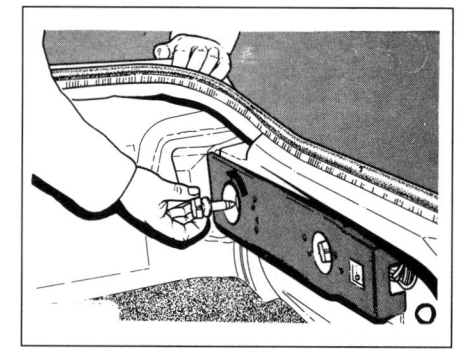

Fig. 13.65 Turning rear lamp bulbholder toggle (Sec 16)

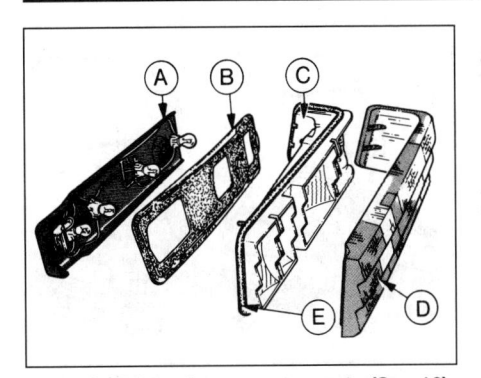

Supplement: Revisions and information on later models

Fig. 13.67 Rear lamp components (Sec 16)

A Bulbholder B Gasket C Reflector
D Lens E Gasket

39 The direction indicator reflector, secured by a further two screws, can be renewed independently of the main reflector.
40 Reassemble in the reverse order to dismantling. Do not overtighten the lens securing screws, or the threads in the lens may be stripped. The multi-plug can only be connected one way round.
41 Refit the light cluster in the reverse order to removal.

Rear foglamp removal and refitting - Estate models

42 Disconnect the wiring plug from the lamp.
43 Remove the two securing screws to free the lamp.
44 Remove the two lens securing screws to gain access to the bulb. (This can be done without removing the lamp from the car.)
45 Refitting is a reversal of removal.

Number plate light bulb renewal - 1982 on

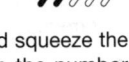

46 Reach under the bumper and squeeze the spring clips together to release the number plate light assembly. Push it upwards and unplug its connector.
47 Twist the bulbholder anti-clockwise and pull it to separate it from the lens and cover.
48 Pull the bulb from the holder and push in a new one.
49 Reassemble the unit, plug in its connector and gently press it back into its location.

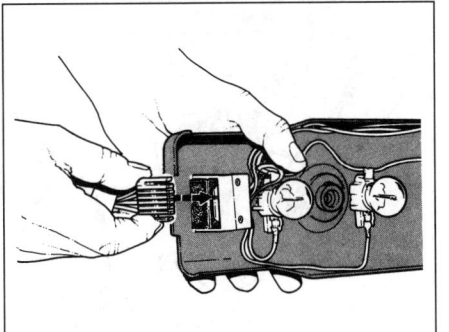

Fig. 13.68 Alignment ridge for rear lamp multi-plug (Sec 16)

Interior light bulb renewal - 1982 on

Rear interior light

50 Release the light unit by turning it anti-clockwise. If the unit is reluctant to come out, lever it free with a flat-bladed knife.
51 Pull the bulbholder from the back of the light unit and renew the bulb. Refit the bulbholder.
52 Press the light unit back into position.

Map reading light

53 Make sure that the light switch is off, then carefully prise the light from the roof console with a screwdriver.
54 Pull out the bulbholder and renew the bulb. Refit the bulbholder.
55 Press the light unit back into position.

Ashtray light

56 Open the ashtray and remove the drawer.
57 Pull the bulbholder out of its housing and renew the bulb.
58 Refit the bulbholder into its housing and reassemble the ashtray.

Glove compartment light

59 Open the ashtray and remove the drawer.
60 Pull the bulb out of the holder and push in the new one.
61 Refit the bulbholder by pushing it into its bracket and turning it clockwise.

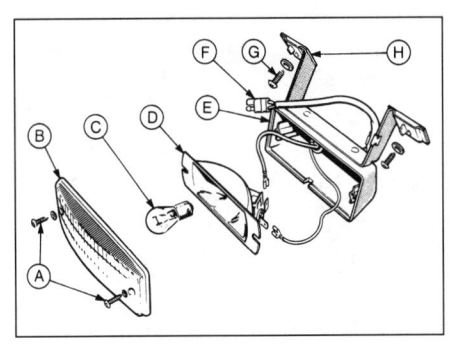

Fig. 13.69 Rear fog lamp components (Estate) (Sec 16)

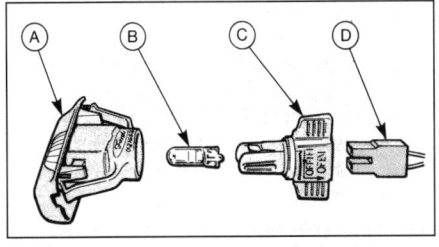

A Lens retaining screws   E Body
B Lens   F Connector
C Bulb   G Screw
D Reflector   H Bracket

Fig. 13.70 Later type rear number plate lamp (Sec 16)

A Lens/body   C Bulbholder
B Bulb   D Connecting plug

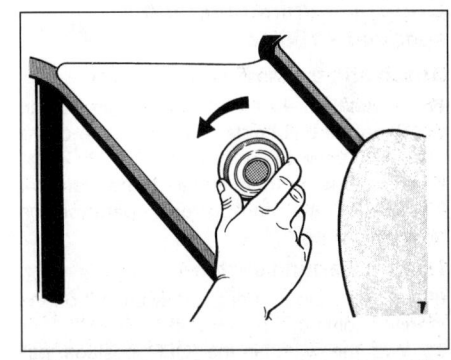

Fig. 13.71 removing interior lamp (Sec 16)

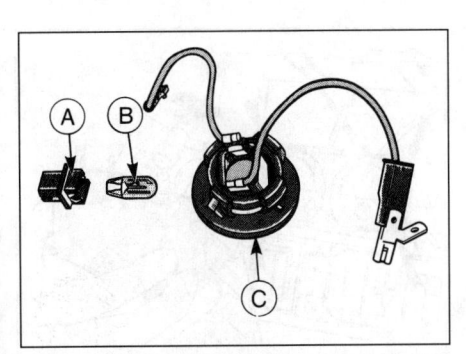

Fig. 13.72 Map reading lamp components (Sec 16)

A Bulbholder B Bulb C Body

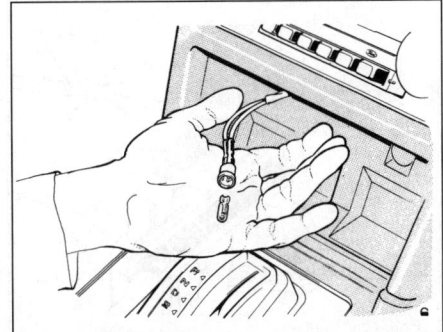

Fig. 13.73 Ashtray bulb and holder (Sec 16)

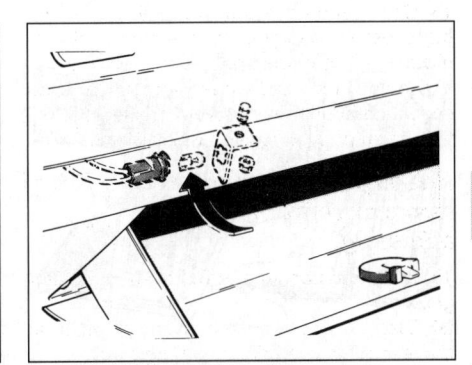

Fig. 13.74 Glove compartment lamp (Sec 16)

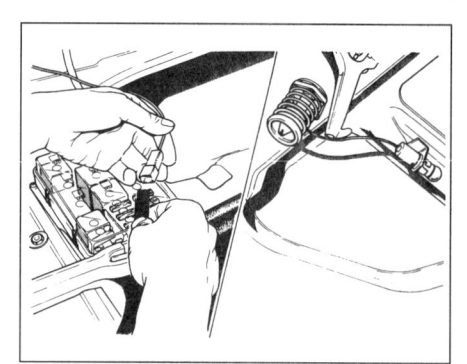

13•22 Supplement: Revisions and information on later models

Fig. 13.75 Engine compartment lamp and connector (Sec 16)

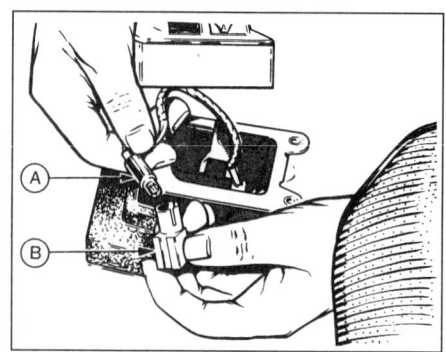

Fig. 13.76 Automatic transmission selector index plate illuminating bulb and holder (A) and bulb housing (B) (Sec 16)

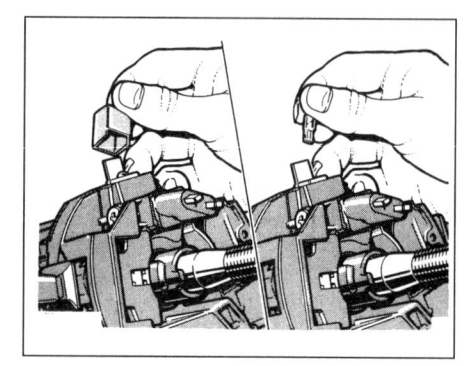

Fig. 13.77 Hazard warning switch lamp lens and bulb (Sec 16)

Engine compartment light

62 The engine compartment light bulb is a simple bayonet fit. Should it be necessary to renew the light unit itself, disconnect its electrical lead at the bonnet hinge end and tie a piece of string to the lead. Draw the lead and string through the bonnet panel before removing the light unit so that the string can be used to draw the new lead through the panel.

Vanity mirror light

63 Lower the sun visor and carefully prise off the mirror and diffuser.
64 Extract the festoon bulb(s) from the spring contacts and press in the new bulb(s).
65 Press the mirror back into position.

Control illumination bulb renewal - 1982 on

Automatic transmission selector

66 Unclip the selector housing and lift it clear.
67 Slide the bulbholder from the base of the lever and renew the bulb.
68 Refit the bulbholder and the selector housing. Check for correct operation on completion.

Hazard warning switch

69 Remove the securing screw and lift off the steering column upper shroud.
70 With the switch in the "OFF" position, pull the cap off the switch to expose the bulb.

71 Pull out the old bulb and press in a new one.
72 Refit the switch cap and the steering column shroud.

Heater controls

73 The heater fan switch bulb is simply renewed after pulling off the switch knob. Twist and pull the bulb to remove it.
74 The heater slide control bulbs are only accessible after the facia trim panel has been removed (see Section 18, this Supplement).
75 Pull the control knobs from their levers and carefully prise off the bulb cover to expose the bulb.
76 Renew the bulb and refit the control and trim components.

Warning light bulb renewal - 1982 on

77 Warning lights located in the main instrument cluster can only be renewed after removing the instrument cluster.
78 The oil pressure warning light is accessible after the facia/trim panel has been removed (see Section 18, this Supplement).
79 To renew an auxiliary warning system light bulb, carefully prise the warning light panel out of the overhead console.
80 Unplug the panel connector. Twist the appropriate bulbholder to remove it from the panel.
81 Fit the new bulb, reconnect the panel

multi-plug and carefully press the panel back into position.

Oil pressure gauge and/or ammeter - removal and refitting (1982 on)

82 Remove the facia trim panel as described in Section 18, this Supplement.
83 Remove the four screws which secure the gauge assembly and pull the gauges forwards slightly.
84 Unscrew the pipe union from the oil pressure gauge, disconnect the wires from the ammeter and extract the bulbholder.
85 The oil pressure gauge can be removed from the housing after undoing the two securing screws. The ammeter can be removed after undoing the terminal stud nuts and washers.
86 Refitting is a reversal of removal. The ammeter terminals are different sizes so it should not be possible to connect it incorrectly.

Digital clock - removal and refitting

87 Carefully prise the clock out of the centre console.
88 Disconnect the multi-plug and remove the clock.
89 Refitting is a reversal of removal. Rest the clock on completion.

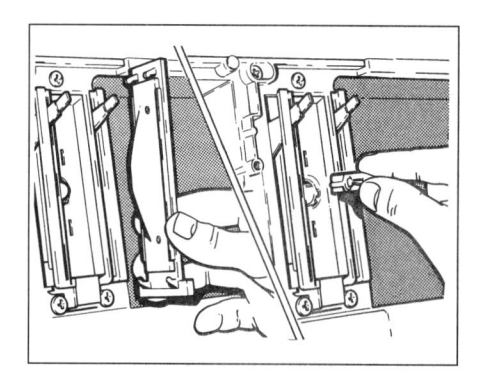

Fig. 13.78 Access to heater control illumination bulbs (Sec 16)

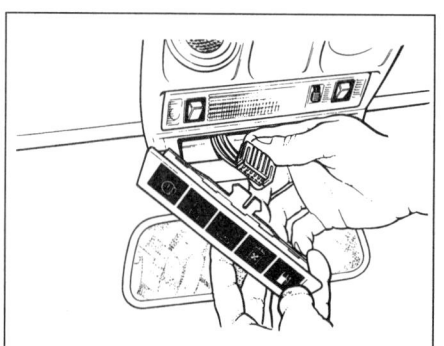

Fig. 13.79 Removing auxiliary warning lamp panel (Sec 16)

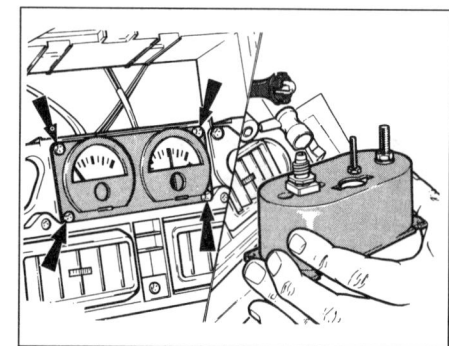

Fig. 13.80 Oil pressure gauge and ammeter fixing screws and rear connections (Sec 16)

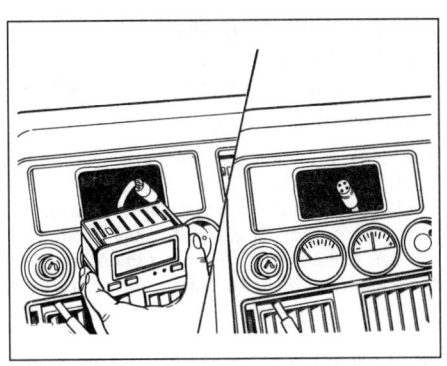

Supplement: Revisions and information on later models 13•23

Fig. 13.81 Removing digital clock (Sec 16)

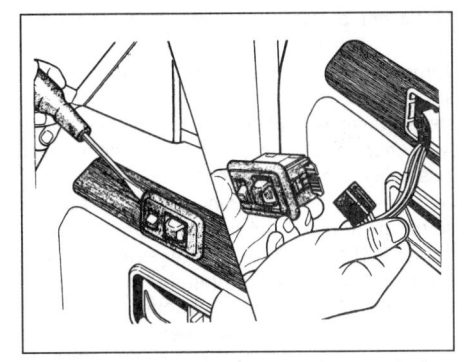

Fig. 13.82 Removing door mirror switch (Sec 16)

Fig. 13.83 Seat control (movement) wiring harness and plug (Sec 16)

Minor switches - removal and refitting (1982 on)

Door mirror control switch

90 Carefully prise the switch out of its location in the door. Disconnect the multi-plug and remove the switch.
91 Refit by reconnecting the multi-plug and pressing the switch into position. Check for correct operation on completion.

Seat movement switches

92 The seat tilt switch is simply removed by prising it out and unplugging it.
93 To remove the multi-function seat control switch, first remove the two bolts from the front of the seat frame and tip the seat backwards.
94 Remove the seat side trim and the switch housing, and unclip the wiring loom from the seat. Disconnect the multi-plugs.
95 Free the switch from its housing and withdraw the switch and loom together.
96 Refitting is a reversal of removal.

Heater fan switch

97 Remove the facia trim panel as described in Section 18, this Supplement.
98 Remove the facia bracket screws, squeeze the lugs on the switch together and withdraw the switch. Disconnect the multi-plug.
99 Refitting is a reversal of removal.

Boot light switch (Saloon)

100 Open the boot and disconnect the wire from the switch.
101 Remove the screw to free the switch.
102 Align the switch when refitting so that it is operated by the opening and closing of the boot lid.

Load compartment light switch (Estate)

103 Open the tailgate and remove its trim panel.
104 Disconnect the wire from the switch. Note the angle at which the switch is fitted, then unscrew and remove it.
105 Refit the switch at approximately the same angle and check for correct operation before refitting the trim panel.

Seat operating motor - removal and refitting

106 Remove the two bolts from the front of the seat and tip the seat backwards.
107 Remove the motor mounting bolts (two or three) and disconnect the multi-plug. Detach the bracket from the tilt or rake motor.
108 Withdraw the motor, making sure that the drive cables are not trapped in the worm drives.
109 Refitting is a reversal of removal. Check for correct operation of the motor on completion.

Sliding roof motor - removal and refitting

110 Disconnect the battery earth lead.
111 Open the boot and remove the spare wheel, the jack and the wheelbrace.
112 Open the sliding roof by turning the white hexagonal drive nut (the wheelbrace fits this nut).
113 Remove the sliding roof guides and plates. Pull the roof panel forwards far enough to expose the cable and bracket. Remove the two screws which secure the end bracket to the roof panel.
114 Grip the cable end bracket and pull the cable out of the guide tube.
115 Remove the two nuts which hold the motor mounting plate to the body, and the three bolts which hold the motor to the mounting plate. Disconnect the multi-plug.

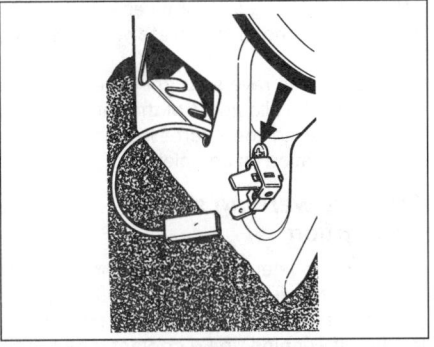

Fig. 13.84 Luggage boot lamp switch. Fixing screw (arrowed) (Sec 16)

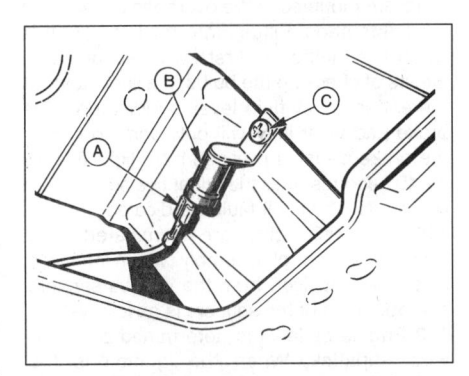

Fig. 13.85 Estate load compartment lamp switch (Sec 16)

A Connector B Switch C Fixing screw

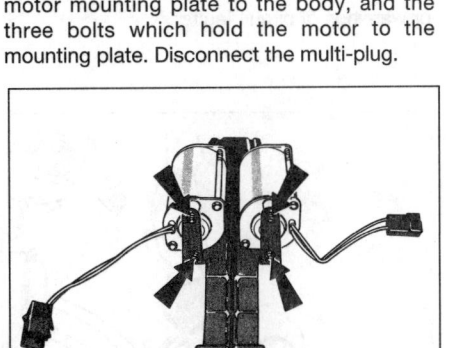

Fig. 13.86 Seat rake and tilt electric motors (Sec 16)

Mounting bolts (arrowed)

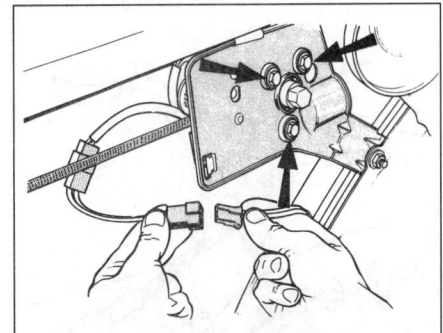

Fig. 13.87 Sliding roof motor wiring plug (Sec 16)

Motor to mounting plate bolts (arrowed)

13

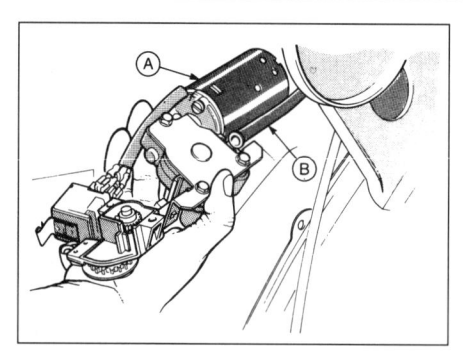

13•24 Supplement: Revisions and information on later models

Fig. 13.88 Removing sliding roof motor (Sec 16)

A Motor   B Guide tube

116 Remove the guide tube securing clip and withdraw the motor complete with relay and microswitch.

117 Refit in the reverse order to removal. **Do not** disturb the alignment of the cam and gear drive relative to the microswitch. Check for correct operation on completion.

Auxiliary warning system - description

118 Fitted to later Granada models with high level trim, the auxiliary warning system (AWS) provides the driver with information on the levels of engine oil, coolant, fuel and windscreen washer fluid; warning of front brake pad wear is also provided. The AWS lights are mounted in the overhead console. All the lights should illuminate for five seconds when the ignition is first switched on, as a means of checking the bulbs. If a light stays on thereafter, a low fluid level is indicated. (The brake pad wear light will only come on when the brake pedal is pressed.) If a light flashes for 40 seconds after the initial five second test illumination, a circuit fault is indicated.

119 The AWS lights are illuminated by a "black box" control assembly, which interprets messages received from the various sensors. The operation of the sensors is as follows.

120 Engine oil level is determined by a "hot wire" dipstick. When the ignition is first switched on, a small electric current is passed through a wire in the tip of the dipstick. The resistance of this wire varies with temperature. If the dipstick is immersed in oil, the temperature change caused by the electric current will be small. If the oil level is low, the sensing element will heat up and its resistance will change, thus signalling to the control assembly that the oil level is low.

121 The oil level warning only operates at initial switch-on, since once the engine is running the oil level in the sump will fall considerably. It will not operate again until the ignition has been switched off for approximately three minutes, this being the time necessary for the oil in circulation to drain back into the sump.

122 False readings of low oil level are possible if the dipstick is incorrectly inserted, or if the car is parked on a steep slope.

123 The low fuel level warning light operates when the fuel remaining in the tank is approximately 1½ gallons (7 litres) or less. The sensor is operated by the fuel gauge sender unit float.

124 Coolant level and washer fluid level are both monitored by reed switches, which are activated by magnets mounted on floats in the respective reservoirs.

125 Low fluid level warnings (including fuel and coolant) are ignored by the control assembly until they have persisted for 6 to 10 seconds. In this way spurious warnings due to fluid movement when cornering etc are avoided.

126 Brake pad wear is registered by means of a wire buried in the lining of one brake pad on each side of the car. When the lining is worn down far enough, the wire contacts the brake disc and completes the warning light circuit.

127 No repairs are possible to the AWS components. Testing of the control assembly and sensors is by substitution of known good units. Before doing this, however, check the security and condition of all electrical connectors in the system, especially the control assembly multi-plug. The last item should be cleaned carefully with a typewriter eraser or a folded sheet of clean writing paper if troubles are experienced with false readings or warnings of circuit faults.

Auxiliary warning system components - removal and refitting

Coolant level switch

128 Drain the cooling system expansion tank, taking precautions against scalding if the coolant is hot. Save the coolant for re-use.

129 Disconnect the multi-plug from the switch. Unscrew the switch retainer and pull the switch out of its sealing grommet. If a non-pressurised overflow tank is fitted, the switch is simply prised out.

130 Refit in the reverse order to removal and refill the expansion tank.

Washer fluid level switch

131 Drain or syphon out the contents of the washer fluid reservoir.

132 Disconnect the switch multi-plug and carefully prise the switch out of its location in the reservoir.

133 Push the new switch into position, making sure that its sealing grommet is correctly located. Reconnect the multi-plug and refill the reservoir.

Control assembly

134 Open the glovebox lid, reach up inside the glovebox and disconnect the multi-plug.

135 Undo the four plastic securing nuts and remove the control assembly.

136 Refit in the reverse order to removal. The unit is fitted with the multi-plug towards the front of the vehicle. Check for correct operation on completion.

Trip computer - description

137 Fitted as standard equipment to top line Granada models, the trip computer incorporates a clock with alarm facility and monitors mileage, speed, fuel consumption and fuel level. Warning is automatically given of low fuel level (in terms of distance remaining before the tank is empty!) and another alarm can be programmed to warn the driver if a selected speed is being exceeded. Integration of the clock and stopwatch with the other functions enables the display of information relating to average speed and journey time.

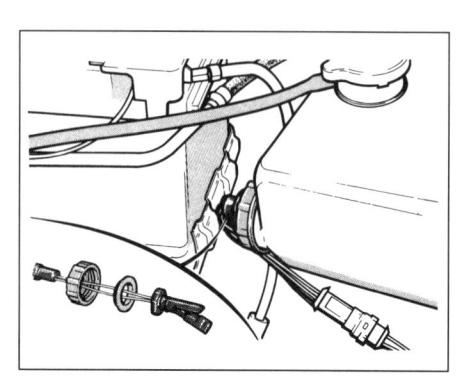

Fig. 13.89 Low coolant level warning reed switch (Sec 16)

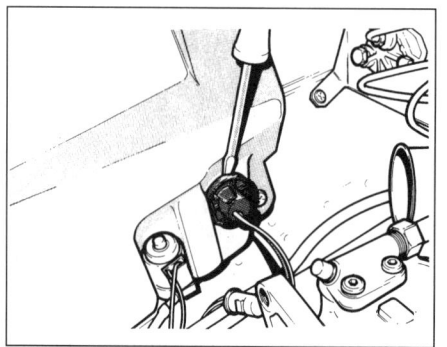

Fig. 13.90 Prising out the washer fluid level warning switch (Sec 16)

Fig. 13.91 Removing the auxiliary warning system control unit (Sec 16)

Supplement: Revisions and information on later models 13•25

Fig. 13.92 Removing the trip computer module (Sec 16)

Fixing screws (arrowed)

Fig. 13.93 Trip computer speed sender unit (Sec 16)

A Fixing nuts   B Multi-plug connector

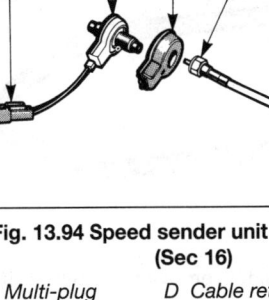

Fig. 13.94 Speed sender unit components (Sec 16)

A Multi-plug   D Cable retaining nut
B Sender unit   E Speedometer drive clip
C Cover

138 Use of the various functions of the trip computer is fully described in the owner's handbook supplied with the car. Copies of this publication should be available from your Ford dealer. The information given here is confined to removal and refitting of the computer module and its associated sender units. Testing of the computer should be entrusted to a Ford dealer.

Trip computer components - removal and refitting

Computer module

139 Disconnect the battery earth lead.
140 Remove the steering column upper and lower shrouds and the facia panel.
141 Undo the two bracket retaining screws and carefully withdraw the module from its location. Disconnect the three multi-plugs.
142 If a new module is being fitted, transfer the mounting brackets from the old module. Note that the left-hand and right-hand brackets are not interchangeable
143 Refitting is a reversal of removal.

Speed sender unit

144 Disconnect the battery earth lead.
145 Remove the under-dash cover. It is secured by three screws or clips.
146 Disconnect the sender unit multi-plug,

undo the two securing nuts and remove the sender unit. Note which way round the unit is fitted. The shouldered side with a brass collar (visible when the cover is removed) faces towards the speedometer.
147 Refitting is a reversal of removal.

Fuel flow sensor (fuel injection models)

148 The fuel sender unit is located in the fuel line serving No 2 cylinder injector.
149 Disconnect the battery earth lead.
150 Disconnect the sensor multi-plug from the wiring loom.
151 Remove the banjo connector bolts from each end of the sensor unit. Be prepared for some fuel spillage. Retrieve the washers.
152 Remove the sensor bracket securing screws and withdraw the sensor complete with bracket. Transfer the bracket to the new sensor if applicable.
153 Refit in the reverse order to removal, noting that the line from the fuel distributor goes to the "IN" connection and the line to No 2 injector goes to the "OUT" connection. Tighten the banjo bolts to the specified torque (see *Fuel Injection Specifications* at the beginning of this Chapter).

Fuel flow sensor (carburettor models)

154 Disconnect the battery earth lead.
155 Remove the plug and three fuel lines from the sender unit. Be prepared for some fuel

spillage. If the fuel lines are damaged by removal of the clips, they must be renewed.
156 Disconnect the multi-plug and remove the bracket securing screws. Remove the sensor complete with bracket. Transfer the bracket to the new sensor if applicable.
157 Refit in the reverse order to removal. Note that there are three arrows on the sensor: the arrows next to the unions indicate the direction of fuel flow, and the arrow on the side face indicates which way up the unit is to be mounted.

Fuel tank sender unit

158 The procedure is as described for the fuel gauge sender unit in Chapter 3, but there are considerably more electrical connections to consider. Take careful note of all connections before removing the sender unit.

Radio removal - 1984 models

159 Special tools are required for removing radio/cassette players equipped with DIN standard fittings (Fig. 13.98). These tools are used to release the retaining lugs at the back of the set.
160 Suitable tools should be available from in-car entertainment specialists or from a Ford dealer, or can be made up from pieces of stiff (welding) wire.

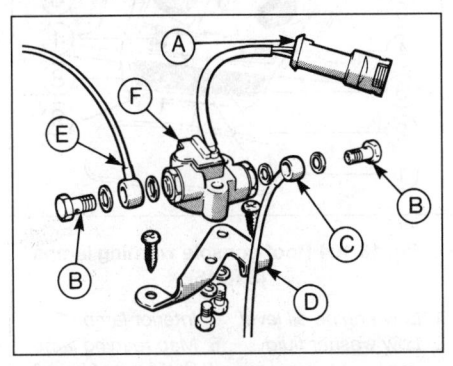

Fig. 13.95 Fuel flow sensor components (fuel injection type) (Sec 16)

A Multi-plug   D Bracket
B Banjo bolts   E Outlet connector
C Inlet connector   F Sensor

Fig. 13.96 Fuel flow sensor (carburettor type) (Sec 16)

Fuel hoses (arrowed)

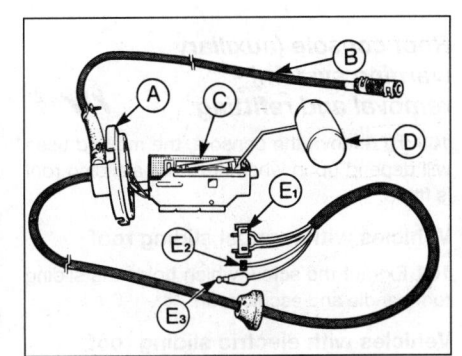

Fig. 13.97 Fuel tank sender unit used in conjunction with trip computer (fuel injection type) (Sec 16)

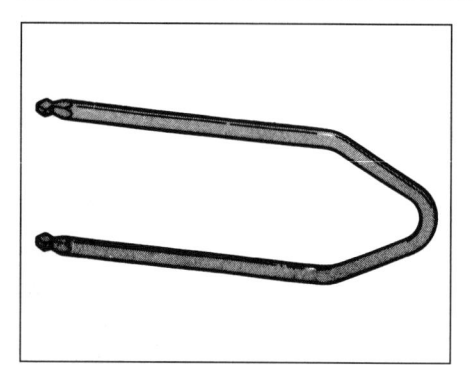

13•26 Supplement: Revisions and information on later models

Fig. 13.98 Tool for removing radio (Sec 16)

Fig. 13.99 Tools inserted and pressed outwards to release radio fixing clips (Sec 16)

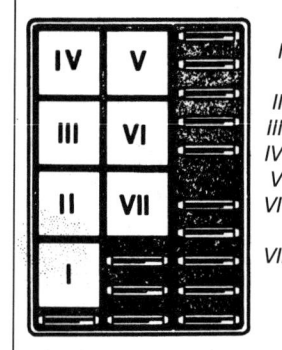

Fig. 13.100 Relay identification - main fuse block in engine compartment (Sec 16)

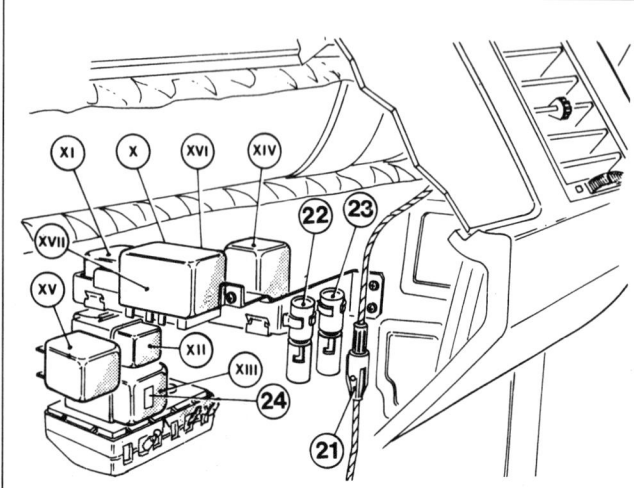

I   Ignition switch
II Spot lamps
III Foglamps
IV Horn
V   Flasher unit
VI Headlamp wash
VII Wiper (intermittent)

X   Power windows
XI   Sliding rood
XII   Air conditioner
XIII Heated mirror
XIV   Interior lamp (delay)
XV   Daytime running lamps
XVI   Tailgate release
XVII Central door locking

Fuses
21 Trip computer
22 Power-operated seat (LH)
23 Power-operated seat (RH)
24 heated rear view mirror

Fig. 13.101 Relay identification - bracket on RH A body pillar (Sec 16)

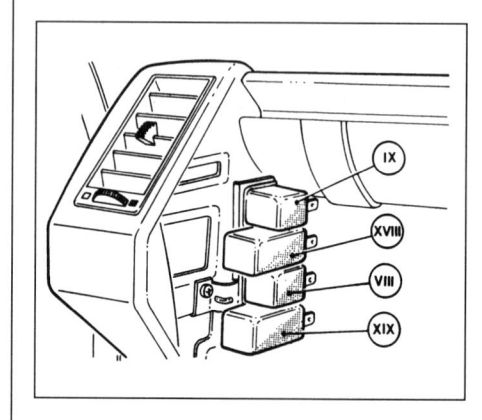

Fig. 13.102 Relay identification - bracket on LH A body pillar (Sec 16)

VIII   Automatic transmission
IX   Heated seats
XV III Fuel injection
XIX   Fuel injection (hot start)

Fuses and relays - 1982 on

161 Fuse functions, locations and ratings are largely unchanged from the details given in Chapter 10. Precise details of fuse application will be found on the fuse box lid.

162 More relays are used than previously, at least on fully equipped vehicles. Their locations are shown typically in Figs 13.100 to 13.106.

Roof console (auxiliary warning system) - removal and refitting

163 To remove the console, the method used will depend upon whether or not a sliding roof is fitted.

Vehicles with manual sliding roof

164 Extract the screw which holds the sliding roof handle and escutcheon.

Vehicles with electric sliding roof

165 Remove the operating switch by unclipping its rear edge, releasing its retaining lugs and wiring connector plug.

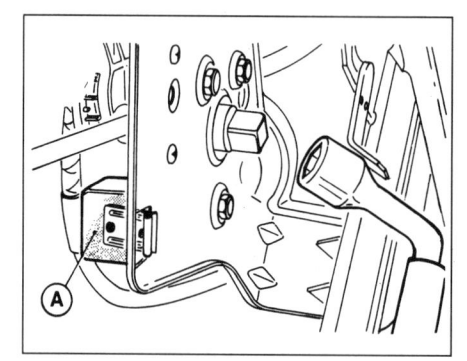

Fig. 13.103 Sliding roof relay (A) located on left-hand side of luggage boot (Sec 16)

Note hand crank for sliding roof

All models

166 Detach the warning lamp assembly and disconnect the multi-plugs.

167 Extract the fixing screws and remove the console. As this is done, disconnect the leads from the interior lamp and the map reading lamp. Note the two retaining lugs used on the console fitted to models without a sliding roof. Refitting is a reversal of removal.

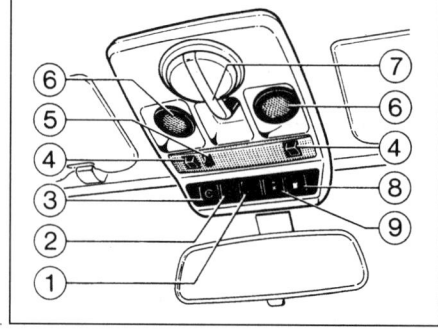

Fig. 13.104 Roof console warning lamps (Sec 16)

1 Low engine oil level
2 Low washer fluid level
3 Brake disc pad wear
4 Map reading lamp switch
5 Interior lamp
6 Map reading lamp
7 Sliding roof handle
8 Low fuel level
9 Low engine coolant level

Supplement: Revisions and information on later models 13•27

Fig. 13.105 Roof console with sliding roof fitted (Sec 16)

Fig. 13.106 Roof console without sliding roof (Sec 16)

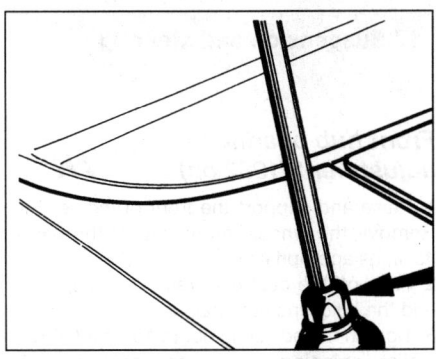

Fig. 13.107 Aerial upper retaining nut (arrowed) (Sec 16)

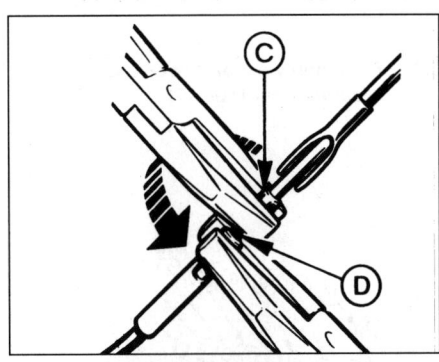

Fig. 13.108 Disconnecting drive cable (Sec 16)

C Inner section locknut
D Drive cable thread

Electrically-operated aerial mast - renewal

168 The extending aerial tube (mast) can be removed and refitted without the need to remove the complete aerial assembly.
169 Operate the aerial electrically until it is extended fully. Switch off the radio to begin retracting the aerial. When the aerial reaches an extended length of 400 mm (16.0 in) switch off the ignition.
170 Disconnect the battery.
171 Unscrew the aerial upper retaining nut Fig. 13.107. Once the nut has been unscrewed do not operate the aerial.
172 Pull the aerial tubular sections upwards from the wing.
173 Clean the plastic drive cable.
174 Wind masking tape round the aerial knob and aerial upper section. Using two pairs of pliers, grip the knob and aerial, then unscrew the knob.
175 Slide the upper section of the aerial into the tubular section until the drive cable locknut is exposed.
176 Again using pliers, release the screw connection between the aerial rod and the drive cable. Remove the aerial mast. Reassembly is a reversal of dismantling.

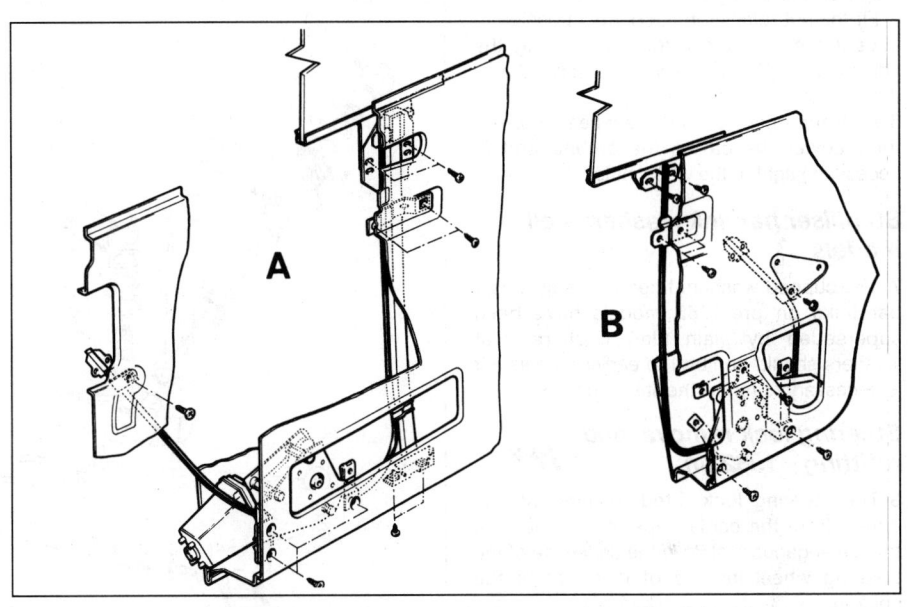

Fig. 13.109 Electrically-operated window mechanism (Sec 16)

A Front Door   B Rear Door

Electrically-operated windows

177 Electrically-operated windows were available as an optional fitment on Granada models. Two centrally-mounted switch panels control all of the individual windows, with additional switches being fitted for the rear doors. The centrally-mounted switch panel on the driver's side also has an isolating switch which overrides the rear door switches.

Switch panel - removal and refitting

178 Disconnect the battery earth lead.
179 Using a thin-bladed screwdriver prise the appropriate switch panel from the centre console.
180 Taking care that the multi-plug does not fall back into the aperture, detach it from the base of the switch.
181 Refitting is the reverse of the removal procedure; check the operation on completion.

Window operating motor - removal and refitting

182 Remove the door trim panel as described in Chapter 12, Section 13 or Section 18 of this Chapter.
183 Peel the plastic weathershield from the inside of the door panel, then disconnect the wiring multi-plug.
184 Remove the three cross head screws securing the motor and regulator bracket to the door panel.
185 Remove the three bolts, and separate the motor from the regulator bracket. If necessary, remove the drive gear from the motor (one screw).
186 Refitting is basically the reverse of the removal procedure, but test the operation of the window at the earliest opportunity once the motor has been refitted.

13•28 Supplement: Revisions and information on later models

17 Suspension and steering

Front hub bearings - adjustment (1982 on)

1 Raise and support the front of the vehicle. Remove the wheel centre trim or the wheel itself, as appropriate.
2 Prise off the dust cap, remove the split pin and the adjusting nut retainer.
3 Tighten the adjusting nut to 15 lbf ft (20 Nm) whilst spinning the wheel or hub anti-clockwise.
4 Slacken the nut by half a turn (180°), then retighten it finger tight.
5 Fit the nut retainer in the correct position to accept the split pin without disturbing the adjusting nut. Fit a new split pin and bend over the ends.
6 Refit the dust cap and the wheel or wheel trim. Lower the car to the ground and (if necessary) tighten the wheel nuts.

Stabiliser bar link washers - all models

7 The cupped washers fitted to the stabiliser bar links on pre-1982 models have been superseded by plain (flat) washers. Flat washers should be fitted to earlier models if it is necessary to renew the link insulators.

Steering lock removal and refitting - 1982 on

8 The steering lock fitted to later models differs from the earlier type in that the lock tongue engages in slots in the underside of the steering wheel instead of in a slot in the column.
9 To remove the lock, the steering wheel and steering column must first be removed (see Chapter 11). The single shear head retaining bolt is then drilled out or otherwise extracted to remove the lock (Fig. 13.111).
10 When fitting the new lock, check for correct operation and then tighten the new shear head bolt until its head breaks off.

Steering wheel removal - 1982 on

11 On models with the horn push in the steering wheel, disconnect the battery earth lead before commencing removal. Disconnect the horn lead from the centre terminal on models with a centre push.

Manual steering gear - lubrication (1982 on)

12 If it is necessary to lubricate the steering rack (eg after bellows renewal), note that the total lubricant capacity is now 0.25 pint (0.14 litre).
13 The specified quantity of lubricant should be distributed so that 0.16 pint (0.09 litre) is placed in the pinion end of the housing and 0.09 pint (0.05 litre) is placed in the other end.

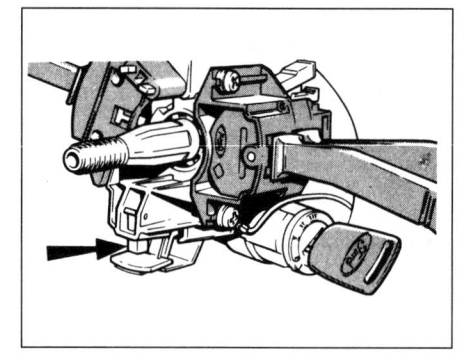

Fig. 13.110 Steering lock pawl (arrowed) on later models (Sec 17)

Fig. 13.111 Steering column lock shear head bolt (A) (Sec 17)

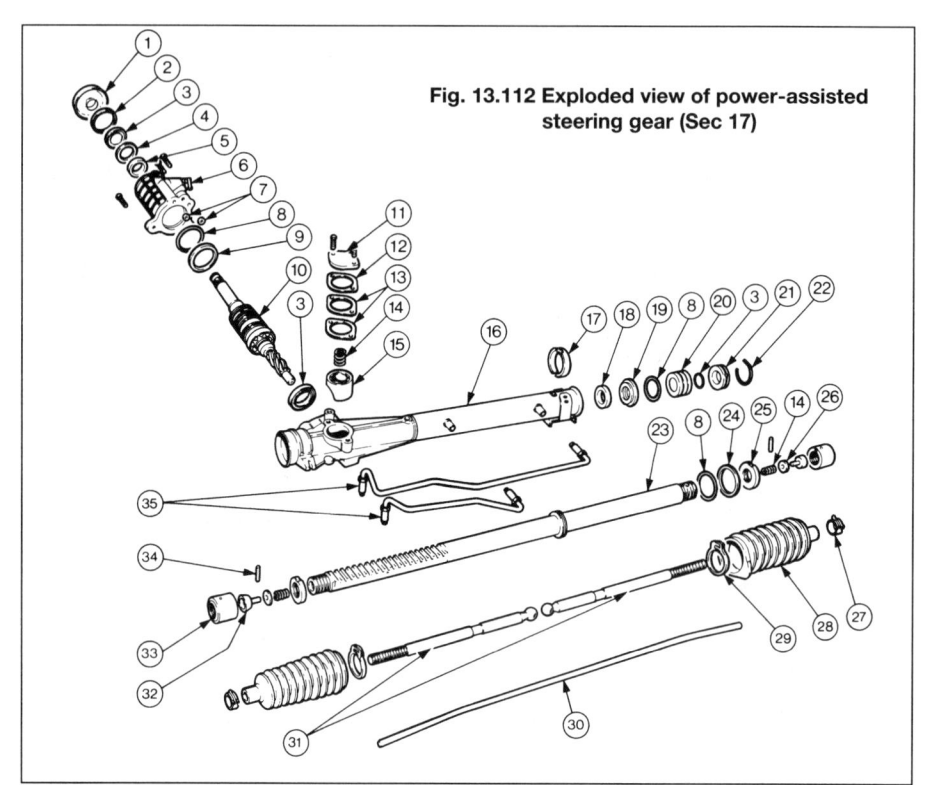

Fig. 13.112 Exploded view of power-assisted steering gear (Sec 17)

1 Dust cap	10 Valve assembly with shaft and pinion	18 Spacer	27 Clip
2 Circlip		19 Seal	28 Gaiter
3 Seal	11 Yoke cover	20 Bearing	29 Clip
4 Spacer	12 Paper shims	21 Locking collar	30 Air bleed tube
5 Bearing	13 Metal shims	22 Locking clip	31 Tie-rods (track-rods)
6 Valve housing	14 Spring	23 Rack	32 Ball seat
7 Transfer port seals	15 Yoke	24 Piston ring	33 Ball housing
8 O-ring	16 Housing	25 Locknut	34 Pin
9 Spacer	17 Clip	26 Washer	35 Ram feed pipes

Power-assisted steering gear (1982 on) - summary of changes

14 An exploded view of the Cam Gear power steering gear fitted to later models is given in Fig. 13.112. Comparison with Fig. 11.27 in Chapter 11 will show that the main difference is in the rack and pinion housings, which are in three separate pieces in the older version but are of one-piece construction in the later version. Less apparent is a change in the design and function of the control valve assembly, which cannot now be dismantled or adjusted.
15 "Quick fit" hydraulic pipe connectors are used at certain unions in the power steering system. The union nuts cannot be separated from the pipes with this type of coupling, though the O-ring seals may be renewed by pushing the nut up the pipe to expose the seal.
16 When a "quick fit" union is tightened to its specified torque, it is still possible to move the

pipe relative to the union. Do not tighten the union further in an attempt to eliminate this movement.

17 In view of the delicate nature of the overhaul work required, and the potentially serious consequences of error, the home mechanic is not advised to attempt to overhaul the power-assisted steering assembly. An overhauled or new unit, which may be available on an exchange basis, should be fitted if the old unit is badly worn or damaged.

Power-assisted steering system - bleeding (1982 on)

18 Open the bonnet and top up the steering fluid reservoir to the maximum level. Disconnect the coil negative lead.
19 Have an assistant operate the starter motor in short bursts (2 or 3 seconds), at the same time slowly turning the steering wheel from lock to lock. Keep the reservoir topped up with fluid during this operation.
20 When air bubbles cease to appear in the fluid in the reservoir, bleeding is complete. Top up the reservoir to the correct level and refit its cap. Reconnect the coil lead to complete.

Power steering pump (1982 on) - removal and refitting

21 Refer to Chapter 11, Section 39. The same procedure applies, but special bolts are no longer used to secure the pump.

Power steering pump (1982 on) - overhaul

22 Thoroughly clean the outside of the pump and reservoir.
23 Clamp the pump in a vice with protected jaws, shaft side downwards. **Do not** clamp on the reservoir.
24 Remove the mounting stud, the union and the O-ring from the back of the pump. Remove the reservoir from the pump housing by carefully pulling and rocking it free. Retrieve the O-rings from the housing.
25 Insert a pin punch or nail into the hole in the pump housing opposite the flow control valve. Press the punch to compress the endplate retaining ring and lever the ring out with a screwdriver.
26 Remove the endplate, spring and O-ring. If the endplate is not free rock it gently to release it and the spring will push it up. Remove the pressure plate.
27 Take the pump out of the vice and turn it over. Remove the flow control valve and its spring.
28 Remove the circlip from the pump shaft and push out the shaft. Do not attempt to separate the pulley flange from the shaft.
29 Remove the pump ring, vanes, rotor and thrust plate. Extract the pressure plate O-ring.
30 Extract the two dowel pins from the housing and remove the shaft seal, taking care not to damage the seal housing.
31 Remove the magnet and clean it, noting its

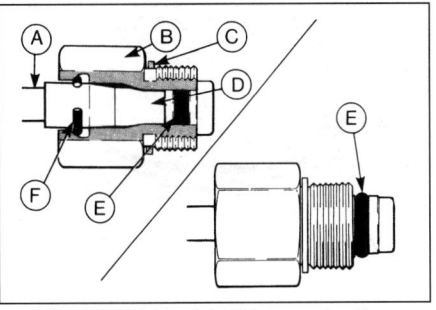

Fig. 13.113 "Quick-fit" type pipeline connector (Sec 17)

A Pipe
B Union nut
C PTFE seal
D Pre-formed end
E O-ring
F Snap-ring

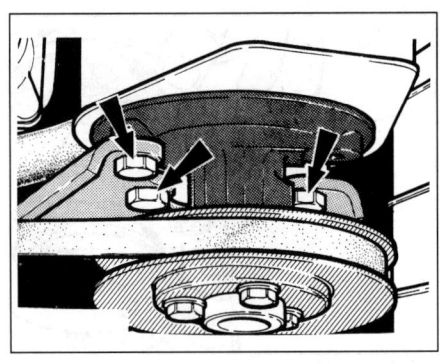

Fig. 13.114 Power steering pump mounting bolts (arrowed) (Sec 17)

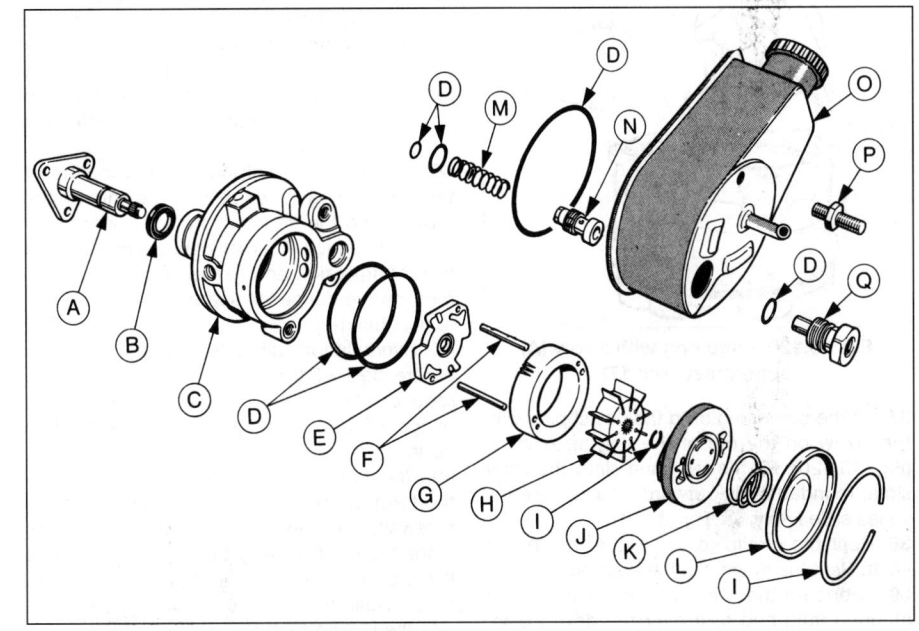

Fig. 13.115 Exploded view of power steering pump (Sec 17)

A Shaft
B Seal
C Housing
D O-rings
E Thrust plate
F Dowel pins
G Pump ring
H Vanes and rotor
I Retaining ring
J Pressure plate
K Spring
L Endplate
M Flow control spring
N Flow control valve
O Reservoir
P Stud
Q Union

location for reassembly. Keep it away from ferrous metal filings.
32 Clean all metal components in a suitable solvent and dry them. Scrupulous cleanliness must be observed from now on.
33 Fit a new pump shaft seal to the housing, using a socket or tube and a hammer to drive it home. Refit the shaft.
34 Fit a new pressure plate O ring into the third groove from the rear of the housing (Fig. 13.119) Lubricate the O-ring with power steering fluid.
35 Mount the housing in the vice and fit the two dowel pins.
36 Fit the thrust plate and rotor onto the shaft, with the countersunk side of the rotor facing the thrust plate. Make sure the thrust plate holes engage with the dowel pins.

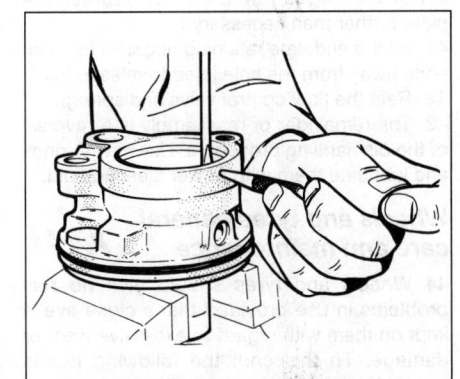

Fig. 13.116 Removing power steering pump endplate retaining ring (Sec 17)

Fig. 13.117 Removing endplate (A) and spring (B) from power steering pump (Sec 17)

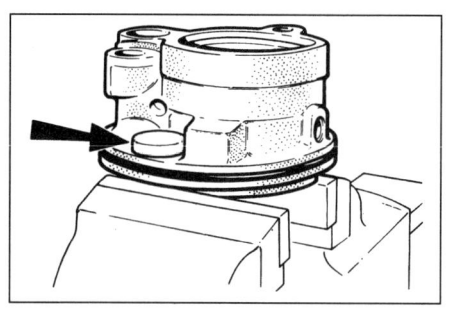

Fig. 13.118 Power steering pump magnet (arrowed) (Sec 17)

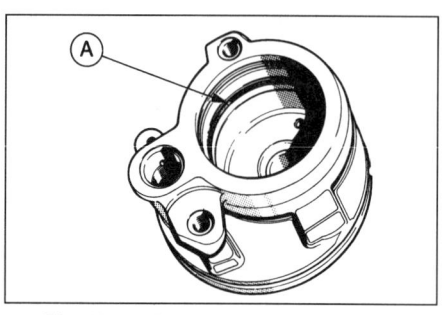

Fig. 13.119 Pressure plate O-ring (A) located in third groove (Sec 17)

Fig. 13.120 Pump ring with arrow (A) uppermost (Sec 17)

37 Fit the pump ring onto the dowel pins with the arrow on the rim of the ring uppermost (Fig. 13.120). Fit the ten vanes into the rotor slots, rounded ends outwards. Make sure the vanes slide freely.
38 Refit the circlip to the end of the pump shaft. Refit the magnet in the housing.
39 Lubricate the pressure plate with power steering fluid and fit it over the dowel pins. Make sure the spring seat is uppermost. Press the plate a little way below the O-ring to seat it.
40 Lubricate and fit the endplate O-ring in the second groove from the end. Fit the endplate spring, lubricate the endplate and press it into position so that the retaining ring can be fitted. (It may be necessary to use a G-clamp or similar device to keep the endplate depressed whilst the ring is fitted. Do not depress the plate further than necessary.)
41 Fit the endplate retaining ring with its open ends away from the hole used to release it.
42 Refit the flow control valve and spring.
43 The remainder of reassembly is a reversal of the dismantling procedure. Use new O-rings and lubricate them with power steering fluid.

Wheels and tyres general care and maintenance

44 Wheels and tyres should give no real problems in use provided that a close eye is kept on them with regard to excessive wear or damage. To this end, the following points should be noted.
45 Ensure that tyre pressures are checked regularly and maintained correctly. Checking should be carried out with the tyres cold and not immediately after the vehicle has been in use. If the pressures are checked with the tyres hot, an apparently high reading will be obtained owing to heat expansion. Under no circumstances should an attempt be made to reduce the pressures to the quoted cold reading in this instance, or effective underinflation will result.
46 Underinflation will cause overheating of the tyre owing to excessive flexing of the casing, and the tread will not sit correctly on the road surface. This will cause a consequent loss of adhesion and excessive wear, not to mention the danger of sudden tyre failure due to heat build-up.
47 Overinflation will cause rapid wear of the centre part of the tyre tread coupled with reduced adhesion, harsher ride, and the danger of shock damage occurring in the tyre casing.
48 Regularly check the tyres for damage in the form of cuts or bulges, especially in the sidewalls. Remove any nails or stones embedded in the tread before they penetrate the tyre to cause deflation. If removal of a nail *does* reveal that the tyre has been punctured, refit the nail so that its point of penetration is marked. Then immediately change the wheel and have the tyre repaired by a tyre dealer. Do not drive on a tyre in such a condition. In many cases a puncture can be simply repaired by the use of an inner tube of the correct size and type. If in any doubt as to the possible consequences of any damage found, consult your local tyre dealer for advice.
49 Periodically remove the wheels and clean any dirt or mud from the inside and outside surfaces. Examine the wheel rims for signs of rusting, corrosion or other damage. Light alloy wheels are easily damaged by "kerbing" whilst parking, and similarly steel wheels may become dented or buckled. Renewal of the wheel is very often the only course of remedial action possible.
50 The balance of each wheel and tyre assembly should be maintained to avoid excessive wear, not only to the tyres but also to the steering and suspension components. Wheel imbalance is normally signified by vibration through the vehicle's bodyshell, although in many cases it is particularly noticeable through the steering wheel. Conversely, it should be noted that wear or damage in suspension or steering components may cause excessive tyre wear. Out-of-round or out-of-true tyres, damaged wheels and wheel bearing wear/maladjustment also fall into this category. Balancing will not usually cure vibration caused by such wear.
51 Wheel balancing may be carried out with the wheel either on or off the vehicle. If balanced on the vehicle, ensure that the wheel-to-hub relationship is marked in some way prior to subsequent wheel removal so that it may be refitted in its original position.
52 General tyre wear is influenced to a large degree by driving style - harsh braking and acceleration or fast cornering will all produce more rapid tyre wear. Interchanging of tyres may result in more even wear, but this should only be carried out where there is no mix of tyre types on the vehicle. However, it is worth bearing in mind that if this is completely effective, the added expense of replacing a complete set of tyres simultaneously is incurred, which may prove financially restrictive for many owners.
53 Front tyres may wear unevenly as a result of wheel misalignment. The front wheels should always be correctly aligned according to the settings specified by the vehicle manufacturer.
54 Legal restrictions apply to the mixing of tyre types on a vehicle. Basically this means that a vehicle must not have tyres of differing construction on the same axle. Although it is not recommended to mix tyre types between front axle and rear axle, the only legally permissible combination is crossply at the front and radial at the rear. An obvious disadvantage of such mixing is the necessity to carry two spare tyres to avoid contravening the law in the event of a puncture.

> **HAYNES HINT** *When mixing radial ply tyres, textile braced radials must always go on the front axle, with steel braced radials at the rear.*

55 In the UK, the Motor Vehicles Construction and Use Regulations apply to many aspects of tyre fitting and usage. It is suggested that a copy of these regulations is obtained from your local police if in doubt as to the current legal requirements with regard to tyre condition, minimum tread depth, etc

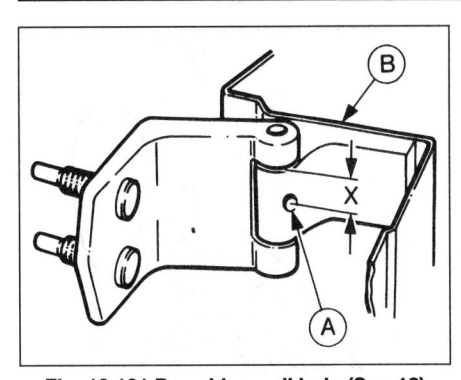

Supplement: Revisions and information on later models 13•31

Fig. 13.121 Door hinge oil hole (Sec 18)
A Hole B Door X 0.5 in (12.5 mm)

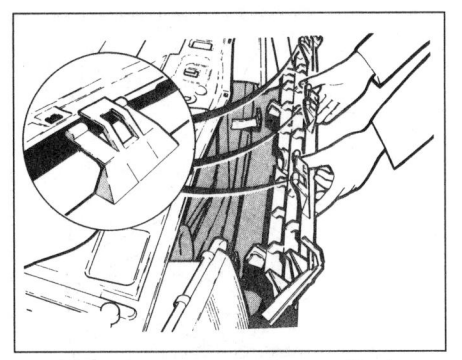

Fig. 13.122 Removing radiator grille. Inset shows fixing clip (Sec 18)

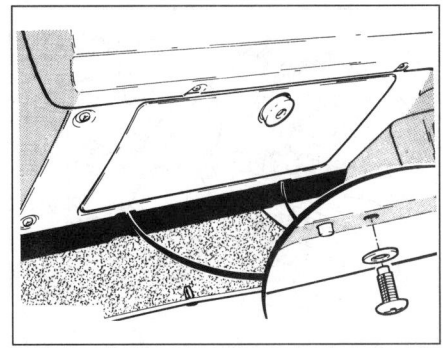

Fig. 13.123 Glovebox fixing screws (Sec 18)

18 Bodywork and fittings

Door hinge lubrication (pre- 1982 models)

1 If the door hinges seem to be suffering from lack of lubrication on pre-1982 models, a lubrication hole 0.08 in (2 mm) in diameter can be drilled as shown in Fig. 13.121 and oil injected using an oil gun. Take care not to drill into the hinge pin itself

2 Later models have hinges with built-in lubricating grooves and the drilling of a lubrication hole should not be necessary.

Bumper - removal and refitting (1982 on)

3 Proceed as described in Chapter 12, Section 8 or 9, but note that on models with plastic quarter bumpers, the quarter sections are retained by clips instead of by nuts and bolts.

Radiator grille removal and refitting (1982 on)

4 Open the bonnet and release the three clips which hold the top of the grille in place. Tilt the grille outwards.
5 Release the two lower clips and remove the grille.
6 Refitting is a reversal of removal.

Glovebox - removal and refitting (all models)

7 Free the under-dash trim panel in the region of the glovebox in order to expose the glovebox hinge securing screws (Fig. 13.123).
8 Remove the screws and withdraw the glovebox complete with lid. Disconnect the glovebox lamp leads, where fitted.
9 Refitting is a reversal of removal.

Front spoiler - removal and refitting

10 Drill out the two rivets on each side which secure the sides of the spoiler to the wheel arch.
11 Drill out the ten rivets which secure the spoiler to the lower valance. Remove the spoiler.
12 Align the rivet holes when refitting and secure the spoiler with proprietary snap-off rivets. Note that washers are used under the heads of the rivets securing the spoiler to the valance.

Rear spoiler - removal and refitting

13 Open the boot lid and remove the four bolts (two at each end) which secure the ends of the spoiler.
14 Remove the four nuts which secure the centre section of the spoiler and remove the spoiler.

15 Refitting is a reversal of removal. Fit all the nuts and bolts loosely at first to allow for any slight misalignment, then tighten them when all are in place.

Seat heating pads - removal and refitting

16 Remove the seat, disconnecting the wiring harness multi-plugs when so doing. Remove the cushion trim and/or backrest trim as necessary.
17 Remove the adhesive tape strips and the wire clips which secure the pad. Withdraw the pad and extract the support rod from its slot.
18 Fit the support rod to the slot in the new pad. Fit the new pad in the reverse order to removal, with the thermostat facing the foam padding or cushion. Make sure the pad is flat before securing it with adhesive tape and allow some slack so that the pad can flex when the seat is in use.
19 Refit the seat, reconnect the wiring plugs and check for correct operation.

Air conditioning compressor drivebelt - 1982 on

20 The drivebelt adjustment arrangement described in Chapter 12, involving an adjustable idler pulley, has been superseded in favour of moving the air conditioning compressor itself.
21 To adjust or renew the drivebelt, slacken

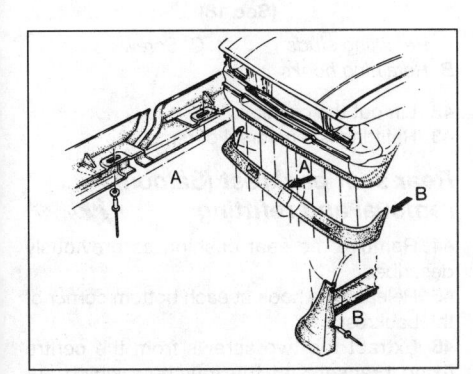

Fig. 13.124 Front spoiler (Sec 18)
A Front fitting detail B Side fitting detail

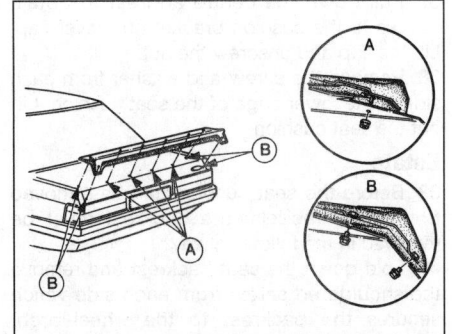

Fig. 13.125 Rear spoiler (Sec 18)
A Studs and nuts B Nuts and bolts

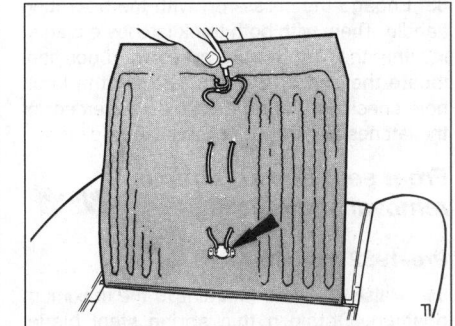

Fig. 13.126 Seat heating thermostat (arrowed) (Sec 18)

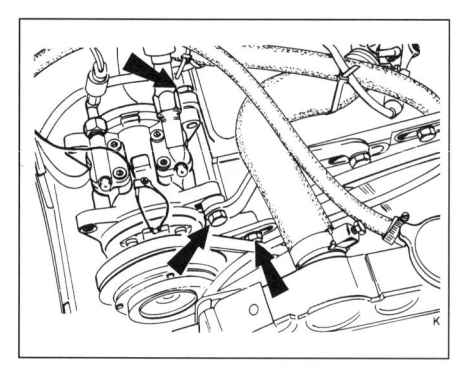

Fig. 13.127 Air conditioner compressor mounting and adjuster link bolts (arrowed) (Sec 18)

the three bolts (Fig. 13.127) and pivot the compressor towards the engine to slacken or remove the belt, away from the engine to tighten it. Tighten the bolts when adjustment is correct.

Front seat - removal and refitting

22 If heated seats are fitted, disconnect the wiring plugs.
23 Slide the seat forward as far as it will go. Remove the seat rear mounting bolts.
24 Now slide the seat fully rearwards and remove the front mounting bolts.
25 Lift the seat from the vehicle interior.
26 Refitting is a reversal of removal.

Front seat slide - removal and refitting

27 With the seat removed from the vehicle as previously described, invert it and take off the cross-rod clips and the cross-rod.
28 Unscrew the bolts which hold the slides to the underside of the seat and remove the slides.
29 When refitting, place the slide on the underside of the seat so that the rear hole in the slide aligns with the one in the seat frame.
30 Insert the bolt finger-tight and then push the slide track fully to the rear.
31 Insert the front bolt finger-tight.
32 Move the track fully forwards and rearwards, checking for smooth operation and tighten the fixing bolts.
33 Engage the cross-rod with the operating handle. Then, with both track latches engaged in either the fully forward or rearward position, locate the end of the cross-rod in the latch hole checking that no partial displacement of the latches occurs. Fit new cross-rod clips.

Front seat head restraints - removal and refitting

Pre-1982 models

34 Raise the head restraint to the maximum position. Obtain a thin spring steel blade, approximately 5 in (130 mm) long x 1.4 in (35 mm) wide x 0.020 in (0.5 mm) thick, and press it down between the front face of the

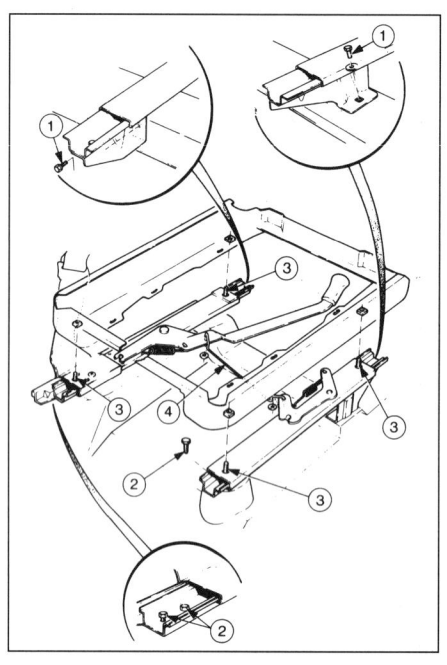

Fig. 13.128 Front seat mechanism (Sec 18)

1 Seat bolts (front)   3 Slide fixing bolts
2 Seat bolts (rear)   4 Cross-rod

rectangular pillar and the plastic guide sleeve (it may be necessary to break the top of the sleeve to obtain sufficient clearance). This will prevent the spring restraint from engaging, and the head restraint can be pulled out. Take care not to lose the blade inside the seat back.

1982-on

35 Turn the right-hand sleeve bezel clockwise, and pull the restraint upwards. If the sleeves are to be removed, push them in and turn them through 45° anti-clockwise, then pull them up and out of the seat back.

All models

36 Refitting is a straightforward operation. Renew or repair the guide sleeve on pre-1982 models, if necessary; note that the notch is towards the rear.

Rear seat cushion - removal and refitting

Saloon

37 Pull down the centre armrest to obtain access to the cushion bracket nut cover flap. Lift the flap and unscrew the nut.
38 Remove the screw and washer from each side at the lower edge of the seat cushion. Lift out the seat cushion.

Estate

39 Before the seat cushion can be removed from these models, the backrest must be removed in the following way.
40 Fold down the seat backrest and remove the shouldered screw from each side which secures the backrest to the wheel arch. Remove the backrest.
41 Now remove the two screws which retain the seat cushion retaining bracket.

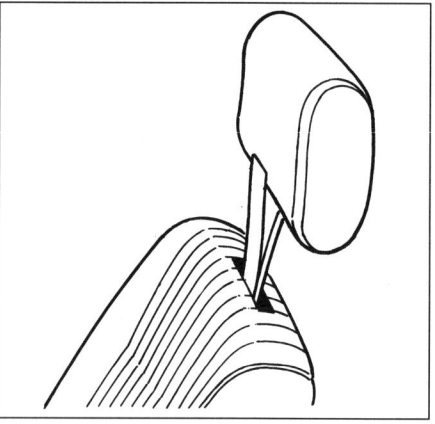

Fig. 13.129 Head restraint removal (pre-1982) (Sec 18)

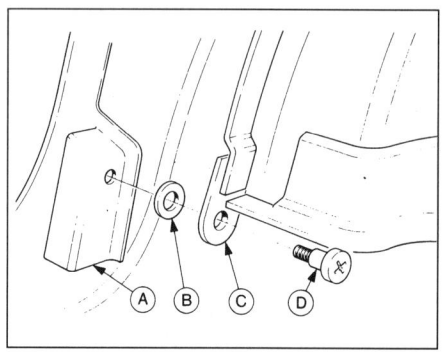

Fig. 13.130 Bracket pivot (Estate) (Sec 18)

A Bracket   C Seat back panel
B Washer   D Shouldered screw

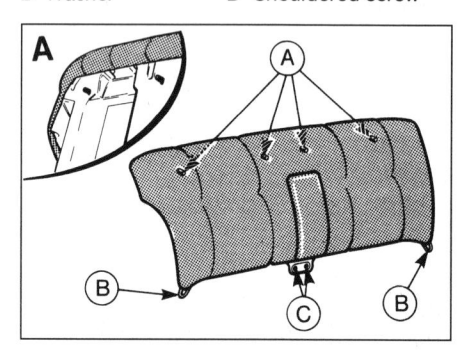

Fig. 13.131 Rear seat backrest (Saloon) (Sec 18)

A Retaining studs   C Screws
B Retaining hooks

42 Lift out the cushion.
43 Refitting is a reversal of removal.

Rear seat backrest (Saloon) - removal and refitting

44 Remove the seat cushion as previously described.
45 Release the hook at each bottom corner of the backrest.
46 Extract the two screws from the centre fixing bracket and the four nuts which are accessible from within the luggage boot.
47 Refitting is a reversal of removal.

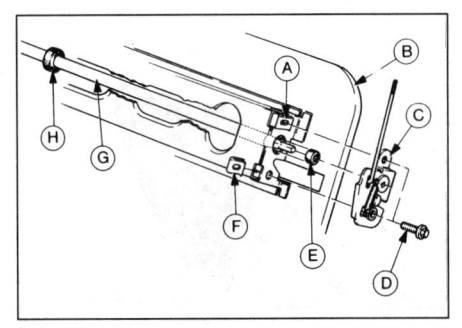

Supplement: Revisions and information on later models 13•33

Fig. 13.132 Rear seat backrest lock (Estate) (Sec 18)

A Spring nut   D Bolt   G Connecting tube
B Seat back   E Bush   H Insulator
C Catch   F Spring nut

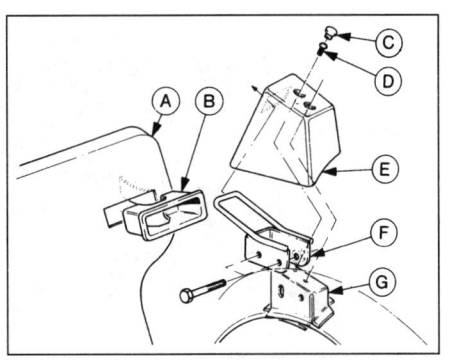

Fig. 13.133 Rear seat striker (Estate) (Sec 18)

A Seat panel   E Striker cover
B Socket   F Striker
C Screw cover   G Striker bracket
D Screw

Fig. 13.134 Front seat belt Torx bolts (arrowed) (Sec 18)

Rear seat back locking mechanism (Estate) - dismantling and reassembly

48 Fold down the seat back and unscrew the release knobs.
49 Lift the edges of the rear seat cover from the flange of the seat panel. Pull the cushion away from the panel to gain access to the seat catch.
50 Remove the two bolts which hold the seat catch and pull the catch from the connecting tube.
51 Withdraw the connecting tube from the connecting rod housing. Remove the tube end bushes and centre grommet. Reassembly is a reversal of dismantling.

Rear seat catch striker (Estate) - removal and refitting

52 Fold down the seat back, extract the two screws and remove the striker cover.
53 Mark around the striker to facilitate refitting.
54 Unscrew the two fixing bolts and remove the striker.
55 Refitting is a reversal of removal.

> **HAYNES HiNT** *Do not fully tighten the striker bolts until the seat back has been checked for smooth, positive locking.*

Seat belts - maintenance

56 Inspect the inertia reel type seat belts regularly for fraying or other damage.
57 Any cleaning should only be carried out using warm water and household detergent. Never use any kind of solvent.

Seat belts (front) - removal and refitting

58 The front seat belt stalk is bolted to the seat frame. If it is to be removed therefore, the seat must first be taken out. The bolts are of Torx type.
59 The upper anchor plate is bolted to the body pillar, the securing bolt being covered by a plastic cap which can be removed by prising it off.
60 The reel is secured to the body pillar by a single bolt. Note the alignment pin.
61 A trim panel covers the reel. This must be removed to reach the mounting bolt.

Seat belts (rear) Saloon - removal and refitting

62 The belt anchor plates are bolted to the floor pan. The seat cushion must be removed to gain access to them.
63 The upper belt anchor plate is bolted to

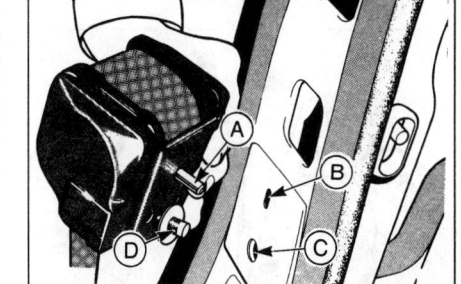

Fig. 13.135 Seat belt upper anchorage (Sec 18)

Fig. 13.136 Front seat belt reel (Sec 18)

A Locating peg   C Mounting bolt hole
B Peg slot   D Mounting bolt

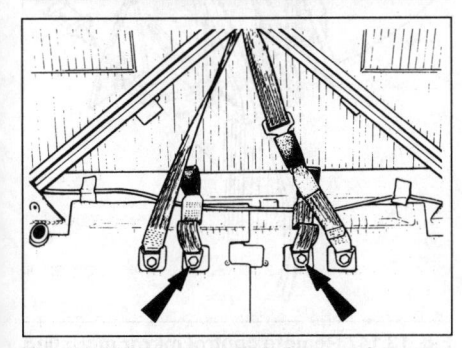

Fig. 13.137 Rear seat belt floor anchorages (Sec 18)

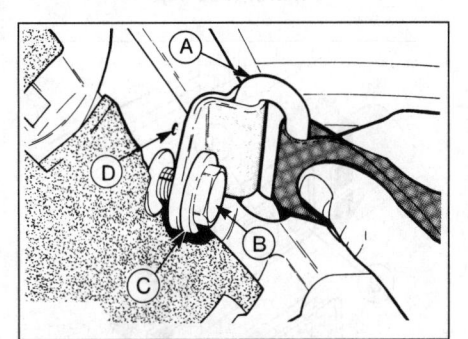

Fig. 13.138 Rear belt upper anchorage (Sec 18)

A Plate   C Washer
B Bolt   D Locating peg hole

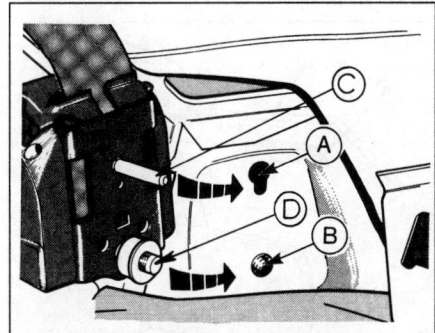

Fig. 13.139 Rear belt reel (Sec 18)

A Locating pin hole   C Locating pin
B Bolt hole   D Mounting bolt

13•34 Supplement: Revisions and information on later models

Fig. 13.140 Belt guide at rear parcels shelf (Sec 18)

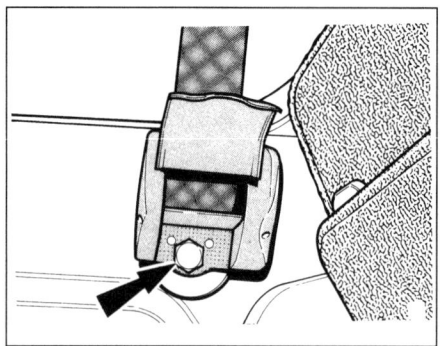

Fig. 13.141 Rear belt reel (Estate) (Sec 18)

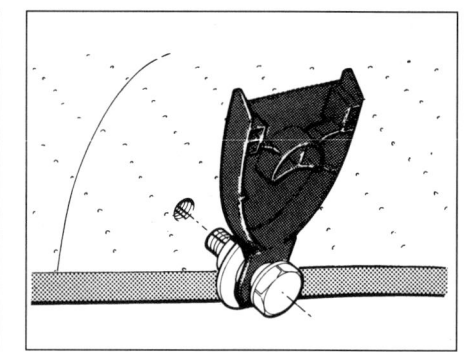

Fig. 13.142 Hinged belt anchor bolt cap (Estate) (Sec 18)

the body panel just inside the rear window. The trim panel must be removed for access. The anchor plate has an alignment pin.

64 The reel is located within the luggage boot under a cover. If the reel is to be removed, then the belt guide runner will have to be released from the rear parcels shelf so that the belt and anchor plates can be fed through the shelf aperture.

65 Refitting is a reversal of removal. Make sure that the alignment pins locate correctly in their holes.

Seat belts (rear) Estate - removal and refitting

66 The general arrangement is similar to that for Saloon models, but note the hinged flaps which cover the bolt on the reel and the supporting strap anchor plate.

67 The reel has a spacer located between it and the load space panel.

Seat belts - general

68 When removing or refitting a belt anchor plate or reel, always retain the original fitted sequence of the components (washers, spacer and anchor plate), otherwise their correct operation will be impaired.

69 Always tighten bolts to the specified torque (see Specifications in this Supplement).

Rear view mirrors removal and refitting

Interior mirror (1982 on)

70 This mirror is stuck directly to the glass. If it must be removed, grip the mirror head and twist as shown in Fig. 13.144.

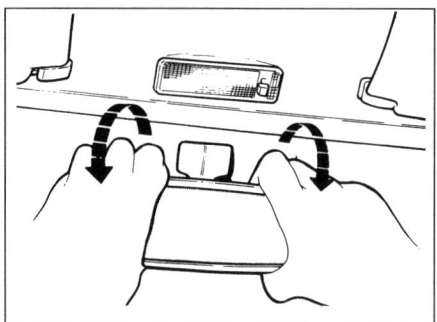

71 Clean the residue of adhesive from the glass using solvent. Also clean the mirror contact area if it is to be refitted. Use a new adhesive patch and avoid touching it with the fingers. Sustain pressure for two minutes once the mirror is pressed onto the adhesive patch.

72 New windscreens are supplied with a "black patch". Peel the protective tape from it to expose the adhesive surface.

73 when sticking a mirror to the windscreen, make sure that the screen is at a temperature not lower than 68°F (20°C) and warm the mirror base and adhesive patch to a temperature of between 122 and 158°F (50 and 70°C).

Exterior mirror

74 Remove the door trim panel as described in Chapter 12.

75 If the mirror is electrically-operated, disconnect the multi-plug from the mirror control switch.

76 If the mirror is manually operated, extract the small retaining screw and pull off the knob.

77 Prise off the triangular inner trim panel inserting a screwdriver at its rear edge.

78 Unscrew the three fixing screws, remove the inner plate and withdraw the mirror.

79 On electrically operated mirrors, the withdrawal of the mirror will necessitate detaching the multi plug.

80 Refitting is a reversal of removal.

Fig. 13.143 Belt reel with mounting spacer (Estate) (Sec 18)

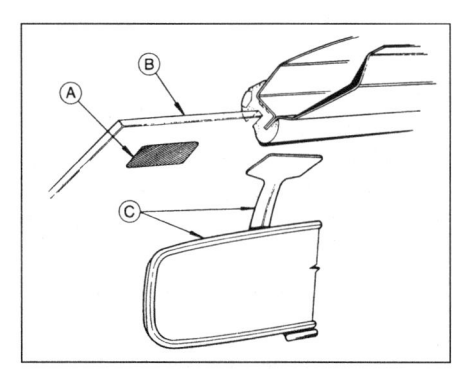

Fig. 13.144 Removing interior rear view mirror (Sec 18)

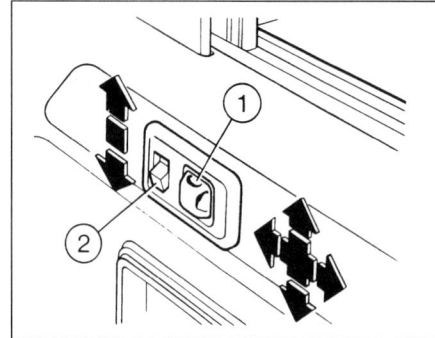

Fig. 13.145 Interior mirror ready for fitting (Sec 18)

A Black patch   B Windscreen   C Mirror

Fig. 13.146 Electric mirror control (Sec 18)

1 Operating switch
2 Passenger/driver selector switch

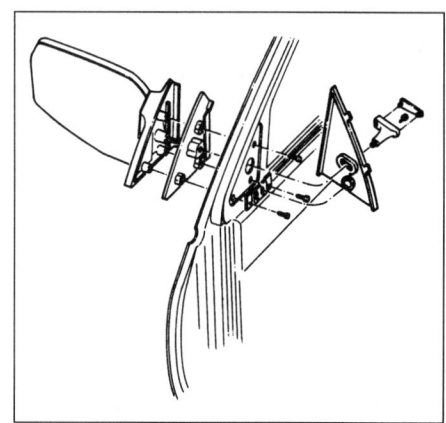

Fig. 13.147 Remote control mirror mounting components (Sec 18)

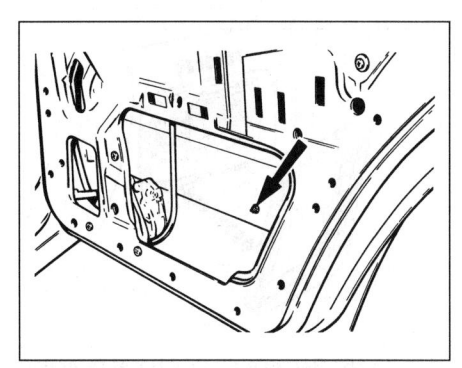

Supplement: Revisions and information on later models 13•35

Fig. 13.148 Door moulding stud (Sec 18)

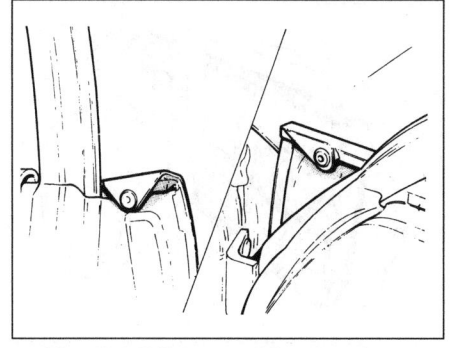

Fig. 13.149 Door moulding pop rivets (Sec 18)

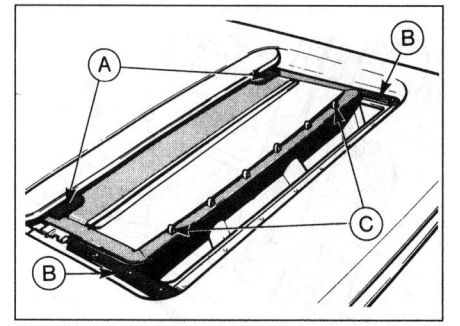

Fig. 13.150 Removing sliding roof trim panel (Sec 18)

A Rear trim clips   C Front trim clips
B Slide rails

Body mouldings - removal and refitting

Side door moulding

81 Remove the door trim panel as described in Chapter 12. Peel away the waterproof sheet.
82 Unscrew the moulding fixing nuts and then detach the moulding from its door clips.

Door belt moulding

83 Remove the exterior rear view mirror as described earlier in this Section.
84 Carefully drill out the "pop" rivets from both ends of the moulding.
85 Wind masking tape around a screwdriver to prevent damage to the paintwork and then prise the door belt moulding from its clips.

Window frame moulding

86 To remove this moulding, start with the rear section working from the inside edge of the frame. Drive off the moulding using a block of wood as an impact tool.
87 Refitting all mouldings is the reversal of removal.

> **HAYNES HiNT** When fitting the window frame moulding, use a rubber mallet to avoid damage.

Sliding roof - adjustment

88 Open the sliding roof to the half-way position. Unclip the trim from the front edge of the sliding roof panel (Fig. 13.150).
89 On manually operated roofs, the control lever should be in the lower or normal operating position.
90 Fully open the sliding roof and lift the front edge of the panel trim over the slide rails and slide it out from the front.
91 If necessary to facilitate trim removal, depress the rear trim retaining clips with the fingers.
92 Close the sliding roof and set it in the tilt position. On manually-operated roofs, make sure that the control lever is in the upper position.
93 Move the roof panel from side to side across the aperture to check for wear or incorrect adjustment of the rear slides or front guides.

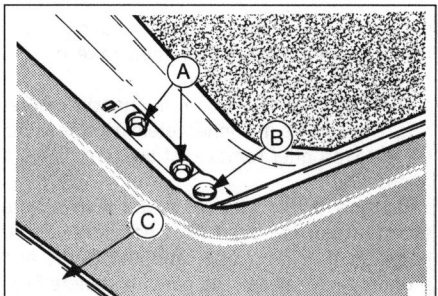

Fig. 13.151 Sliding roof front guides (Sec 18)

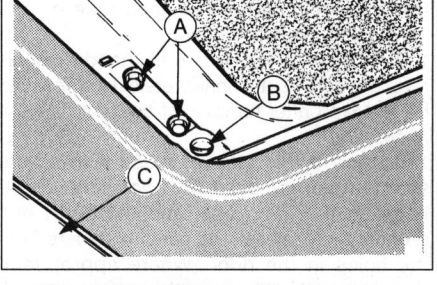

A Fixing bolts   C Sun visor
B Height adjusting screw

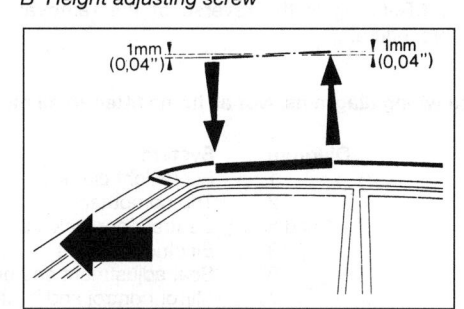

Fig. 13.153 Sliding roof panel alignment (Sec 18)

94 Close the roof and loosen the front guide retaining bolts (Fig. 13.151).
95 Loosen the three nuts on the rear adjusting plates (Fig. 13.152).
96 Centralise the sliding roof panel in its aperture and check for an even gap all round. Tighten the rear adjusting plate nuts.
97 Slide the front guides fully inwards, but do not tighten the fixing bolts at this stage.
98 Check that the sliding roof panel is flush with the surrounding roof, or at least not exceeding the tolerance shown in Fig.13.153.
99 Where necessary, check that the front guides are still loose and turn the adjusting screw (B) Fig.13.151 to achieve the correct setting.
100 Hold the front guides fully inwards and tighten the fixing bolts.
101 If the rear edge of the roof panel requires

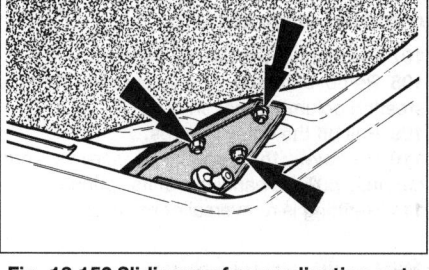

Fig. 13.152 Sliding roof rear adjusting nuts (Sec 18)

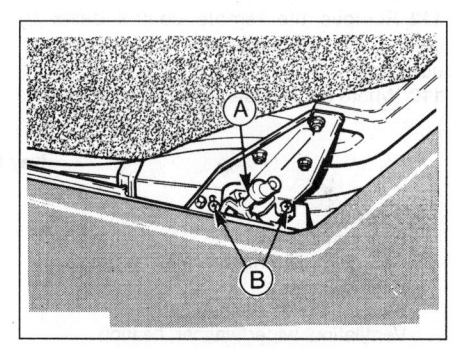

Fig. 13.154 Sliding roof rear height adjustment (Sec 18)

A Link assembly   B Height adjusting screws

adjustment, tilt the panel until both the link adjusting screws are accessible, (Fig. 13.154). The roof panel should be approximately in the half tilted position.
102 Loosen the adjusting screws on each side.
103 Close the panel and adjust by lowering or raising it until the setting is as shown in Fig. 13.153. Tighten the adjusting screws which are still accessible
104 Tilt the roof panel and tighten the remaining adjusting screws.
105 Check the operation of the sliding roof then fully open it and clip the trim to the front edge.
106 In the event of failure of an electrically-operated sliding roof, it can be operated manually using the wheel brace supplied in the car tool kit on the nut located in the luggage boot or load space (Estate) see Fig. 13.103.

13•36 Supplement: Revisions and information on later models

Fig. 13.155 Facia trim panel fixing screws (Sec 18)

Fig. 13.156 Removing facia trim panel (Sec 18)

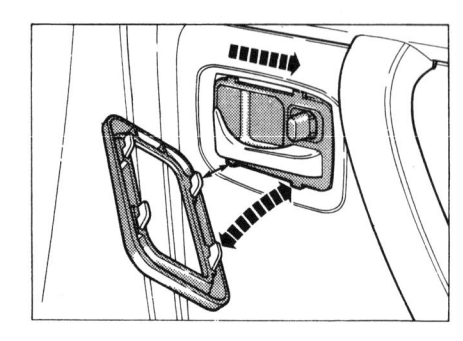

Fig. 13.157 Remove control bezel removal (Sec 18)

Facia trim panel (1982 on) - removal and refitting

107 Disconnect the battery earth lead.
108 Remove the securing screw and lift off the steering column upper shroud.
109 Pull off the control knobs.
110 Remove the securing screws and carefully pull the panel out of its spring clips.
111 Refitting is a reversal of removal.

Door trim - removal and refitting

112 Remove the remote control bezel by raising the rear edge and lifting the bezel outwards.
113 Where appropriate, carefully prise out the centre cap from the window winder handle. Remove the single screw then take off the handle and the escutcheon.
114 Remove the two pull handle outer retaining screws (the inner retaining screw can be removed after removal of the trim panel, if necessary).
115 Insert a thin strip of metal with all the sharp edges removed or a flat screwdriver between the trim panel and the door, and progressively release the panel fasteners from the holes in the door. Where appropriate, disconnect the plugs to the mirror control, loudspeaker and electric window switch. Lift the trim panel up to disengage it from the inner belt weatherstrip.
116 Refitting is the reverse of the removal procedure.

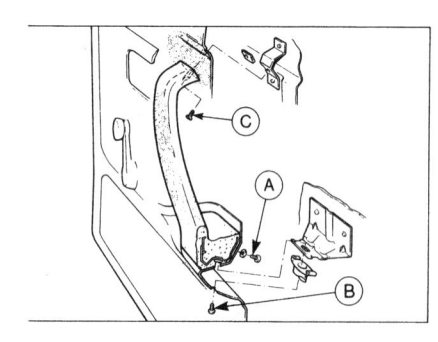

Fig. 13.158 Door pull handle - typical (Sec 18)

A Inner retaining screw
C Outer retaining screw
B Outer retaining screw

Key to wiring diagrams. Not all items fitted to all models

System	Diagram	System	Diagram
Explanation	1	Econolight circuit	4
Charging, starting, fuel injection	2	Trip computer	9
Exterior lighting	4 and 5	Central door locking	10
Indicators and warning lamps	3	Electric windows	11
Ventilation, wiper/washer, heating	6	Seat adjustment and heating	12
Interior lighting	7	Mirror control and heating	13
Instrument illumination	8	Accessories	7
Auxiliary warning system	9		

Colour code

Black	SW	Pink	RS
Blue	BL	Red	RT
Brown	BR	Violet	VI
Green	GN	White	WS
Grey	GR	Yellow	GE

Component	Diagram	Location	Component	Diagram	Location	Component	Diagram	Location
Actuator	10	F4	Battery	6	D1	Cigar lighter	8	B6
Actuator	10	F5	Battery	7	D1	Clock	7	C11
Actuator	10	F8	Battery	8	D1	Clock	8	B7
Actuator	10	D7	Battery	9	D1	Gauge		
Luggage compartment release	10	F9	Battery	10	D1	Ammeter	2	C13
Alternator	2	C2	Battery	11	D1	Auxiliary warn	1	C8
Alternator	10	E11	Battery	12	D1	Auxiliary warn	9	B14
Alternator	11	D3	Battery	13	D1	Trip computer tone generator	9	E5
Battery	2	A1	Cigar lighter	7	C9	Instrument cluster	1	C6
Battery	3	D1	Cigar lighter	7	10	Instrument cluster	2	A14
Battery	4	D9	Cigar lighter	8	B5	Instrument cluster	3	C14

Wiring diagrams WD•1

Key to wiring diagrams (continued)

Component	Diagram	Location	Component	Diagram	Location	Component	Diagram	Location
Instrument cluster	4	B4	Starter	12	E1	Luggage compartment	7	C16
Instrument cluster	5	D15	Starter	13	E1	Luggage compartment rel	8	B12
Instrument cluster	7	E14	Electric window	11	F5	Luggage compartment rel	10	D10
Instrument cluster	8	E8	Electric window	11	F7	Door cont	7	E4
Oil pressure	2	C14	Electric window	11	F9	Door cont`	7	E5
Horn	3	D5	Electric window	11	F12	Door cont	7	E6
Horn	3	E5	Electric window	11	F14	Door locking	10	D4
Ignition			Electric window	11	F16	Front fog lamp	5	D5
Coil	2	B5	Windshi. washer	6	B12	Glove box lamp	8	D7
Control trans.	2	C10	Wiper	6	E10	Handbrake warn ind	3	E11
Distributor contactless	2	A9	Radio	7	E3	Heat. seat	8	B13
Illumination			Relay			Heat. seat	12	C4
Ashtray	8	B5	Luggage compartment			Heat. seat	12	E4
Heater regul.	8	F14	release	10	A11	Heat. seat	12	F4
Mirror	7	D8	Delay Int. Light	7	D4	Heated rear window	8	B4
Oil pres. + Amm	8	F15	Door locking	10	C6	Heated rear window	13	C3
Switch stage ind.	8	A15	Flasher	3	C12	Heater blower	6	B5
Lamp			Fog lamp	5	D8	Heater blower	8	F11
Add. interior	7	F12	Fuel injection	2	E8	Horn	3	F4
Add. interior	7	F11	Headl. wash	6	B14	Light/wiper	4	E12
Cargo space	7	B16	Heat. seats	12	A5	Light/wiper	6	E13
Engine comp	8	E4	Heated mirror	13	D10	Light/wiper	7	B2
Glove box	8	C7	Horn	3	B4	Light/wiper	8	E2
Interior, reading	7	C6	Hot start	2	A11	Light/wiper	9	E7
LH combined rear	1	B6	Ignition switch	3	C2	Long ran. lamp	5	F11
LH combined rear	3	F7	Ignition switch	4	C10	Low fluid level warning	3	E12
LH combined rear	5	F2	Ignition switch	6	C2	Luggage compt	7	C15
LH flash	3	A5	Ignition switch	10	C2	Mirror adjustm	13	C5
LH fog	5	A5	Ignition switch	11	C2	Multi function	3	C9
LH long range	5	A3	Ignition switch	12	C2	Multi function	4	E15
LH main head beam	5	A2	Ignition switch	13	C2	Oil pressure	2	E14
LH side mark. flash	3	B5	Power roof	10	B14	Power roof	10	D13
Licence plate	5	F9	Electric window	11	A4	Power seat	12	B9
Luggage compt	7	B12	Spot lamp	5	C9	Power seat	12	B13
Rear interior	7	E7	Starter autom.transm	2	E5	Rear fog lamp	5	C5
RH combined rear	3	F15	Wiper interm	6	E10	Rear window w/w	6	E6
RH combined rear	5	F15	Resistor			Rear window w/w	8	B2
RH flash	3	A15	Heat. rear win	13	E4	Seat inclin	12	C12
RH fog	5	A13	Heat. seat back	12	C8	Seat inclin	12	C16
RH long range	5	A14	Heat. seat back	12	E8	Steering ign	2	E3
RH main head beam	5	A15	Heat. seat pad	12	D8	Steering ign	3	E2
RH side mark, flash	3	A15	Heat. seat pad	12	F8	Steering ign	4	E10
Module box	9	C16	Ign.coil series	2	C5	Steering ign	6	E2
Motor			Illumination poti	8	E6	Steering ign	7	E2
Fuel pump	2	E9	Illumination poti	9	B7	Steering ign	9	E2
Headlamp wash	6	B16	Wiper 1. poti	6	E12	Steering ign	10	E2
Heater blower	6	B6	Wiper 1. poti	8	C5	Steering ign	11	E2
Outside mirror	13	F6	Sender			Steering ign	12	E2
Outside mirror	13	F11	Low screen wash warn	9	E11	Steering ign	13	E2
Power roof	10	E14	Coolant warn	9	E10	Stop light	3	E14
Power seat	12	E10	Fuel flow	9	C5	Thermal time	2	D12
Power seat	12	E12	LH brake pad	9	E13	Vacuum	2	E10
Power seat	12	E11	Oil warn	9	E9	Econolight	4	B1
Power seat	12	E14	RH brake pad	9	E15	Econolight	4	B2
Power seat	12	E16	Speed	9	E4	Electric window	11	D5
Power seat	12	E15	Transm. fuel	2	E16	Electric window	11	D7
Rear w/washer	6	E8	Transm. fuel	2	D15	Electric window	11	D11
Rear window wiper	6	E5	Transm. fuel	9	E12	Electric window	11	D14
Starter	2	E1	Transm. water temp	2	E15	Electric window	11	E15
Starter	3	E1	Switch			Trip computer	9	E6
Starter	4	E9	Add. interior lamp	7	F10	Valve		
Starter	6	E1	Autom. gear blocking	2	F6	Air dumper	2	E12
Starter	10	E1	Autom. gear blocking	4	A15	Cold starting	2	E13
Starter	11	E1	Reverse lamp	4	A13	EGR system	2	E11
						Hot running governor	2	E11

WD•2 Wiring diagrams

FIGURE 1

- SEPARATION BETWEEN CONNECTORS OR CONNECTOR AND COMPONENT
- BASIC PART NUMBER OF THE WIRE
- LOCATION IN CAR FOR CONNECTION S
- DRV.SIDE B PILLAR

LH COMBINED REAR LAMP

1 = FLASHER LAMP
2 = BRAKE+SIDE MARKER LAMP
3 = BACK UP LAMP
4 = REAR FOG LAMP

INSTRUMENT CLUSTER GAUGE

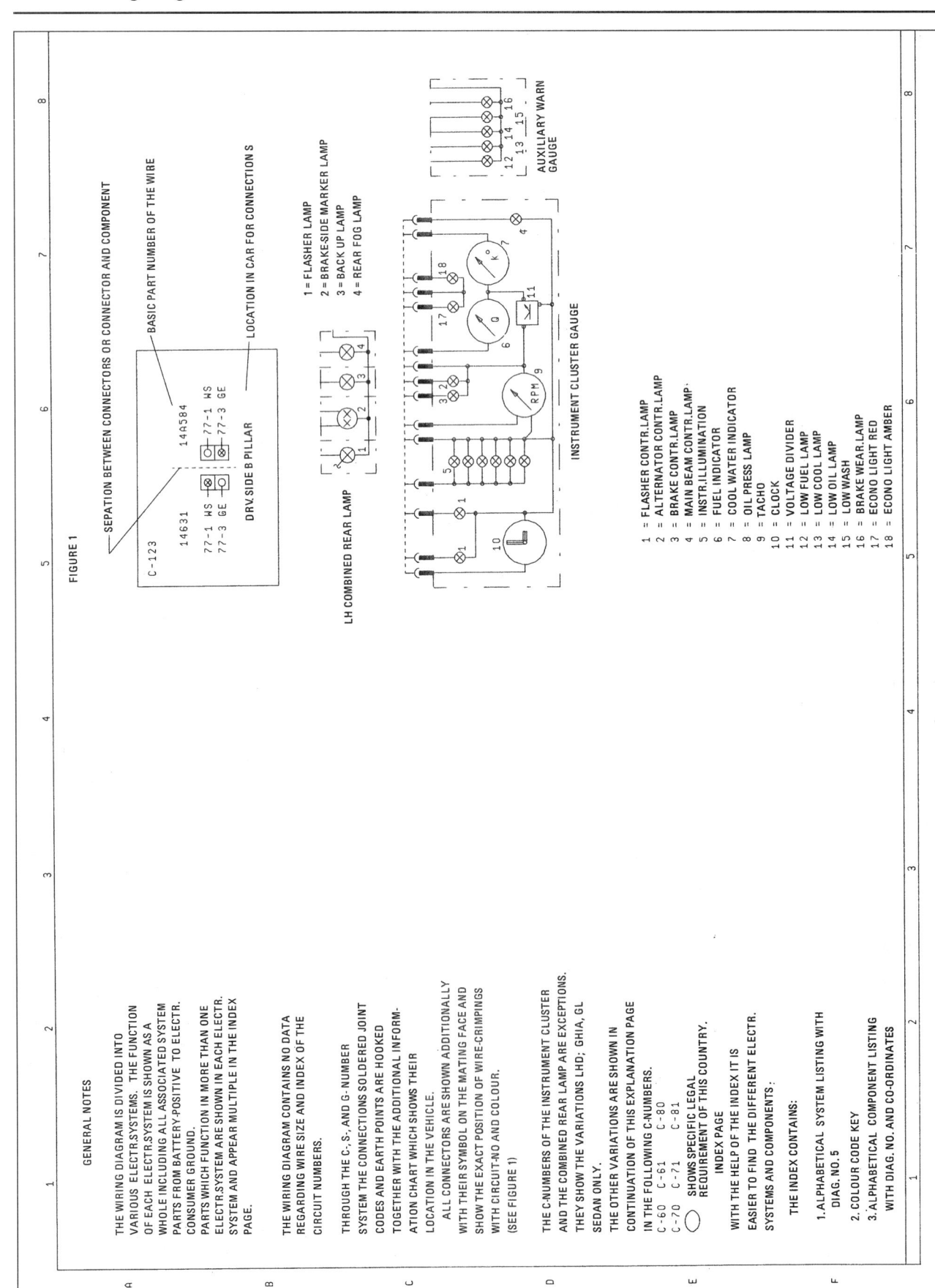

AUXILIARY WARN GAUGE

1 = FLASHER CONTR.LAMP
2 = ALTERNATOR CONTR.LAMP
3 = BRAKE CONTR.LAMP
4 = MAIN BEAM CONTR.LAMP
5 = INSTR.ILLUMINATION
6 = FUEL INDICATOR
7 = COOL WATER INDICATOR
8 = OIL PRESS LAMP
9 = TACHO
10 = CLOCK
11 = VOLTAGE DIVIDER
12 = LOW FUEL LAMP
13 = LOW COOL LAMP
14 = LOW OIL LAMP
15 = LOW WASH
16 = BRAKE WEAR.LAMP
17 = ECONO LIGHT RED
18 = ECONO LIGHT AMBER

GENERAL NOTES

THE WIRING DIAGRAM IS DIVIDED INTO VARIOUS ELECTR.SYSTEMS. THE FUNCTION OF EACH ELECTR.SYSTEM IS SHOWN AS A WHOLE INCLUDING ALL ASSOCIATED SYSTEM PARTS FROM BATTERY-POSITIVE TO ELECTR. CONSUMER GROUND.
PARTS WHICH FUNCTION IN MORE THAN ONE ELECTR.SYSTEM ARE SHOWN IN EACH ELECTR. SYSTEM AND APPEAR MULTIPLE IN THE INDEX PAGE.

THE WIRING DIAGRAM CONTAINS NO DATA REGARDING WIRE SIZE AND INDEX OF THE CIRCUIT NUMBERS.

THROUGH THE C-, S-, AND G- NUMBER SYSTEM THE CONNECTIONS SOLDERED JOINT CODES AND EARTH POINTS ARE HOOKED TOGETHER WITH THE ADDITIONAL INFORMATION CHART WHICH SHOWS THEIR LOCATION IN THE VEHICLE.
ALL CONNECTORS ARE SHOWN ADDITIONALLY WITH THEIR SYMBOL ON THE MATING FACE AND SHOW THE EXACT POSITION OF WIRE-CRIMPINGS WITH CIRCUIT-NO AND COLOUR.
(SEE FIGURE 1)

THE C-NUMBERS OF THE INSTRUMENT CLUSTER AND THE COMBINED REAR LAMP ARE EXCEPTIONS. THEY SHOW THE VARIATIONS LHD; GHIA, GL SEDAN ONLY.
THE OTHER VARIATIONS ARE SHOWN IN CONTINUATION OF THIS EXPLANATION PAGE IN THE FOLLOWING C-NUMBERS.
C-60   C-61   C-80
C-70   C-71   C-81

◯ SHOWS SPECIFIC LEGAL REQUIREMENT OF THIS COUNTRY.

INDEX PAGE
WITH THE HELP OF THE INDEX IT IS EASIER TO FIND THE DIFFERENT ELECTR. SYSTEMS AND COMPONENTS :

THE INDEX CONTAINS:

1. ALPHABETICAL SYSTEM LISTING WITH DIAG. NO. 5
2. COLOUR CODE KEY
3. ALPHABETICAL COMPONENT LISTING WITH DIAG. NO. AND CO-ORDINATES

Diagram 1. Explanatory notes for wiring diagrams

FUSE PANEL SHOWN IN C-1 6 WAY FUSE BOX SHOWN IN C-11

LOCATION RELAY

	FUSE PANEL		BRACKET A PILLAR RHS
I	IGNTION SWITCH	X	POWER WINDOW
II	SPOT LAMPS	XI	POWER ROOF
III	FOG LAMPS	XII	AIR CONDITION
IV	HORN	XIII	HEAT. MIRROR
V	FLASHER UNIT	XIV	DELAY INTERIOR LIGHT
VI	HEADL. WASH	XV	DAYTIME RUN LIGHT
VII	WIPER INTERMITTENT	XVI	TAILGATE RELEASE
		XVII	DOOR LOCKING

BRACKET A PILLAR LHS

VIII	AUTOM.GEAR	XVIII	FUEL INJECTION 1
IX	HEATED SEAT	XIX	HOT START (FUEL INJ.)

	ENGINE COMPARTMENT
XX	PRE GLOW
XXI	GLOW PLUGS
XXII	TAXI

FUSE	AMPERE	FUSE CONNECTIONS
1	16	HEATER BLOWER + ECONO LIGHT
2	16	STOP LIGHT + WIPER + WIPER REAR + BACK UP LAMP + FLASHER
3	8	LICENCEPLATE LAMP + GLOVE BOX LAMP+ ILLUMIN, INSTRUMENT CLUSTER
4	8	SIDE LIGHTS
5	8	SIDE LIGHTS
6	8	LONG RANGE LAMP
7	8	RH HIGH BEAM
8	8	LH HIGH BEAM
9	8	FOG LAMP
10	8	LH LOW BEAM
11	8	RH LOW BEAM + REAR FOG LAMP
12	16	HEATED REAR WINDOW
13	16	HAZARD + INTERIOR LIGHTING + CLOCK +MIRROR + CIGAR LIGHTER
14	16	WINDSHIELD WASHER REAR WASHER
15	25	DOOR LOCKING/LUGGAGE COMPARTMENT RELEASE/POWER ROOF
16	16	HEATED SEAT } LOCATED ON BRACKET A-PILLAR
17	16	FUEL PUMP } LOCATED ON BRACKET A-PILLAR
18	16	AIR CONDITION
19	16	LH/RH } 2 DOOR LH/RH } 4 DOOR ELECTRIC WINDOWS
20	16	LH/RH } 2 DOOR LH/RH } 4 DOOR ELECTRIC WINDOWS
21	1	TRIP COMPUTER LOCATED NEAR FUSE BOX
22	25	LH/RH } POWER SEAT LOCATED ON BRACKET A-PILLAR
23	25	LH/RH } POWER SEAT LOCATED ON BRACKET A-PILLAR
24	16	HEATED MIRROR ON RELAY XIII
25	16	TAXI LOCATED ON MAIN LOOM NEAR BATTERY
26	8	REAR FOG LAMP LOCATED ON A-PILLAR

Diagram 1. Explanatory notes for wiring diagrams (continued)

WD

WD•4 Wiring diagrams

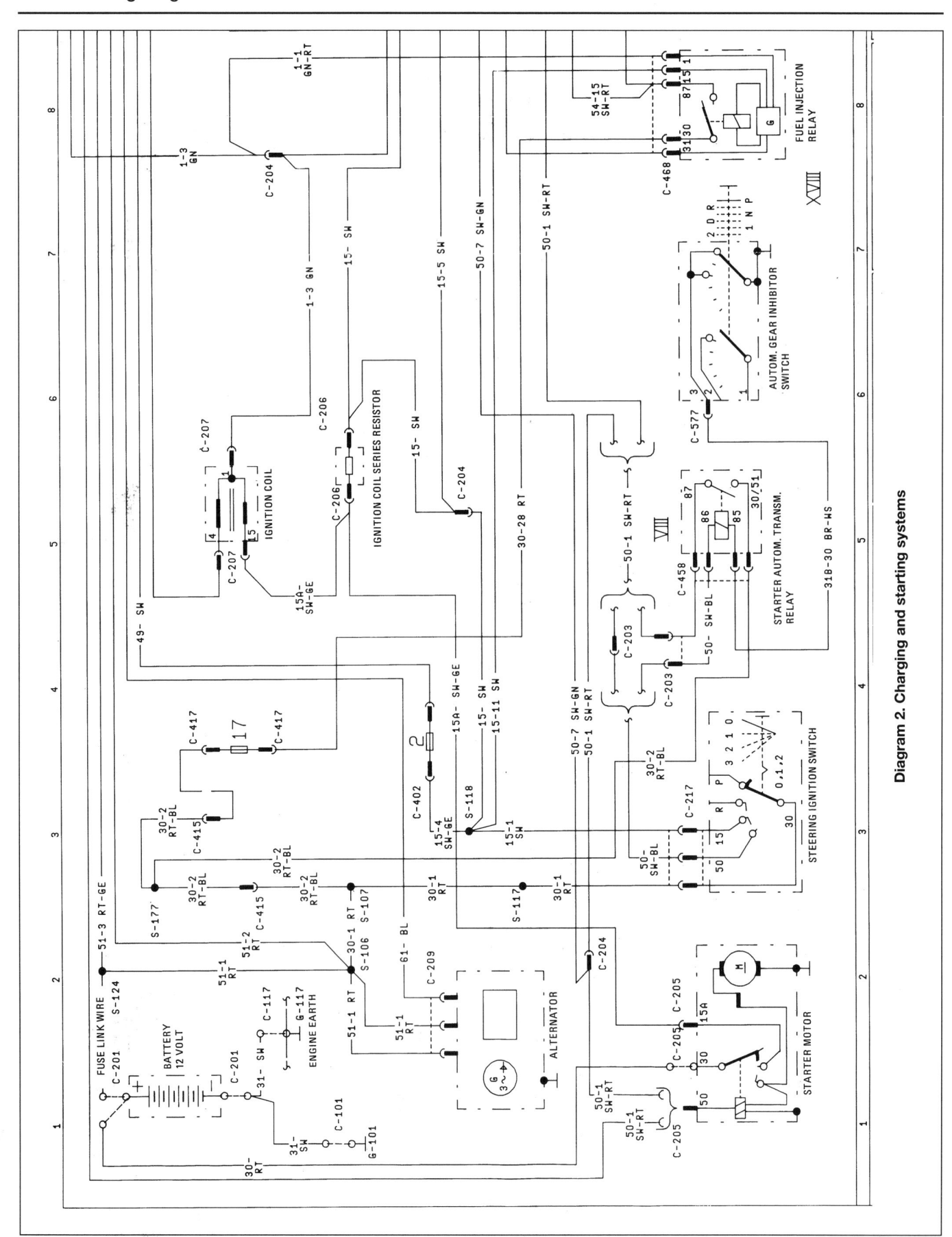

Diagram 2. Charging and starting systems

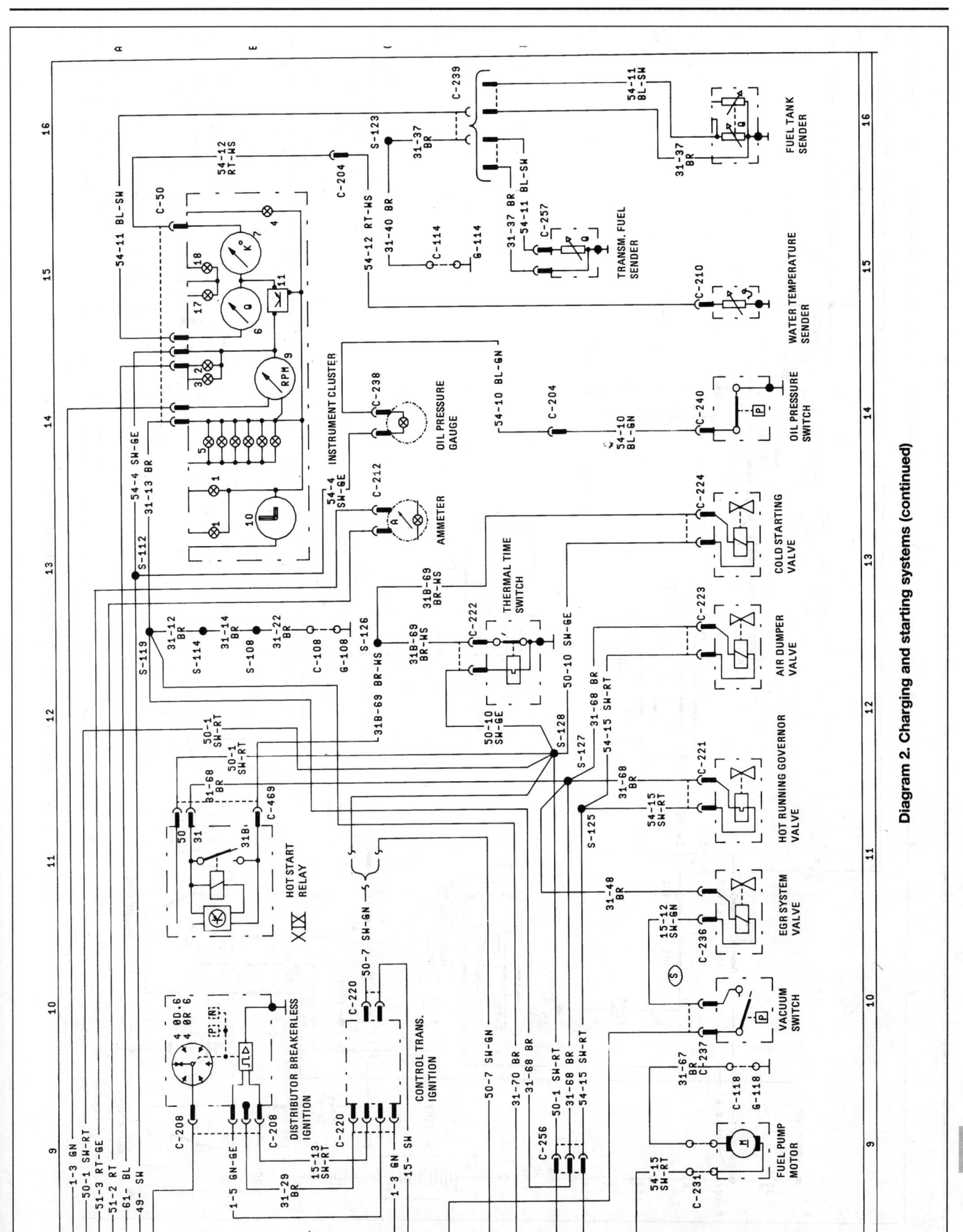

Wiring diagrams WD•5

Diagram 2. Charging and starting systems (continued)

WD•6 Wiring diagrams

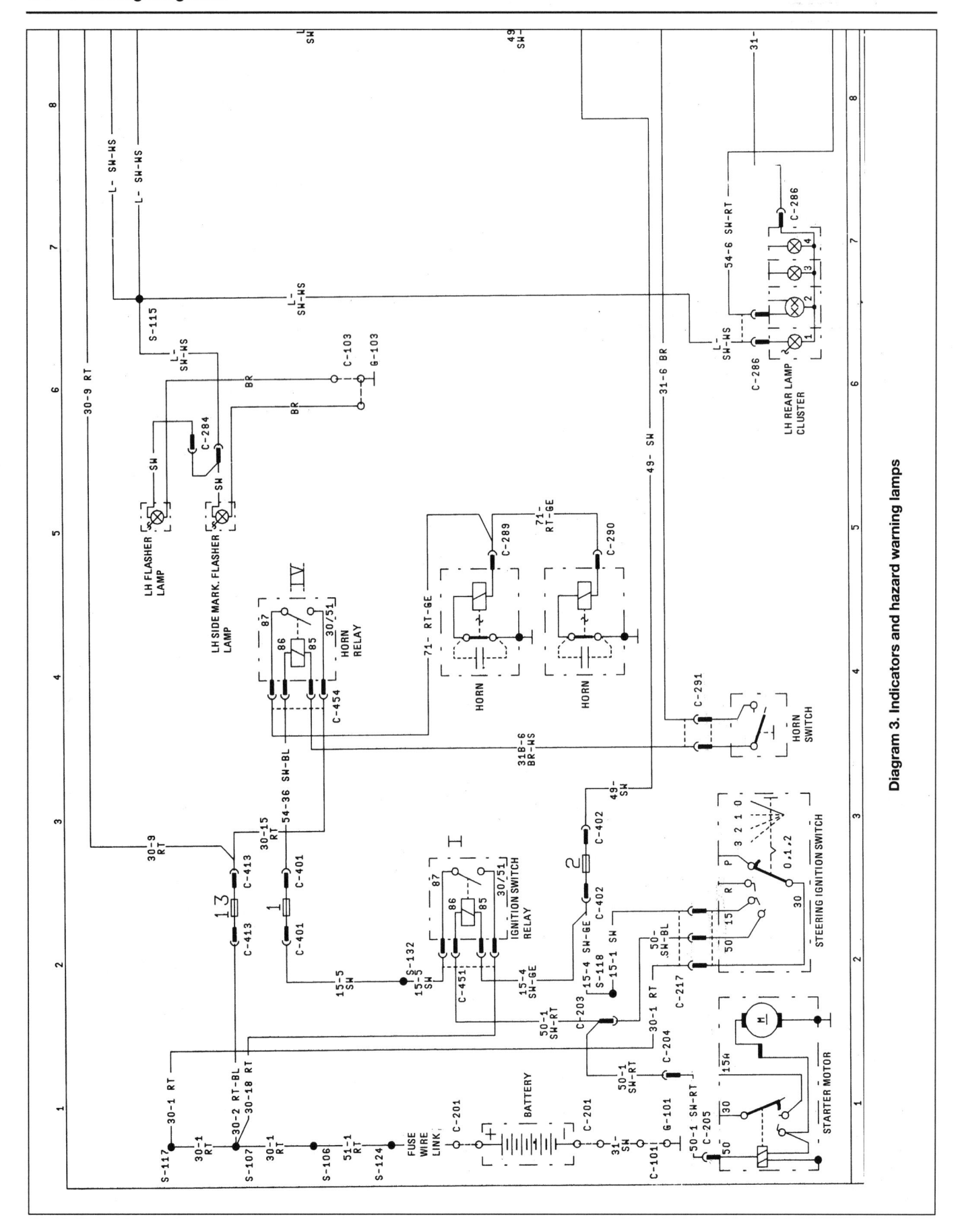

Diagram 3. Indicators and hazard warning lamps

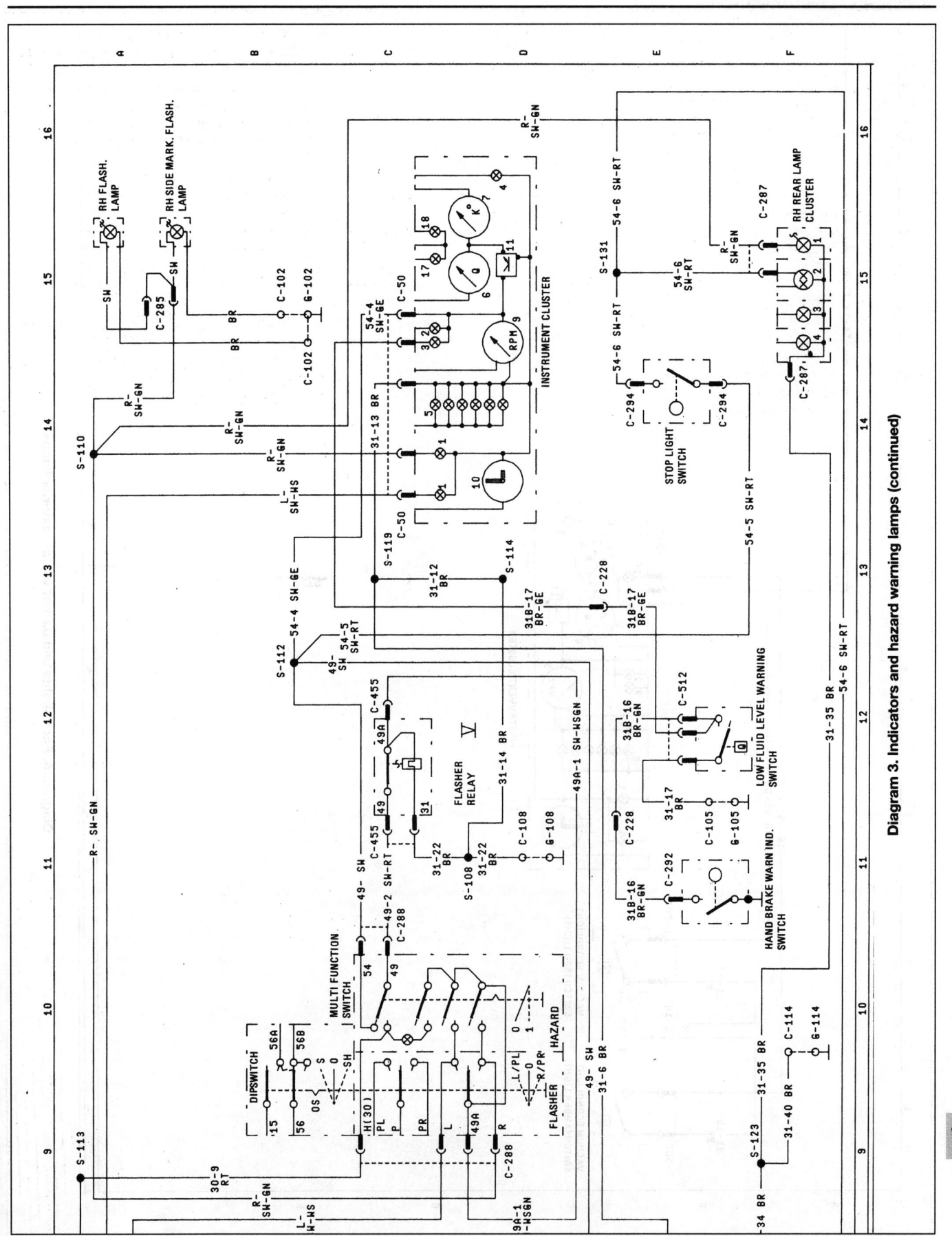

Wiring diagrams WD•7

Diagram 3. Indicators and hazard warning lamps (continued)

WD•8 Wiring diagrams

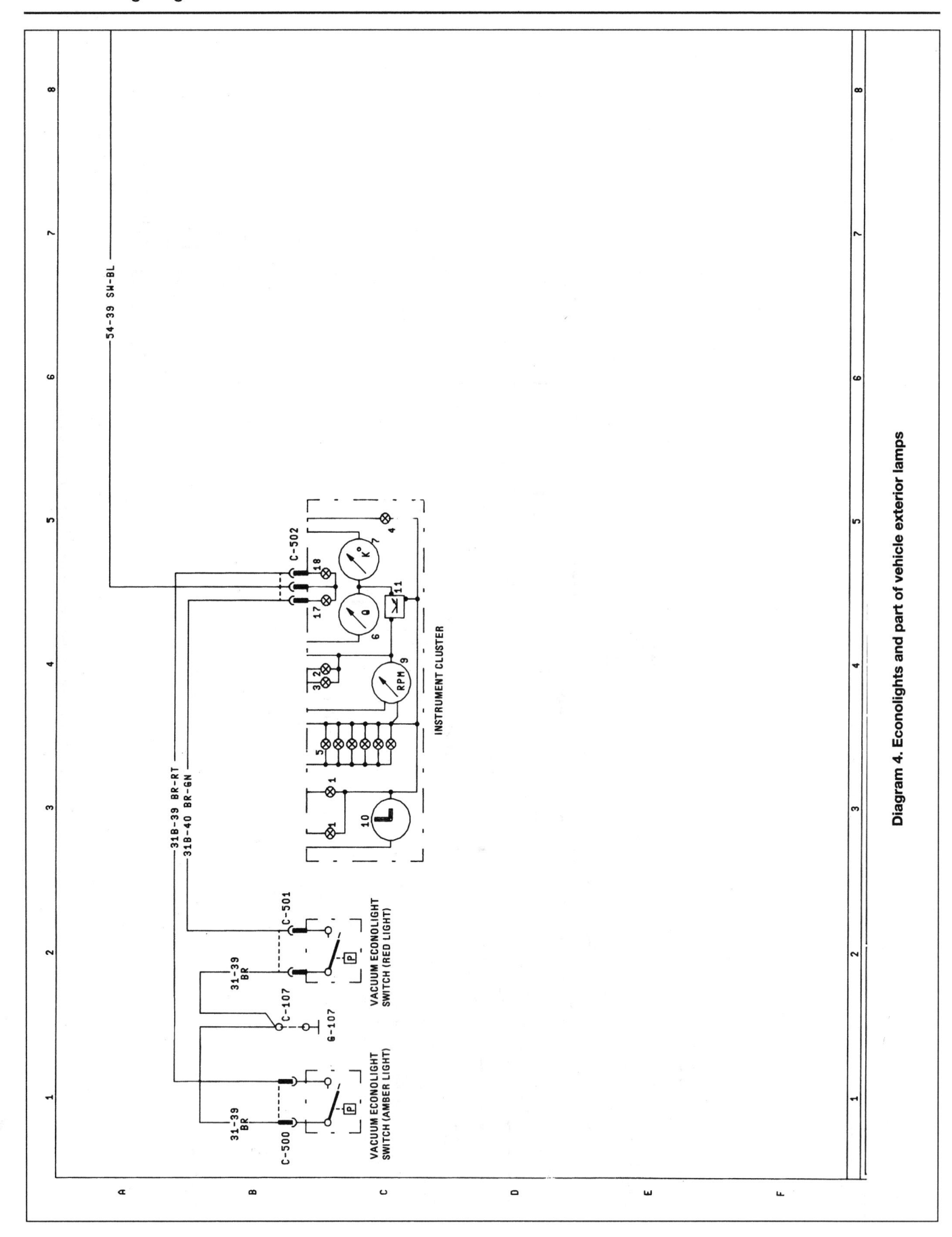

Diagram 4. Econolights and part of vehicle exterior lamps

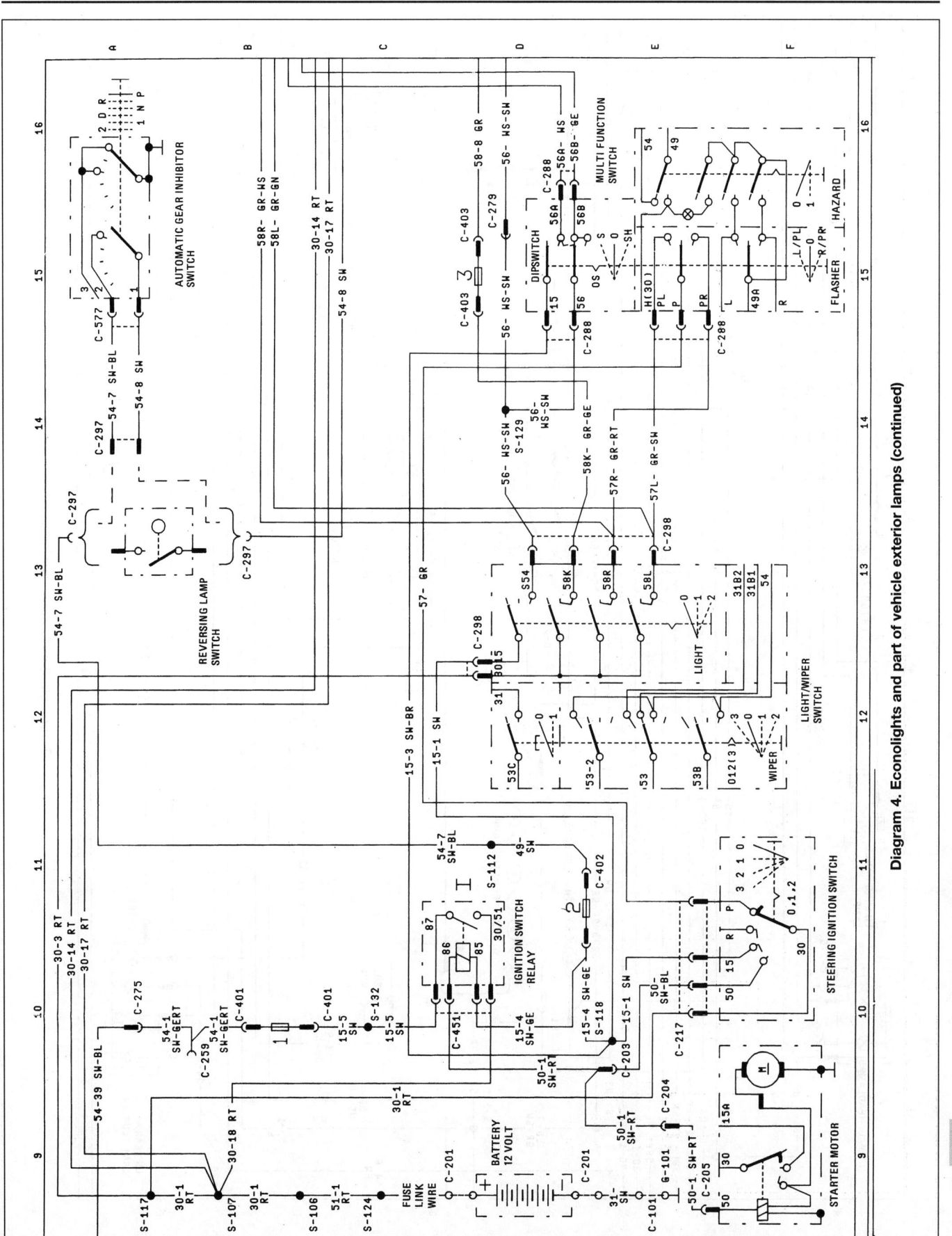

Wiring diagrams WD•9

Diagram 4. Econolights and part of vehicle exterior lamps (continued)

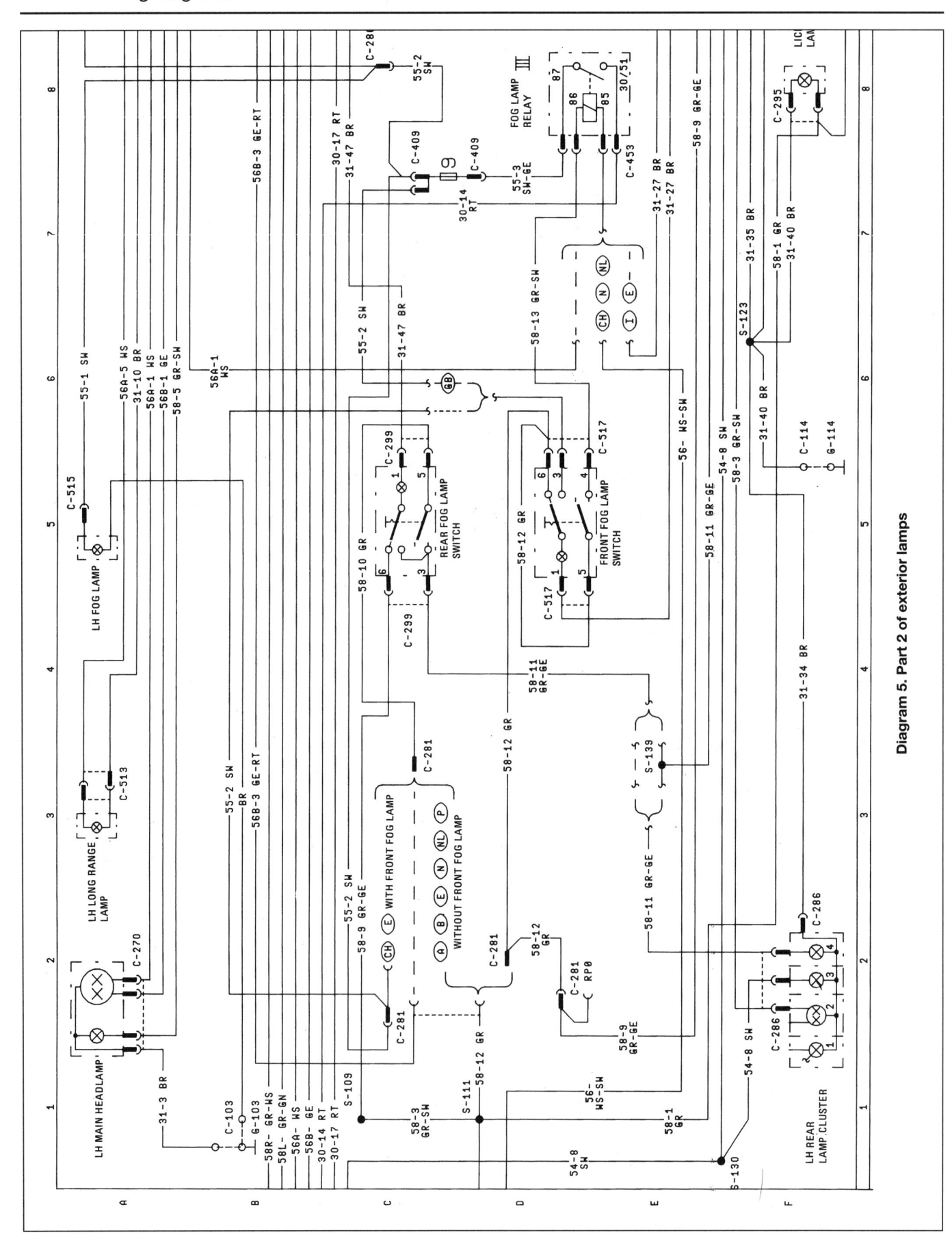

WD•10 Wiring diagrams

Diagram 5. Part 2 of exterior lamps

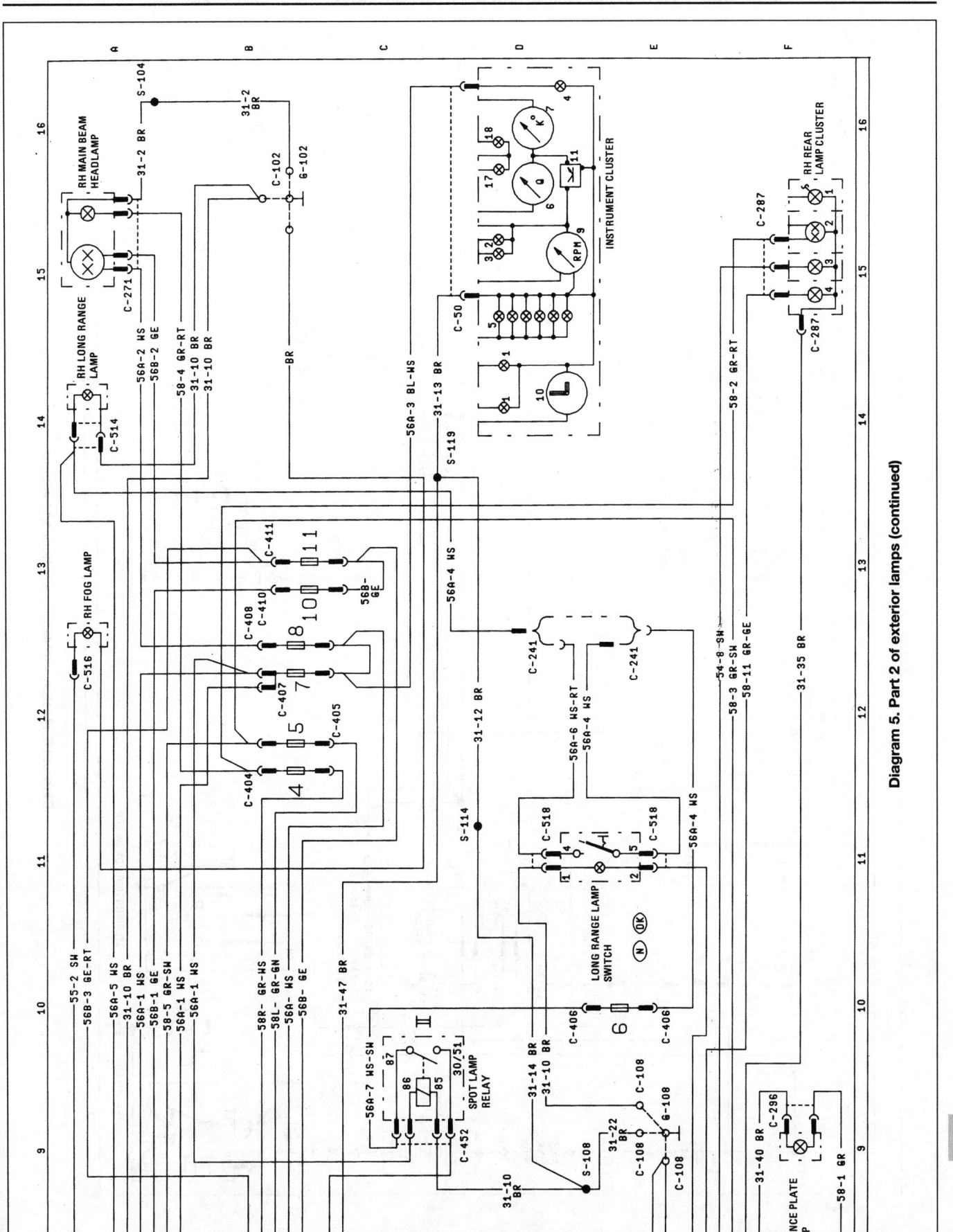

Wiring diagrams WD•11

Diagram 5. Part 2 of exterior lamps (continued)

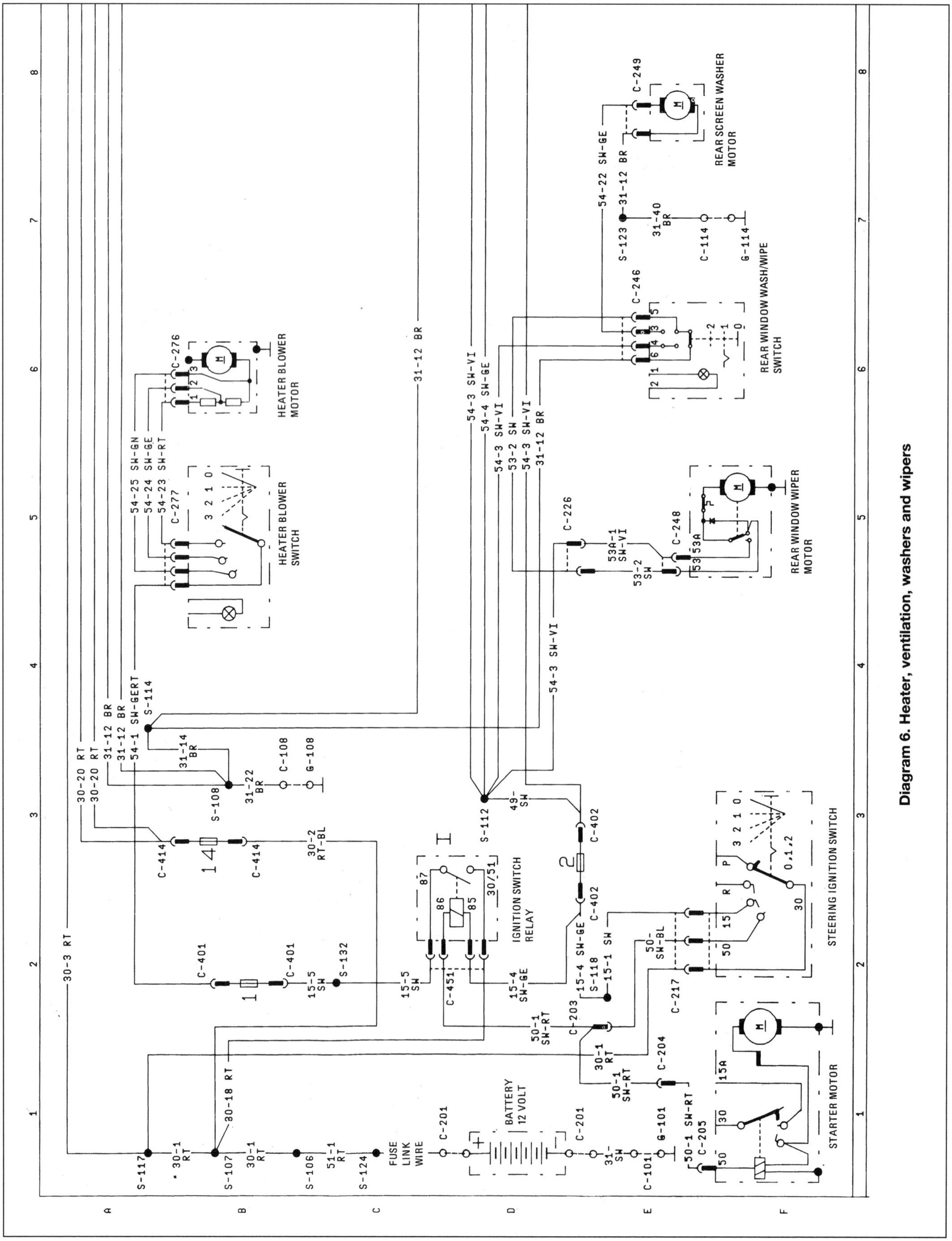

WD•12 Wiring diagrams

Diagram 6. Heater, ventilation, washers and wipers

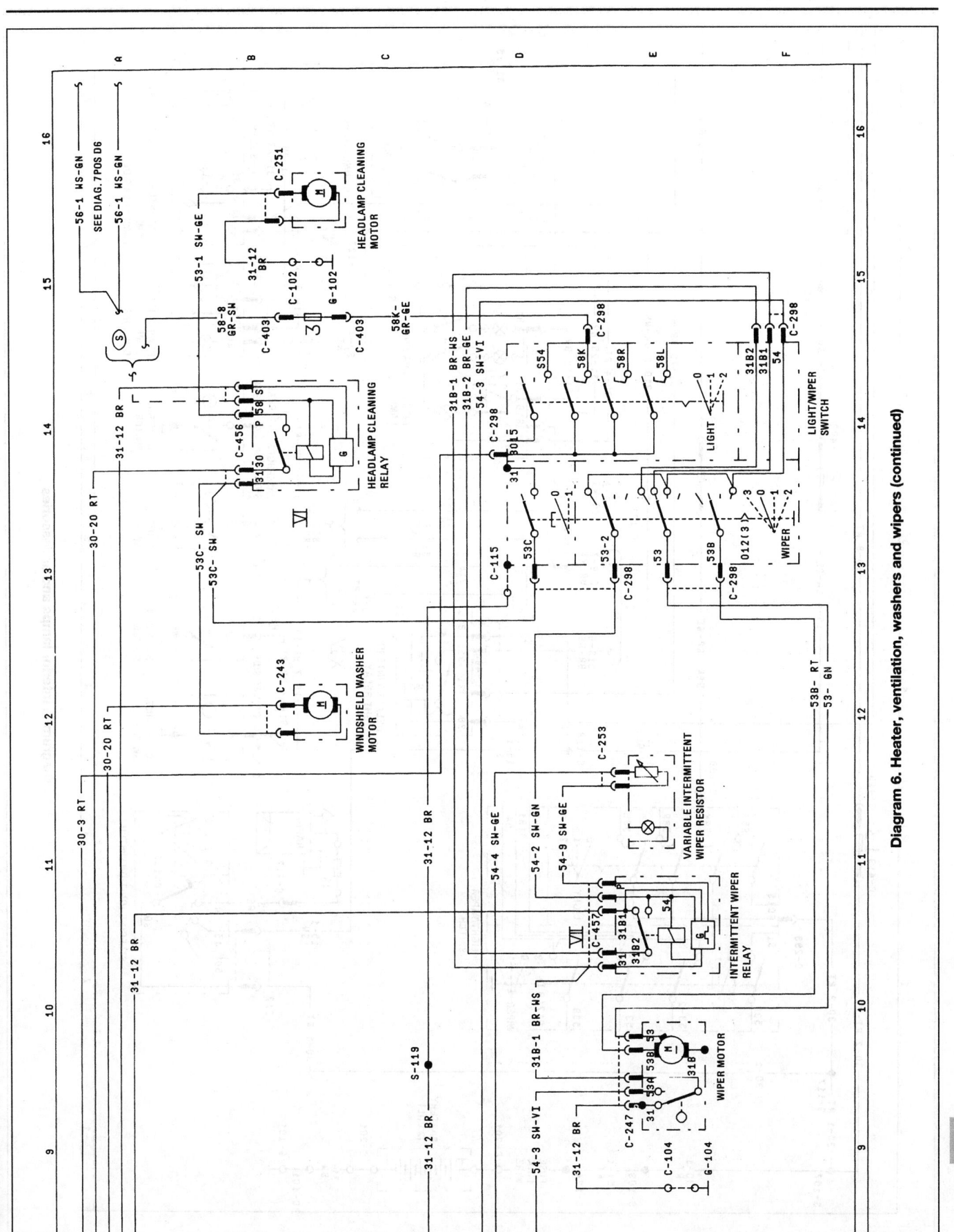

Wiring diagrams WD•13

Diagram 6. Heater, ventilation, washers and wipers (continued)

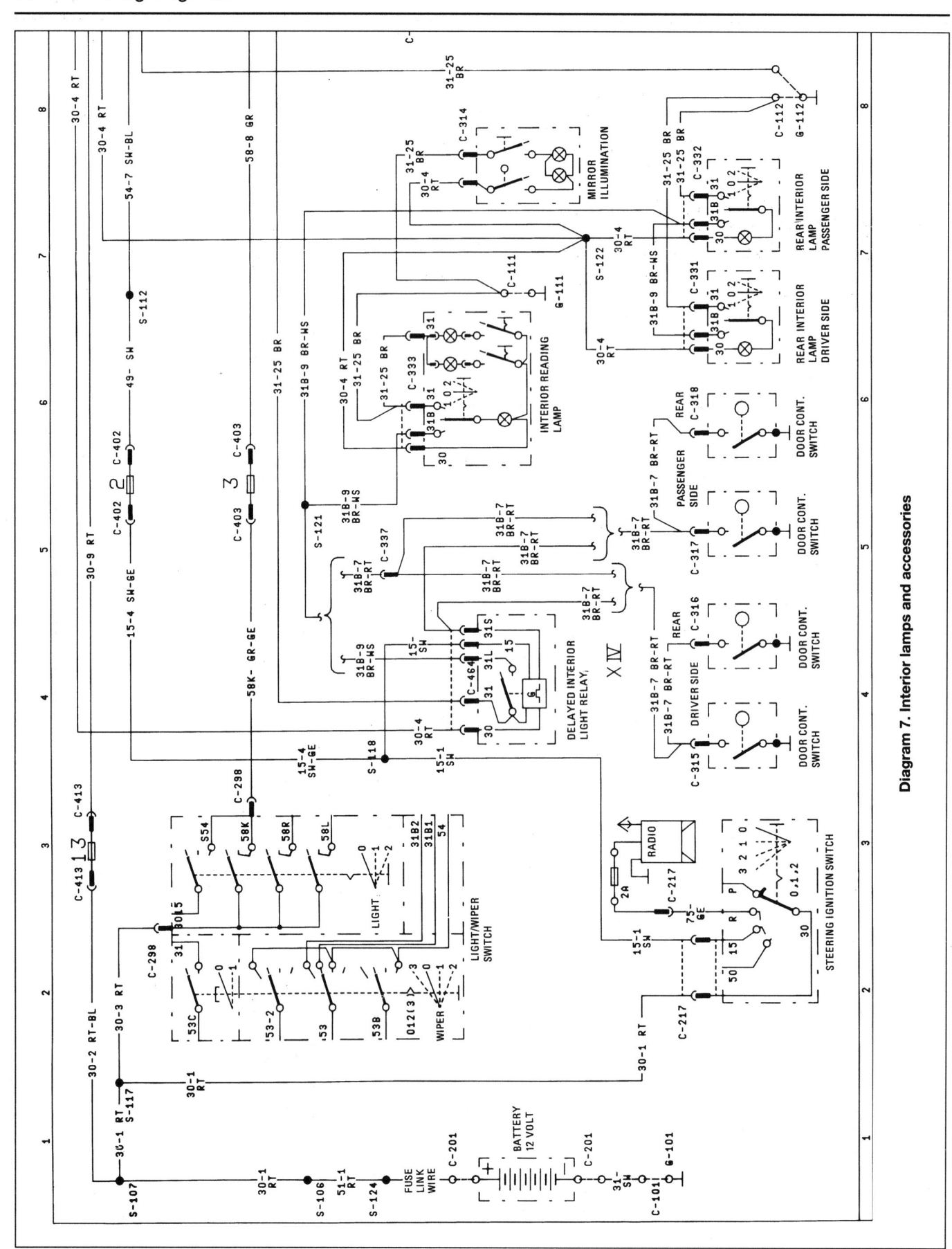

WD•14 Wiring diagrams

Diagram 7. Interior lamps and accessories

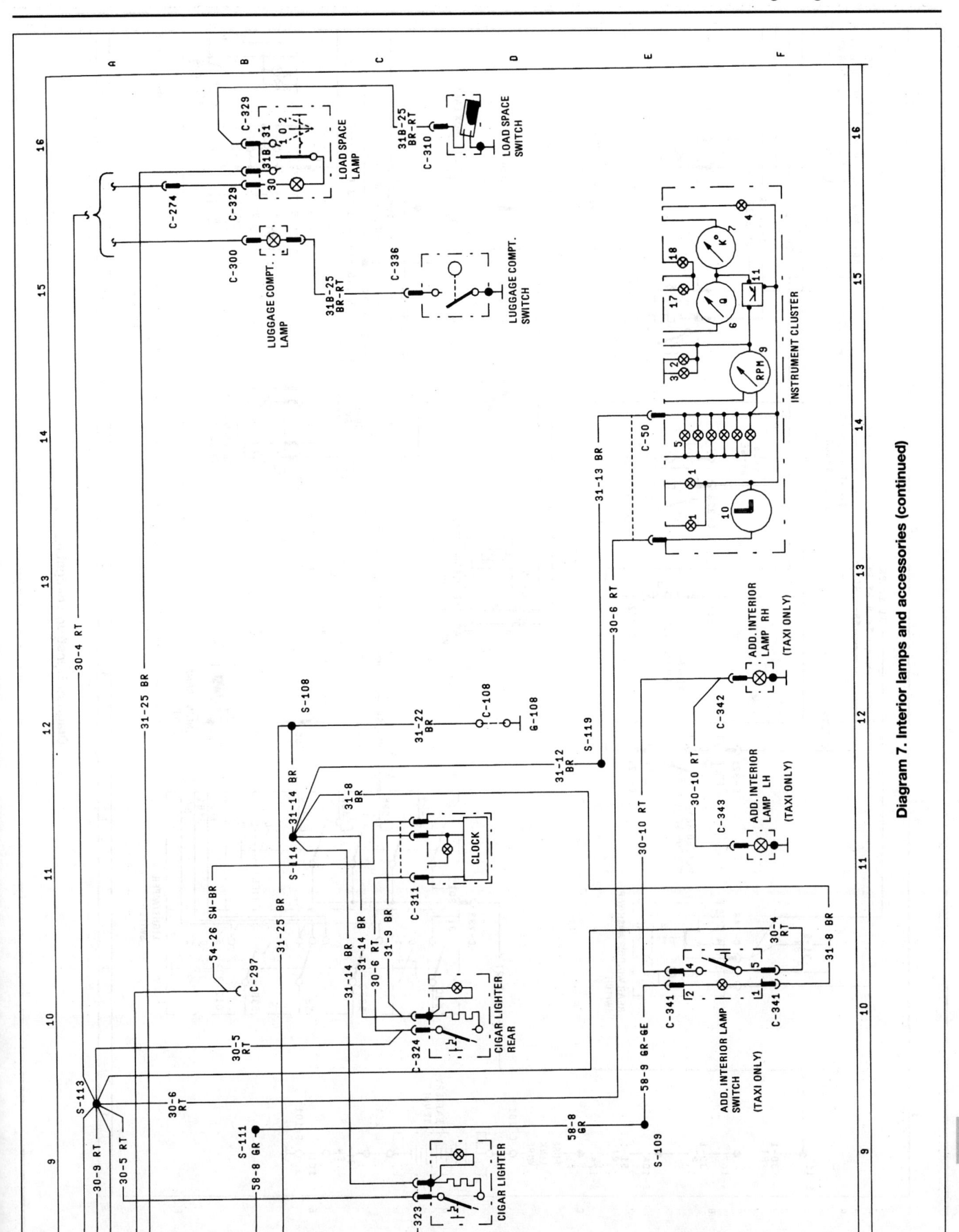

Wiring diagrams WD•15

Diagram 7. Interior lamps and accessories (continued)

Diagram 8. Instrument lighting

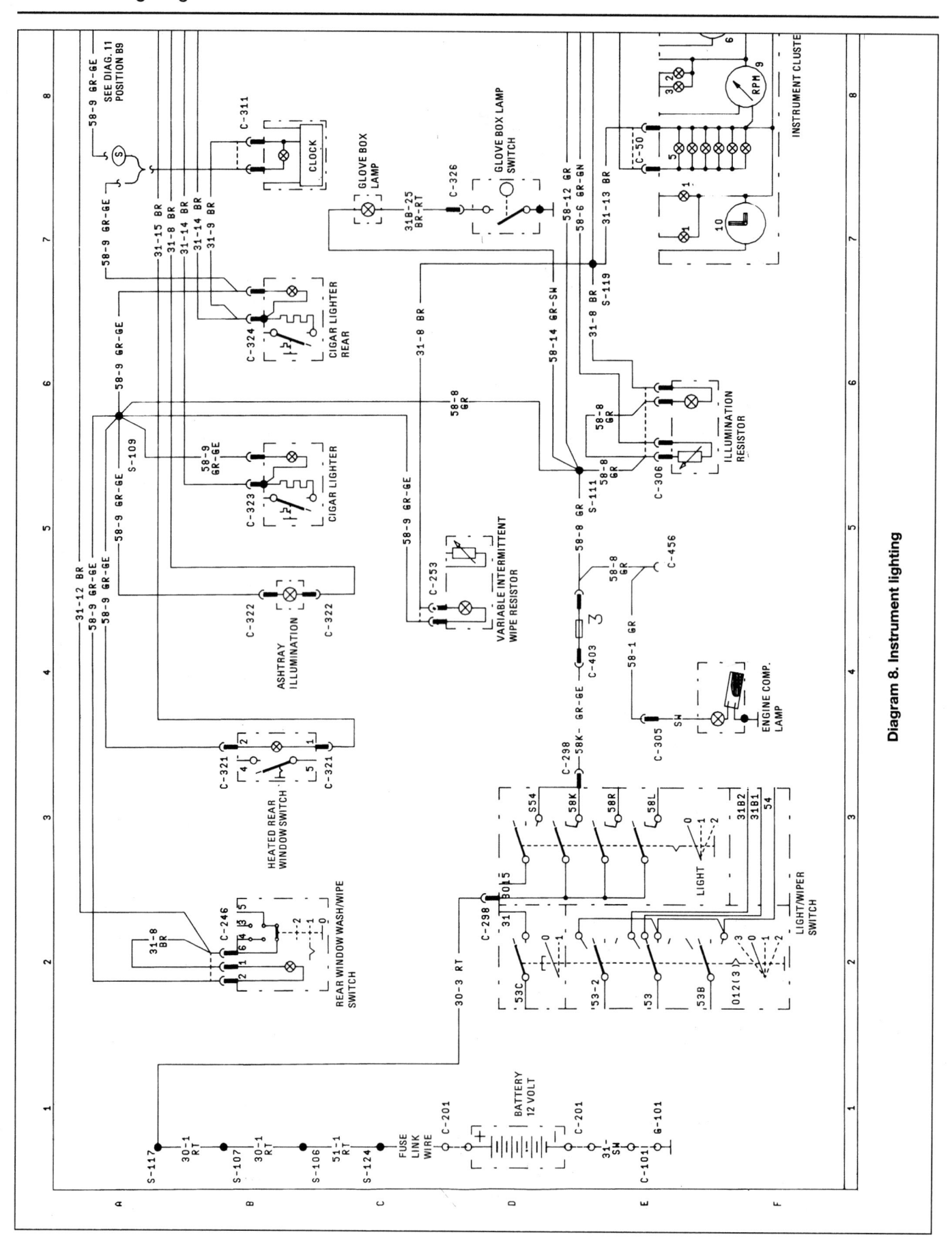

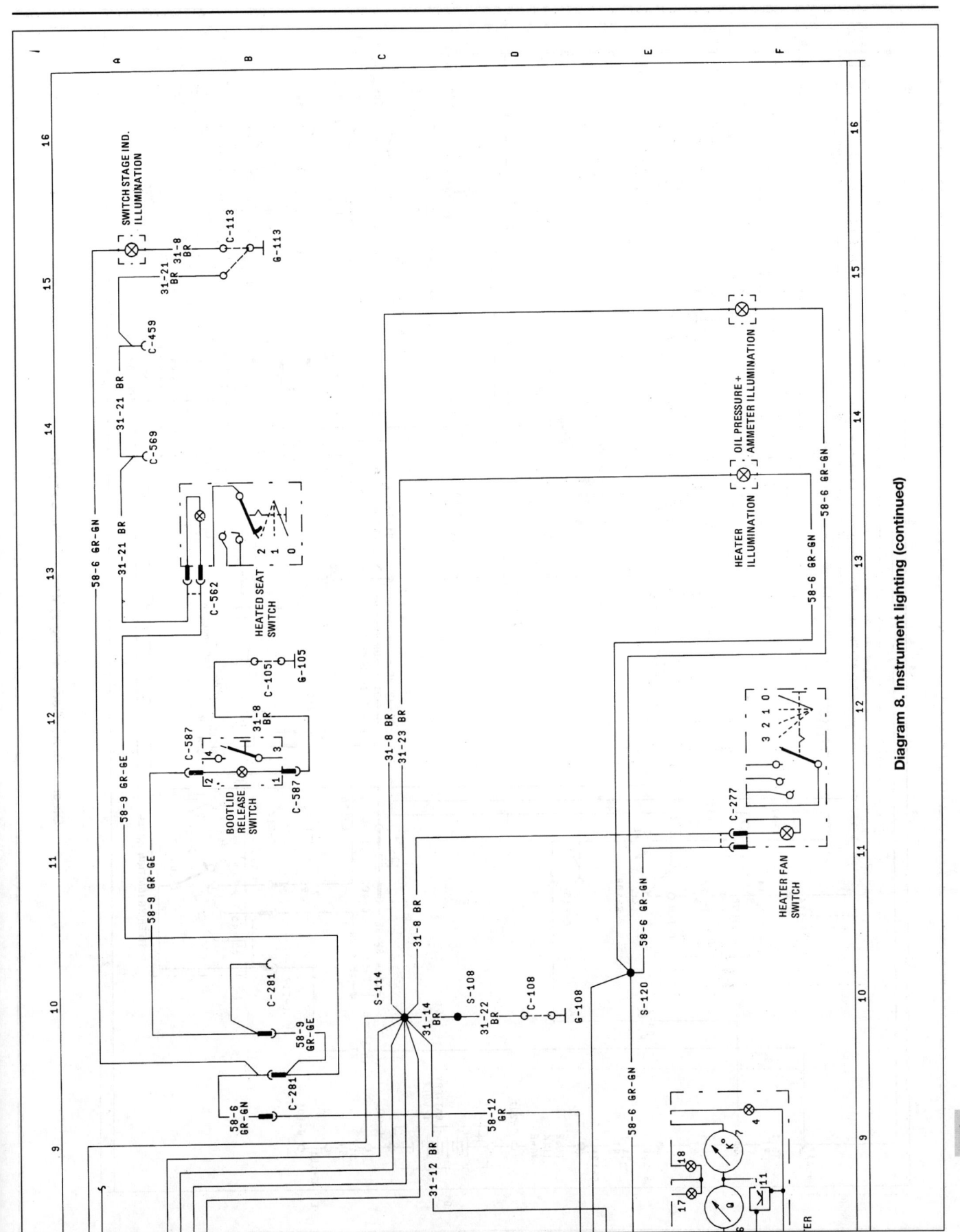

Wiring diagrams WD•17

Diagram 8. Instrument lighting (continued)

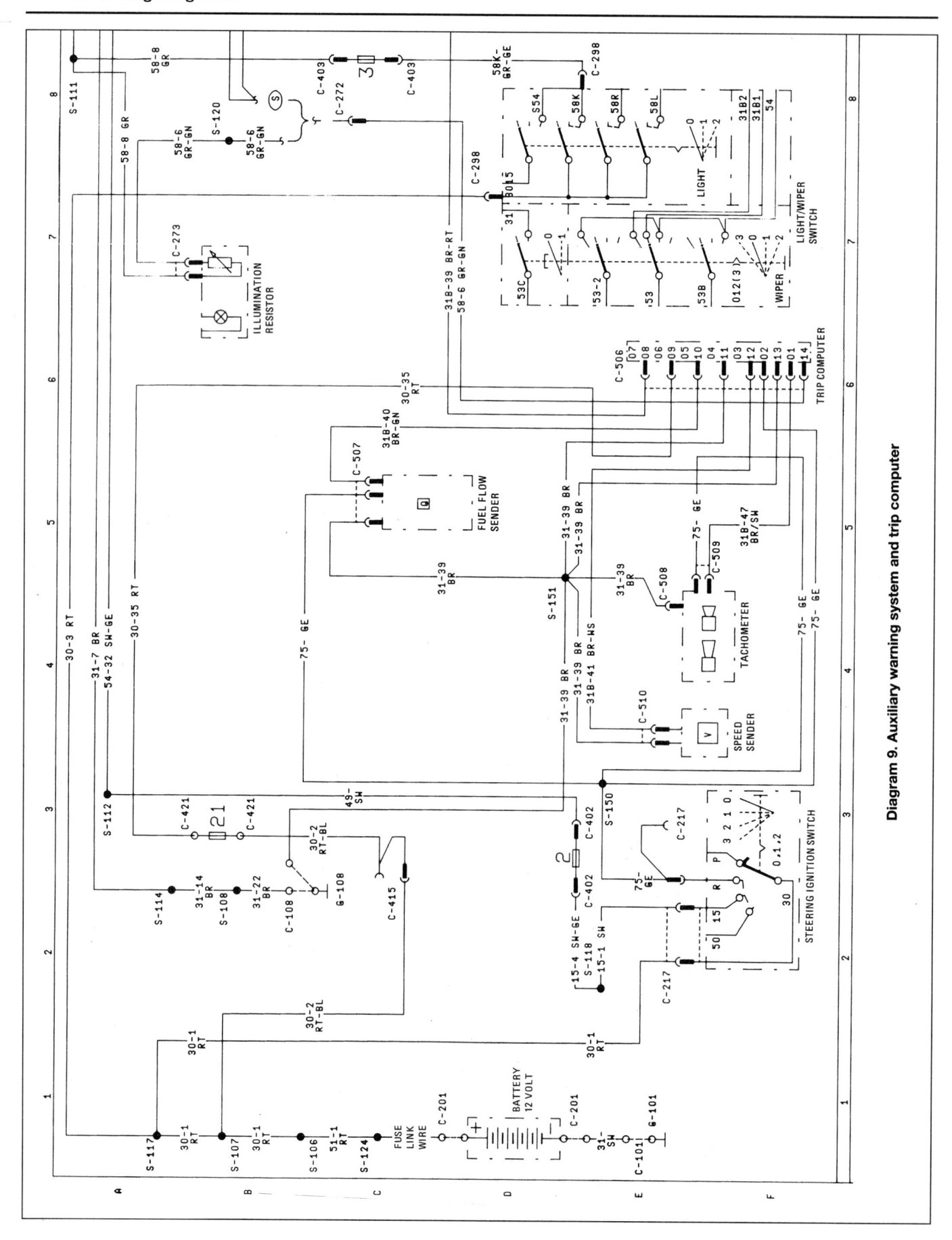

WD•18 Wiring diagrams

Diagram 9. Auxiliary warning system and trip computer

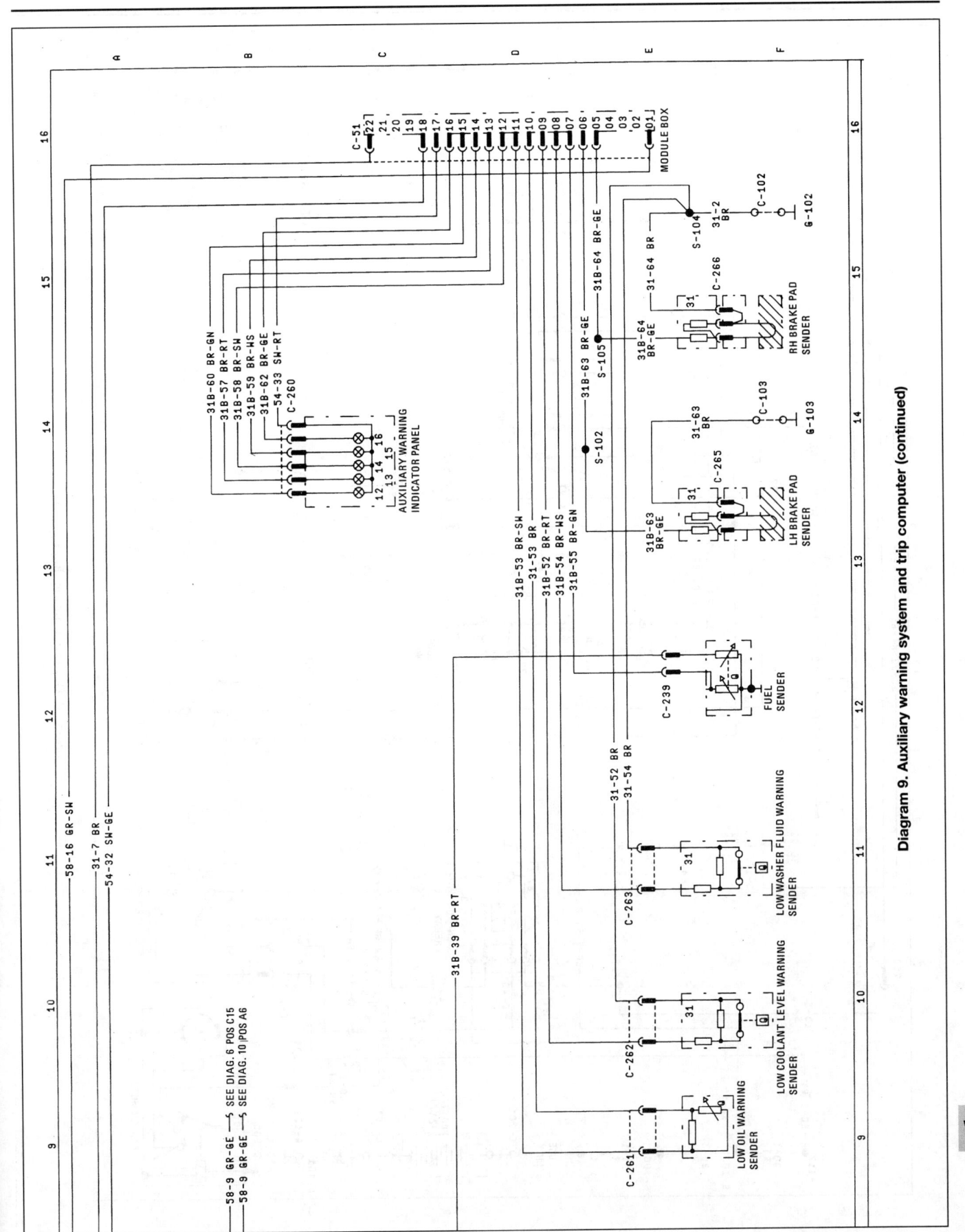

Wiring diagrams WD•19

Diagram 9. Auxiliary warning system and trip computer (continued)

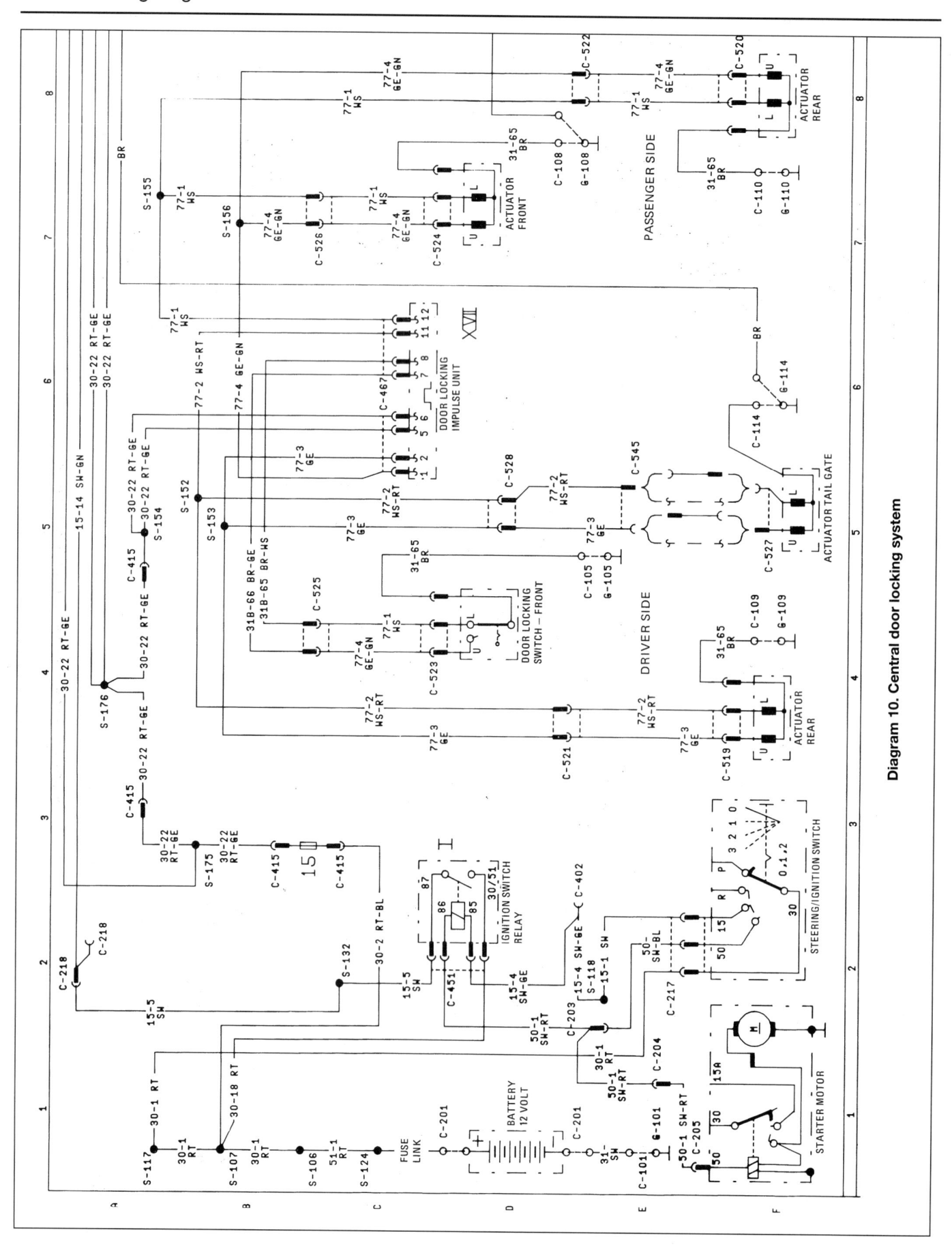

WD•20 Wiring diagrams

Diagram 10. Central door locking system

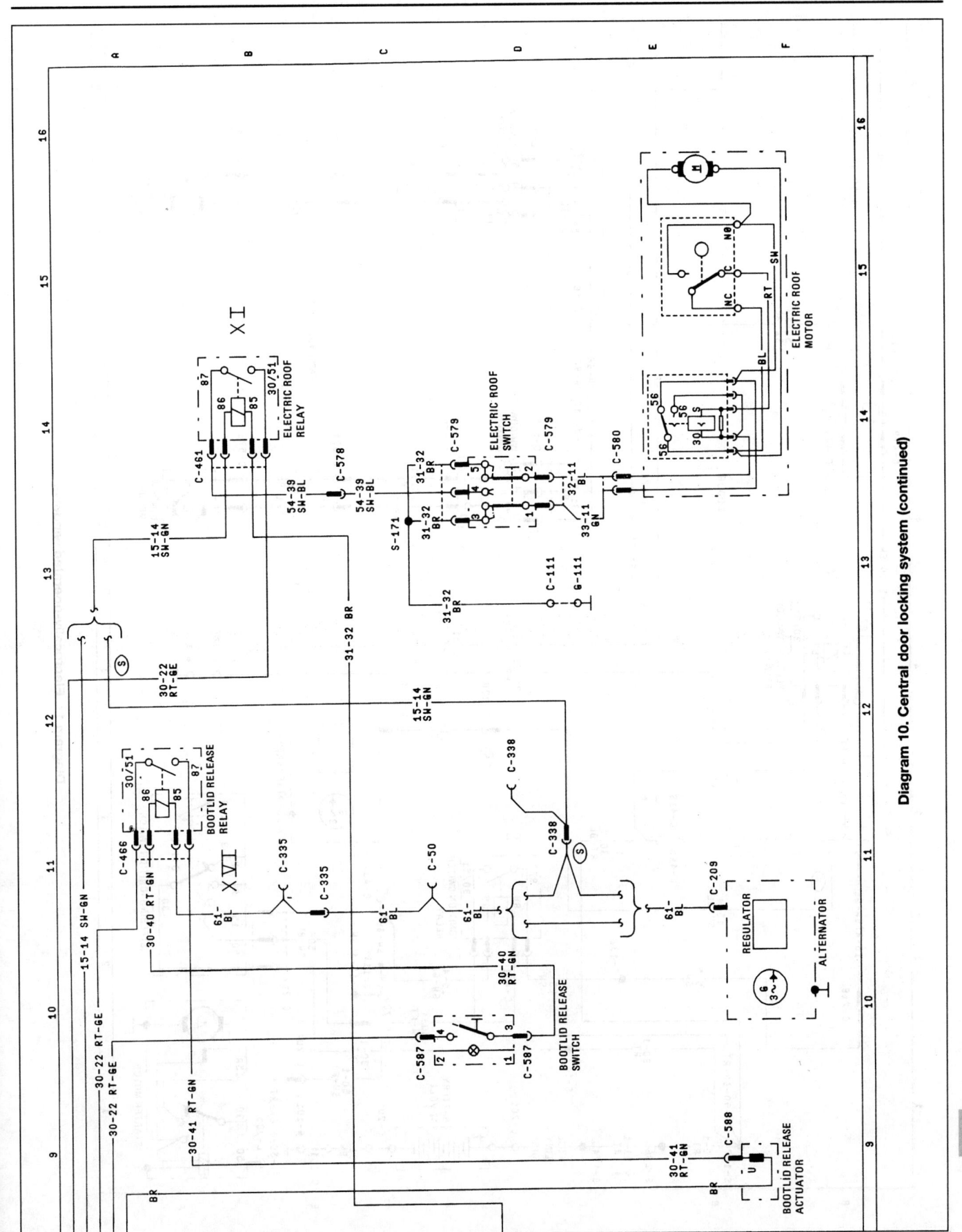

Wiring diagrams WD•21

Diagram 10. Central door locking system (continued)

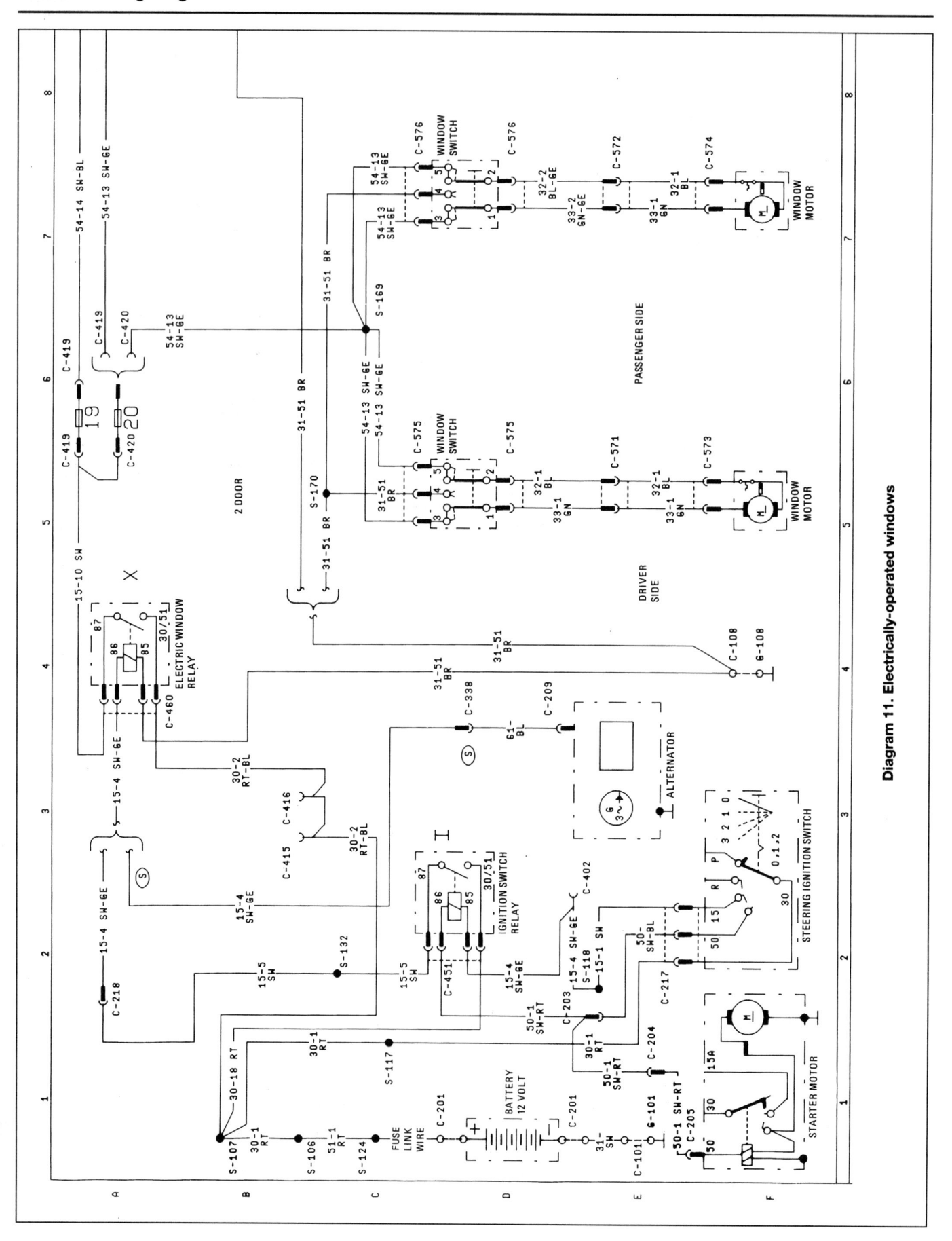

WD•22 Wiring diagrams

Diagram 11. Electrically-operated windows

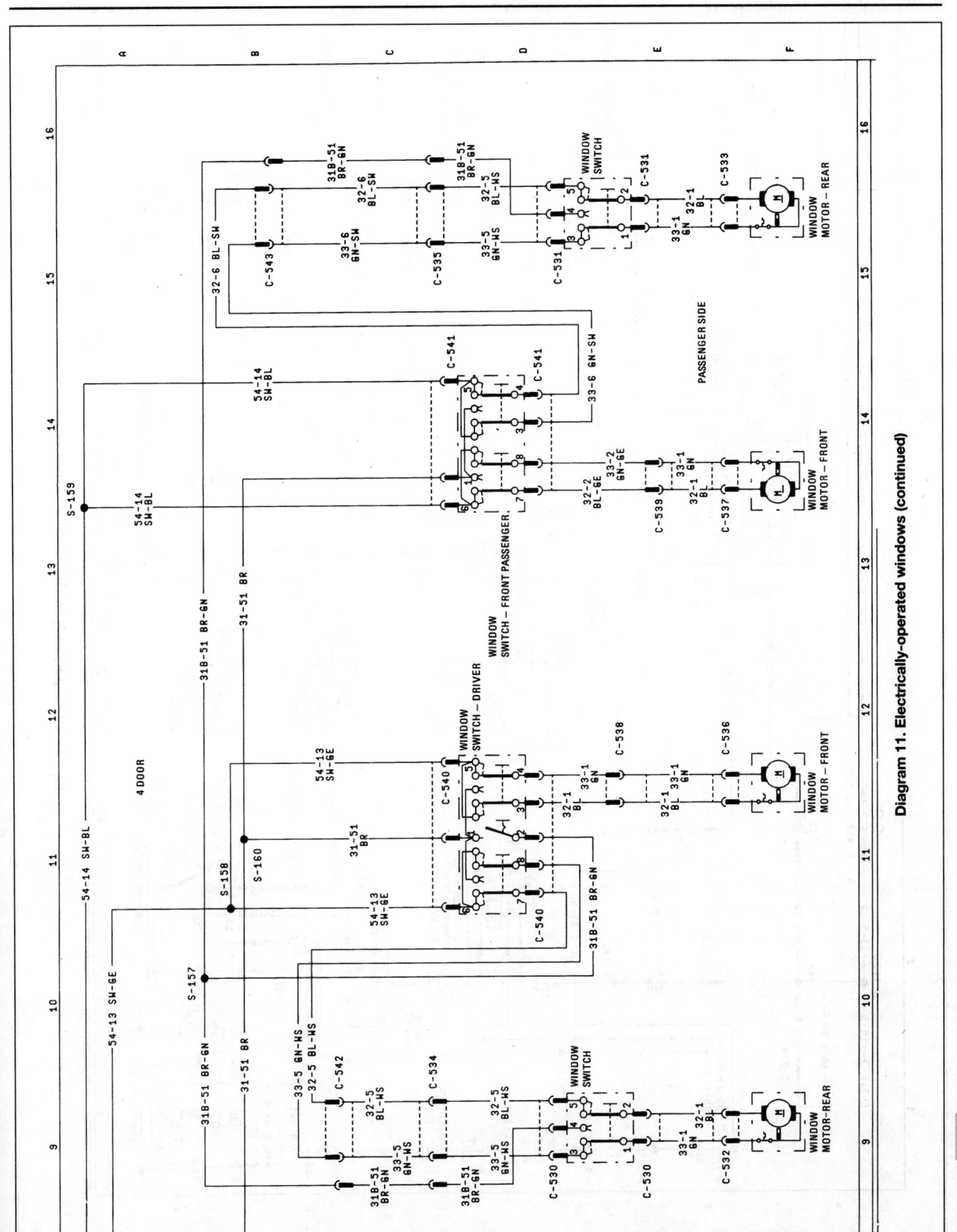

Wiring diagrams WD•23

Diagram 11. Electrically-operated windows (continued)

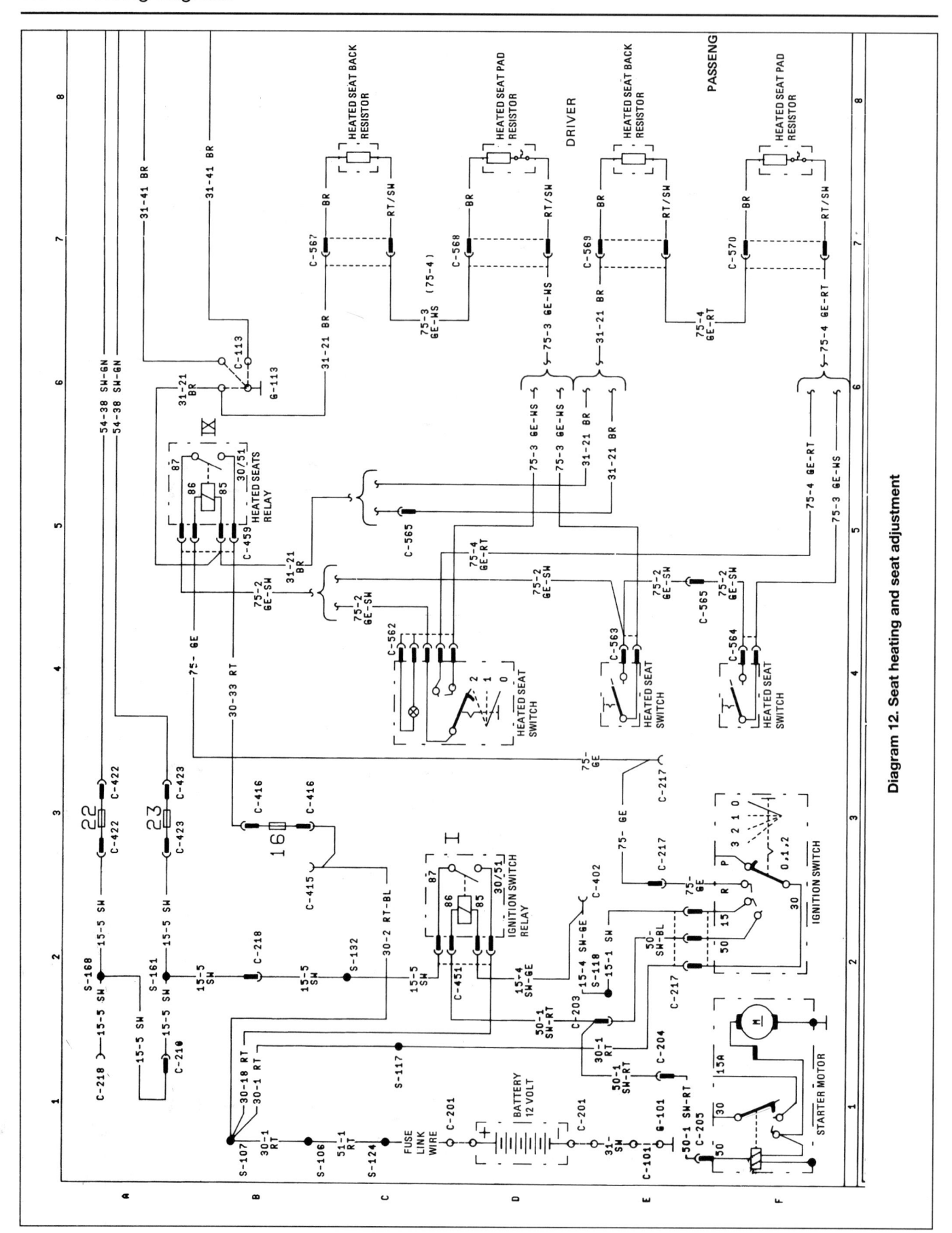

WD•24 Wiring diagrams

Diagram 12. Seat heating and seat adjustment

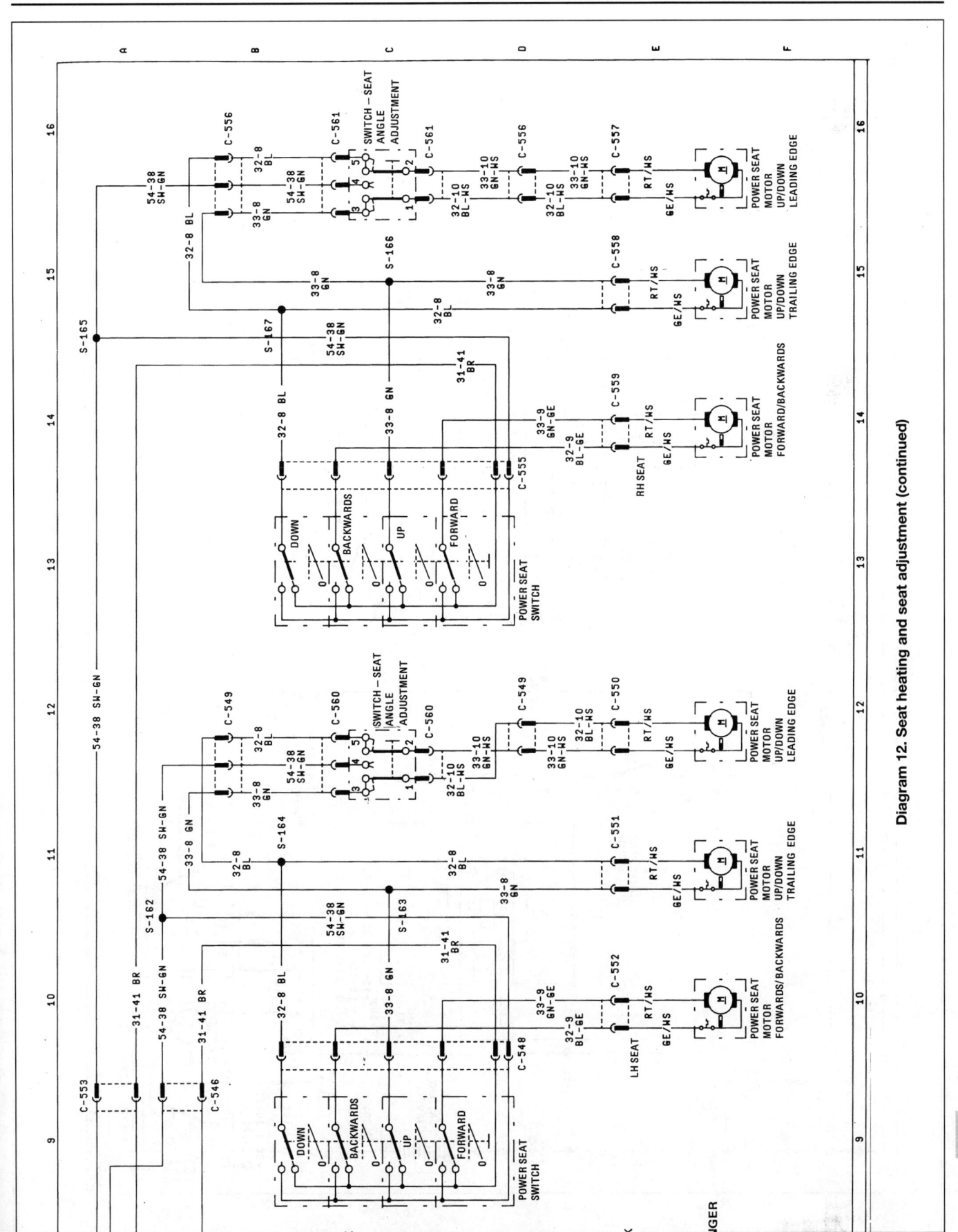

Wiring diagrams WD•25

Diagram 12. Seat heating and seat adjustment (continued)

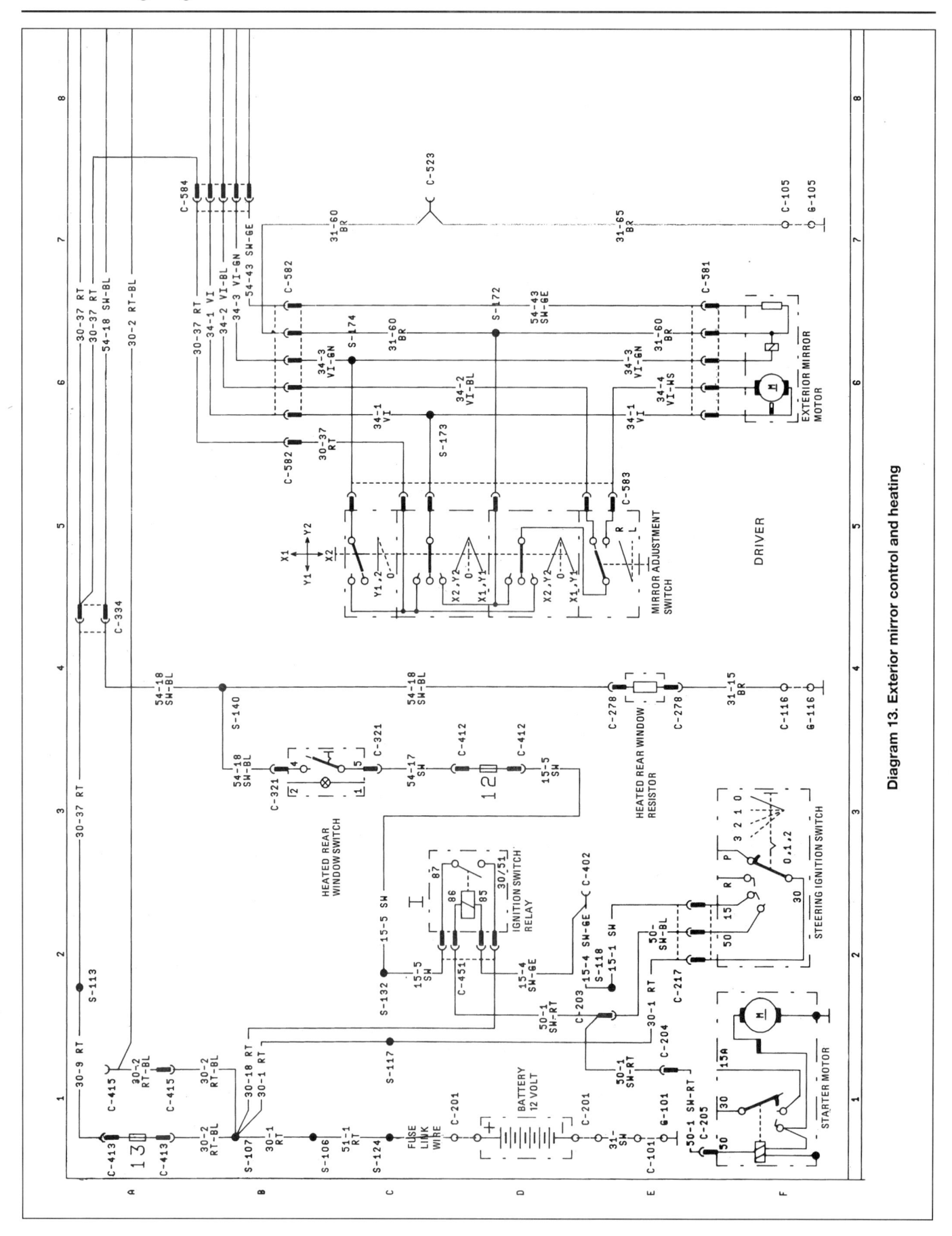

WD•26 Wiring diagrams

Diagram 13. Exterior mirror control and heating

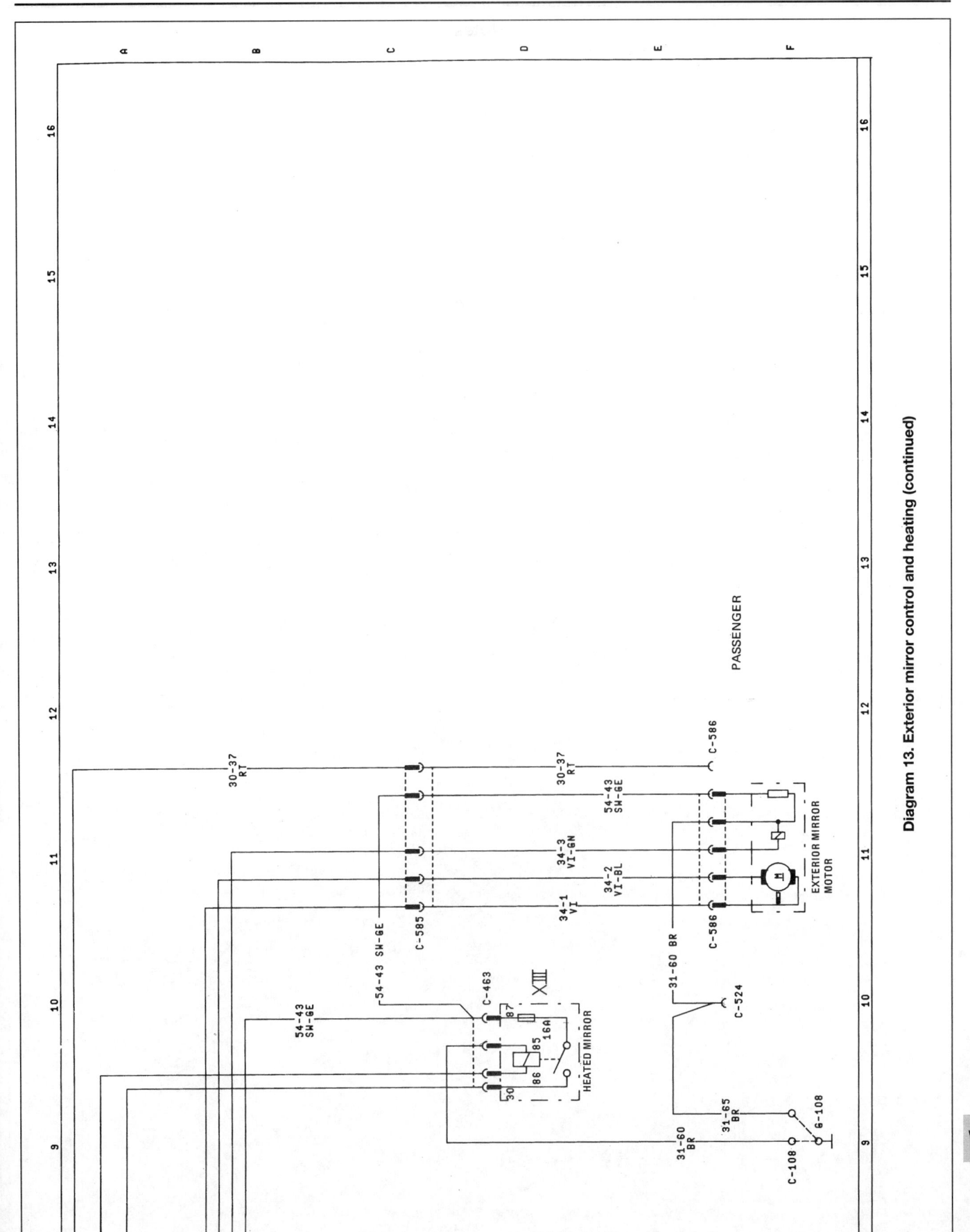

Wiring diagrams WD•27

Diagram 13. Exterior mirror control and heating (continued)

Notes

Fault Finding

Introduction

The car owner who does his or her own maintenance according to the recommended schedules should not have to use this section of the manual very often. Modern component reliability is such that, provided those items subject to wear or deterioration are inspected or renewed at the specified intervals, sudden failure is comparatively rare. Faults do not usually just happen as a result of sudden failure, but develop over a period of time. Major mechanical failures in particular are usually preceded by characteristic symptoms over hundreds or even thousands of miles. Those components which do occasionally fail without warning are often small and easily carried in the car.

With any fault finding, the first step is to decide where to begin investigations. Sometimes this is obvious, but on other occasions a little detective work will be necessary. The owner who makes half a dozen haphazard adjustments or replacements may be successful in curing a fault (or its symptoms), but he will be none the wiser if the fault recurs and he may well have spent more time and money than was necessary. A calm and logical approach will be found to be more satisfactory in the long run. Always take into account any warning signs or abnormalities that may have been noticed in the period preceding the fault - power loss, high or low gauge readings, unusual noises or smells, etc - and remember that failure of components such as fuses or spark plugs may only be pointers to some underlying fault.

The pages which follow here are intended to help in cases of failure to start or breakdown on the road. There is also a Fault Diagnosis Section at the end of each Chapter which should be consulted if the preliminary checks prove unfruitful. Whatever the fault, certain basic principles apply. These are as follows:

Verify the fault. This is simply a matter of being sure that you know what the symptoms are before starting work. This is particularly important if you are investigating a fault for someone else who may not have described it very accurately.

Don't overlook the obvious. For example, if the car won't start, is there petrol in the tank? (Don't take anyone else's word on this particular point, and don't trust the fuel gauge either!) If an electrical fault is indicated, look for loose or broken wires before digging out the test gear.

Cure the disease, not the symptom. Substituting a flat battery with a fully charged one will get you off the hard shoulder, but if the underlying cause is not attended to, the new battery will go the same way, Similarly, changing oil-fouled spark plugs for a new set will get you moving again, but remember that the reason for the fouling (if it wasn't simply an incorrect grade of plug) will have to be established and corrected.

Don't take anything for granted. Particularly, don't forget that a "new" component may itself be defective (especially if it's been rattling round in the boot for months), and don't leave components out of a fault diagnosis sequence just because they are new or recently fitted. When you do finally diagnose a difficult fault, you'll probably realise that all the evidence was there from the start.

Electrical faults

Electrical faults can be more puzzling than straightforward mechanical failures, but they are no less susceptible to logical analysis if the basic principles of operation are understood. Car electrical wiring exists in extremely unfavourable conditions - heat, vibration and chemical attack - and the first things to look for are loose or corroded connections and broken or chafed wires, especially where the wires pass through holes in the bodywork or are subject to vibration.

All metal-bodied cars in current production have one pole of the battery "earthed", ie connected to the car bodywork, and in nearly all modern cars it is the negative (-) terminal. The various electrical components - motors, bulb holders etc - are also connected to earth, either by means of a lead or directly by their mountings. Electric current flows through the component and then back to the battery via the car bodywork. If the component mounting is loose or corroded, or if a good path back to the battery is not available, the circuit will be incomplete and malfunction will result. The engine and/or gearbox are also earthed by means of flexible metal straps to the body or subframe; if these straps are loose or missing, starter motor, generator and ignition trouble may result.

Assuming the earth return to be satisfactory, electrical faults will be due either to component malfunction or to defects in the current supply. Individual components are dealt with in Chapter 10. If supply wires are broken or cracked internally this results in an open-circuit, and the easiest way to check for this is to bypass the suspect wire temporarily with a length of wire having a crocodile clip or suitable connector at each end. Alternatively, a 12V test lamp can be used to verify the presence of supply voltage at various points along the wire and the break can be thus isolated.

If a bare portion of a live wire touches the car bodywork or other earthed metal part, the electricity will take the low-resistance path thus formed back to the battery: this is known as a short-circuit. Hopefully a short-circuit will blow a fuse, but otherwise it may cause burning of the insulation (and possibly further short-circuits) or even a fire. This is why it is inadvisable to bypass persistently blowing fuses with silver foil or wire.

Spares and tool kit

Most cars are only supplied with sufficient tools for wheel changing; the Maintenance and minor repair tool kit detailed in Tools and working facilities, with the addition of a hammer, is probably sufficient for those repairs that most motorists would consider attempting at the roadside. In addition a few items which can be fitted without too much trouble in the event of a breakdown should be carried. Experience and available space will modify the list below, but the following may save having to call on professional assistance:

- [ ] *Spark plugs, clean and correctly gapped*
- [ ] *HT lead and plug cap - long enough to reach the plug furthest from the distributor*
- [ ] *Distributor rotor, condenser and contact breaker points (as applicable)*
- [ ] *Drivebelt(s) - emergency type may suffice*
- [ ] *Spare fuses*
- [ ] *Set of principal light bulbs*
- [ ] *Worm drive hose clips*
- [ ] *Tin of radiator sealer and hose bandage*
- [ ] *Tube of filler paste*
- [ ] *Exhaust bandage*
- [ ] *Roll of insulating tape*
- [ ] *Length of soft iron wire*
- [ ] *Length of electrical flex*
- [ ] *Torch or inspection lamp (can double as test lamp)*
- [ ] *Battery jump leads*
- [ ] *Tow-rope*
- [ ] *Ignition water dispersant aerosol*
- [ ] *Litre of engine oil*
- [ ] *Sealed can of hydraulic fluid*
- [ ] *Emergency windscreen*
- [ ] *Tyre valve core*

If spare fuel is carried, a can designed for the purpose should be used to minimise risks of leakage and collision damage. A first aid kit and a warning triangle, whilst not at present compulsory in the UK, are obviously sensible items to carry in addition to the above.

When touring abroad it may be advisable to carry additional spares which, even if you cannot fit them yourself, could save having to wait while parts are obtained. The items below may be worth considering:

Cylinder head gasket
Alternator brushes

One of the motoring organisations will be able to advise on availability of fuel etc in foreign countries.

REF•2 Fault Finding

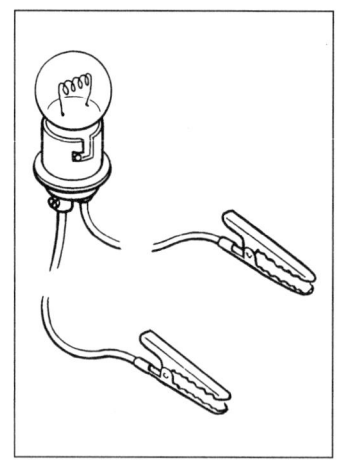

Engine will not start

Engine fails to turn when starter operated
- ☐ Flat battery (recharge, use jump leads, or push start)
- ☐ Battery terminals loose or corroded
- ☐ Battery earth to body defective
- ☐ Engine earth strap loose or broken
- ☐ Starter motor (or solenoid) wiring loose or broken
- ☐ Automatic transmission selector in wrong position, or inhibitor switch faulty
- ☐ Ignition/starter switch faulty
- ☐ Major mechanical failure (seizure) or long disuse (piston rings rusted to bores)
- ☐ Starter or solenoid internal fault (see Chapter 10)

Starter motor turns engine slowly
- ☐ Partially discharged battery (recharge, use jump leads, or push start)
- ☐ Battery terminals loose or corroded
- ☐ Battery earth to body defective
- ☐ Engine earth strap loose
- ☐ Starter motor (or solenoid) wiring loose
- ☐ Starter motor internal fault (see Chapter 10)

Starter motor spins without turning engine
- ☐ Flywheel gear teeth damaged or worn
- ☐ Starter motor mounting bolts loose

Engine turns normally but fails to start
- ☐ Damp or dirty HT leads and distributor cap (crank engine and check for spark) - try moisture dispersant
- ☐ Dirty or incorrectly gapped contact breaker points (if applicable)
- ☐ No fuel in tank (check for delivery at carburettor)
- ☐ Excessive choke (hot engine) or insufficient choke (cold engine)
- ☐ Fouled or incorrectly gapped spark plugs (remove, clean and regap)
- ☐ Other ignition system fault (see Chapter 4)
- ☐ Other fuel system fault (see Chapter 3)
- ☐ Poor compression (see Chapter 1)
- ☐ Major mechanical failure (eg camshaft drive)

Engine fires but will not run
- ☐ Insufficient choke (cold engine)
- ☐ Air leaks at carburettor or inlet manifold
- ☐ Fuel starvation (see Chapter 3)
- ☐ Ballast resistor defective, or other ignition fault (see Chapter 4)

Engine cuts out and will not restart

Engine cuts out suddenly - ignition bulb
- ☐ Loose or disconnected LT wires
- ☐ Wet HT leads or distributor cap (after traversing water splash)
- ☐ Coil or condenser failure (check for spark)
- ☐ Other ignition fault (see Chapter 4)

A simple test lamp is useful for investigating electrical faults

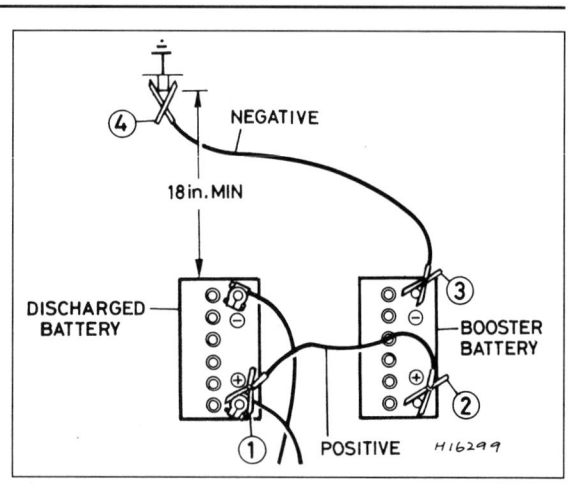

Engine misfires before cutting out - fuel fault
- ☐ Fuel tank empty
- ☐ Fuel pump defective or filter blocked (check for delivery)
- ☐ Fuel tank filler vent blocked (suction will be evident on releasing cap)
- ☐ Carburettor needle valve sticking
- ☐ Carburettor jets blocked (fuel contaminated)
- ☐ Other fuel system fault (see Chapter 3)

Engine cuts out - other causes
- ☐ Serious overheating
- ☐ Major mechanical failure (eg camshaft drive)

Engine overheats

Ignition (no-charge) warning light illuminated
- ☐ Slack or broken drivebelt - retension or renew (Chapter 2)

Ignition warning light not illuminated
- ☐ Coolant loss due to internal or external leakage (see Chapter 2)
- ☐ Thermostat defective
- ☐ Low oil level
- ☐ Brakes binding
- ☐ Radiator clogged externally or internally
- ☐ Electric cooling fan not operating correctly (if fitted)
- ☐ Engine waterways clogged
- ☐ Ignition timing incorrect or automatic advance malfunctioning
- ☐ Mixture too weak

Note: Do not add cold water to an overheated engine or damage may result

Low engine oil pressure

Gauge reads low or warning light illuminates with engine running
- ☐ Oil level low or incorrect grade
- ☐ Defective gauge or sender unit
- ☐ Wire to sender unit earthed

Jump start lead connection for negative earth vehicles – connect leads in order shown

- ☐ Engine overheating
- ☐ Oil filter clogged or bypass valve defective
- ☐ Oil pressure relief valve defective
- ☐ Oil pick-up strainer clogged
- ☐ Oil pump worn or mountings loose
- ☐ Worn main or big-end bearings

Note: Low oil pressure in a high-mileage engine at tickover is not necessarily a cause for concern. Sudden pressure loss at speed is far more significant. In any event, check the gauge or warning light sender before condemning the engine.

Engine noises

Pre-ignition (pinking) on acceleration
- ☐ Incorrect grade of fuel
- ☐ Ignition timing incorrect
- ☐ Distributor faulty or worn
- ☐ Worn or maladjusted carburettor
- ☐ Excessive carbon build-up in engine

Whistling or wheezing noises
- ☐ Leaking vacuum hose
- ☐ Leaking carburettor or manifold gasket
- ☐ Blowing head gasket

Tapping or rattling
- ☐ Incorrect valve clearances
- ☐ Worn valve gear
- ☐ Worn timing chain or belt
- ☐ Broken piston ring (ticking noise)

Knocking or thumping
- ☐ Unintentional mechanical contact (eg fan blades)
- ☐ Worn fanbelt
- ☐ Peripheral component fault (generator, water pump etc)
- ☐ Worn big-end bearings (regular heavy knocking, perhaps less under load)
- ☐ Worn main bearings (rumbling and knocking, perhaps worsening under load)
- ☐ Piston slap (most noticeable when cold

Tools and Working Facilities

Introduction

A selection of good tools is a fundamental requirement for anyone contemplating the maintenance and repair of a motor vehicle. For the owner who does not possess any, their purchase will prove a considerable expense, offsetting some of the savings made by doing-it-yourself. However, provided that the tools purchased are of good quality, they will last for many years and prove an extremely worthwhile investment.

To help the average owner to decide which tools are needed to carry out the various tasks detailed in this manual, we have compiled three lists of tools under the following headings: Maintenance and minor repair, Repair and overhaul, and Special. The newcomer to practical mechanics should start off with the Maintenance and minor repair tool kit and confine himself to the simpler jobs around the vehicle. Then, as his confidence and experience grows, he can undertake more difficult tasks, buying extra tools as, and when, they are needed. In this way, a Maintenance and minor repair tool kit can be built-up into a Repair and overhaul tool kit over a considerable period of time without any major cash outlays. The experienced do-it-yourselfer will have a tool kit good enough for most repair and overhaul procedures and will add tools from the Special category when he feels the expense is justified by the amount of use to which these tools will be put.

Maintenance and minor repair tool kit

The tools given in this list should be considered as a minimum requirement if routine maintenance, servicing and minor repair operations are to be undertaken. We recommend the purchase of combination spanners (ring one end, open-ended the other); although more expensive than open-ended ones, they do give the advantages of both types of spanner.

- ☐ Combination spanners - 7/16, 1/2, 9/16, 5/8, 11/16, 3/4 in AF
- ☐ Combination spanners - 10, 11, 12, 13, 14, 15, 16, 17, 18, 19 mm
- ☐ Adjustable spanner - 9 inch
- ☐ Engine sump/gearbox/rear axle drain plug key (where applicable)
- ☐ Spark plug spanner (with rubber insert)
- ☐ Spark plug gap adjustment tool
- ☐ Set of feeler gauges
- ☐ Screwdriver - 4 in long x 1/4 in dia (flat blade)
- ☐ Screwdriver - 4 in long x 1/4 in dia (cross blade)
- ☐ Combination pliers - 6 inch
- ☐ Hacksaw, junior
- ☐ Tyre pump
- ☐ Tyre pressure gauge
- ☐ Oil can
- ☐ Fine emery cloth (1 sheet)
- ☐ Wire brush (small)
- ☐ Funnel (medium size)

Repair and overhaul tool kit

These tools are virtually essential for anyone undertaking any major repairs to a motor vehicle, and are additional to those given in the Maintenance and minor repair list. Included in this list is a comprehensive set of sockets. Although these are expensive they will be found invaluable as they are so versatile - particularly if various drives are included in the set. We recommend the _ in square-drive type, as this can be used with most proprietary torque wrenches. If you cannot afford a socket set, even bought piecemeal, then inexpensive tubular box spanners are a useful alternative.

The tools in this list will occasionally need to be supplemented by tools from the Special list.

- ☐ Sockets (or box spanners) to cover range in previous list
- ☐ Reversible ratchet drive (for use with sockets)
- ☐ Extension piece 10 inch (for use with sockets)
- ☐ Universal joint (for use with sockets)
- ☐ Torque wrench (for use with sockets)
- ☐ "Mole" wrench - 8 inch
- ☐ Ball pein hammer
- ☐ Soft-faced hammer plastic or rubber
- ☐ Screwdriver - 6 in long x 5/16 in dia (flat blade)
- ☐ Screwdriver - 2 in long x 5/16 in square (flat blade)
- ☐ Screwdriver - 11/8 in long x 1/4 in dia (cross blade)
- ☐ Screwdriver - 3 in long x 1/8 in dia (electricians)
- ☐ Pliers - electricians side cutters
- ☐ Pliers - needle nosed
- ☐ Pliers - circlip (internal and external)
- ☐ Cold chisel - 1/2 inch
- ☐ Scriber
- ☐ Scraper
- ☐ Centre punch
- ☐ Pin punch
- ☐ Hacksaw
- ☐ Brake hose clamp
- ☐ Valve grinding tool
- ☐ Steel rule/straight edge
- ☐ Allen keys
- ☐ Selection of files
- ☐ Wire brush (large)
- ☐ Axle-stands
- ☐ Jack (strong scissor or hydraulic type)

Special tools

The tools in this list are those which are not used regularly, are expensive to buy, or which need to be used in accordance with their manufacturers' instructions. Unless relatively difficult mechanical jobs are undertaken frequently, it will not be economic to buy many of these tools. Where this is the case, you could consider clubbing together with friends (or a motorists' club) to make a joint purchase, or borrowing the tools against a deposit from a local garage or tool hire specialist.

The following list contains only those tools and instruments freely available to the public, and not those special tools produced by the vehicle manufacturer specifically for its dealer network. You will find occasional references to these manufacturers' special tools in the text of this manual. Generally, an alternative method of doing the job without the vehicle manufacturer's special tool is given. However, sometimes, there is no alternative to using them. Where this is the case and the relevant tool cannot be bought or borrowed you will have to entrust the work to a franchised garage.

- ☐ Valve spring compressor
- ☐ Piston ring compressor
- ☐ Balljoint separator
- ☐ Universal hub/bearing puller
- ☐ Impact screwdriver
- ☐ Micrometer and/or vernier gauge
- ☐ Carburettor flow balancing device (where applicable)
- ☐ Dial gauge
- ☐ Stroboscopic timing light
- ☐ Dwell angle meter/tachometer
- ☐ Universal electrical multi-meter
- ☐ Cylinder compression gauge
- ☐ Lifting tackle
- ☐ Trolley jack
- ☐ Light with extension lead

Buying tools

For practically all tools, a tool factor is the best source since he will have a very comprehensive range compared with the average garage or accessory shop. Having said that, accessory shops often offer excellent quality tools at discount prices, so it pays to shop around.

There are plenty of good tools around at reasonable prices, but always aim to purchase items which meet the relevant national safety standards. If in doubt, ask the proprietor or manager of the shop for advice before making a purchase.

Care and maintenance of tools

Having purchased a reasonable tool kit, it is necessary to keep the tools in a clean

Tools and Working Facilities

serviceable condition. After use, always wipe off any dirt, grease and metal particles using a clean, dry cloth, before putting the tools away. Never leave them lying around after they have been used. A simple tool rack on the garage or workshop wall, for items such as screwdrivers and pliers is a good idea. Store all normal spanners and sockets in a metal box. Any measuring instruments, gauges, meters, etc, must be carefully stored where they cannot be damaged or become rusty.

Take a little care when tools are used. Hammer heads inevitably become marked and screwdrivers lose the keen edge on their blades from time to time. A little timely attention with emery cloth or a file will soon restore items like this to a good serviceable finish.

Working facilities

Not to be forgotten when discussing tools, is the workshop itself. If anything more than routine maintenance is to be carried out, some form of suitable working area becomes essential.

It is appreciated that many an owner mechanic is forced by circumstances to remove an engine or similar item, without the benefit of a garage or workshop. Having done this, any repairs should always be done under the cover of a roof.

Wherever possible, any dismantling should be done on a clean flat workbench or table at a suitable working height.

Any workbench needs a vice: one with a jaw opening of 4 in (100 mm) is suitable for most jobs. As mentioned previously, some clean dry storage space is also required for tools, as well as the lubricants, cleaning fluids, touch-up paints and so on which become necessary.

Another item which may be required, and which has a much more general usage, is an electric drill with a chuck capacity of at least 5/16 in (8 mm). This, together with a good range of twist drills, is virtually essential for fitting accessories such as wing mirrors and reversing lights.

Last, but not least, always keep a supply of old newspapers and clean, lint-free rags available, and try to keep any working area as clean as possible.

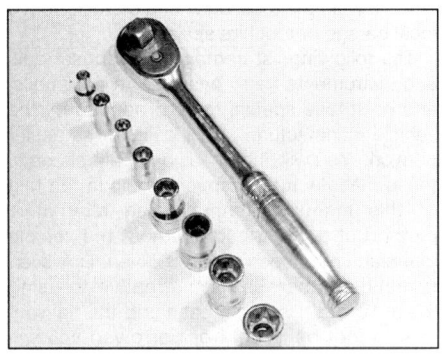

Sockets and reversible ratchet drive

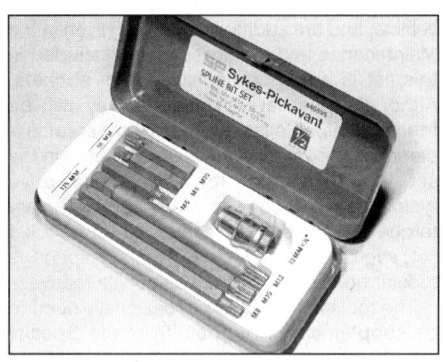

Spline bit set

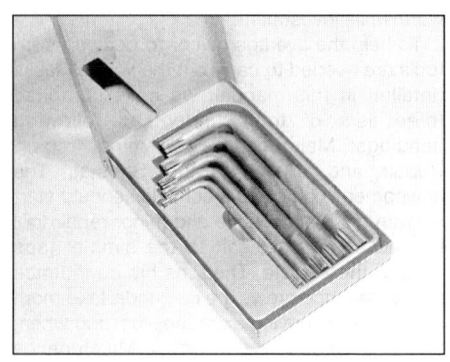

Spline key set

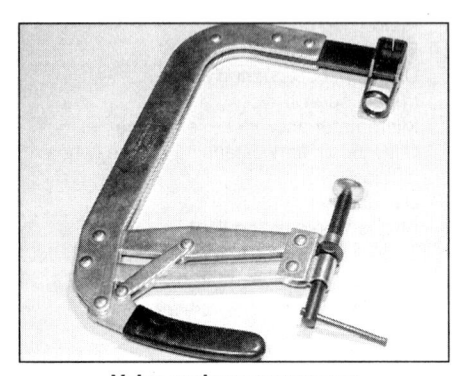

Valve spring compressor

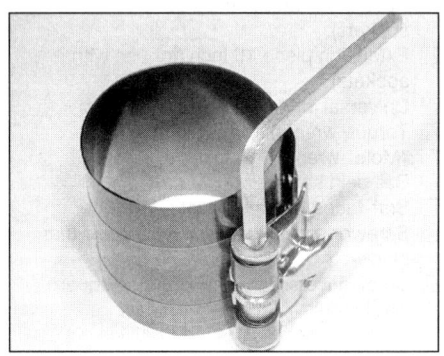

Piston ring compessor

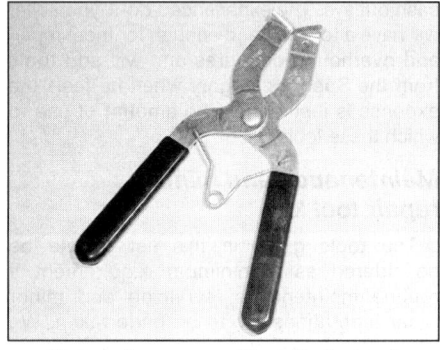

Piston ring removal/installation tool

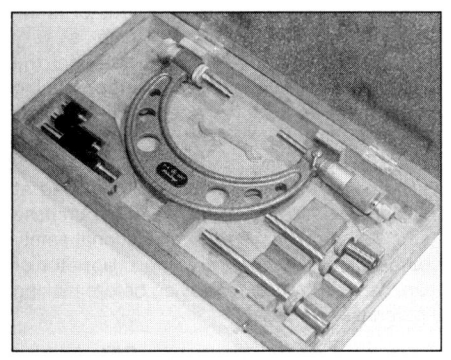

Micrometer set

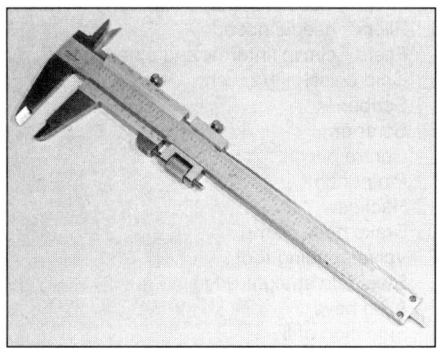

Vernier calipers

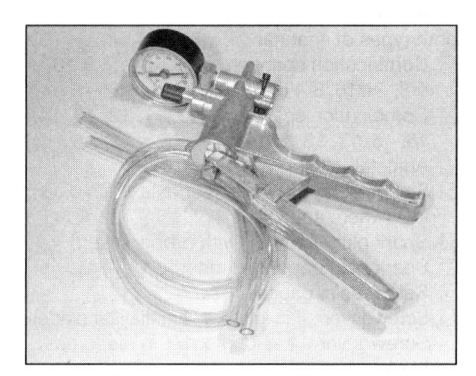

Vacuum pump and gauge

General Repair Procedures

Whenever servicing, repair or overhaul work is carried out on the car or its components, it is necessary to observe the following procedures and instructions. This will assist in carrying out the operation efficiently and to a professional standard of workmanship.

Joint mating faces and gaskets

When separating components at their mating faces, never insert screwdrivers or similar implements into the joint between the faces in order to prise them apart. This can cause severe damage which results in oil leaks, coolant leaks, etc upon reassembly. Separation is usually achieved by tapping along the joint with a soft-faced hammer in order to break the seal. However, note that this method may not be suitable where dowels are used for component location.

Where a gasket is used between the mating faces of two components, ensure that it is renewed on reassembly, and fit it dry unless otherwise stated in the repair procedure. Make sure that the mating faces are clean and dry, with all traces of old gasket removed. When cleaning a joint face, use a tool which is not likely to score or damage the face, and remove any burrs or nicks with an oilstone or fine file.

Make sure that tapped holes are cleaned with a pipe cleaner, and keep them free of jointing compound, if this is being used, unless specifically instructed otherwise.

Ensure that all orifices, channels or pipes are clear, and blow through them, preferably using compressed air.

Oil seals

Oil seals can be removed by levering them out with a wide flat-bladed screwdriver or similar implement. Alternatively, a number of self-tapping screws may be screwed into the seal, and these used as a purchase for pliers or some similar device in order to pull the seal free.

Whenever an oil seal is removed from its working location, either individually or as part of an assembly, it should be renewed.

The very fine sealing lip of the seal is easily damaged, and will not seal if the surface it contacts is not completely clean and free from scratches, nicks or grooves.

Protect the lips of the seal from any surface which may damage them in the course of fitting. Use tape or a conical sleeve where possible. Lubricate the seal lips with oil before fitting and, on dual-lipped seals, fill the space between the lips with grease.

Unless otherwise stated, oil seals must be fitted with their sealing lips toward the lubricant to be sealed.

Use a tubular drift or block of wood of the appropriate size to install the seal and, if the seal housing is shouldered, drive the seal down to the shoulder. If the seal housing is unshouldered, the seal should be fitted with its face flush with the housing top face (unless otherwise instructed).

Screw threads and fastenings

Seized nuts, bolts and screws are quite a common occurrence where corrosion has set in, and the use of penetrating oil or releasing fluid will often overcome this problem if the offending item is soaked for a while before attempting to release it. The use of an impact driver may also provide a means of releasing such stubborn fastening devices, when used in conjunction with the appropriate screwdriver bit or socket. If none of these methods works, it may be necessary to resort to the careful application of heat, or the use of a hacksaw or nut splitter device.

Studs are usually removed by locking two nuts together on the threaded part, and then using a spanner on the lower nut to unscrew the stud. Studs or bolts which have broken off below the surface of the component in which they are mounted can sometimes be removed using a proprietary stud extractor. Always ensure that a blind tapped hole is completely free from oil, grease, water or other fluid before installing the bolt or stud. Failure to do this could cause the housing to crack due to the hydraulic action of the bolt or stud as it is screwed in.

When tightening a castellated nut to accept a split pin, tighten the nut to the specified torque, where applicable, and then tighten further to the next split pin hole. Never slacken the nut to align the split pin hole, unless stated in the repair procedure.

When checking or retightening a nut or bolt to a specified torque setting, slacken the nut or bolt by a quarter of a turn, and then retighten to the specified setting. However, this should not be attempted where angular tightening has been used.

For some screw fastenings, notably cylinder head bolts or nuts, torque wrench settings are no longer specified for the latter stages of tightening, "angle-tightening" being called up instead. Typically, a fairly low torque wrench setting will be applied to the bolts/nuts in the correct sequence, followed by one or more stages of tightening through specified angles.

Locknuts, locktabs and washers

Any fastening which will rotate against a component or housing in the course of tightening should always have a washer between it and the relevant component or housing.

Spring or split washers should always be renewed when they are used to lock a critical component such as a big-end bearing retaining bolt or nut. Locktabs which are folded over to retain a nut or bolt should always be renewed.

Self-locking nuts can be re-used in non-critical areas, providing resistance can be felt when the locking portion passes over the bolt or stud thread. However, it should be noted that self-locking stiffnuts tend to lose their effectiveness after long periods of use, and in such cases should be renewed as a matter of course.

Split pins must always be replaced with new ones of the correct size for the hole.

When thread-locking compound is found on the threads of a fastener which is to be re-used, it should be cleaned off with a wire brush and solvent, and fresh compound applied on reassembly.

Special tools

Some repair procedures in this manual entail the use of special tools such as a press, two or three-legged pullers, spring compressors, etc. Wherever possible, suitable readily-available alternatives to the manufacturer's special tools are described, and are shown in use. Unless you are highly-skilled and have a thorough understanding of the procedures described, never attempt to bypass the use of any special tool when the procedure described specifies its use. Not only is there a very great risk of personal injury, but expensive damage could be caused to the components involved.

Environmental considerations

When disposing of used engine oil, brake fluid, antifreeze, etc, give due consideration to any detrimental environmental effects. Do not, for instance, pour any of the above liquids down drains into the general sewage system, or onto the ground to soak away. Many local council refuse tips provide a facility for waste oil disposal, as do some garages. If none of these facilities are available, consult your local Environmental Health Department for further advice.

With the universal tightening-up of legislation regarding the emission of environmentally-harmful substances from motor vehicles, most current vehicles have tamperproof devices fitted to the main adjustment points of the fuel system. These devices are primarily designed to prevent unqualified persons from adjusting the fuel/air mixture, with the chance of a consequent increase in toxic emissions. If such devices are encountered during servicing or overhaul, they should, wherever possible, be renewed or refitted in accordance with the vehicle manufacturer's requirements or current legislation.

Note: It is antisocial and illegal to dump oil down the drain. To find the location of your local oil recycling bank, call this number free.

OIL BANK LINE
0800 66 33 66

Preserving Our Motoring Heritage

The Model J Duesenberg Derham Tourster. Only eight of these magnificent cars were ever built – this is the only example to be found outside the United States of America

Almost every car you've ever loved, loathed or desired is gathered under one roof at the Haynes Motor Museum. Over 300 immaculately presented cars and motorbikes represent every aspect of our motoring heritage, from elegant reminders of bygone days, such as the superb Model J Duesenberg to curiosities like the bug-eyed BMW Isetta. There are also many old friends and flames. Perhaps you remember the 1959 Ford Popular that you did your courting in? The magnificent 'Red Collection' is a spectacle of classic sports cars including AC, Alfa Romeo, Austin Healey, Ferrari, Lamborghini, Maserati, MG, Riley, Porsche and Triumph.

A Perfect Day Out

Each and every vehicle at the Haynes Motor Museum has played its part in the history and culture of Motoring. Today, they make a wonderful spectacle and a great day out for all the family. Bring the kids, bring Mum and Dad, but above all bring your camera to capture those golden memories for ever. You will also find an impressive array of motoring memorabilia, a comfortable 70 seat video cinema and one of the most extensive transport book shops in Britain. The Pit Stop Cafe serves everything from a cup of tea to wholesome, home-made meals or, if you prefer, you can enjoy the large picnic area nestled in the beautiful rural surroundings of Somerset.

John Haynes O.B.E., Founder and Chairman of the museum at the wheel of a Haynes Light 12.

Graham Hill's Lola Cosworth Formula 1 car next to a 1934 Riley Sports.

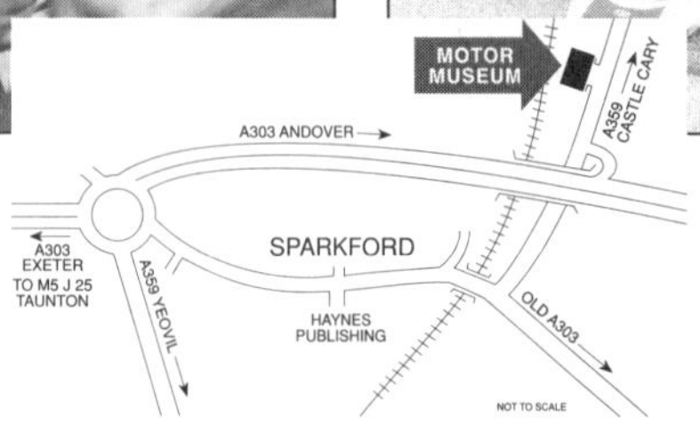

The Museum is situated on the A359 Yeovil to Frome road at Sparkford, just off the A303 in Somerset. It is about 40 miles south of Bristol, and 25 minutes drive from the M5 intersection at Taunton.
Open 9.30am - 5.30pm (10.00am - 4.00pm Winter) 7 days a week, *except Christmas Day, Boxing Day and New Years Day*
Special rates available for schools, coach parties and outings Charitable Trust No. 292048

Index

Note: *References throughout this index relate to Chapter•page number*

A

Accelerator cable - 3•4, 3•10
Accelerator cable adjustment - 13•9
Accelerator linkage - 3•5, 3•10
Accelerator pedal assembly - 3•4, 3•10
Acknowledgements - 0•4
Additional bleeding methods - 13•17
Air cleaner - 3•3, 3•9, 3•17
Air cleaner thermostatically controlled valve - 3•9
Air conditioning compressor drivebelt - 13•31
Air conditioning system - 12•11
Alternator - 10•3, 10•5
Alternator brushes - 10•5, 10•6, 13•18
Ancillary components - 1•48
Antifreeze precautions - 2•6
Ashtray light - 13•21
Automatic transmission - 1•33, 6•2, 6•6, 13•15
Automatic transmission selector - 6•8, 6•9, 13•22
Automatic transmission vacuum diaphragm unit - 6•10
Auxiliary air device - 3•18
Auxiliary light bulb renewal (typical) - 13•20
Auxiliary shaft - 1•14, 1•26
Auxiliary warning system - 13•24
Axle and suspension assembly - 11•7
Axle casing rubber mounting - 11•9

B

Battery - 10•2, 10•3
Battery will not hold charge for more than a few days - 10•20
Bleeding - 9•2, 13•17
Body mouldings - 13•35
Bodywork and fittings - 12•1 *et seq*, 13•31
Bonnet - 12•7
Boot lid - 12•7
Boot lid lock - 12•8
Boot light switch - 13•23
Bosch distributor (in-line ohc engine) - 4•5
Bosch short frame starter motor - 13•20
Brake band adjustment - 13•15
Brake caliper - 9•4
Brake disc - 9•4
Brake master cylinder - 9•8, 9•9
Brake pads - 9•3, 13•16
Brake pedal - 9•9
Brake pedal travel excessive - 9•10
Brake pressure control valve - 9•10, 13•17
Brake shoe renewal - 13•16
Brakes binding, juddering, overheating - 9•10
Brakes judder when applied - 9•10

Braking system - 0•9, 9•1 *et seq*, 13•5 13•16
Breakerless distributor - 13•12, 13•13
Breakerless ignition - 13•12
Bumper - 12•4, 13•31
Buying spare parts - 0•12
Buying tools - REF•4

C

Cajavec starter motor - 13•20
Cam followers - 1•30
Camshaft - 1•17, 1•29
Camshaft and camshaft bearings - 1•39
Camshaft and front intermediate plate - 1•43
Camshaft drivebelt - 1•16, 1•32, 13•6
Camshaft drivebelt tensioner and thermostat housing - 1•31
Camshaft, camshaft bearings and cam followers - 1•21
Capacities - 0•11
Carburation and ignition faults - 1•50
Care and maintenance of tools - REF•4
Central door locking system - 10•18
Central door locks fail to operate - 10•20
Central door locks will lock but not unlock - 10•20
Central door locks will unlock but not lock - 10•20
Central locking system, boot lock solenoid - 12•8
Central locking system, door lock solenoid - 12•6
Central locking system, tailgate lock solenoid - 12•9
Centre console - 10•17, 12•11
Choke phasing - 3•13
Cleaning - 0•13
Clunk on acceleration or deceleration - 8•3
Clutch - 5•1 *et seq*, 13•14
Clutch cable noise (ohc models) - 13•14
Clutch judder - 5•4
Clutch pedal and self adjusting mechanism - 5•3
Clutch release bearing - 5•3
Clutch slip - 5•4
Clutch spin - 5•4
Clutch squeal - 5•4
CO emissions (mixture) - 0•10
Computer module - 13•25
Condenser (in-line ohc engine) - 4•4
Connecting rods and gudgeon pin - 1•21, 1•40
Connecting rods to crankshaft - 1•25
Console - 12•11
Contact breaker points (in-line ohc engine) - 4•3, 4•4

Contents - 0•2
Control assembly - 13•24
Control illumination bulb renewal - 13•22
Control modules - 13•10
Conversion factors - 0•6
Cool running - 2•6
Coolant hoses chafing - 13•8
Coolant level switch - 13•24
Cooling system - 2•1 *et seq*, 13•8
Corrosion - 0•10
Crankcase ventilation system - 1•41
Crankshaft - 1•20, 1•22, 1•40, 1•41
Crankshaft and main bearings - 1•15
Crankshaft main and big-end bearings - 1•40
Crankshaft pulley, sprocket and timing belt cover - 1•15
Crankshaft rear oil seal - 1•26
Crankshaft reconditioning - 13•8
Crankshaft sprocket and pulley and auxiliary shaft sprocket - 1•27
Current at starter motor - 1•50
Cylinder bores - 1•20, 1•40
Cylinder head - 1•12, 1•14, 1•31, 1•39
Cylinder head and piston crowns - 1•22
Cylinder head bolts - 13•7
Cylinder heads, rocker shafts and inlet manifold - 1•46

D

Depressurizing the fuel injection system - 13•11
Difficulty in engaging gear - 6•10
Digital clock - 13•22
Dimensions - 0•11
Direction indicator, stop and tail light bulbs - 10•14, 10•15
Distributor (in-line ohc engine) - 4•4, 4•5, 4•6
Distributor (V6 engines) - 4•6, 4•7
Door belt moulding - 13•35
Door glass and regulator - 12•6
Door hinge lubrication - 13•31
Door lock assembly - 12•5, 12•6
Door mirror control switch - 13•23
Door mounted loudspeaker - 10•18
Door rattles - 12•4
Door striker plate - 12•6
Door trim - 12•5, 13•36
Door window regulator - 12•12
Doors - 0•8, 12•4
Downshift cable - 6•9
Drive shafts - 8•2
Drum brake backplate - 9•7
Drum brake shoes - 9•5
Drum brake wheel cylinder - 9•7

Index

E

Electric windows fail to operate - 10•20
Electric windows work front or rear only - 10•20
Electrical equipment - 0•8
Electrical faults - REF•1
Electrical system - 10•1 et seq, 13•18
Electrically operated aerials - 10•18, 13•27
Electrically operated windows - 10•18, 13•27
Engine - 1•1 et seq
Engine compartment light - 13•22
Engine components - 1•20
Engine cuts out - REF•2
Engine dismantling - 1•11, 1•36, 13•7
Engine fails to start - 4•8
Engine fails to turn over when starter operated - 1•50
Engine fails to turn when starter operated - REF•2
Engine fires but will not run - REF•2
Engine misfires - 4•8
Engine misfires before cutting out - REF•2
Engine misfires on road - 3•19
Engine misfires or idles unevenly - 1•50
Engine modifications - 1•49
Engine noises - REF•2
Engine oil pump water sump, and crankshaft pulley - 1•45
Engine overheats - REF•2
Engine reassembly - 1•22, 1•41
Engine removal - 13•7
Engine runs on - 3•19
Engine stalls and will not start - 1•50
Engine turns normally but fails to start - REF•2
Engine turns over but will not start - 1•50
Engine will not start - REF•2, 3•19
Environmental considerations - REF•5
Excess of petrol in cylinder or carburettor - 1•50
Exhaust emission checks - 0•10
Exhaust system - 0•9, 0•13, 3•8, 3•14
Exterior mirror - 13•34

F

Facia crash pad - 12•9
Facia mounted loudspeaker - 10•17
Facia trim panel - 13•36
Fan - 2•5, 13•8
Fan belt - 2•5
Fault diagnosis - automatic transmission - 6•10
Fault diagnosis - braking system - 9•10
Fault diagnosis - clutch - 5•4
Fault diagnosis - cooling system - 2•6
Fault diagnosis - electrical system - 10•20
Fault diagnosis - engine - 1•50
Fault diagnosis - fuel and exhaust systems - 3•19
Fault diagnosis - ignition system - 4•8
Fault diagnosis - manual gearbox - 6•10
Fault diagnosis - propeller shaft - 7•3
Fault diagnosis - rear axle - 8•3
Fault diagnosis - suspension and steering - 11•19
Fault finding - REF•1

Final drive - 8•3
Five-speed gearbox (type N) - 13•15
Fixed centre fans - 2•5
Flasher unit - 10•11
Flexible hoses - 9•3
Fluid level checking - 13•15
Flywheel and clutch - 1•28, 1•44
Flywheel and sump - 1•14
Flywheel ring gear - 1•22, 1•41
Foglamp removal and refitting - 13•21
Footbrake - 0•7, 0•8
Fuel accumulator - 3•17, 13•11
Fuel and exhaust systems - 0•10, 3•1 et seq, 13•3
Fuel consumption excessive - 3•19
Fuel consumption high - 3•19
Fuel distributor - 3•17
Fuel filler neck - 3•4
Fuel filter - 3•17, 13•12
Fuel flow sensor - 13•25
Fuel gauge gives no reading - 10•20
Fuel gauge registers full all the time - 10•20
Fuel gauge sender unit - 3•4, 3•10
Fuel injection system - 3•2
Fuel injectors - 3•18
Fuel pump - 3•1, 3•3, 3•9, 3•17, 13•11
Fuel pump one-way valve - 13•12
Fuel shortage at engine - 1•50
Fuel start valve - 3•18, 13•10
Fuel system - 13•9, 13•10
Fuel tank - 3•3, 3•4, 3•10, 13•10, 13•12
Fuel tank indicator unit - 10•16
Fuel tank sender unit - 13•25
Fuel/air mixture and adjustment - 3•19
Fuel/air mixture leaking from cylinder - 1•50
Full length console - 12•11
Fuses and relays - 10•18, 13•26

G

Gauge reads low or warning light illuminates with engine running - REF•2
Gearbox input shaft spigot diameter - 13•14
Glove compartment light - 13•21
Glovebox - 13•31
Gudgeon pin - 1•18

H

Handbrake - 0•7, 9•7
Handbrake adjustment - 13•16
Handbrake cable - 9•8
Handbrake lever - 9•8
Hazard warning switch - 13•22
HC emissions - 0•10
Headlight and auxiliary light alignment - 10•13
Headlight and sidelight bulbs - 10•13
Headlight assembly - 10•13
Headlight washer jets - 13•20
Headlight washer pump - 10•13
Heater assembly - 12•9, 12•10
Heater controls - 12•10, 13•22
Heater fan switch - 13•23
Heater motor - 12•11
Horn - 10•12
Horn emits intermittent or unsatisfactory noise - 10•20

Horn fails to operate - 10•20
Horn operates all the time - 10•20
Hub - 11•4
Hub bearings - 11•3, 11•4, 13•28

I

Ignition (no-charge) warning light illuminated - REF•2
Ignition amplifier (all types) - 13•14
Ignition coil (all types) - 13•13
Ignition lights fails to go out, battery runs flat in a few days - 10•20
Ignition switch and lock - 10•17
Ignition system - 4•1 et seq, 4•8, 13•12
Ignition timing (in-line ohc engine) - 4•6
Ignition timing (V6 engines) - 4•7
Ignition timing - 13•13
Ignition warning light not illuminated - REF•2
Instrument cluster and printed circuit - 10•15
Instrument panel and centre console switches - 10•17
Instrument voltage regulator - 10•16
Instruments, warning lights and illumination light - 10•16
Insufficient fuel delivery or weak mixture due to air leaks - 3•19
Interior light - 10•15, 13•21
Intermittent spark at spark plugs - 1•50
Introduction to the Ford Granada - 0•4

J

Jacking and towing - 0•12
Joint mating faces and gaskets - REF•5
Jumps out of gear - 6•10

K

Kerb weight (nominal) - 0•11
Knock, or clunk when taking up drive - 7•3
Knocking or thumping - REF•2

L

Lack of power and poor compression - 1•50
Lack of power assistance - 11•19
Leaks from radiator blanking plug - 13•8
Light bulb renewal - 13•20
Light cluster unit - 13•20
Lights come on but fade out - 10•20
Lights do not come on - 10•20
Lights give very poor illumination - 10•20
Lights work erratically - especially over bumps - 10•20
Limited-slip differential - 13•16
Load compartment light switch - 13•23
Locknuts, locktabs and washers - REF•5
Loss of cooling water - 2•6
Loudspeakers - 10•17
Low engine oil pressure - REF•2
Lower suspension swinging arm - 11•6
Lubrication and crankcase ventilation systems - 1•18
Lubrication system - 1•41
Lucas M35J and 9M90 starter motor - 13•19

Index IND•3

M

Main and big-end bearings - 1•20
Main bearing caps and bolts - 13•8
Maintenance - 12•4
Maintenance - bodywork and underframe - 12•2
Maintenance - upholstery and carpets - 12•2
Maintenance and minor repair tool kit - REF•3
Maintenance intervals - 13•6
Maintenance-free battery - 13•18
Major body damage - 12•3
Manual and automatic transmission - 6•1 et seq, 13•14
Manual steering - 11•9, 13•28
Manually operated aerials - 10•18
Map reading light - 13•21
Mechanical wear - 1•50
Minor body damage - 12•2
Minor switches - 13•23
Mirror- 13•34
Modulator spring gap - 3•13
MOT test checks - 0•7
Motorcraft distributor (in-line ohc engine) - 4•5

N

Nippondenso starter motor - 13•20
No current at starter motor - 1•50
No fuel at carburettor or fuel injection system - 1•50
No fuel getting to engine - 1•50
No spark at spark plug - 1•50
Noise - 6•10, 8•3
Noisy operation of steering pump - 11•19
Non-sealed bearing type - 7•2
Number plate light assembly and bulb - 10•15, 13•21

O

Oil being burnt by engine - 1•50
Oil being lost due to leaks - 1•50
Oil consumption high - 1•50
Oil cooler - 13•7
Oil filter - 1•20
Oil leakage - 8•3
Oil pressure gauge and/or ammeter - 13•22
Oil pump - 1•15, 1•19, 1•25, 1•40, 1•44
Oil seals - REF•5
One individual lock does not operate - 10•20
Overheating - 2•6

P

Parcel shelf mounted loudspeaker - 10•18
Pedal feels spongy when the brakes are applied - 9•10
Pedal travel excessive with little or no resistance and brakes are virtually non-operative - 9•10
Piston rings - 1•18, 1•25, 1•40
Pistons - 1•25
Pistons and connecting rods - 1•25, 1•43
Pistons and piston rings - 1•21, 1•40
Pistons, connecting rods and big-end bearings - 1•15
Poor stering self centring - 11•19
Power assisted steering control valve - 11•17
Power assisted steering fluid cooler - 11•18
Power assisted steering gear - 11•14, 11•15, 11•17, 11•18, 13•28, 13•29
Power assisted steering pump - 11•18, 13•29
Power assisted steering pump drivebelt - 11•17
Pre-ignition (pinking) on acceleration - REF•2
Propeller shaft - 7•1 et seq, 13•15
Propeller shaft noise - 13•15

R

Rack and pinion steering gear - 11•11, 11•12, 11•13
Radiator - 2•3, 2•4, 13•9
Radiator grille - 12•9, •31
Radiator repairs - 13•8
Radio - 10•17, 13•25
Radio aerials - 10•18
Rear axle - 8•1 et seq, 13•16
Recommended lubricants and fluids - 0•14
Relays - 10•18
Repair and overhaul tool kit - REF•3
Repair procedures - REF•5
Repairs of dents in bodywork - 12•3
Repairs of minor scratches in bodywork - 12•2
Repairs of rust holes or gashes in bodywork - 12•3
Rocker ball stud locknuts - 13•7
Rocker covers and gaskets - 13•8
Rocker shaft springs - 13•7
Roof console (auxiliary warning system) - 13•26
Rough idling- 3•19
Routine maintenance - 0•13

S

Safety control module - 13•10, 13•11
Safety first! - 0•5
Screw threads and fastenings - REF•5
Seat - 13•32
Seat back locking mechanism (estate) - 13•33
Seat backrest (saloon) - 13•32
Seat belts and seats - 0•8
Seat belts- 13•33, 13•34
Seat catch striker (estate) - 13•33
Seat cushion - 13•32
Seat head restraints - 13•32
Seat heating pads - 13•31
Seat movement switches - 13•23
Seat operating motor - 13•23
Seat slide - 13•32
Shock absorbers - 0•9, 11•7, 11•9
Side door moulding - 13•35
Sidelight bulb renewal - 13•20
Sliding roof - 12•9, 13•35
Sliding roof motor - 13•23
Solex carburettor - 3•10, 3•11
Solex carburettor automatic choke - 3•13, 3•14
Solex carburettor fast idle speed - 3•13
Solex dual venturi carburettor - 3•10
Solex/Pierburg carburettor - 13•9, 13•10
Solex/Pierburg carburettor electrically-heated choke - 13•9
Solex/Pierburg carburettor throttle damper - 13•9
Spares and tool kit - REF•1
Spark plugs and HT leads - 4•7, 13•14
Special tools - REF•3, REF•5
Speed sender unit - 13•25
Speedometer inner and outer cable - 10•16
Spoiler - 13•31
Springs and shock absorbers - 0•9
Stabiliser bar - 11•6, 11•7, 13•28
Starter inhibitor/reverse lamp switch - 6•10
Starter motor - 10•6, 10•7, 10•8, 10•9, 10•10, 13•19
Starter motor fails to turn engine - 10•20
Starter motor noisy or excessively rough engagement - 10•20
Starter motor operates without turning engine - 10•20
Starter motor spins without turning engine - REF•2
Starter motor turns engine slowly - REF•2
Starter motor turns engine very slowly - 10•20
Steering column assembly - 11•10
Steering column flexible coupling and universal joint assembly - 11•11
Steering column shaft - 11•11
Steering column switches - 10•16
Steering lock - 10•17, 13•28
Steering mechanism - 0•9
Steering rack rubber gaiter - 11•13
Steering wheel - 0•7, 11•10, 13•28
Stop light switch - 9•10
Stopping ability poor, even though pedal pressure is firm - 9•10
Stub axle - 11•6
Stub axle bearing and seals - 8•2
Stub axle shaft - 8•2
Sub-frame mounting bushes - 11•6
Sump - 1•22, 1•27
Supplement: revisions and information on later models - 13•1 et seq
Suspension and steering - 0•8, 0•9, 11•1 et seq, 13•28
Switch panel - 13•27

T

Tailgate assembly - 12•8
Tailgate lock - 12•8
Tailgate lock striker plate - 12•9
Tapping or rattling - REF•2
Temperature gauge and sender unit - 2•6
Thermostat - 2•4
Thermostat housing and belt tensioner - 1•18
Tie-bar - 11•7
Timing gears and belt - 1•22
Timing gears and timing cover - 1•45
Tools and working facilities - REF•3
Top cover gasket - 13•7
Towing points - 0•12
Track-rod end - 11•14
Trip computer - 13•24, 13•25

Index

U

Unbalanced steering - 11•19
Universal joints - 7•3
Unsatisfactory road performance - 3•19
Unusual noises from engine - 1•50
Using this manual - 0•5

V

Vacuum pull-down - 3•13
Vacuum servo unit - 9•10
Valve clearances - 1•32, 1•47
Valve guides - 1•22
Valve rockers - 1•39
Valve tappets and pushrods - 1•39
Valves - 1•16, 1•29
Valves and valve seats - 1•21
Vanity mirror light - 13•22
Variation of clutch pedal operation - 5•4
Vehicle identification - 0•8
Vehicle identification numbers - 0•12
Vibration - 7•3, 8•3
View mirrors - 13•34
Viscous drive fans - 2•5

W

Warm up regulator - 3•18
Warning light bulb renewal - 13•22
Washer fluid level switch - 13•24
Water pump - 1•28, 2•5
Weak or ineffective synchromesh - 6•10
Weber carburettor - 3•5
Weber carburettor automatic choke - 3•7
Weber carburettor top cover securing screws - 13•9
Wheel alignment and steering angles - 11•9
Wheel wobble and vibration - 11•19
Wheels and tyres - 0•10, 11•9, 11•10, 11•18, 13•30
Whistling or wheezing noises - REF•2
Window and tailgate glass - 12•9
Window frame moulding - 13•35
Window operating motor - removal and refitting - 13•27
Window washer pump - 10•13
Windscreen and mirrors - 0•7
Windscreen glass - 12•4
Windscreen washer pump and reservoir - 10•12
Wiper arms - 10•11
Wiper blades - 10•11
Wiper mechanism - 10•11
Wiper motor - 10•12
Wiper motor and linkage - 10•12
Wiper motor fails to work - 10•20
Wiper motor works but wiper blades remain static - 10•20
Wiper motor works slowly and takes little current - 10•20
Wiper motor works very slowly and takes excessive current - 10•20
Wiring diagrams - WD•1 *et seq*
Working facilities - REF•4